FUNCTIONAL REYNOLDS STRESS MODELING

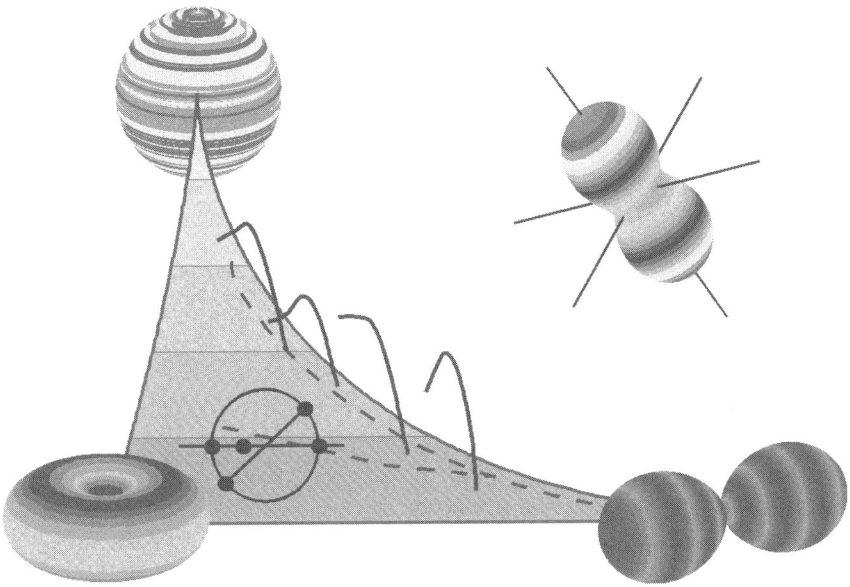

JOAN G. MOORE and JOHN MOORE

Pocahontas Press, Inc., Blacksburg, Virginia

Functional Reynolds Stress Modeling
by Joan G. Moore and John Moore

Published by Pocahontas Press, Inc., P.O. Drawer F, Blacksburg, VA 24063-1020.
Printed in the United States of America by McNaughton & Gunn, Saline, Michigan.

First printing 2006

Cover/title page illustration: The World of Turbulence Glyphs. The isotropic sun rises over the field nourished by the 2-d lake. The Mohr's circle plow harvests 1-d potatoes in the shade of spectral trees. Shear flow paths wend through the woods and a satellite glyph provides internet communications. The illustration is a collage from Figs. 1.2.1, 1.2.2, 1.2.5 and 17.6.5.

Library of Congress Cataloging-in-Publication Data

Moore, Joan G., 1943-
Functional Reynolds stress modeling / Joan G. Moore and John Moore.
p. cm.
Summary: "Develops models for turbulence in fluid flow using the Reynolds stress equations. Applies the models to stationary and rotating flows including boundary layers, laminar to turbulent transition, and rapid distortion. Uses glyphs to visualize turbulence stress and spectrum tensors"--Provided by publisher.
Includes bibliographical references and index.
ISBN 0-926487-21-3 (alk. paper)
1. Turbulence--Mathematical models. 2. Reynolds stress--Mathematical models.
I. Moore, John, 1943- II. Title. TA357.5.T87M66 2006
532'.0527015118--dc22
2006025242

CONTENTS

LIST OF TABLES

PREFACE

This book presents contributions to the understanding of turbulence centered around the development and application of a family of Reynolds stress models for incompressible rotating flows. The book format provides a forum in which a coherent and complete story may be told, giving the 'why' and 'how' and 'based on what', as well as describing the new Reynolds stress models themselves.

The model development is mostly concerned with determining algebraic models for pressure-strain which is a major contributor to the differential equations for the Reynolds stresses. 'Functional' in the title is meant to have two meanings. With respect to the pressure-strain modeling, it refers to the choices or functionality included in the pressure-strain model coefficients. Are they constants, do they depend on turbulence anisotropy invariants, are they different for rapidly distorting turbulence, or for flows with and without vorticity? While complex models can be more accurate, simpler models are easier to implement in 3-d CFD codes. Thus, four models, varying in complexity, are summarized in the last chapter.

This brings us to the second meaning of 'functional'. We hope the models will be practical and useful. This is reflected in the basic choice of variables, and our insistence that all the models pass basic tests and return reasonable wall skin friction coefficients without the use of wall functions. We found we could meet this goal without wall-reflection terms, except that, for the complex model, an additional pressure-strain term is added near rotating walls, to make the near-wall turbulence materially frame indifferent. Of course, we expect (and get) from the Reynolds stress models all of the turbulence stresses, not just the component which controls skin friction.

<u>What's new or different?</u>

The emphasis on boundary layer flows, including: fully developed pipe and channel flow over a broad range of Reynolds number, rotating channel and (axially) rotating pipe flow for a broad range in rotation rates, and flat plate transition for a broad range of turbulence Reynolds number and intensity.

Consideration of the similarity of homogeneous shear flow with high dimensionless strain rate, and near-wall turbulence (Lee, Kim and Moin DNS).

Analysis of DNS data (Rogers, Moin and Reynolds) to confirm the strong Reynolds number dependence of slow pressure-strain (Lumley and Newman).

The use of rapid distortion theory: as an inexpensive alternative to DNS, to set rapid pressure-strain model coefficients, to understand the limitations of algebraic rapid pressure-strain models, and to understand the effects of vorticity and rotation.

Recognition of the importance of rapid distortion effects to the calculation of flat plate transition.

A model for turbulent diffusion based on the square root of the Reynolds stress tensor, giving the correct near-wall asymptotic behavior of the turbulent diffusion term.

Glyphs to visualize the Reynolds stress tensor (Moore, Schorn and Moore), and to visualize the effects of terms in the Reynolds stress equations (both actual and modeled) on the development of the tensor.

A model for the spectrum tensors (which cover 3-d wave-number space) based on the Reynolds stresses (any anisotropy, isotropic to 1-d) and the turbulence dissipation. This uses a new model for the anisotropy of dissipation based on DNS data.

Using glyphs to visualize spectrum tensors, including the high-rotation Taylor-Proudman limit.

For completeness, the book puts Reynolds stress models in context within turbulence modeling in general, and derives the equations for the Reynolds stresses and for algebraic forms for the pressure-strain models. Several related subjects, which did not directly contribute to the final family of Reynolds stress models, are also examined: in particular, non-linear return-to-isotropy, wall reflection, elliptical instabilities in high rotation flows, and turbulence spectra.

Following this preface is a section defining the common nomenclature used throughout the book. We have tried to use commonly accepted choices for variables, but have departed from that to avoid confusing double meanings. We have converted older models (e.g. Donaldson, and Rotta) to this common nomenclature so that they may be compared with more modern models. The nomenclature also allows the option of ‘reading in the book’, piecemeal, without the daunting task of finding definitions hidden in paragraphs.

An extensive body of data was analyzed to determine model coefficients, to estimate errors, and to put the data in an appropriate form for comparison with model results. We have relegated much of this interpretation and analysis of the data in the literature to appendices on CD, so that the book text may concentrate on model development. Correspondingly, where the derivation of equations requires much algebra, the details are given in appendices on CD.

The appendices on CD serve two purposes. They allow the reader to more fully understand what we have done, and to check its accuracy. We have tried to check all of the book carefully, but should we have missed anything, it is our intent to place on a website information about the book (including corrections) and a contact email address.

We wish to thank the Virginia Tech Library, and the Commonwealth of Virginia for making this library possible. We were able to find nearly all of the publications we needed in the library, and those that were not directly available the library obtained for us through the interlibrary loan program. We also wish to thank all the authors who have published their data in tabulated form and especially those who have made it available over the internet. A list of the internet resources we have used is given at the end of the References.

Joan and John Moore, Blacksburg, Virginia, March, 2006

NOMENCLATURE

Vectors and tensors are indicated by their Cartesian subscripts. When the subscript is repeated in an expression, the sum from 1 to 3 is implied. In parentheses is the section, figure or equation where the variable is first mentioned.

a_i	Fourier transform of the fluctuating velocity, Eqs. 9.1.4, 9.1.6
a_i^*	complex conjugate of a_i. $a_i^*(\mathbf{K},t) = a_i(-\mathbf{K},t)$, since u_i is real, Eq. 9.2.6
A_1^2, A_2^2, A_3^2	coefficients in the spectrum tensor model, Eqs. 17.5.28, 17.5.38 (Eq. 17.5.24)
A_{ij}	slow pressure-strain coefficient tensor, Eq. 3.1.8
b_{ij}	turbulence anisotropy, $b_{ij} = \overline{u_i u_j}/(2k) - \delta_{ij}/3$, Eq. 1.2.3
B	dimensionless rate of change of anisotropy, Eq. 10.4.1
c_f	skin friction coefficient, $c_f = \tau_w/(0.5\rho U_{mean}^2)$ for channels and pipes, Eq. 8.1.9, or $c_f = \tau_w/(0.5\rho U_\infty^2)$ for flat plates, Eq. 11.1.2
$c_{t\psi}$	diffusion coefficient in Eq. 8.1.2, for turbulence property ψ
c_{vq}	coefficient for new v_{ij} diffusion model (Eq. 7.2.10)
$c_1,, c_{12}$	coefficients in algebraic pressure-strain model, Eqs. 3.2.7 and 3.2.17
c_{1d}, c_{2d}	coefficients in the model for ε_{ij}, Eq. 3.2.3
c_{1s}, c_{2s}	coefficients in the model for Π_{ij}^S, Eq. 3.2.2
$c_{1,D}$	D1971 model for c_1, Eq. 3.5.2
$c_{1,L}$	SL1985 model for c_1, Eq. 3.5.5
$c_{1\infty}$	high R_t limit for c_1 (Eq. 8.2.2)
$c_{\varepsilon 1}, c_{\varepsilon 2}$	coefficients in dissipation equation (Eq. 1.3.12)
c_η	parameter in 3-d spectrum model, Eq. 17.3.11, Table 17.4.1
c_μ	turbulent viscosity coefficient (Eq. 1.3.9)
$c_{\mu\psi}$	diffusion coefficient for turbulence property ψ (Eq. 7.1.2)
C_{ij}	turbulent diffusion coefficient tensor, Eq. 8.1.12
C_1	coefficient in spectrum tensor model, Eqs. 17.5.31, 17.5.38
$C_{1,\min}$	minimum value of C_1, occurs at $K = 0$ (Section 17.6)
C_B	$C_B(K)$, coefficient in the model for $B_{ij}(K)$, Eq. 17.5.14
$C_{B,\max}$	upper limit of C_B for realizability of Φ_{ij}^N, Eq. 17.5.19
$C_{B,0}$	coefficient in model for C_B, Eq. 17.5.20
dA	area of integration (Eq. 3.1.4)

D	bar or pipe diameter, or characteristic width. Two-equation analysis variable, Eq. 5.3.1
D_{ij}	pressure diffusion of Reynolds stresses, Eq. 2.2.19
D/Dt	substantial derivative, $D/Dt = \partial/\partial t + U_i \partial/\partial x_i$ (Eq. 6.4.6)
e_i	vector used in formation of spectrum tensor model (Eq. 17.5.21)
E	3-d turbulence energy spectrum, $k = \int_0^\infty E(K)dK$ (Eq. 6.2.5)
E_{ij}	1-d spectra, $\overline{u_i u_j} = \int_0^\infty E_{ij}(K_1)dK_1$, Eq. 17.1.12
E_{ij}^Z	1-d spectra in the x_3 direction, Eq. 17.6.1, $\overline{u_i u_j} = \int_0^\infty E_{ij}^Z(K_3)dK_3$
E_ψ	extra source term for property ψ (Eq. 6.1.5)
f_B	rapid distortion parameter for pressure-strain model coefficients, Eq. 10.4.6 (Eq. 10.1.1)
f_{c1}	Reynolds number dependence of c_1, Eq. 8.2.2
f_{g0}, f_{g1}	quadratic functions of g, Eqs. 5.2.7, 5.2.5
f_{g3}	cubic function of g, Eq. 5.3.9
f_{lin}	step-wise linear function of g, Eq. 5.3.10
f_R	rotating-wall parameter (Eq. 12.5.3)
f_w	vorticity parameter for pressure-strain model coefficients, Eq. 9.6.7
$F(Z)$	Fourier transform of Z (Eq. 9.1.4)
F_i	centrifugal and Coriolis force, Eq. 2.1.3
F_{ij}	Coriolis term in Reynolds stress equations, Eq. 2.2.13
g	two-dimensionality parameter for turbulence, $g = 1 - 4.5II + 9III$, Eqs. 1.2.8-9
g_{eff}	effective g in inhomogeneous turbulence, Eq. 8.3.3
G	inhomogeneity of anisotropy, Eq. 8.3.2
i	$\sqrt{-1}$ (Eq. 9.1.5)
II, III	turbulence anisotropy invariants, $II = b_{ij}b_{ij}$, $III = b_{ij}b_{jk}b_{ki}$, Eq. 1.2.4
k	turbulence kinetic energy, Eq. 1.2.2
K	magnitude of the wave number, $K = \sqrt{K_i K_i}$ (Eq. 6.2.5)
K,θ,ψ	spherical wave-number coordinates, Fig. 9.5.1
K_i, $\mathbf{K}$	wave-number vector (Eq. 9.1.5)
L	length scale, mixing length (Eq. 1.3.3)
L_{ij}	laminar diffusion of Reynolds stresses, Eq. 2.2.2
$L_{ij,k}$	integral length scales, Eq. 17.2.2
L_p	length scale in Pope's spectrum model, Eq. 17.3.11
L_{vk}	length scale in von Kármán spectrum, Eq. 17.3.9
L_x	longitudinal integral length scale, $L_x = L_{11,1}$ (Section 16.3)

M_{ijmn}	rapid pressure-strain coefficient tensor, Eq. 3.2.12
n	coordinate locally normal to the surface (Eq. 3.1.4)
n_i	unit vector normal to the wall (16.2.2)
$\mathbf{n}_k$	unit vector in the x_k direction (17.2.2)
n_t	model constant for turbulent diffusion (Eq. 8.1.2)
N	dimensionless rotation rate of pipe about its axis, $N = V_{\theta,wall}/U_{mean}$, Eq. 14.1.1
$O(Z)$	order of magnitude of Z
p	fluctuating static pressure (Eq. 2.1.6)
$\overline{P}$	time-mean static pressure (Eq. 1.3.1)
$\tilde{P}$	static pressure, $\tilde{P} = \overline{P} + p$ (Eq. 2.1.2)
P	production rate of turbulence kinetic energy, $P_{ii}/2$, Eq. 2.2.6
P_ψ	production rate of turbulence property ψ (Eq. 1.3.10)
$\mathbf{P}_i$	unit vectors in the principal stress directions (Section 1.2)
P_{ij}	production rate of Reynolds stresses, Eq. 2.2.3
$\overline{P}_r$	reduced time-mean static pressure, $\overline{P}_r = \overline{P} - \frac{1}{2}\rho\Omega^2 r^2$, Eq. 13.1.4
q	turbulence velocity scale, $q = \sqrt{k}$, Eq. 1.3.17
r,θ,z	cylindrical coordinates
R	rotation parameter, Eqs. 4.3.10, 4.3.13.
R_D	Reynolds number based on diameter or channel width and mean velocity, $R_D = U_{mean}D/\nu$, Eqs. 8.1.13, 8.1.8
R_i	Richardson number, Eq. 4.3.14
R_{ij}	the two-point velocity correlation, Eq. 9.2.1
R_o	rotation number for channel flow, Eq. 13.1.1
$R_t, \tilde{R}_t$	turbulence Reynolds number, $R_t = k^2/(\nu\varepsilon)$, $\tilde{R}_t = k^2/(\nu\tilde{\varepsilon})$, Eq. 1.3.16
R_{t1}	R_t evaluated from longitudinal spectrum, Eq. 17.4.8
R_x	dimensionless distance, $R_x = x_1 U_\infty/\nu$, Eq. 11.1.1
R_y	Reynolds number based on wall distance, $R_y = yq/\nu$ (Section 1.3)
R_θ	momentum thickness Reynolds number, $R_\theta = \theta U_\infty/\nu$, Eq. 11.1.7
R_λ	Taylor-microscale Reynolds number, $R_\lambda = \lambda\sqrt{\overline{u_1 u_1}}/\nu$, Eq. 17.2.9
$R_{\lambda 1}$	R_λ evaluated from longitudinal spectrum, $R_{\lambda 1} = \lambda_1\sqrt{\overline{u_1 u_1}}/\nu$, Eq. 17.4.9
R_τ	boundary layer or channel Reynolds number, $R_\tau = U_\tau\delta/\nu$ (Fig. 1.3.2); for rotating channel, Eq. 13.1.7
$s, \tilde{s}$	dimensionless strain rate parameter, $s = Sk/\varepsilon$, $\tilde{s} = Sk/\tilde{\varepsilon}$, Eq. 1.3.23
S	strain rate parameter, $S = \sqrt{2S_{ij}S_{ij}}$, Eq. 4.3.19
S_{ij}	strain rate, $S_{ij} = \frac{1}{2}\left(\partial U_i/\partial x_j + \partial U_j/\partial x_i\right)$, Eq. 1.3.20

S_ψ	source term for property ψ (Eq. 6.1.6)
t	time (Eq. 1.3.10)
T_{ij}	turbulent diffusion of Reynolds stresses, Eq. 2.2.16
T_u	turbulence intensity, Eqs. 11.1.8-9
T_ψ	turbulent diffusion term for property ψ (Eq. 6.1.4)
u_i	fluctuating velocity (Section 1.0)
u,v,w	fluctuating velocities, u_1,u_2,u_3, Eq. 12.1.1
$\overline{u_i u_j}$	Reynolds stress tensor, Eq. 1.2.1
U_i	relative time-mean velocity (in rotating coordinate system) (Section 1.0)
$\tilde{U}_i$	relative velocity, $\tilde{U}_i = U_i + u_i$ (Section 1.0)
U_{mean}	area-mean velocity, Eq. 8.1.7
U_τ	wall friction velocity, $U_\tau = \sqrt{\tau_w/\rho}$ (Eq. 1.3.4)
$U_{\tau,a}$	average friction velocity for rotating channel, Eq. 13.1.6
U^+	dimensionless velocity in a boundary layer, $U^+ = U/U_\tau$ (Fig. 1.3.1)
V, Vol	volume (Eq. 9.2.5, Eq. 3.1.4)
V_i	absolute mean velocity, Eq. 2.1.7
v_{ij}	square root of the Reynolds stress tensor, $v_{ik}v_{kj} = \overline{u_i u_j}$, Eq. 7.2.7
W	absolute vorticity parameter, $W = \sqrt{2W_{ij}W_{ij}}$, Eq. 4.3.19
W_{ij}	absolute vorticity tensor, $W_{ij} = \frac{1}{2}\left(\partial V_i/\partial x_j - \partial V_j/\partial x_i\right)$, Eq. 3.2.16
x_i, $\mathbf{x}$	Cartesian coordinates, position vector (Section 1.2)
$x_{1,s}$	location of the start of transition (Fig. 11.3.4)
X_{ijkm}	rapid pressure-strain coefficient tensor, Eq. 3.1.7
y	coordinate normal to the wall, distance to the wall (Eq. 1.3.4)
y^+	dimensionless distance to the wall, $y^+ = yU_\tau/\nu$, Eq. 1.3.8
α	Kolmogorov constant (Eq. 6.2.6), or parameter (Table 17.4.1)
"α"	α determined from compensated spectra, Eq. 17.4.11
β	dimensionless time, $\beta = St$ (Eq. 3.3.3); or coefficient in 3-d spectrum models, Eqs. 17.3.5, 17.3.11
β_B	coefficient in model for C_B, Eq. 17.5.20
B_{ij}	$B_{ij}(K)$ describes the average anisotropy of Φ_{ij}^N at K, Eq. 17.5.10
Γ	the gamma or factorial function, $\Gamma(n+1) = n\Gamma(n)$ (Eq. 17.3.10)
δ	boundary layer thickness (Eq. 1.3.7)
δ^*	boundary layer displacement thickness, Eq. 11.1.5
δ_{ij}	Kronecker delta, $\delta_{ij} = 1$ for $i = j$; $\delta_{ij} = 0$ for $i \neq j$ (Eq. 1.2.3)
ε	dissipation rate of turbulence kinetic energy, $\varepsilon = \varepsilon_{ii}/2$, Eq. 2.1.18

ε_1	ε evaluated from longitudinal spectrum, Eq. 17.4.6
$\tilde{\varepsilon}$	homogeneous part of dissipation rate, $\tilde{\varepsilon} = \varepsilon - 2\nu(\partial q/\partial x_i)^2$, Eq. 1.3.15
ε_ψ	dissipation rate of turbulence property ψ (Eq. 1.3.10)
ε_{ij}	dissipation rate of Reynolds stresses, Eq. 2.2.21
ε_{ijk}	permutation or cross-product tensor, $(\mathbf{a} \times \mathbf{b})_i = \varepsilon_{ijk} a_j b_k$ (Eq. 2.1.7)
η	Kolmogorov length scale, $\eta = \nu^{3/4}/\varepsilon^{1/4}$, Eq. 17.2.1
η_1	η evaluated from longitudinal spectrum, $\eta_1 = \nu^{3/4}/\varepsilon_1^{1/4}$, Eq. 17.4.7
θ	boundary layer momentum thickness, Eq. 11.1.6
ρ	density (Eq. 1.3.1)
κ	von Kármán constant ($\kappa \sim 0.418$) (Eq. 1.3.4)
λ	Taylor microscale, Eq. 17.2.5
λ_1	longitudinal microscale, Eq. 17.2.10
μ	laminar viscosity (Eq. 2.1.2)
ν	kinematic laminar viscosity, $\nu = \mu/\rho$ (Eq. 1.3.1)
ν_t	turbulent viscosity (Eq. 1.3.2)
ξ	von Mises transverse (streamline) variable, Eq. 11.2.1
ξ,η	invariant coordinates for turbulence anisotropy, Eq. 15.3.10
Π_{ij}	pressure-strain, Eq. 2.2.20
σ_ψ	turbulent Prandtl number of turbulence property ψ (Eq. 1.3.10)
τ_w	wall shear stress, Eq. 8.1.10
ϕ_{ij}	effective pressure-strain including anisotropy of dissipation, Eq. 2.2.22
$\hat{\phi}_{ij}$	generalized effective pressure-strain, Eq. 12.3.10
Φ_{ij}	Fourier transform of the two-point velocity correlation, Eq. 9.2.2, also referred to as the spectrum tensor
Φ_{ij}^N	dimensionless part of the spectrum tensor, Eq. 17.5.6
ω	specific turbulence dissipation rate, $\omega = \tilde{\varepsilon}/k$, Eq. 1.3.18
ω_i	mean relative vorticity vector, $\omega_i = \varepsilon_{ijk}\, \partial U_k/\partial x_j$, Eq. 2.2.10
ω_{ij}	mean relative vorticity, $\omega_{ij} = \frac{1}{2}(\partial U_i/\partial x_j - \partial U_j/\partial x_i)$, Eq. 1.3.25
Ω_E	rotation parameter for elliptic flow, Eq. 18.5.8
Ω_i	coordinate system rotation rate (Eq. 2.1.3)
ψ	a turbulence property governed by a transport equation, $\psi = q, k, \varepsilon, b_{ij}$, etc. (Eq. 1.3.10)
—	overbar, time or ensemble average

Subscripts

fdf	fully-developed non-rotating flow
lam	laminar
p	pressure side of rotating channel (Eq. 13.1.2)
P_i	components in the principal stress directions (Eq. 1.2.5)
r,θ,z	cylindrical components
rdt	rapid distortion theory
s	suction side of rotating channel (Eq. 13.1.2)
0,o	initial value, or non-rotating value
∞	free stream value

Superscripts

R	rapid
S	slow
W	wall
+	dimensionless quantities based on wall shear (Appendix A)

Abbreviations

DNS	direct numerical simulations
MFI	materially frame indifferent
IP	"isotropization of production" model (Eq. 3.5.4)
RDT	rapid distortion theory
rms	root mean square

Chapter 1

INTRODUCTION

The velocity in a flow can be decomposed into a mean or ensemble average velocity, and a fluctuating or turbulent velocity, $\tilde{U}_i = U_i + u_i$. By definition the means, or ensemble averages, of the components of the turbulent fluctuating velocity are zero, $\overline{u_i} = 0$. Turbulence can be seen in the irregular movement of snow as it falls on a stormy day, or in the churning of water at the bottom of a waterfall. Five hundred years ago, Leonardo da Vinci (1510) sketched and described the transient swirls and eddies he observed in turbulent flow.

Osborne Reynolds, in 1895, used an energy approach to analyze turbulence. Energy is taken out of the kinetic energy of the mean flow and transferred into turbulence kinetic energy, which then dissipates to internal energy. The energy transfer occurs through the interaction between the Reynolds stresses, $\overline{u_i u_j}$, which bear his name, and the mean velocity gradient. This is in a sense a thermodynamic approach to turbulence. The effect of the 'black box' called turbulence on mean momentum and energy is considered without detailed knowledge of the individual transient swirls or other motions in the turbulence.

It is many decades since Chou (1945) suggested that differential equations be solved for the Reynolds stresses and the triple velocity correlations. At the time the numerical solutions to these equations were impractical. Instead turbulence was modeled by the Prandtl (1925) algebraic mixing length model, which is still used today in generalized form. Currently, two-equation models, which solve two differential equations for turbulence properties, for example $k-\varepsilon$ or $q-\omega$ models, are also widely used.

Rotta, in 1951, completed a Reynolds stress model, with an algebraic model for the pressure-velocity correlation term in the Reynolds stress equations, using a combination of theoretical constraints and empirical data. Numerous Reynolds stress models have followed, with different ideas on what is important.

Recent direct numerical simulations (DNS) of low Reynolds number turbulent flows have dramatically increased the body of knowledge regarding Reynolds stresses and the terms in the Reynolds stress equations. With increased computer power, the practical implementation of Reynolds stress models is now widely feasible. Now is therefore an appropriate time to stand back, in a sense start again, consider the wide body of data which is now available, and see what modeling choices may or should be made differently to model the terms in the differential equations for the Reynolds stresses, and therefore the Reynolds stresses themselves.

1.1 Goals of This Work

A central goal of this work is the development of practical models of the Reynolds stress equations, which may be implemented in three-dimensional Reynolds-averaged Navier-Stokes solvers. Considering a seven-equation model as an upgrade from two-equation turbulence modeling, we require from it the same generality, and obviously more, than has been obtained with two-equation models. In particular, we require low Reynolds number capability so that flows may be calculated to the walls without the use of wall-function boundary conditions. The equations are developed for constantly rotating reference frames as are used in the calculation of flows in, for example, turbomachinery applications. The effects of curvature and rotation on the turbulence are considered, as is the difference between rapidly distorting and nearly fully developed turbulence. Although compressibility, heat transfer and buoyancy are important in many applications, their effects on turbulence are not considered here. The new Reynolds stress models are based on incompressible flow, the basis of most Reynolds stress models in the literature.

The most important modeling aspect is the algebraic model for the pressure-strain. Considerations such as isotropic limits and realizability are included. In developing the model, the data in the literature is reanalyzed. For example, many models for pressure-strain are based on single, or fixed-point homogeneous shear data. Here, no litmus test is applied, but instead a broad group of data is considered. Early modelers did not have this luxury.

To complete the model it is necessary to choose the differential equations to be solved, determine coefficients for the dissipation or time-scale equation, and model the turbulent diffusion in the equations. All these are reconsidered based on practicality, theoretical limits and available data.

Testing on examples is important to determine if the goals have been reached. Indeed the choice of examples and comparison with data tends to drive the direction model development takes. The current model has been developed within a fairly flexible framework where capabilities can be added.

Some history is included and comparisons with other models will be made. But the purpose is to understand the basis of the various models. If other modelers had other objectives and limited available data in forming and selecting their models, as well as very limited computing power, it should not be surprising if their models do not satisfy the goals of this work.

1.2 The Reynolds Stress Tensor, Invariants, and Visualization

The Reynolds stresses are the components of a 3 by 3 Cartesian tensor,

$$\overline{u_i u_j} = \begin{bmatrix} \overline{u_1 u_1} & \overline{u_1 u_2} & \overline{u_1 u_3} \\ \overline{u_2 u_1} & \overline{u_2 u_2} & \overline{u_2 u_3} \\ \overline{u_3 u_1} & \overline{u_3 u_2} & \overline{u_3 u_3} \end{bmatrix} , \tag{1.2.1}$$

which is symmetric, $\overline{u_i u_j} = \overline{u_j u_i}$. The turbulence kinetic energy is one half of the trace of the Reynolds stress tensor,

$$k = \frac{1}{2}\left(\overline{u_1 u_1} + \overline{u_2 u_2} + \overline{u_3 u_3}\right) = \frac{1}{2}\overline{u_i u_i} \ . \tag{1.2.2}$$

(Note the summation convention, sum $i = 1,3$, implied by the repeated subscript, i.)

The dimensionless anisotropy of the Reynolds stresses is defined by

$$b_{ij} = \frac{\overline{u_i u_j}}{2k} - \frac{1}{3}\delta_{ij} , \tag{1.2.3}$$

so that $b_{ii} = 0$. Like the Reynolds stress tensor, the b_{ij} tensor is symmetric, $b_{ij} = b_{ji}$.

The Reynolds stresses and the turbulence anisotropy are obviously dependent on the coordinate system chosen. The turbulence kinetic energy is an invariant of the Reynolds stress tensor, i.e., it is a scalar, which is independent of the choice of coordinate system. Two other independent scalar invariants of the Reynolds stress tensor may be defined by

$$II = b_{ij} b_{ji} \qquad \text{and} \qquad III = b_{ij} b_{jk} b_{ki} . \tag{1.2.4}$$

Coordinate Transformation and Principal Coordinates

The component of the velocity in any direction is found by the dot product of a unit vector in the direction of interest with the velocity vector. For unit vector $\mathbf{P}_1$,

$$u_{P_1} = u_i P_{1,i} \ . \tag{1.2.5}$$

Because the Reynolds stress tensor is symmetric, in may be transformed into principal directions, as used in mechanics. In the principal-direction coordinate system, the shear, or off-diagonal, stresses are zero and only the normal stresses remain. With $\mathbf{P}_i$ as the unit vectors in the principal directions,

$$\overline{u_{P_i}u_{P_j}} = \begin{bmatrix} \overline{u_{P_1}u_{P_1}} & 0 & 0 \\ 0 & \overline{u_{P_2}u_{P_2}} & 0 \\ 0 & 0 & \overline{u_{P_3}u_{P_3}} \end{bmatrix} . \tag{1.2.6}$$

Reynolds stresses differ from their mechanical counterparts in that the normal stresses are always positive, regardless of the coordinate system chosen to represent them.

$$u_\alpha u_\alpha \geq 0 \qquad \text{(no sum on } \alpha\text{)}. \tag{1.2.7}$$

Insuring that the results of a model always give positive normal stresses, no matter the coordinate system chosen, is defined as the condition of realizability. (This is sometimes referred to as 'weak realizability', Pope, 2000, p. 402)

A dimensionless product of the principal normal stresses,

$$g = \frac{\overline{u_{P_1}u_{P_1}}}{\frac{2}{3}k} \frac{\overline{u_{P_2}u_{P_2}}}{\frac{2}{3}k} \frac{\overline{u_{P_3}u_{P_3}}}{\frac{2}{3}k} , \tag{1.2.8}$$

is a useful invariant for tagging the limit for which one or more of the principal normal stresses goes to zero. When the turbulence is isotropic, each of the principal normal components is $2k/3$, and $g = 1$. In terms of the invariants II and III,

$$g = 1 - \frac{9}{2} II + 9III . \tag{1.2.9}$$

The turbulence is called two-dimensional when only two of the principal normal stresses are non-zero, and one-dimensional when only one of the principal normal stresses is non-zero.

Mohr's Circles

In mechanics, Mohr's circles (see e.g. Beer and Johnston, 1981) are used to determine the shear stresses in the principal planes from the principal normal stresses. They are particularly useful as a way of viewing coordinate transformation when one of the principal directions coincides with one of the chosen Cartesian coordinate directions. For example, consider a typical Reynolds stress tensor, which may occur in a two-dimensional turbulent boundary layer, where the x_1 direction is the flow direction and the x_2 direction is normal to the wall (Appendix A),

$$\text{B.L. example:} \qquad \frac{\overline{u_i u_j}}{k} = \begin{bmatrix} 1.07 & -0.3 & 0 \\ -0.3 & 0.39 & 0 \\ 0 & 0 & 0.54 \end{bmatrix}. \tag{1.2.10}$$

The x_3 direction is a principal direction since the shear stresses $\overline{u_1 u_3}$ and $\overline{u_2 u_3}$ are zero. The Mohr's circle for this tensor is shown in Fig. 1.2.1. In the Mohr circle representation, the normal stresses in the x_1 and x_2 directions (points d and e) are 180 degrees apart, twice the physical angle difference of the axes. Note that the principal normal stresses in the x_1-x_2 plane (points a and c) are the minimum and maximum normal stresses in the plane.

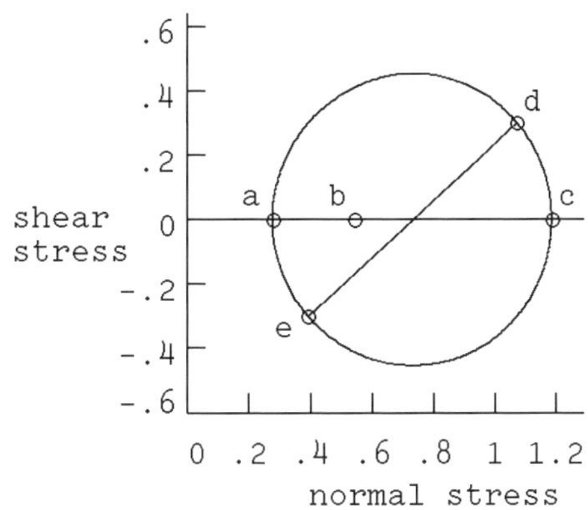

Fig. 1.2.1. Mohr's circle for Reynolds stresses in Eq. 1.2.10. Points a and c, principal normal stresses in x_1-x_2 plane. Point b, principal normal stress, $\overline{u_3 u_3}/k$. Point d, $\overline{u_1 u_1}/k$, $-\overline{u_1 u_2}/k$. Point e, $\overline{u_2 u_2}/k$, $\overline{u_1 u_2}/k$.

A Reynolds Stress Tensor Glyph

Using the property that the normal stresses are always positive, Moore et al. (MSM1996) devised a stress glyph to visualize the Reynolds stress tensor in three dimensions. The glyphs are defined such that the distance from the glyph origin to any point on its surface is the magnitude of the normal stress in that direction. The 3-d object created is independent of arbitrary choices of Cartesian coordinate directions.

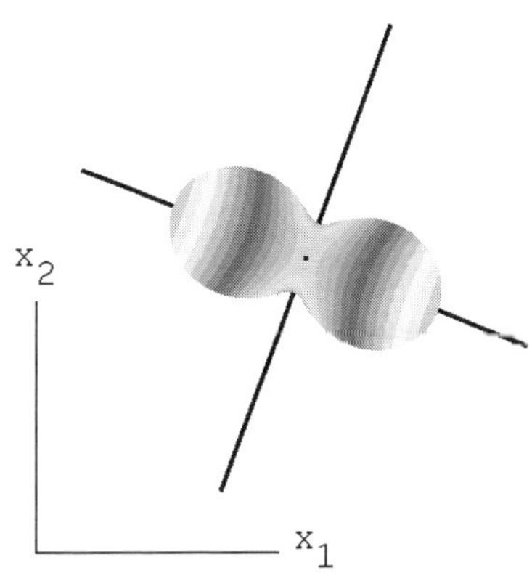

Fig. 1.2.2. Three dimensional glyph for Reynolds stresses in Eq. 1.2.10.

An example of the glyph is given in Fig. 1.2.2 for the Reynolds stress tensor given in Eq. 1.2.10. Contours of constant normal stress are marked on the surface of the glyph to help define its three-dimensional shape. The principal directions are marked, but are also easily seen based on the symmetry of the glyph shape.

The two-dimensional outline of the glyph, Fig 1.2.3, as seen in the $x_1 - x_2$ plane shows how the glyph was constructed. The surface of a cube was transformed into the glyph by moving each point along the direction from the center of the glyph, so that the distance from the center to the point on the surface is the normal stress in that direction.

Fig. 1.2.3. Sketch of glyph construction.

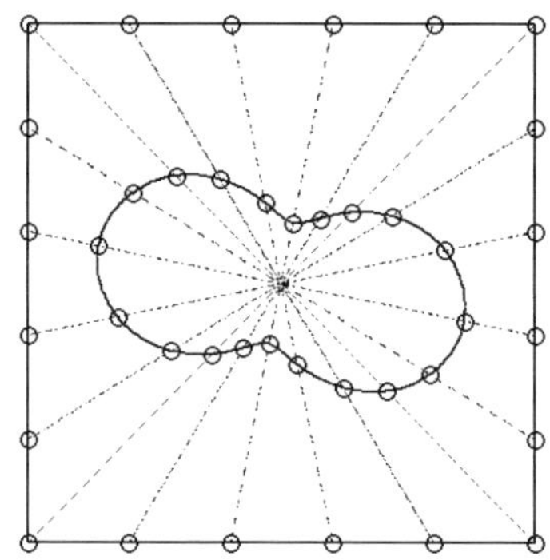

Figure 1.2.4 shows a glyph for a more general Reynolds stress tensor where none of the principal directions coincide with the Cartesian directions. The length of the principal axes in each direction was taken to be the maximum value the normal stress could take if the glyph were shaped differently, but the point had the same turbulence kinetic energy. For this dimensionless glyph, the principal axes have a radial length of 2, since $\overline{u_i u_j}/k \le 2$. Reynolds stress glyphs will be used throughout the book to visualize the shape and orientation of the Reynolds stress tensor.

Fig. 1.2.4. Three-dimensional glyph for the Reynolds stress tensor,

$$\frac{\overline{u_i u_j}}{k} = \begin{bmatrix} 0.610 & -0.332 & -0.235 \\ -0.332 & 0.855 & 0.240 \\ -0.235 & 0.240 & 0.535 \end{bmatrix}.$$

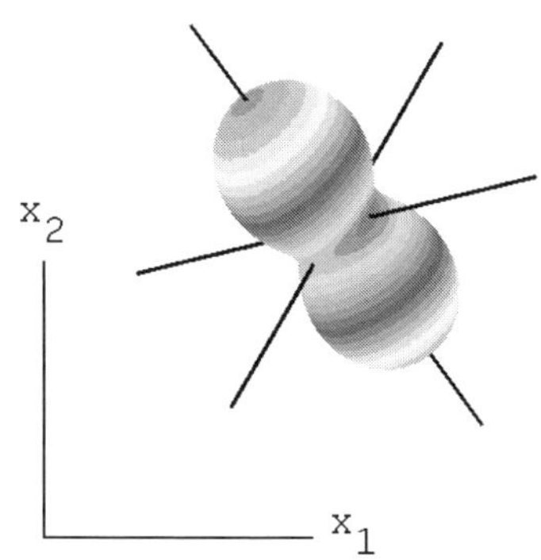

Approaching Two-Dimensionality

Figure 1.2.5 outlines the range of anisotropy that turbulence can have in terms of the invariants g and III.

Shown at the corners of the permissible range are glyphs of the turbulence stresses. For $g = 1$, the turbulence is isotropic with the same normal stress in all directions, and the glyph is a sphere. Turbulence generated by axisymmetric contraction has a low normal stress in the direction of the flow and equal normal stresses in the directions perpendicular to the flow. In the 2-d limit, $g = 0$, the stress glyph is doughnut shaped. For axisymmetric expansion, the normal stress in the direction of the flow is large and the normal stresses in the perpendicular plane are small. The limit at $g = 0$ is one-dimensional turbulence. The corresponding glyph is peanut shaped.

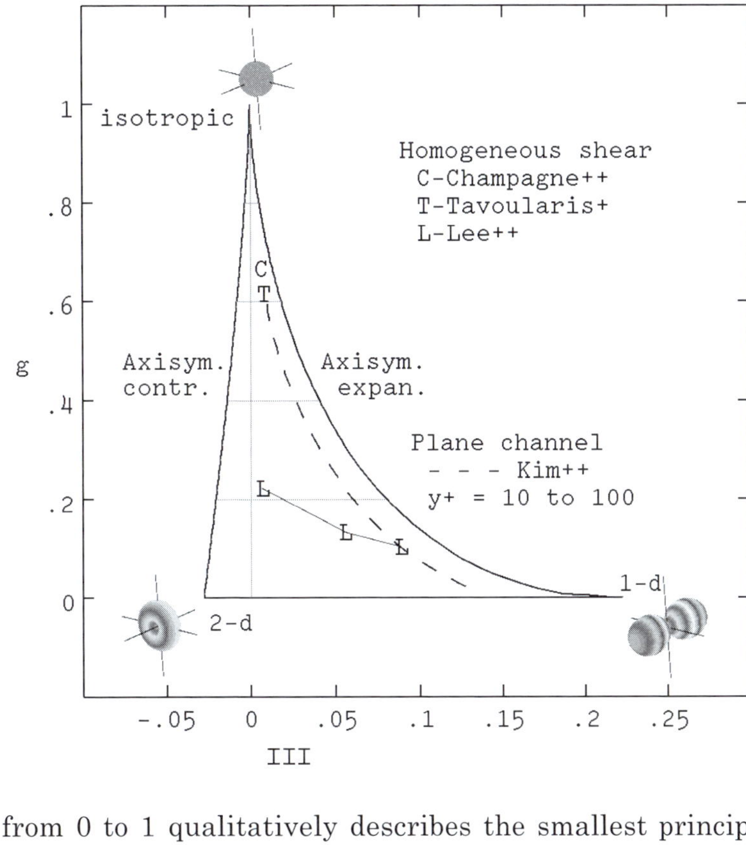

Fig. 1.2.5. Range of turbulence anisotropy in terms of g and III.

While g varying from 0 to 1 qualitatively describes the smallest principal stress varying between 0 and the average value $2k/3$, III describes whether most of the principal stresses are above average, $III < 0$, or below average, $III > 0$. When $III = 0$, one principal normal stress equals $2k/3$.

Also shown on the plot are stresses for some simple shear flows. Kim et al. (KMM1987) performed direct numerical simulations (DNS) for fully developed channel flow. At $y^+ = 10$, g is very low and the turbulence is nearly two-dimensional and approaching one-dimensional. The end of the curve for $y^+ = 100$ is near the point C on the plot, and has turbulence stresses corresponding to the log-law region of a turbulent boundary layer.

The points C, T and L are from nearly homogeneous shear experiments where the only component of the velocity gradient, $\partial U_1 / \partial x_2$, is uniform. Champagne et al. (CHC1970), point C, achieved nearly fully-developed turbulence anisotropy for a dimensionless strain rate, $s = (k/\varepsilon)\partial U_1 / \partial x_2$, which corresponds closely to the log-law region of a turbulent boundary layer with s approximately 3.33 (Appendix A). Tavoularis and Corrsin (TC1981), with nearly double the shear rate ($s = 6.1$), achieved a more fully developed state for turbulence anisotropy. Lee et al. (LKM1990), performed DNS calculations for large shear rates with s approximately 17. The three points marked L on Fig. 1.2.5 follow the development of the anisotropy for their calculations, with the lowest g point being the closest to fully developed.

It is important to note, as did Lee et al., that the turbulence for high Reynolds number homogeneous shear flow, in the limit approaching fully developed anisotropy, is similar to the turbulence in moderate-to-low Reynolds number near-wall flows. Both can have turbulence which is approaching one-dimensional with low g values. It seems reasonable then, that any Reynolds stress model which will at some time be extended to low Reynolds number near-wall flows, should first be able to model high Reynolds number homogeneous shear flows. Since the g values for the near-wall flow are very low, having a model which ensures that realizability will be satisfied as g tends to zero, is also likely to be important.

1.3 Levels of Turbulence Modeling

The level of a turbulence model may be defined by the number of differential equations which are used to specify the properties of the turbulence. Turbulence models in the literature range from zero-equation, Prandtl mixing length, models, to a model developed by Poroseva et al. (PKLR2000) which uses 25 equations. The k-ε model uses two differential equations. This book primarily considers the development of Reynolds stress models which solve seven differential equations: six equations corresponding to each of six independent Reynolds stresses, and one equation for a length, time, or dissipation scale.

Before embarking on the task of considering Reynolds stress models, and the development of new Reynolds stress models, it is useful to review briefly simpler turbulence models, and what they do, and do not do. In developing a Reynolds stress model, the goal is to have it perform as well as simpler models, on the tasks the simpler models have been designed to accomplish, and then do more.

Zero-Equation Models

For a simple two-dimensional turbulent boundary layer, with flow in the x_1 direction, and with y or x_2 being the direction normal to the wall, the only component of the Reynolds stress tensor to significantly influence the mean velocity is $\overline{u_1 u_2}$. For fully developed pipe or channel flow, the momentum equation reduces to

$$0 = -\frac{1}{\rho}\frac{\partial \overline{P}}{\partial x_1} - \frac{\partial \overline{u_1 u_2}}{\partial x_2} + \frac{\partial}{\partial x_2}\nu\left(\frac{\partial U_1}{\partial x_2} + \frac{\partial U_2}{\partial x_1}\right) . \tag{1.3.1}$$

Boussinesq (1877) suggested that the turbulent shear stress could be modeled in an analogous way to the laminar shear stress using an eddy viscosity model.

$$-\overline{u_1 u_2} = \nu_t\left(\frac{\partial U_1}{\partial x_2} + \frac{\partial U_2}{\partial x_1}\right) . \tag{1.3.2}$$

The turbulent viscosity, ν_t, which can be considered as the product of a length scale times a velocity scale, is specified algebraically. Prandtl (1925) suggested a mixing length model where the velocity scale is modeled as the mixing length times the mean velocity gradient, so that,

$$\nu_t = L^2\left|\frac{\partial U_1}{\partial x_2}\right| . \tag{1.3.3}$$

Using the empirical relation for the velocity gradient in the log-law region (Appendix A) of a turbulent boundary layer, and with the wall and local shear stresses being approximately equal,

$$\frac{\partial U_1}{\partial x_2} = \frac{U_\tau}{\kappa y} = \frac{\sqrt{\tau_w/\rho}}{\kappa y} \approx \frac{\sqrt{-\overline{u_1 u_2}}}{\kappa y} . \tag{1.3.4}$$

This gives the mixing length as $L = \kappa y$, where κ is the von Kármán constant.

The Boussinesq model, in a generalized form gives all the components of the Reynolds stress tensor,

$$\overline{u_i u_j} = -\nu_t\left(\frac{\partial U_i}{\partial x_j} + \frac{\partial U_j}{\partial x_i}\right) + \frac{2}{3}k\delta_{ij} . \tag{1.3.5}$$

This is used in zero-, one-, and two-equation models, with the last term being combined with the static pressure and therefore omitted in zero-equation models. Correspondingly, the turbulent viscosity can be generalized,

$$\nu_t = L^2 \sqrt{\frac{1}{2}\left(\frac{\partial U_i}{\partial x_j} + \frac{\partial U_j}{\partial x_i}\right)\left(\frac{\partial U_i}{\partial x_j} + \frac{\partial U_j}{\partial x_i}\right)} \;, \tag{1.3.6}$$

and a more general relation for the mixing length used, e.g.,

$$L = \min\left(0.41y(1 - e^{-y^+/26}),\;\; 0.08\delta\right), \tag{1.3.7}$$

where

$$y^+ = \frac{yU_\tau}{\nu} \;. \tag{1.3.8}$$

An important feature of this model is the exponential van Driest (1956) correction to the mixing length. This allows the model to calculate $\overline{u_1 u_2}$ through the buffer layer (approx. $5 < y^+ < 40$) and the laminar sublayer (approx. $y^+ < 5$), and therefore to successfully calculate the velocity profile and pressure drop in fully developed turbulent pipe or channel flows, Fig. 1.3.1. To complete the model for developing flows, a search procedure is needed to determine the boundary layer thickness, δ.

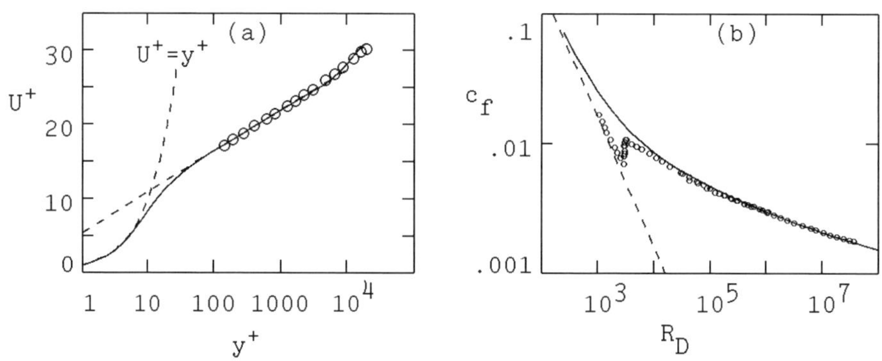

Fig. 1.3.1. Results of mixing length model for fully developed pipe flow. (a) Velocity profile for R_D=10^6, —— model; o, data of Zagarola (1996); - - - $U^+=y^+$, U^+=5.45+ln(y^+)/0.418. (b) Wall skin friction, —— model; o, data from McKeon et al. (MSZDS2004); - - - laminar limit.

The mixing length model has been successfully employed in performance evaluation of turbomachinery flows (e.g., Moore et al., MMT1984, Moore and Moore, 1985). The success in highly three-dimensional calculations relied strongly on the ability of the underlying code to accurately calculate inviscid flow. In other words, the inviscid convection terms must be evaluated accurately, so that the turbulence model governs the mixing in the flow, without the corrupting influence of numerical mixing (Moore and Moore, 1984).

Using mixing length models, both Moore and Moore (1991) and Cleak et al. (CGB1991) found that some details of the computed tip leakage or passage vortices in turbine cascades could be improved by "turning off" the turbulent viscosity in parts of the flow. Moore and Moore also noted that the low static pressure measured in the vortex core was not reproduced by the calculations. Reynolds stress modeling might be able to shed light on the flow physics in these regions.

One-Equation Models

In one-equation models, the velocity scale for the eddy viscosity is calculated using a differential equation for either the turbulence kinctic energy, k, or for the velocity scale, $q=\sqrt{k}$. The mixing length is still specified algebraically. Near the wall, a van Driest type correction based on a turbulence/wall-distance Reynolds number, $R_y = yq/\nu$, may be used to allow calculations through the buffer layer and sublayer. Additional (algebraic) models for the turbulent diffusion and dissipation of either q or k complete the model. In addition to the flows calculated by zero-equation models, because they include some information on the history of the turbulence, one-equation models can be extended to calculate transitional flows (Moore and Moore, 1995).

Two-Equation Models

The difficulty of determining the mixing length for flows other than simple boundary layers, has led to the widespread use of two-equation turbulence models. The most well known is the standard k-ε model (Launder and Spalding, 1974), where ε is the dissipation rate of the turbulence. The turbulent viscosity is given by

$$\nu_t = c_\mu \frac{k^2}{\varepsilon} \ . \tag{1.3.9}$$

The generic model equation for the transport of a turbulence property, ψ, includes convection, laminar and turbulent diffusion, production and dissipation,

$$\frac{\partial\psi}{\partial t} + U_k \frac{\partial\psi}{\partial x_k} = \frac{\partial}{\partial x_k}\nu\frac{\partial\psi}{\partial x_k} + \frac{\partial}{\partial x_k}\frac{\nu_t}{\sigma_\psi}\frac{\partial\psi}{\partial x_k} + P_\psi - \varepsilon_\psi \ . \tag{1.3.10}$$

For the k equation, $P_k = P = -\overline{u_i u_j}\,\partial U_i/\partial x_j$, and $\varepsilon_k = \varepsilon$.

In the log-law region of simple high Reynolds number turbulent boundary layers, it has been observed that the turbulence kinetic energy is essentially uniform (Patel et al., PRS1985). With convection and diffusion negligible, turbulence production is equal to turbulence dissipation. Therefore,

$$-\overline{u_1 u_2} = \nu_t \frac{\partial U_1}{\partial x_2} = c_\mu \frac{k^2}{\varepsilon} \frac{\partial U_1}{\partial x_2} = c_\mu \frac{k^2 \, \partial U_1 / \partial x_2}{-\overline{u_1 u_2} \, \partial U_1 / \partial x_2} \, . \qquad (1.3.11)$$

With the structure coefficient, $\overline{u_1 u_2}/k$, being around -0.3 in this region, the standard value, $c_\mu = 0.09$ is obtained.

For the standard k-ε model, the production and dissipation of ε are modeled as being proportional to the production and dissipation of k,

$$P_\varepsilon = c_{\varepsilon 1} P \frac{\varepsilon}{k}, \qquad \varepsilon_\varepsilon = c_{\varepsilon 2} \varepsilon \frac{\varepsilon}{k}. \qquad (1.3.12)$$

The diffusion of ε is not negligible. To obtain the correct slope for the velocity in the log-law region, Eq. 1.3.4, the coefficients in the ε equation must satisfy,

$$c_{\varepsilon 1} = c_{\varepsilon 2} - \frac{\kappa^2}{\sigma_\varepsilon \sqrt{c_\mu}} \, . \qquad (1.3.13)$$

The coefficients for the standard model are:

$$c_\mu = 0.09, \ \sigma_k = 1, \ c_{\varepsilon 1} = 1.44, \ c_{\varepsilon 2} = 1.92, \text{ and } \sigma_\varepsilon = 1.3. \qquad (1.3.14)$$

Low Reynolds number modifications are needed to represent the buffer layer and laminar sublayer near walls with the same accuracy as zero- and one-equation models. Several low-Reynolds-number models were reviewed by Patel et al. (PRS1985). Some of these models separate the dissipation into a transported part, $\tilde{\varepsilon}$, and an algebraic term,

$$\varepsilon = \tilde{\varepsilon} + 2\nu \left(\frac{\partial \sqrt{k}}{\partial y} \right)^2 \, . \qquad (1.3.15)$$

Additional modifications include reductions to the turbulent viscosity near walls, i.e. van Driest type corrections, based on the turbulence Reynolds number,

$$R_t = \frac{k^2}{\varepsilon \nu}, \text{ or } \tilde{R}_t = \frac{k^2}{\tilde{\varepsilon} \nu} \, . \qquad (1.3.16)$$

These modifications are preferred by some code developers over ones which are based on y^+, since R_t and $\tilde{R}_t$ may be computed independent of the distance to the wall. Figure 1.3.2 shows typical turbulence Reynolds number variations near a wall, with R_t and $\tilde{R}_t$ zero at the wall and increasing to about 100 at $y^+ = 10$.

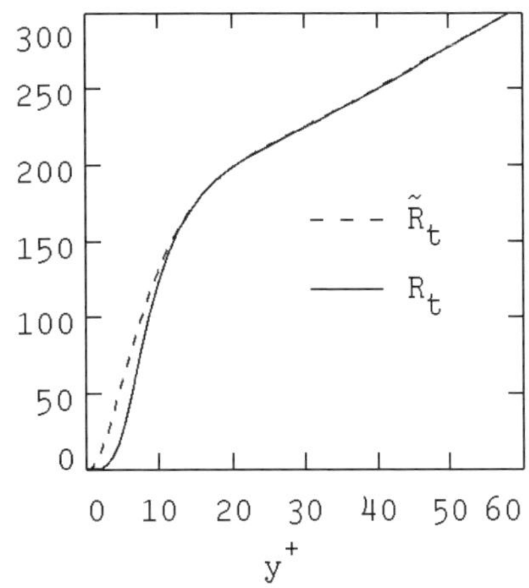

Fig. 1.3.2. Turbulence Reynolds numbers from the fully developed channel flow DNS data of Moser et al. (MKM1999), for $R_\tau = 590$.

Coakley considerably simplified the calculation of near-wall flows by choosing different primary turbulence variables. He chose the velocity scale,

$$q = \sqrt{k}, \tag{1.3.17}$$

which varies linearly near the wall, rather than k, which varies as y^2. And, instead of $\tilde{\varepsilon}$, he chose the specific dissipation rate,

$$\omega = \tilde{\varepsilon}/k, \tag{1.3.18}$$

which has less variation than $\tilde{\varepsilon}$ near the wall. The Coakley and Hsieh (1985) model includes a van Driest type correction in the turbulent viscosity. The functions and constants are:

$$\begin{aligned} &\nu_t = c_\mu f_v q^2/\omega, \quad f_v = 1 - \exp(-\alpha y q/\nu), \\ &c_\mu = 0.09, \quad \alpha = 0.02, \quad P_q = P/(2q), \quad \varepsilon_q = q\omega/2, \quad \sigma_q = 2, \\ &P_\omega = (0.555 + 0.055/f_v)P\omega/k, \quad \varepsilon_\omega = 0.833\omega^2 \text{ and } \sigma_\omega = 2. \end{aligned} \tag{1.3.19}$$

Two-equation models are considered superior to the simpler models for the calculation of highly turbulent flows in, for example, combustors. They are also *potentially* better than the simpler models when the convection history of the turbulence is important.

Realizability. However, Moore and Moore (1997) found that in order to reach that potential, a realizability correction needed to be added to two-equation models. In terms of the strain rate,

$$S_{ij} = \frac{1}{2}\left(\frac{\partial U_i}{\partial x_j} + \frac{\partial U_j}{\partial x_i}\right), \tag{1.3.20}$$

Eq. 1.3.5 may be combined with Eq. 1.3.8 to give the dimensionless stresses,

$$\frac{\overline{u_i u_j}}{2k} = -c_\mu \frac{k}{\varepsilon} S_{ij} + \frac{1}{3}\delta_{ij} . \tag{1.3.21}$$

Clearly if, for example, $c_\mu k S_{11}/\varepsilon$ is greater than 1/3, then $\overline{u_1 u_1}$ will be less than zero and not realizable. Durbin (1994) corrected his model by limiting c_μ, using

$$c_\mu = \min\left(0.09,\ 1/(\sqrt{3}\, s)\right), \tag{1.3.22}$$

where the dimensionless strain rate, s, is given by

$$s = \frac{k}{\varepsilon}\sqrt{2 S_{ij} S_{ij}} . \tag{1.3.23}$$

Durbin (1996), as well as Moore and Moore, found that unacceptable overproduction of turbulence at blade leading edges could be corrected with realizability modifications to c_μ. Several formulae for c_μ are in the literature, see e.g., Moore and Moore (1999). Some, e.g. Pope (1975), attempt to do more than simply correct for realizability with an algebraic model for c_μ.

Moore and Moore (1998) calculated the flow in a compressor rotor using a $q-\omega$ model with a realizability constraint. The distributions of k and ε downstream of the rotor were used to obtain turbulence spectra using a Chapman/Pao model. The results agreed well with hot-wire spectra measurements. Therefore, it may be possible to obtain some spectra information from a Reynolds stress model. This is considered further in Chapter 17.

Boussinesq Limitation. All of the zero-, one-, and two-equation models are limited by the Boussinesq model of the Reynolds stresses,

$$\overline{u_i u_j} = -2\nu_t S_{ij} + \frac{2}{3} k \delta_{ij} . \tag{1.3.24}$$

In the Boussinesq model, the turbulence stress tensor, $\overline{u_i u_j}$ has the same principal directions as the strain rate tensor, S_{ij}. For simple two-dimensional boundary layer flow, with $\partial U_1/\partial x_2$ the only non-zero component of the velocity gradient, these principal directions are at 45 degrees to the flow (x_1) and wall normal (x_2) directions. The modeled normal stresses are all equal in the x_1, x_2 and x_3 directions.

Glyphs for the Boussinesq model are compared with Reynolds stress glyphs for DNS data for fully developed channel flow in Fig. 1.3.3. The glyphs and correspondingly the normal stresses are quite different. For the Boussinesq model, the shear stress, $\overline{u_1 u_2}/k$, was taken to be the same as for the DNS data, and is shown as a dashed line. Therefore, for both sets of glyphs, $\overline{u_1 u_2}/k$ increases to zero as the wall is approached. However, this occurs in two different ways. In the Boussinesq model, $\overline{u_1 u_2}/k$ approaches zero because the turbulence becomes isotropic as the wall is approached. In the DNS data, the turbulence becomes more two-dimensional, and $\overline{u_1 u_2}/k$ approaches zero because the glyph rotates so that at the wall the principal directions coincide with the x_1 and x_2 directions. Correspondingly the DNS data shows $\overline{u_1 u_1} >> \overline{u_2 u_2}$ near the wall.

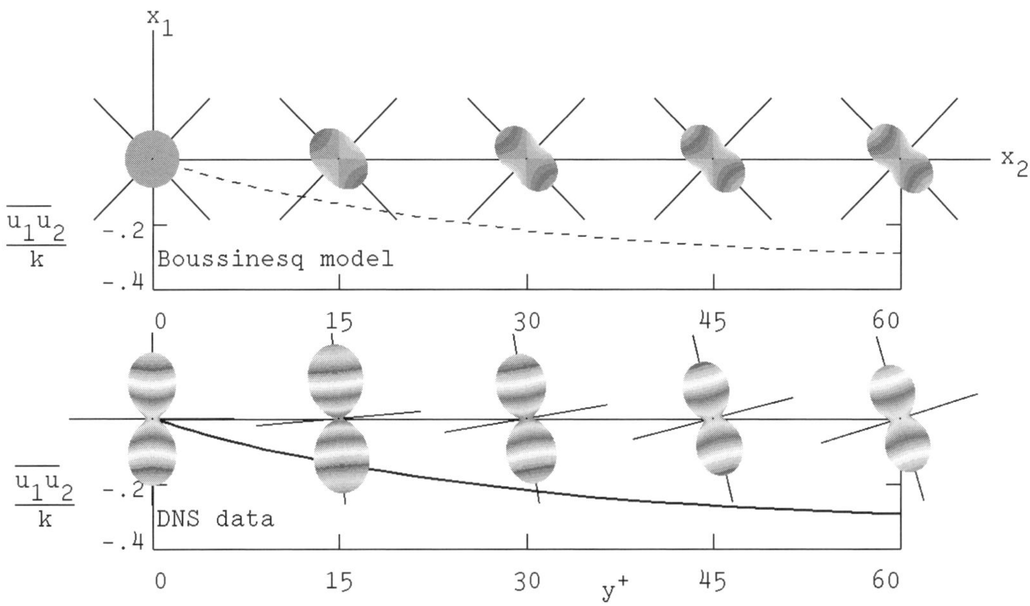

Fig. 1.3.3. Glyphs of the dimensionless Reynolds stresses, $\overline{u_i u_j}/k$, at y^+=0, 15, 30, 45 and 60, for fully developed channel flow. Bottom: DNS data, Moser et al., (MKM1999), R_τ=180. Top: Boussinesq model with the same $\overline{u_1 u_2}/k$ as the DNS data.

Algebraic Reynolds Stress Models

Algebraic Reynolds Stress Models (ARSM), sometimes referred to as non-linear two-equation models, use a more complex algebraic model for the Reynolds stresses than the simple Boussinesq approximation. Some ARSM are based on the stresses observed in boundary layers (e.g. Wilcox, 1993, Eq. 6.13). Some ARSM are fully developed limits or subsets of seven-equation Reynolds stress models (e.g. Rodi, 1984, Eq. 2.80).

As discussed above, for the Boussinesq model the principal directions of the Reynolds stress tensor are the same as for the strain rate tensor, S_{ij}. ARSM may have different orientations for the Reynolds stress tensor due to the influence of the mean-vorticity tensor,

$$\omega_{ij} = \frac{1}{2}\left(\frac{\partial U_i}{\partial x_j} - \frac{\partial U_j}{\partial x_i}\right). \tag{1.3.25}$$

As an example, Craft et al. (CLS1996) proposed a model, which is cubic in the strain-rate and vorticity tensors, and begins with the Boussinesq term,

$$b_{ij} = -C_o \frac{k}{\tilde{\varepsilon}} S_{ij} + C_1 \frac{k^2}{\tilde{\varepsilon}^2}(S_{ik}S_{kj} - \frac{1}{3} S_{mn}S_{mn}\delta_{ij}) + C_2 \frac{k^2}{\tilde{\varepsilon}^2}(\omega_{ik}S_{jk} + \omega_{jk}S_{ik}) + ...$$

Of particular note in the Craft paper are the observations that in homogeneous shear flow, $\overline{u_1u_2}/k$ is a function of the dimensionless strain rate, $s = (k/\varepsilon)\sqrt{2S_{ij}S_{ij}}$, and that the magnitude of $\overline{u_1u_2}/k$ *decreases* at large s. As will be seen, this has important implications for near-wall Reynolds stress modeling.

Chapter 2

FUNDAMENTAL EQUATIONS

At a theoretical level, it is simply a matter of choice in what reference frame a flow is analyzed. The results, with post-processing changes for the change in reference frame, should be the same. At a practical level, it is usually obvious what choice should be made. For example, the flow in a rotor may be steady in the rotating frame, except for the presence of turbulence, but unsteady in the stationary frame. In this chapter, the continuity and momentum equations in rotating coordinates are used to derive the equations for the Reynolds stresses, and thence the turbulence kinetic energy equation. The derivation shows the difference between the thermodynamic definition of dissipation and the modeler's definition. Terms in the Reynolds stress equations are discussed briefly.

2.1 ***Basic Equations for Flow in a Rotating Reference Frame***

Continuity and Momentum Equations

The final Reynolds stress equations to be derived use the assumption that the flow is incompressible. However, in order to aid the understanding of turbulence dissipation, the starting point here is the continuity and Navier-Stokes equations for a compressible fluid,

$$\frac{\partial \rho}{\partial t} + \frac{\partial}{\partial x_i}(\rho \tilde{U}_i) = 0 \ , \qquad (2.1.1)$$

$$\rho \frac{\partial \tilde{U}_i}{\partial t} + \rho \tilde{U}_k \frac{\partial \tilde{U}_i}{\partial x_k} = -\frac{\partial \tilde{P}}{\partial x_i} + \frac{\partial}{\partial x_k}\left\{\mu\left(\frac{\partial \tilde{U}_i}{\partial x_k} + \frac{\partial \tilde{U}_k}{\partial x_i} - \frac{2}{3}\delta_{ik}\frac{\partial \tilde{U}_n}{\partial x_n}\right)\right\} + F_i \ . \qquad (2.1.2)$$

Here, $\tilde{U}_i$ is the unsteady relative velocity in the rotating reference frame and $\tilde{P}$ is the unsteady static pressure. For flows with constant rotation rate, Ω_k, the Coriolis and centrifugal force terms are

$$F_i = -2\rho\varepsilon_{ikm}\Omega_k\tilde{U}_m + \rho\frac{\partial}{\partial x_i}\left(\frac{\Omega^2 r^2}{2}\right) , \qquad (2.1.3)$$

where r is the distance from the axis of rotation and $\Omega^2 = \Omega_n\Omega_n$.

For incompressible flow with constant density, ρ, the continuity equation reduces to

$$\frac{\partial \tilde{U}_i}{\partial x_i} = 0. \tag{2.1.4}$$

Combining the above equations, the momentum equation may be written as

$$\frac{\partial \tilde{U}_i}{\partial t} + \tilde{U}_k \frac{\partial \tilde{U}_i}{\partial x_k} = -2\varepsilon_{ikm}\Omega_k \tilde{U}_m + \frac{\partial}{\partial x_i}\left(\frac{\Omega^2 r^2}{2}\right) - \frac{1}{\rho}\frac{\partial \tilde{P}}{\partial x_i} + \frac{\partial}{\partial x_k}\nu\frac{\partial \tilde{U}_i}{\partial x_k} + \frac{\partial}{\partial x_k}\nu\frac{\partial \tilde{U}_k}{\partial x_i}, \tag{2.1.5}$$

where $\nu = \mu/\rho$. The analysis now proceeds assuming incompressible flow with ν constant. The last term in Eq. 2.1.5 is then zero, but will be retained for mathematical convenience.

Reynolds-Averaged and Fluctuating Momentum Equations

To form the Reynolds-averaged and fluctuating momentum equations, the velocity and pressure are decomposed into mean and turbulent fluctuating components,

$$\tilde{U}_i = U_i + u_i\,, \qquad \tilde{P} = \overline{P} + p\,, \tag{2.1.6}$$

where U_i and $\overline{P}$ are the mean or ensemble average values, and u_i and p are the turbulent fluctuating components which have zero ensemble average values.

The ensemble averaging process is to average out only the turbulence. For example, if we consider steady turbulent flow in a rotor, when viewed in the frame rotating with the rotor, the only unsteadiness is the turbulence and the ensemble average is simply a time average in the rotating frame. If this same flow is analyzed in a stationary frame, the flow just downstream of the rotor for example will have time dependence or unsteadiness due to the passing of blade wakes. This unsteadiness is not part of the turbulence. The ensemble average of the velocity, which filters out only the turbulence, is a once per revolution, or once per blade passage, average of the velocity at a point.

The absolute mean velocity is given by (Appendix B)

$$V_i = U_i + \varepsilon_{ikm}\Omega_k x_m\,, \tag{2.1.7}$$

when the origin of the Cartesian, x_m, coordinate system is on the axis of rotation. The turbulent fluctuating velocities are the same in both the stationary and rotating frames.

In other situations, other criteria may be used to separate out turbulence from other unsteadiness. For example when unsteady terms in the average equations are retained, it is sometimes assumed that there is a separation of time scales so that u_i and p fluctuate on time scales corresponding to the turbulence and U_i and $\overline{P}$ vary comparatively slowly in time.

The continuity equations for the mean and fluctuating velocities are

$$\frac{\partial U_i}{\partial x_i} = 0 \qquad \text{and} \qquad \frac{\partial u_i}{\partial x_i} = 0. \tag{2.1.8}$$

Substituting Eq. 2.1.6 into Eq. 2.1.5 yields

$$\begin{aligned} &\frac{\partial U_i}{\partial t} + \frac{\partial u_i}{\partial t} + U_k \frac{\partial U_i}{\partial x_k} + u_k \frac{\partial U_i}{\partial x_k} + U_k \frac{\partial u_i}{\partial x_k} + u_k \frac{\partial u_i}{\partial x_k} \\ &= -2\varepsilon_{ikm}\Omega_k U_m - 2\varepsilon_{ikm}\Omega_k u_m + \frac{\partial}{\partial x_i}\left(\frac{\Omega^2 r^2}{2}\right) - \frac{1}{\rho}\frac{\partial \overline{P}}{\partial x_i} - \frac{1}{\rho}\frac{\partial p}{\partial x_i} \\ &+ \frac{\partial}{\partial x_k}\nu\frac{\partial U_i}{\partial x_k} + \frac{\partial}{\partial x_k}\nu\frac{\partial U_k}{\partial x_i} + \frac{\partial}{\partial x_k}\nu\frac{\partial u_i}{\partial x_k} + \frac{\partial}{\partial x_k}\nu\frac{\partial u_k}{\partial x_i} \quad . \end{aligned} \tag{2.1.9}$$

Ensemble averaging Eq. 2.1.9 yields the Reynolds-averaged Navier-Stokes equations,

$$\begin{aligned} \frac{\partial U_i}{\partial t} + U_k \frac{\partial U_i}{\partial x_k} + \frac{\partial \overline{u_i u_k}}{\partial x_k} &= -2\varepsilon_{ikm}\Omega_k U_m + \frac{\partial}{\partial x_i}\left(\frac{\Omega^2 r^2}{2}\right) - \frac{1}{\rho}\frac{\partial \overline{P}}{\partial x_i} \\ &+ \frac{\partial}{\partial x_k}\nu\frac{\partial U_i}{\partial x_k} + \frac{\partial}{\partial x_k}\nu\frac{\partial U_k}{\partial x_i} \quad , \end{aligned} \tag{2.1.10}$$

where the overbars indicate ensemble averages and $\overline{u_i u_k}$ are the Reynolds stresses which need to be modeled.

The difference between Eqs. 2.1.9 and 2.1.10, yields the fluctuating momentum equation

$$\begin{aligned} &\frac{\partial u_i}{\partial t} + u_k \frac{\partial U_i}{\partial x_k} + U_k \frac{\partial u_i}{\partial x_k} + \frac{\partial u_i u_k}{\partial x_k} - \frac{\partial \overline{u_i u_k}}{\partial x_k} \\ &= -2\varepsilon_{ikm}\Omega_k u_m - \frac{1}{\rho}\frac{\partial p}{\partial x_i} + \frac{\partial}{\partial x_k}\nu\frac{\partial u_i}{\partial x_k} + \frac{\partial}{\partial x_k}\nu\frac{\partial u_k}{\partial x_i} \quad . \end{aligned} \tag{2.1.11}$$

Reynolds Stress Equations

The Reynolds stress equations are obtained by taking u_j times the fluctuating momentum equation for u_i, plus u_i times the corresponding equation for u_j, and ensemble averaging (Appendix EQ.2.1.12). The result may be written as

$$\begin{aligned}
&\frac{\partial\overline{u_iu_j}}{\partial t}+U_k\frac{\partial\overline{u_iu_j}}{\partial x_k}=-\left(\overline{u_ju_k}\frac{\partial U_i}{\partial x_k}+\overline{u_iu_k}\frac{\partial U_j}{\partial x_k}\right)\\
&-2\Omega_k\left(\varepsilon_{ikm}\overline{u_ju_m}+\varepsilon_{jkm}\overline{u_iu_m}\right)-\frac{\partial\overline{u_iu_ju_k}}{\partial x_k}-\frac{1}{\rho}\left(\overline{u_j\frac{\partial p}{\partial x_i}}+\overline{u_i\frac{\partial p}{\partial x_j}}\right)\\
&+\frac{\partial}{\partial x_k}\nu\frac{\partial\overline{u_iu_j}}{\partial x_k}+\frac{\partial}{\partial x_k}\nu\overline{u_j\frac{\partial u_k}{\partial x_i}}+\frac{\partial}{\partial x_k}\nu\overline{u_i\frac{\partial u_k}{\partial x_j}}\\
&-2\nu\overline{\frac{\partial u_i}{\partial x_k}\frac{\partial u_j}{\partial x_k}}-\nu\overline{\frac{\partial u_j}{\partial x_k}\frac{\partial u_k}{\partial x_i}}-\nu\overline{\frac{\partial u_i}{\partial x_k}\frac{\partial u_k}{\partial x_j}}\quad.
\end{aligned}\tag{2.1.12}$$

Note that the third and fourth lines of Eq. 2.1.12 (all the terms with ν) originated as the laminar diffusion term in the momentum equation. The third line still consists of diffusion terms. The fourth line represents the dissipation rate of the Reynolds stresses.

For incompressible flow, with ν constant, four of the terms in Eq. 2.1.12 sum to zero. When these four terms are omitted, the more familiar form of the Reynolds stress equations is obtained,

$$\begin{aligned}
&\frac{\partial\overline{u_iu_j}}{\partial t}+U_k\frac{\partial\overline{u_iu_j}}{\partial x_k}=-\left(\overline{u_ju_k}\frac{\partial U_i}{\partial x_k}+\overline{u_iu_k}\frac{\partial U_j}{\partial x_k}\right)-2\Omega_k\left(\varepsilon_{ikm}\overline{u_ju_m}+\varepsilon_{jkm}\overline{u_iu_m}\right)\\
&-\frac{\partial\overline{u_iu_ju_k}}{\partial x_k}-\frac{1}{\rho}\left(\overline{u_j\frac{\partial p}{\partial x_i}}+\overline{u_i\frac{\partial p}{\partial x_j}}\right)+\frac{\partial}{\partial x_k}\nu\frac{\partial\overline{u_iu_j}}{\partial x_k}-2\nu\overline{\frac{\partial u_i}{\partial x_k}\frac{\partial u_j}{\partial x_k}}\quad.
\end{aligned}\tag{2.1.13}$$

In terms of modeling the Reynolds stress equations, the last term in Eq. 2.1.13 is taken as the dissipation rate of the Reynolds stresses which needs to be modeled. When the turbulence is homogeneous, the last two terms in Eq. 2.1.12 are zero and the definitions of dissipation rate are consistent.

Turbulence Kinetic Energy Equation

Contracting Eq. 2.1.12 yields an equation for the turbulence kinetic energy, $k=\overline{u_iu_i}/2$,

$$\begin{aligned}
&\frac{\partial k}{\partial t}+U_k\frac{\partial k}{\partial x_k}=-\overline{u_iu_k}\frac{\partial U_i}{\partial x_k}-\frac{1}{2}\frac{\partial\overline{u_iu_iu_k}}{\partial x_k}-\frac{1}{\rho}\frac{\partial\overline{pu_i}}{\partial x_i}+\frac{\partial}{\partial x_k}\nu\frac{\partial k}{\partial x_k}\\
&+\frac{\partial}{\partial x_k}\nu\frac{\partial\overline{u_iu_k}}{\partial x_i}-\nu\overline{\frac{\partial u_i}{\partial x_k}\frac{\partial u_i}{\partial x_k}}-\nu\overline{\frac{\partial u_i}{\partial x_k}\frac{\partial u_k}{\partial x_i}}\quad.
\end{aligned}\tag{2.1.14}$$

The dissipation rate for turbulence kinetic energy is the last two terms in Eq. 2.1.14. This gives the correct form for the thermodynamic representation of dissipation as a viscous stress times a strain rate,

$$\varepsilon_{th} = \frac{1}{\rho}\overline{\mu\left(\frac{\partial u_i}{\partial x_j} + \frac{\partial u_j}{\partial x_i}\right)\frac{1}{2}\left(\frac{\partial u_i}{\partial x_j} + \frac{\partial u_j}{\partial x_i}\right)} , \tag{2.1.15}$$

or

$$\varepsilon_{th} = 2\nu\overline{s_{ij}s_{ij}} , \qquad \text{where} \qquad s_{ij} = \frac{1}{2}\left(\frac{\partial u_i}{\partial x_j} + \frac{\partial u_j}{\partial x_i}\right) . \tag{2.1.16}$$

Modelers, however, find it more convenient to note that (with ν constant) the first and last terms on the second line of Eq. 2.1.14 sum to zero, giving the familiar form for the k equation,

$$\frac{\partial k}{\partial t} + U_k\frac{\partial k}{\partial x_k} = -\overline{u_i u_k}\frac{\partial U_i}{\partial x_k} - \frac{1}{2}\frac{\partial \overline{u_i u_i u_k}}{\partial x_k} - \frac{1}{\rho}\frac{\partial \overline{p u_i}}{\partial x_i} + \frac{\partial}{\partial x_k}\nu\frac{\partial k}{\partial x_k} - \nu\overline{\frac{\partial u_i}{\partial x_k}\frac{\partial u_i}{\partial x_k}} . \tag{2.1.17}$$

The dissipation rate, which will need to be modeled, is then

$$\varepsilon = \nu\overline{\frac{\partial u_i}{\partial x_k}\frac{\partial u_i}{\partial x_k}} , \tag{2.1.18}$$

and for homogeneous turbulence, $\varepsilon = \varepsilon_{th}$.

2.2 *Terms in the Reynolds Stress Equations*

The Reynolds stress equations, Eq. 2.1.13, may be written as,

$$\frac{\partial \overline{u_i u_j}}{\partial t} + U_k\frac{\partial \overline{u_i u_j}}{\partial x_k} = L_{ij} + P_{ij} + F_{ij} + T_{ij} + D_{ij} + \Pi_{ij} - \varepsilon_{ij} . \tag{2.2.1}$$

The left hand side represents a (slow) time variation and convection of the Reynolds stresses. The terms on the right hand side correspond to laminar diffusion, production, Coriolis, turbulent diffusion, pressure diffusion, pressure-strain and dissipation, respectively. The first three of these are simple functions of the Reynolds stresses themselves and do not need modeling. The latter four will require models.

Laminar Diffusion

The laminar diffusion term,

$$L_{ij} = \frac{\partial}{\partial x_k} \nu \frac{\partial \overline{u_i u_j}}{\partial x_k} , \tag{2.2.2}$$

is in general important only near walls. Mathematically, it provides the "glue" which makes the equations elliptic and spatially links the Reynolds stresses so that they are continuous.

Production

The Reynolds stress production term is of primary importance in determining how the Reynolds stresses develop,

$$P_{ij} = -\overline{u_j u_k} \frac{\partial U_i}{\partial x_k} - \overline{u_i u_k} \frac{\partial U_j}{\partial x_k} . \tag{2.2.3}$$

It may be rewritten in terms of the turbulence anisotropy,

$$P_{ij} = -2k\left(b_{jk} \frac{\partial U_i}{\partial x_k} + b_{ik} \frac{\partial U_j}{\partial x_k} \right) - \frac{2}{3} k \left(\frac{\partial U_i}{\partial x_j} + \frac{\partial U_j}{\partial x_i} \right) , \tag{2.2.4}$$

and further broken down in terms of the mean strain rate, S_{ij}, and the mean relative vorticity, ω_{ij}, using $\partial U_i / \partial x_k = S_{ik} + \omega_{ik}$, to give

$$P_{ij} = -\frac{4}{3} k S_{ij} - 2k\left(b_{jk} S_{ik} + b_{ik} S_{jk}\right) - 2k\left(b_{jk}\omega_{ik} + b_{ik}\omega_{jk}\right) . \tag{2.2.5}$$

When these are contracted to give the production of turbulence kinetic energy,

$$P = \frac{1}{2} P_{ii} = -2k b_{ik} \frac{\partial U_i}{\partial x_k} = -k b_{ik} \left(\frac{\partial U_i}{\partial x_k} + \frac{\partial U_k}{\partial x_i} \right) = -2k b_{ik} S_{ik} , \tag{2.2.6}$$

it is found that only the second term in Eq. 2.2.5 contributes to turbulence production. Therefore turbulence production is an interaction between the mean strain rate and the anisotropy of the turbulence, and is zero for isotropic turbulence ($b_{ik} = 0$).

The initial effect of the Reynolds stress production term on isotropic turbulence is then to produce anisotropy of the turbulence, negatively proportional to the mean flow strain rate, through the first term on the right hand side of Eq. 2.2.5. This effect can be demonstrated by integrating the simplified equation,

$d\overline{u_i u_j}/dt = -(4/3)kS_{ij}$. Here, we use the example of simple shear flow where $\partial U_1/\partial x_2$ (= 2) is the only non-zero component of the mean velocity gradient. The solution to the equations,

$$\left(\overline{u_i u_j}\right)_{t=0} = \frac{2}{3}k_o \begin{bmatrix} 1 & 0 & 0 \\ 0 & 1 & 0 \\ 0 & 0 & 1 \end{bmatrix} \quad \text{and} \quad d\overline{u_i u_j} = -\frac{4}{3}kS_{ij}dt = -\frac{4}{3}k_o \begin{bmatrix} 0 & 1 & 0 \\ 1 & 0 & 0 \\ 0 & 0 & 0 \end{bmatrix} dt \,, \qquad (2.2.7)$$

is illustrated in Fig. 2.2.1. The principal directions of the strain rate tensor are at 45 degrees to the x_1 and x_2 directions, as illustrated. The glyph develops with a Boussinesq shape (see Fig. 1.3.3, top), the same as if it had been subjected to plane strain, at 45 degrees to the x_1 and x_2 directions. Note that the glyph is stretched in the direction that the strain rate is compressive, and correspondingly the diameter of the glyph is reduced in the direction of positive strain rate. The corresponding change in the Mohr's circles is shown in Fig. 2.2.2.

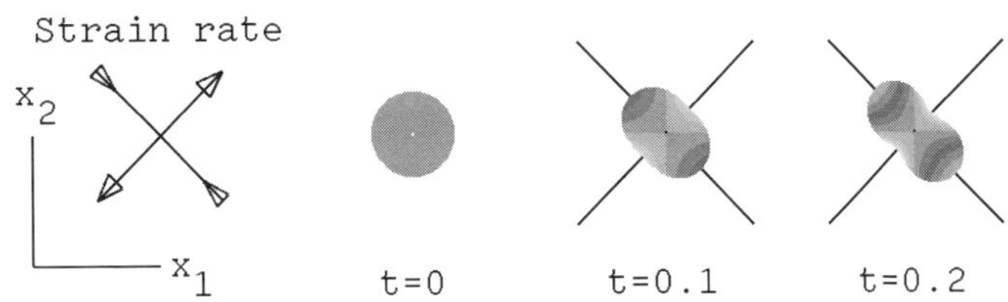

Fig. 2.2.1. Solution to Eq. 2.2.7.

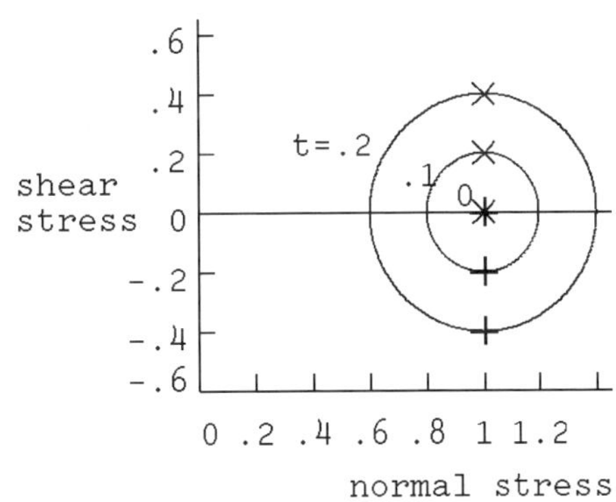

Fig. 2.2.2. Mohr's circles in x_1-x_2 plane for the solution to Eq. 2.2.7 (with k_o=3/2). ×, $\overline{u_1u_1}, -\overline{u_1u_2}$; +, $\overline{u_2u_2}, \overline{u_1u_2}$.

This anisotropy interacts with the mean flow strain rate to produce turbulence kinetic energy through the second term on the right hand side of Eq. 2.2.5. To demonstrate this, we integrate the simplified equation

$$d\overline{u_i u_j}/dt = -2k\left(b_{jk}S_{ik} + b_{ik}S_{jk}\right) \,, \qquad (2.2.8)$$

starting with the results from the solution of Eq. 2.2.7 at $t = 0.2$. An increase in turbulence kinetic energy, of about 10 percent for $t_b (= \delta t) = 0.2$, occurs in the $\overline{u_1 u_1}$ and $\overline{u_2 u_2}$ components, with $\overline{u_3 u_3}$ unchanged. This is shown in Fig 2.2.3 by the increased size of the glyphs.

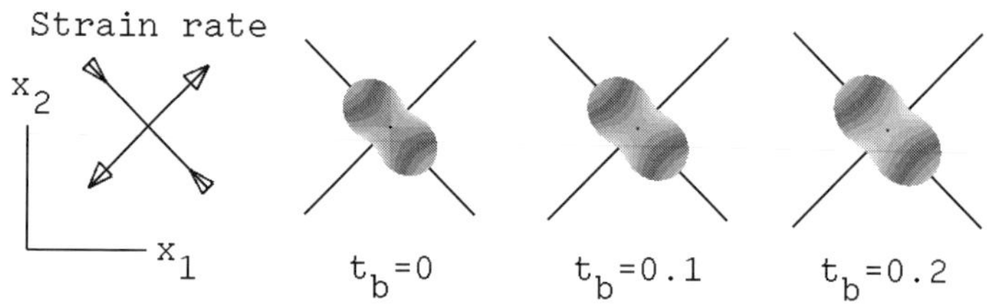

Fig. 2.2.3. Solution to Eq. 2.2.8.

The effect of the vorticity part of the Reynolds stress production term, is to turn the glyph in the direction opposite to the vorticity. This is seen in Fig. 2.2.4, which shows the solution for the simplified equation

$$d\overline{u_i u_j}/dt = -2k\left(b_{jk}\omega_{ik} + b_{ik}\omega_{jk}\right) \ , \tag{2.2.9}$$

again starting with the results from the solution of Eq. 2.2.7 at $t = 0.2$. Vorticity is perhaps more easily visualized in terms of a vorticity vector. The relative vorticity vector in terms of the mean velocity is given by

$$\omega_i = \varepsilon_{ijk} \frac{\partial U_k}{\partial x_j} \ . \tag{2.2.10}$$

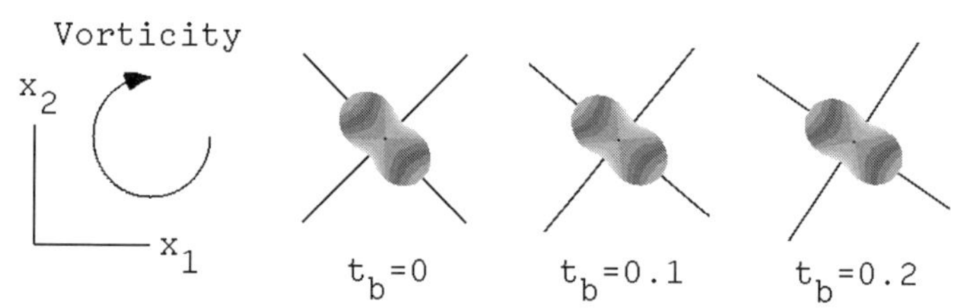

Fig. 2.2.4. Solution to Eq. 2.2.9.

The relations to transform between the vorticity vector and the vorticity tensor are

$$w_{jk} = -\frac{1}{2}\omega_i \varepsilon_{ijk} \qquad \text{and} \qquad \omega_i = -\varepsilon_{ijk}\omega_{jk} \ , \tag{2.2.11}$$

so that Eq. 2.2.9, rewritten in terms of the vorticity vector is

$$d\overline{u_i u_j}/dt = -k\omega_k\left(\varepsilon_{ikm} b_{jm} + \varepsilon_{jkm} b_{im}\right). \tag{2.2.12}$$

With $\partial U_1/\partial x_2 = 2$, for the example in Fig. 2.2.4, $\omega_{12} = 1$, and $\omega_3 = -2$, so the vorticity is clockwise on the figure.

Coriolis

The Coriolis term, which is due to the choice of a rotating coordinate system, is also of primary importance in how the components of the Reynolds stresses develop in the rotating coordinate system,

$$F_{ij} = -2\Omega_k\left(\varepsilon_{ikm}\overline{u_j u_m} + \varepsilon_{jkm}\overline{u_i u_m}\right) , \tag{2.2.13}$$

or

$$F_{ij} = -4k\Omega_k\left(\varepsilon_{ikm}b_{jm} + \varepsilon_{jkm}b_{im}\right) . \tag{2.2.14}$$

It makes no contribution to the turbulence kinetic energy equation ($F_{ii} = 0$).

The effect of the Coriolis term can be seen using the simplified equation,

$$d\overline{u_i u_j}/dt = F_{ij} = -4k\Omega_k\left(\varepsilon_{ikm}b_{jm} + \varepsilon_{jkm}b_{im}\right). \tag{2.2.15}$$

The solution, for a positive rotation about the x_3 axis, $\Omega_3 = 1$, is shown in Fig. 2.2.5. The effect of the Coriolis term is to rotate the glyph in the opposite direction from the coordinate system rotation.

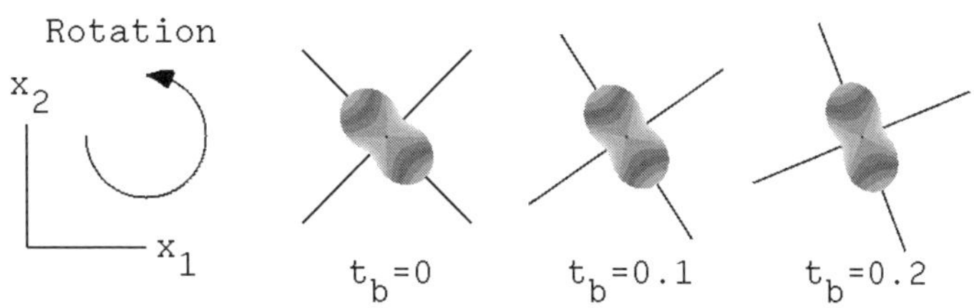

Fig. 2.2.5. Solution to Eq. 2.2.15 with Ω_3=1.

Turbulent Diffusion

The turbulent diffusion of the Reynolds stresses is the gradient of a triple product correlation of the fluctuating velocities,

$$T_{ij} = -\frac{\partial \overline{u_i u_j u_k}}{\partial x_k} . \tag{2.2.16}$$

Chou (1945) suggested that this be handled by solving differential equations for the triple product correlations. This suggestion has not been adopted. Instead, a gradient diffusion type model,

$$T_{ij} = \frac{\partial}{\partial x_k} \nu_{t,kn} \frac{\partial \overline{u_i u_j}}{\partial x_n} \quad , \tag{2.2.17}$$

is usually adopted but with some variations on the tensorial form. The model chosen for turbulent diffusion has not been shown to be a critical component of Reynolds stress models (Chen and Jaw, 1998, p. 27). Models for turbulent diffusion will be discussed in more detail in Chapter 7.

Velocity–Pressure-Gradient Term

The velocity–pressure-gradient term,

$$D_{ij} + \Pi_{ij} = -\frac{1}{\rho}\left(\overline{u_j \frac{\partial p}{\partial x_i}} + \overline{u_i \frac{\partial p}{\partial x_j}} \right) \quad , \tag{2.2.18}$$

is traditionally separated into a pressure-diffusion term,

$$D_{ij} = -\frac{1}{\rho}\left(\frac{\partial \overline{p u_j}}{\partial x_i} + \frac{\partial \overline{p u_i}}{\partial x_j} \right) \quad , \tag{2.2.19}$$

and a pressure-strain term,

$$\Pi_{ij} = \overline{\frac{p}{\rho}\left(\frac{\partial u_i}{\partial x_j} + \frac{\partial u_j}{\partial x_i} \right)} \quad . \tag{2.2.20}$$

When contracted, the pressure-diffusion term may make a contribution to the turbulence kinetic energy equation, but the pressure-strain term does not ($\Pi_{ii} = 0$). The pressure-diffusion term, D_{ij}, is usually either neglected or modeled together with the turbulent diffusion, T_{ij}.

For homogeneous turbulence, D_{ij} is zero, and Π_{ij} is the only pressure influence on the Reynolds stresses. The pressure-strain term is of near equal importance as the production term, and critical to the modeling of Reynolds stresses.

Dissipation

The viscous dissipation of the Reynolds stresses,

$$\varepsilon_{ij} = 2\nu \overline{\frac{\partial u_i}{\partial x_k} \frac{\partial u_j}{\partial x_k}} \quad , \tag{2.2.21}$$

is usually modeled in two parts. First, a differential equation for the turbulence dissipation rate, ε, similar to that employed in $k-\varepsilon$ models may be used. Alternatively, a differential equation for a turbulence frequency, e.g., $\omega = \tilde{\varepsilon}/k$, or a turbulence length scale, e.g., $L = k^{3/2}/\varepsilon$, may be used.

To complete the dissipation model, the anisotropy of the dissipation of the Reynolds stresses is modeled algebraically. It is convenient to model the anisotropy of the dissipation, $\varepsilon_{ij} - \frac{2}{3}\varepsilon\delta_{ij}$, together with the pressure-strain, as an effective pressure-strain,

$$\phi_{ij} = \Pi_{ij} - (\varepsilon_{ij} - \frac{2}{3}\varepsilon\delta_{ij}) \ . \tag{2.2.22}$$

The model used for this effective pressure-strain is the most important modeling component of any Reynolds stress model, and the subject of much of this book.

<u>Exercise 2.2</u>. Show that $P_{ij} + \frac{1}{2}F_{ij} = P_{ij}^{\text{abs}}$, where P_{ij}^{abs} is the Reynolds stress production rate evaluated in the absolute reference frame,

$$P_{ij}^{\text{abs}} = -\overline{u_j u_k}\frac{\partial V_i}{\partial x_k} - \overline{u_i u_k}\frac{\partial V_j}{\partial x_k} \ .$$

Chapter 3

PRESSURE-STRAIN MODEL BACKGROUND

In this chapter the theoretical basis for algebraic pressure-strain modeling is reviewed. A general algebraic form for pressure-strain models for homogeneous turbulence as a series in b_{ij}, truncated using the Cayley-Hamilton theorem, is presented. This format will be used for comparing models in the literature, and later in the book for the presentation of new models. An initial look at some simple pressure-strain models in the literature reveals several, perhaps conflicting, considerations used in their development.

3.1 Theoretical Basis for Pressure-Strain Models

Chou (1945) pioneered the basis for pressure-strain modeling. Here we follow a similar analysis to obtain an integral equation for the fluctuating pressure, and then the beginnings of an algebraic pressure-strain model.

Taking the divergence of the fluctuating momentum equation, 2.1.11, yields

$$\frac{1}{\rho}\frac{\partial^2 p}{\partial x_i^2} = -2\frac{\partial u_m}{\partial x_i}\frac{\partial U_i}{\partial x_m} - 2\varepsilon_{ikm}\Omega_k\frac{\partial u_m}{\partial x_i} - \frac{\partial^2(u_i u_m - \overline{u_i u_m})}{\partial x_i \partial x_m} . \tag{3.1.1}$$

Differentiating Eq. 2.1.7 relates the mean absolute and relative velocity gradients,

$$\frac{\partial V_i}{\partial x_m} = \frac{\partial U_i}{\partial x_m} + \varepsilon_{ikm}\Omega_k . \tag{3.1.2}$$

Therefore, the fluctuating pressure is a function of the mean absolute velocity gradient,

$$\frac{1}{\rho}\frac{\partial^2 p}{\partial x_i^2} = -2\frac{\partial u_m}{\partial x_i}\frac{\partial V_i}{\partial x_m} - \frac{\partial^2(u_i u_m - \overline{u_i u_m})}{\partial x_i \partial x_m} , \tag{3.1.3}$$

with no direct dependence on the relative mean velocity gradient. Using Green's theorem, the general solution to 3.1.3 may be written in the form (Appendix EQ.3.1.4)

$$\frac{1}{\rho}p(\mathbf{x}) = \frac{1}{4\pi}\iiint\left(2\frac{\partial u_m(\mathbf{x}')}{\partial x_i'}\frac{\partial V_i(\mathbf{x}')}{\partial x_m'} + \frac{\partial^2(u_i(\mathbf{x}')u_m(\mathbf{x}') - \overline{u_i(\mathbf{x}')u_m(\mathbf{x}')})}{\partial x_i'\partial x_m'}\right)\frac{dVol'}{|\mathbf{x}-\mathbf{x}'|}$$
$$+\frac{1}{4\pi\rho}\oiint\left(\frac{1}{|\mathbf{x}-\mathbf{x}'|}\frac{\partial p(\mathbf{x}')}{\partial n'} - p(\mathbf{x}')\frac{\partial}{\partial n'}\frac{1}{|\mathbf{x}-\mathbf{x}'|}\right)dA' \quad , \tag{3.1.4}$$

where the integration is over the entire flow domain, Vol'. For points $\mathbf{x}$ far enough away from the boundary, A', the surface integral may be neglected. The effects of walls on turbulence are discussed in Chapters 12 and 16. We continue here, considering only the volume integral term, and contributions to it from homogeneous turbulence.

The mean velocity gradient at $\mathbf{x}'$ may be written as a Taylor series around the point $\mathbf{x}$,

$$\frac{\partial V_k(\mathbf{x}')}{\partial x_m'} = \frac{\partial V_k(\mathbf{x})}{\partial x_m} + (x_n' - x_n)\frac{\partial^2 V_k(\mathbf{x})}{\partial x_n \partial x_m} + \frac{1}{2}(x_j' - x_j)(x_n' - x_n)\frac{\partial^3 V_k(\mathbf{x})}{\partial x_j \partial x_n \partial x_m} + \ldots \quad . \tag{3.1.5}$$

Rotta (1951) retained three terms in this series, but in general only the first term is retained, as suggested by Chou (1945). Retaining just the first term is consistent with the assumption that the pressure-strain may be determined by considering homogeneous flows, in which the mean velocity gradient and turbulence properties are uniform.

With just the first term, we may write

$$\overline{\frac{p}{\rho}\frac{\partial u_i}{\partial x_j}} = 2kX_{ijkm}\frac{\partial V_k}{\partial x_m} + \varepsilon A_{ij} \ , \tag{3.1.6}$$

where X_{ijkm} and A_{ij} are dimensionless, and integrals of two-point correlations,

$$X_{ijkm} = \frac{1}{4\pi k}\iiint \overline{\frac{\partial u_m(\mathbf{x}')}{\partial x_k'}\frac{\partial u_i(\mathbf{x})}{\partial x_j}}\frac{dVol'}{|\mathbf{x}-\mathbf{x}'|} \tag{3.1.7}$$

and

$$A_{ij} = \frac{1}{4\pi\varepsilon}\iiint\left(\overline{\frac{\partial u_m(\mathbf{x}')}{\partial x_k'}\frac{\partial u_k(\mathbf{x}')}{\partial x_m'}\frac{\partial u_i(\mathbf{x})}{\partial x_j}}\right)\frac{dVol'}{|\mathbf{x}-\mathbf{x}'|} \quad . \tag{3.1.8}$$

The pressure-strain term in the Reynolds stress equations may be written as

$$\Pi_{ij} = \overline{\frac{p}{\rho}\left(\frac{\partial u_i}{\partial x_j} + \frac{\partial u_j}{\partial x_i}\right)} = \Pi_{ij}^{\mathrm{R}} + \Pi_{ij}^{\mathrm{S}} \ , \tag{3.1.9}$$

where Π_{ij}^{R} is the rapid part of the pressure-strain, proportional to the mean velocity gradient,

$$\Pi_{ij}^{R} = 2k(X_{ijkm} + X_{jikm})\frac{\partial V_k}{\partial x_m} , \tag{3.1.10}$$

and Π_{ij}^{S} is the slow part, which depends only on the turbulence,

$$\Pi_{ij}^{S} = \varepsilon(A_{ij} + A_{ji}) . \tag{3.1.11}$$

By continuity, the pressure-strain gives no contribution to the equation for turbulence kinetic energy, i.e., $\Pi_{ii} = 0$. Similarly, from Eqs. 3.1.7 and 3.1.8, $X_{iikm} = 0$ and $A_{ii} = 0$, so that individually

$$\Pi_{ii}^{R} = 0 \text{ and } \Pi_{ii}^{S} = 0. \tag{3.1.12}$$

Equation 3.1.12, together with the obvious symmetry,

$$\Pi_{ij}^{R} = \Pi_{ji}^{R} \quad \text{and} \quad \Pi_{ij}^{S} = \Pi_{ji}^{S} , \tag{3.1.13}$$

provide minimum requirements which models for rapid and slow pressure-strain must satisfy.

3.2 An Algebraic Model Form for Pressure-Strain

Slow Pressure-Strain

Equations 3.1.11 and 3.1.8, along with 2.2.21, give the slow pressure-strain, Π_{ij}^{S}, and the Reynolds stress dissipation, ε_{ij}, as functions of properties of the turbulence. Taking Π_{ij}^{S} and ε_{ij} as algebraic functions of the current state of the turbulence, and noting that the terms have the units of dissipation, ε, then

$$\Pi_{ij}^{S} = \Pi_{ij}^{S}(\varepsilon, R_t, b_{mn}) \quad \text{and} \quad \varepsilon_{ij} = \varepsilon_{ij}(\varepsilon, R_t, b_{mn}). \tag{3.2.1}$$

Within the framework of a Reynolds stress model, for homogeneous turbulence, these may be written as a series in b_{mn}, which may be truncated based on the Cayley-Hamilton theorem (Appendix C) without loss of generality, provided the coefficients are allowed to be functions of *II* and *III*. Therefore,

$$\Pi_{ij}^{S} = -c_{1s}(R_t, II, III)\varepsilon b_{ij} + c_{2s}(R_t, II, III)\varepsilon(b_{ik}b_{kj} - \frac{1}{3}b_{mn}b_{mn}\delta_{ij}) \tag{3.2.2}$$

and

$$\varepsilon_{ij} = \frac{2}{3}\varepsilon\delta_{ij} + c_{1d}(R_t, II, III)\varepsilon b_{ij} - c_{2d}(R_t, II, III)\varepsilon(b_{ik}b_{kj} - \frac{1}{3}b_{mn}b_{mn}\delta_{ij}) \ . \tag{3.2.3}$$

Upon contraction these give as they should,

$$\Pi_{ii}^{S} = 0 \text{ and } \varepsilon_{ii} = 2\varepsilon \ . \tag{3.2.4}$$

Lumley and Newman (1977) reviewed a body of experimental work which considers the return to isotropy of homogeneous turbulence in the absence of mean velocity gradients and system rotation. The equations for the Reynolds stresses then reduce to

$$\frac{d\overline{u_i u_j}}{dt} = \Pi_{ij}^{S} - \varepsilon_{ij} \ . \tag{3.2.5}$$

In these experiments only the combined effects of slow pressure-strain and dissipation can be observed. An *effective* slow pressure-strain which combines pressure-strain with the anisotropy of the dissipation,

$$\phi_{ij}^{S} = \Pi_{ij}^{S} - (\varepsilon_{ij} - \frac{2}{3}\varepsilon\delta_{ij}) \ , \tag{3.2.6}$$

may then be determined from experimental measurements.

The algebraic model for the effective slow pressure-strain takes the form

$$\phi_{ij}^{S} = -c_1\varepsilon b_{ij} + c_2\varepsilon(b_{ik}b_{kj} - \frac{1}{3}b_{mn}b_{mn}\delta_{ij}) \ ,$$

with

$$c_1 = c_{1s} + c_{1d}, \qquad c_2 = c_{2s} + c_{2d} \ . \tag{3.2.7}$$

c_1 and c_2 are functions of II, III and R_t.

Rapid Pressure-Strain

Many authors have modeled rapid pressure-strain by modeling the components of X_{ijkm} (Eq. 3.1.7). These include Rotta (1951), Launder et al. (LRR1975), and more recently, Ristorcelli et al. (RLA1995). These authors assumed that the turbulence was nearly homogeneous to obtain restrictions on X_{ijkm}. For homogeneous turbulence, X_{ijkm} can be expressed as an integral in wave-number space, K, of the Fourier transform, Φ_{im}, of the two-point velocity correlation. In particular,

$$X_{ijkm} = \frac{1}{k}\iiint \frac{K_j K_k}{K^2}\Phi_{im} dK_1 dK_2 dK_3 \ . \tag{3.2.8}$$

This approach was used, for example, by Reynolds (1987), and Eq. 3.2.8 is easily seen from the derivation of the rapid distortion theory equations in Section 9.3.

Since $\Phi_{im} = \Phi_{mi}$, and $\overline{u_i u_m} = \iiint \Phi_{im} dK_1 dK_2 dK_3$, models for X_{ijkm} should satisfy

$$X_{ijkm} = X_{ikjm}, \qquad X_{ijkm} = X_{mjki} \qquad \text{and} \qquad X_{ijjm} = \overline{u_i u_m}/k \ , \tag{3.2.9}$$

in addition to the continuity restrictions, apparent from Eq. 3.1.7,

$$X_{ijmm} = 0, \quad \text{and} \quad X_{iikm} = 0 \ . \tag{3.2.10}$$

It is generally assumed that X_{ijkm} can be modeled as a series in the turbulence anisotropy tensor, b_{ij}. The LRR1975 model is a simple example of this type of model, using terms which are at most linear in b_{ij}. The RLA1995 model is a very complex example.

A more general approach to modeling rapid pressure-strain is to take

$$\Pi_{ij}^{\mathrm{R}} = 2kM_{ijkm}\frac{\partial V_k}{\partial x_m} \ . \tag{3.2.11}$$

It follows, for homogeneous turbulence, that

$$M_{ijkm} = \frac{1}{k}\iiint \frac{K_k}{K^2}\left(K_j \Phi_{im} + K_i \Phi_{jm}\right) dK_1 dK_2 dK_3 \ , \tag{3.2.12}$$

and the restrictions on M_{ijkm} are only those that also apply to Π_{ij}^{R}. In particular,

$$M_{ijkm} = M_{jikm} \qquad (\text{from } \Pi_{ij}^{\mathrm{R}} = \Pi_{ji}^{\mathrm{R}}) \tag{3.2.13}$$

and

$$M_{iikm} = 0 \quad (\text{from } \Pi_{ii}^{\mathrm{R}} = 0) \ . \tag{3.2.14}$$

If it is assumed that M_{ijkm} can be modeled as a series in b_{ij}, then a general algebraic expression can be obtained for the rapid pressure-strain with unknown coefficients to be set. Writing the mean velocity gradient as the sum of strain-rate and vorticity terms

$$\frac{\partial V_k}{\partial x_m} = S_{km} + W_{km} \quad , \tag{3.2.15}$$

where

$$S_{km} = \frac{1}{2}\left(\frac{\partial V_k}{\partial x_m} + \frac{\partial V_m}{\partial x_k}\right) \qquad \text{and} \qquad W_{km} = \frac{1}{2}\left(\frac{\partial V_k}{\partial x_m} - \frac{\partial V_m}{\partial x_k}\right) , \tag{3.2.16}$$

it can be shown (Appendix EQ.3.2.17) that a general expression for rapid pressure-strain may be written as

$$\begin{aligned}\frac{\Pi_{ij}^{\mathrm{R}}}{k} &= c_3 S_{ij} + c_4(b_{ik}S_{jk} + b_{jk}S_{ik} - \frac{2}{3}\delta_{ij}b_{mn}S_{mn}) + c_5(b_{ik}W_{jk} + b_{jk}W_{ik}) \\ &+ c_6 b_{ij}b_{mn}S_{mn} + c_7(b_{ik}b_{kn}S_{jn} + b_{jk}b_{kn}S_{in} - 2b_{kj}b_{ni}S_{kn}) \\ &+ c_8(b_{ik}b_{kn}W_{jn} + b_{jk}b_{kn}W_{in}) \\ &+ c_9 b_{ij}b_{kn}b_{mn}S_{mk} + c_{10}(b_{ik}b_{kj} - \frac{1}{3}\delta_{ij}b_{kn}b_{nk})b_{pq}S_{pq} \\ &+ c_{11}(b_{ik}W_{kn}b_{mn}b_{mj} + b_{jk}W_{kn}b_{mn}b_{mi}) \\ &+ c_{12}(b_{ik}b_{kj} - \frac{1}{3}\delta_{ij}b_{kn}b_{nk})b_{mp}b_{pq}S_{qm} \quad .\end{aligned} \tag{3.2.17}$$

The Cayley-Hamilton theorem (Appendix C) has been used to limit this series expression so that it is explicitly 4th order in b_{ij}. Correspondingly, to essentially represent an infinite series, the coefficients c_3 through c_{12} are functions of the turbulence invariants II and III.

Unfortunately, it can be shown using rapid distortion theory (see Section 9.5), that neither X_{ijkm} nor M_{ijkm} can, *in general*, be modeled as a series in b_{ij}. Therefore the underlying assumption used to develop Eq. 3.2.17 is incorrect. However, there appear to be sufficient independent terms in Eq. 3.2.17 to model Π_{ij}^{R} at any point and time in the flow. It is the generality of the model that is compromised by the incorrect assumption.

Equation 3.2.7 for slow pressure-strain, together with Eq. 3.2.17 for rapid pressure-strain, will be used as a consistent framework in which to compare models in the literature.

3.3 Isotropic Turbulence Limits

Crow (1968) used rapid distortion theory to determine the pressure-strain in the limit of isotropic turbulence. He obtained,

$$\Pi_{ij}^{R} = \frac{4}{5} k S_{ij} \qquad \text{when} \qquad b_{mn} = 0. \tag{3.3.1}$$

Comparing with the general model form, Eq. 3.2.17, this yields

$$c_3 = 0.8 \qquad \text{when} \quad g = 1 . \tag{3.3.2}$$

Townsend (1976) used rapid distortion theory and a Taylor series analysis to obtain equations (his Eq. 3.12.5) for the development of the Reynolds stresses in homogeneous shear flow in terms of the dimensionless shear time, $\beta = t\partial U_1 / \partial x_2$. For initially isotropic turbulence, Townsend gives

$$\overline{u_1^2} = \left(1 + \frac{2}{7}\beta^2 + O(\beta^4)\right)(\overline{u_1^2})_{t=0} ,$$

$$\overline{u_2^2} = \left(1 - \frac{4}{35}\beta^2 + O(\beta^4)\right)(\overline{u_1^2})_{t=0} ,$$

$$\overline{u_3^2} = \left(1 + \frac{8}{35}\beta^2 + O(\beta^4)\right)(\overline{u_1^2})_{t=0} , \tag{3.3.3}$$

and

$$-\overline{u_1 u_2} = \left(\frac{2}{5}\beta + O(\beta^3)\right)(\overline{u_1^2})_{t=0} .$$

From these equations, isotropic turbulence limits for c_4 and c_5 can be obtained, as well as the Crow limit for c_3. The analysis (Appendix EQ.3.3.4) gives

$$c_4 = \frac{24}{7} \qquad \text{and} \qquad c_5 = 0 \qquad \text{when} \quad g = 1 . \tag{3.3.4}$$

A general analysis using rapid distortion theory (Section 9.4) confirms that these limits are general, and not particular to shear flow.

3.4 Two-Dimensional Turbulence Limits

When a normal Reynolds stress, for example, $\overline{u_1 u_1}$, goes to zero, the turbulence is two-dimensional. Then the fluctuating velocity, u_1, is zero at that point, and from Eq. 2.2.18,

$$D_{11} + \Pi_{11} = -\frac{2}{\rho}\overline{u_1 \frac{\partial p}{\partial x_1}} = 0 \quad . \tag{3.4.1}$$

With D_{11} zero for homogeneous turbulence, Π_{11} also equals zero.

Similarly the sum of the laminar diffusion and dissipation terms is zero,

$$L_{11} - \varepsilon_{11} = \frac{\partial}{\partial x_k}\nu\frac{\partial \overline{u_1 u_1}}{\partial x_k} - 2\nu\overline{\frac{\partial u_1}{\partial x_k}\frac{\partial u_1}{\partial x_k}} = \overline{2u_1 \frac{\partial}{\partial x_k}\nu\frac{\partial u_1}{\partial x_k}} = 0 \ , \tag{3.4.2}$$

so that with the laminar diffusion term zero for homogeneous turbulence, ε_{11} also equals zero. Then from Eq. 2.2.22,

$$\phi_{11} = \frac{2}{3}\varepsilon \qquad \text{when} \qquad \overline{u_1 u_1} = 0 \ . \tag{3.4.3}$$

Equating this with the model for ϕ_{11}, Eqs. 3.2.7 and 3.2.17, in principal-stress coordinates, and noting that $b_{11} = -1/3$,

$$\begin{aligned}\phi_{11} = \frac{2}{3}\varepsilon = \varepsilon\left(\frac{1}{3}c_1 + c_2\left(\frac{1}{9} - \frac{1}{3}II\right)\right) + kS_{11}\left(c_3 - \frac{2}{3}c_4\right) \\ + kb_{mn}S_{mn}\left(-\frac{2}{3}c_4 - \frac{1}{3}c_6 + c_{10}\left(\frac{1}{9} - \frac{1}{3}II\right)\right) \\ + kb_{kn}b_{mn}S_{mk}\left(-\frac{1}{3}c_9 + c_{12}\left(\frac{1}{9} - \frac{1}{3}II\right)\right) .\end{aligned} \tag{3.4.4}$$

With ε, S_{11}, and the invariants, $b_{mn}S_{mn}$, and $b_{kn}b_{mn}S_{mk}$, all being independent, the two-dimensional limit is satisfied by

$$c_1 = 2 - c_2\left(\frac{1}{3} - II\right) \qquad \text{when} \qquad g = 0 \ , \tag{3.4.5}$$

$$c_4 = \frac{3}{2}c_3 \qquad \text{when} \qquad g = 0 \ , \tag{3.4.6}$$

$$c_6 = -2c_4 + c_{10}\left(\frac{1}{3} - II\right) \qquad \text{when} \qquad g = 0 \ , \tag{3.4.7}$$

and

$$c_9 = c_{12}\left(\frac{1}{3} - II\right) \qquad \text{when} \qquad g = 0 \ . \tag{3.4.8}$$

Realizability

Realizability is a less stringent condition than the two-dimensional turbulence limit. From Eqs. 2.2.1 and 2.2.22, the equation for $\overline{u_1u_1}$ takes the form,

$$\frac{d\overline{u_1u_1}}{dt} = \Pi_{11} - \varepsilon_{11} + \ldots\ldots = \phi_{11} - \frac{2}{3}\varepsilon + \ldots\ldots\ . \tag{3.4.9}$$

Therefore, if $\phi_{11} - \frac{2}{3}\varepsilon \geq 0$ when $\overline{u_1u_1} = 0$, the pressure-strain model will not cause negative values of $\overline{u_1u_1}$ to occur, and the model will satisfy realizability. Correspondingly, Eq. 3.4.4 then becomes the inequality,

$$\begin{aligned}\phi_{11} - \frac{2}{3}\varepsilon = \varepsilon&\left(\frac{1}{3}(c_1 - 2) + c_2(\frac{1}{9} - \frac{1}{3}II)\right) + kS_{11}\left(c_3 - \frac{2}{3}c_4\right)\\ &+kb_{mn}S_{mn}\left(-\frac{2}{3}c_4 - \frac{1}{3}c_6 + c_{10}(\frac{1}{9} - \frac{1}{3}II)\right) \qquad \text{when } b_{11} = -\frac{1}{3}\ .\\ &+kb_{kn}b_{mn}S_{mk}\left(-\frac{1}{3}c_9 + c_{12}(\frac{1}{9} - \frac{1}{3}II)\right) \geq 0\end{aligned} \tag{3.4.10}$$

Since ε is always positive, the two-dimensional limit, Eq. 3.4.5, may be relaxed to the inequality,

$$c_1 \geq 2 - c_2(\frac{1}{3} - II) \qquad \text{when} \qquad g = 0\ , \tag{3.4.11}$$

to satisfy the less stringent condition of realizability.

To relax the equalities 3.4.6 to 3.4.8 requires a greater flexibility in the functionality of the coefficients than just dependence on the turbulence invariants, II, and III. For example, if turbulence production is positive, then $b_{mn}S_{mn}$ is negative, and Eq. 3.4.7 may become the inequality,

$$c_6 \geq -2c_4 + c_{10}(\frac{1}{3} - II) \qquad \text{when } \ g = 0\ , \text{ and } P \geq 0. \tag{3.4.12}$$

3.5 Pressure-Strain Models in the Literature

The coefficients for several pressure-strain models are given in Table 3.5.1. These models use a reduced set of coefficients so that the pressure-strain is explicitly no more than quadratic in the turbulence anisotropy, b_{ij};

$$\begin{aligned}\phi_{ij} = &-c_1 \varepsilon b_{ij} + c_2 \varepsilon (b_{ik} b_{kj} - \frac{1}{3} b_{mn} b_{mn} \delta_{ij}) \\ &+ c_3 k S_{ij} + c_4 k (b_{ik} S_{jk} + b_{jk} S_{ik} - \frac{2}{3} \delta_{ij} b_{mn} S_{mn}) + c_5 k (b_{ik} W_{jk} + b_{jk} W_{ik}) \\ &+ c_6 k b_{ij} b_{mn} S_{mn} + c_7 k (b_{ik} b_{kn} S_{jn} + b_{jk} b_{kn} S_{in} - 2 b_{kj} b_{ni} S_{kn}) \\ &+ c_8 k (b_{ik} b_{kn} W_{jn} + b_{jk} b_{kn} W_{in}) \quad . \end{aligned} \tag{3.5.1}$$

As can be seen in the table, the early models were linear in b_{ij}.

Table 3.5.1. Pressure-strain models for homogeneous turbulence (omitting the wall reflection terms given by S1973, LRR1975 and GL1978).

Model:	D1971	S1973	LRR1975	GL1978	Y1984	SL1985	SSG1991
c_1	$c_{1,\mathrm{D}}(R_t)$	5	3	3.6	6	$c_{1,\mathrm{L}}(II, III, R_t)$	3.4
c_2	0	0	0	0	0	0	4.2
c_3	0	2/3	0.8	0.8	0.4	0.8	$0.8-1.3\sqrt{II}$
c_4	0	1	1.745	1.2	0.6	$1.2+0.96\sqrt{g}$	1.25
c_5	0	1	1.309	1.2	0.6	$1.73-0.75\sqrt{g}$	0.4
c_6	0	0	0	0	0	–2.4	3.6
c_7	0	0	0	0	0	0.8	0
c_8	0	0	0	0	0	0.8	0

D1971, Donaldson; S1973, Shir; LRR1975, Launder, Reece and Rodi; GL1978, Gibson and Launder; Y1984, Younis (see Launder, 1993); SL1985, Shih and Lumley (see Demuren and Sarkar, 1993, or Shih and Lumley, 1993); SSG1991, Speziale, Sarkar and Gatski.

The earliest model in the table, D1971, uses only a slow pressure-strain, with the c_1 coefficient a function of Reynolds number,

$$c_{1,\mathrm{D}} = 2 + \frac{2}{b\left(1 + \sqrt{1 + \dfrac{a}{b^2 R_t}}\right)}, \qquad a = 2.5, \quad b = 0.125 \quad . \tag{3.5.2}$$

This gives limits of $c_1 = 2$ at $R_t = 0$, and $c_1 = 10$ at high R_t. It is interesting that this model, of all the models in the table, comes closest to satisfying the goals of this work (see Section 1.1). Donaldson and Rosenbaum (1969) initially developed the model to calculate turbulent boundary layers. Donaldson (1971) modified the model and showed good agreement with skin friction measurements in turbulent boundary layers, and data on free jets and shear flows. Lewellen and Teske (1973) used the Donaldson model, with added buoyancy terms, to compare with data on atmospheric boundary layers. Donaldson (1973, and Donaldson and Bilanin, 1975) went on to consider the problem of aircraft trailing vortices, showing perhaps the first bifurcation plots for rotating shear flow. Unfortunately he had no data with which to compare.

The LRR1975, GL1978 and SL1985 models, all take $c_3 = 0.8$. This is to satisfy the well-known limit for isotropic turbulence, Eq. 3.3.1 (Crow, 1968),

$$\phi_{ij} = 0.8kS_{ij} \qquad \text{when} \qquad b_{mn} = 0. \tag{3.5.3}$$

The SSG1991 model satisfies the isotropic limit but takes c_3 as a function of II to reduce c_3 for non-isotropic turbulence.

The rapid pressure-strain in the LRR1975 model satisfied the constraints on X_{ijmk}, Eqs. 3.2.9 and 3.2.10. This set relationships between c_3, c_4 and c_5, so they were unable to independently set c_4 and c_5. The slow pressure-strain, c_1 value, was taken to match the homogeneous shear data of Champagne et al. (CHC1970).

The S1973, GL1978 and Y1984 models are all forms of the IP or "isotropization of production" model, proposed by Naot et al. (NSW1970). For these models, the rapid pressure-strain, when considered in the absolute reference frame, may be written as,

$$\phi_{ij}^{\mathrm{R}} = -\frac{c_4}{2}\left(P_{ij} - \frac{2}{3}P\delta_{ij}\right) . \tag{3.5.4}$$

Thus, for these models, the effect of the pressure-strain is to partially counteract the effect production has on the anisotropy of the turbulence. The D1971 model takes $\phi_{ij}^{\mathrm{R}} = 0$, so it also satisfies this form. Launder (1993) noted that for models of this form, with lower values of c_4, higher values of c_1 are needed to get the same results for simple shear flow.

The D1971, Y1984 and SSG1991 models all use low values for c_5. The common theme is swirling or rotational flow. Launder (1993) observed that the Y1984 model performs better on swirling jets than the GL1978 model. The D1971 model was used to analyze vortices, and the SSG1991 model was developed considering large eddy simulation (LES) data for rotating shear flow.

The development of the SL1985 model placed more emphasis on irrotational flows. The slow pressure-strain c_1 coefficient,

$$c_{1,L} = 2 + g\, e^{-11.65/\sqrt{R_t}}\left(12/\sqrt{R_t} + 8.9\ln(1 + 31.2II + 47.8III)\right) , \tag{3.5.5}$$

was developed by Lumley (1978) using data for the return to isotropy following axisymmetric contraction ($III < 0$).

The SL1985 model also has the advantage of satisfying the two-dimensional turbulence limit, Eq. 3.4.3, and the corresponding restrictions on the pressure-strain model coefficients, Eqs. 3.4.5-7. In particular, when $g = 0$, $c_1 = 2$, $c_4 = 1.5c_3$, and $c_6 = -2c_4$. The D1971 model satisfies realizability, but not the two-dimensional limit. While the other models in Table 3.5.1 do not formally satisfy realizability, it would probably take extremely unusual flow circumstances for failure to occur.

Like SL1985, SSG1991 considered the slow pressure-strain terms independent of the rapid pressure-strain model. In particular, they considered return-to-isotropy near the isotropic limit. Besides a c_1 term, they also included a non-linear, c_2 term, and based on realizability took $c_2 = 3(c_1 - 2)$. (See Eq. 3.4.11, and note that the highest value possible for II is 2/3.) Needing more return-to-isotropy, to match the homogeneous shear flow data of Tavoularis and Corrsin (TC1981), they set $c_6 = 3.6$ (note the opposite sign to SL1985). For this data, $P/\varepsilon \approx 1.8$, so that the c_1 and c_6 terms combined give

$$\phi_{ij} = -c_1\varepsilon b_{ij} + c_6 k b_{ij} b_{mn} S_{mn} + \ldots.. = -(3.4 + 1.8P/\varepsilon)\varepsilon b_{ij} + \ldots.. = -6.64\varepsilon b_{ij} + \ldots.. \tag{3.5.6}$$

or an equivalent c_1 coefficient of 6.64 when $P/\varepsilon = 1.8$.

Chapter 4

HOMOGENEOUS SHEAR FLOW

The primary basis for pressure-strain models in the literature is homogeneous turbulence. By this is meant that the ensemble average properties of the turbulence do not vary in space. In this chapter, we derive equations for the development of turbulence anisotropy in homogeneous shear flow with rotation in the plane of the shear, and show the effects that the pressure-strain terms have on the turbulence. The exploration of the effects that rotation has on shear flow gives insight into vortices and curved and rotating boundary layers.

4.1 Equations for Homogeneous Turbulence

For homogeneous turbulence, the convection and diffusion terms in the Reynolds stress equations, Eq. 2.2.1, are zero, and the equations reduce to

$$\frac{d\overline{u_i u_j}}{dt} = P_{ij} + F_{ij} + \Pi_{ij} - \varepsilon_{ij} \ . \tag{4.1.1}$$

These may be used to obtain equations for the turbulence anisotropy, b_{ij}. Noting that,

$$\frac{d\overline{u_i u_j}}{dt} = 2k\frac{db_{ij}}{dt} + 2(b_{ij} + \frac{1}{3}\delta_{ij})\frac{dk}{dt} \ , \tag{4.1.2}$$

and that for homogeneous turbulence, $dk/dt = P - \varepsilon$, the equations for b_{ij} are

$$2k\frac{db_{ij}}{dt} = P_{ij} + F_{ij} + \Pi_{ij} - \varepsilon_{ij} - 2(b_{ij} + \frac{1}{3}\delta_{ij})(P - \varepsilon) \ , \tag{4.1.3}$$

or using Eq. 2.2.22,

$$2k\frac{db_{ij}}{dt} = P_{ij} - \frac{2}{3}\delta_{ij}P + F_{ij} + \phi_{ij} - 2b_{ij}(P - \varepsilon) \ . \tag{4.1.4}$$

When the production term is combined with half of the force term, it is a function of the absolute mean velocity (Exercise 2.2),

$$P_{ij} - \frac{2}{3}\delta_{ij}P + \frac{1}{2}F_{ij} = -\frac{4}{3}kS_{ij} - 2k(b_{ik}S_{jk} + b_{jk}S_{ik} - \frac{2}{3}\delta_{ij}b_{mn}S_{mn}) \\ -2k(b_{ik}W_{jk} + b_{jk}W_{ik}) \quad . \tag{4.1.5}$$

Note that the three terms on the right hand side have the same form as the c_3, c_4 and c_5 terms in the general algebraic rapid pressure-strain model, Eq. 3.2.17. Combining Eqs. 4.1.4 and 4.1.5 with the algebraic pressure-strain model equations, Eq. 3.2.7 and 3.2.17, and Eq. 2.2.14 for F_{ij}, gives a modeled form for the b_{ij} equations for homogeneous turbulence;

$$\begin{aligned} 2k\frac{db_{ij}}{dt} &= -b_{ij}\left((c_1 - 2)\varepsilon + (2 + \frac{1}{2}c_6)P - c_9 k b_{kn} b_{mn} S_{mk}\right) \\ &+(b_{ik}b_{kj} - \frac{1}{3}\delta_{ij}b_{kn}b_{nk})\left(c_2\varepsilon + c_{10}kb_{pq}S_{pq} + c_{12}kb_{mp}b_{pq}S_{qm}\right) \\ &+\left(c_3 - \frac{4}{3}\right)kS_{ij} + (c_4 - 2)k(b_{ik}S_{jk} + b_{jk}S_{ik} - \frac{2}{3}\delta_{ij}b_{mn}S_{mn}) \\ &+c_7 k(b_{ik}b_{kn}S_{jn} + b_{jk}b_{kn}S_{in} - 2b_{kj}b_{ni}S_{kn}) \\ &+(c_5 - 2)k(b_{ik}W_{jk} + b_{jk}W_{ik}) - 2k\Omega_k\left(\varepsilon_{ikm}b_{jm} + \varepsilon_{jkm}b_{im}\right) \\ &+c_8 k(b_{ik}b_{kn}W_{jn} + b_{jk}b_{kn}W_{in}) + c_{11}k(b_{ik}W_{kn}b_{mn}b_{mj} + b_{jk}W_{kn}b_{mn}b_{mi}) \, . \end{aligned} \tag{4.1.6}$$

4.2 Equations for Homogeneous Shear Flow

For homogeneous shear flow, the shear layer has a uniform mean velocity gradient. As shown in Fig. 4.2.1, we define the coordinate system so that

$$\frac{\partial U_i}{\partial x_j} = \delta_{i1}\delta_{j2}\frac{\partial U_1}{\partial x_2} \tag{4.2.1}$$

with $\partial U_1 / \partial x_2$ positive, and we consider only coordinate rotation about the x_3 axis,

$$\Omega_m = \delta_{m3}\Omega_3 \; . \tag{4.2.2}$$

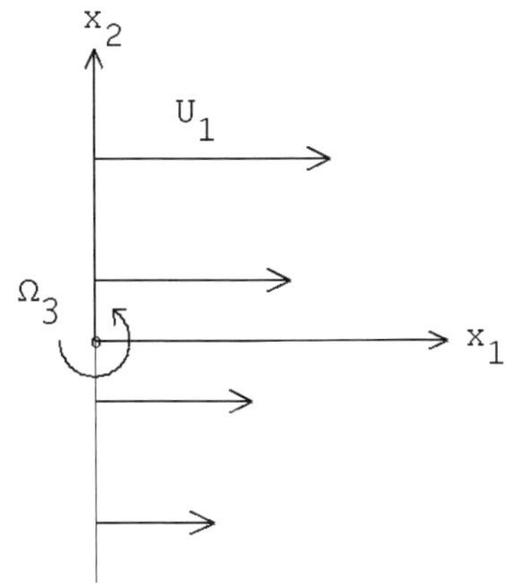

Fig. 4.2.1. Coordinates for rotating homogeneous shear flow.

The non-zero components of the mean strain rate and absolute vorticity tensors are

$$S_{12} = S_{21} = \frac{1}{2}\frac{\partial U_1}{\partial x_2} \qquad \text{and} \qquad W_{12} = -W_{21} = \frac{1}{2}\frac{\partial U_1}{\partial x_2} - \Omega_3 \,. \tag{4.2.3}$$

The turbulence production is

$$P = -2kb_{ik}S_{ik} = -4kb_{12}S_{12} \,. \tag{4.2.4}$$

The algebra to derive the equations for shear flow from Eq. 4.1.6 is given in Appendix EQ.4.2.5. Considering flows in which b_{13} and b_{23} are initially zero, then

$$b_{13} = 0, \qquad b_{23} = 0, \qquad \text{and}$$

$$\begin{aligned} 2k\frac{d}{dt}\begin{bmatrix} b_{11}-b_{22} \\ b_{33} \\ b_{12} \end{bmatrix} &= -\left((c_1-2)\varepsilon + (2+\frac{1}{2}c_6)P - \frac{1}{2}c_9 b_{33}P\right)\begin{bmatrix} b_{11}-b_{22} \\ b_{33} \\ b_{12} \end{bmatrix} \\ &+\left(c_2\varepsilon - \frac{1}{2}c_{10}P + \frac{1}{2}c_{12}b_{33}P\right)\begin{bmatrix} b_{11}^2-b_{22}^2 \\ b_{33}^2 - \frac{1}{3}II \\ -b_{12}b_{33} \end{bmatrix} + \left(c_3 - \frac{4}{3}\right)kS_{12}\begin{bmatrix} 0 \\ 0 \\ 1 \end{bmatrix} - (c_4-2)kS_{12}\begin{bmatrix} 0 \\ \frac{4}{3}b_{12} \\ b_{33} \end{bmatrix} \\ &+\left\{c_7 kS_{12}(b_{22}-b_{11}) + 2k\Omega_3 + kW_{12}\left((c_5-2) - c_8 b_{33} + c_{11}(b_{11}b_{22} - b_{12}^2)\right)\right\}\begin{bmatrix} 4b_{12} \\ 0 \\ b_{22}-b_{11} \end{bmatrix}. \end{aligned} \tag{4.2.5}$$

The rapid pressure-strain terms which are quadratic or higher in b_{ij}, are then tensorially equivalent to the slow pressure-strain terms or the c_5 rapid pressure-strain term. This was noted by Speziale et al. (1991) in the development of the SSG model and it influenced their decision to keep only the slow, and a reduced number of rapid, terms, in particular the c_1 through c_6 pressure-strain terms.

Effects of Pressure-Strain Terms

Equation 4.2.5 shows that there are tensorially five different effects the pressure-strain model terms have on the anisotropy of the turbulence. These can be shown individually using simplified equations. The effects are demonstrated in this section, starting with a typical anisotropy found in turbulent boundary layers (Eq. 1.2.10),

$$\text{Init. Values: } b_{11} = 0.20,\ b_{22} = -0.14,\ b_{33} = -0.06 \text{ and } b_{12} = -0.15\,. \tag{4.2.6}$$

The c_1 term causes the turbulence to return towards isotropy. The simplified equation,

$$\frac{db_{ij}}{dt} = -\frac{1}{2}(c_1 - 2)\frac{\varepsilon}{k} b_{ij}\,, \tag{4.2.7}$$

includes both the c_1 term and the isotropic dissipation term. In general, $c_1 \geq 2$ so that the net effect is a return to isotropy, as shown in Fig. 4.2.2.

Fig. 4.2.2. Effect of the c_1 term on turbulence. Solution of Eq. 4.2.7, with $c_1 = 3.4$, $\varepsilon/k{=}1$.

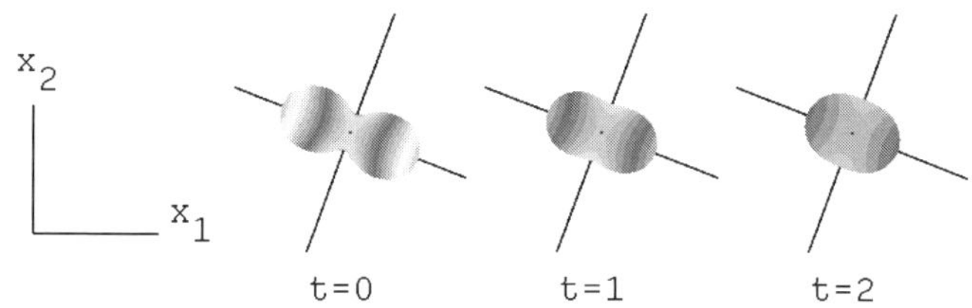

The c_6 and c_9 terms are tensorially equivalent to the c_1 term. Positive values of c_6 will cause a return to isotropy when turbulence production is positive. However, the return occurs more slowly in time as it progresses, since production is reduced as the anisotropy is reduced. The effect of the c_9 term, which is cubic in b_{ij}, depends on the signs of both b_{33} and P. It is usual in boundary layers that P is positive and b_{33} is negative, in which case positive values of c_9 will cause a return to isotropy.

When c_2 is positive, the turbulence anisotropy is enhanced. This is shown in Fig. 4.2.3 using the simplified equation,

$$\frac{db_{ij}}{dt} = \frac{1}{2}c_2\frac{\varepsilon}{k}\left(b_{ik}b_{jk} - \frac{1}{3}\delta_{ij}II\right). \tag{4.2.8}$$

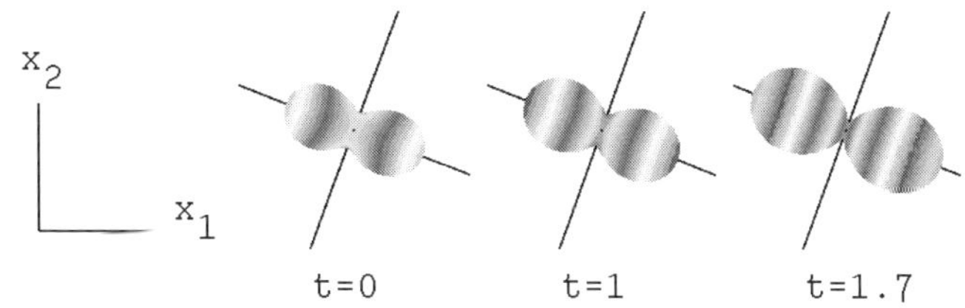

Fig. 4.2.3. Effect of the c_2 term on turbulence. Solution of Eq. 4.2.8, with $c_2 = 4.2$, $\varepsilon/k=1$.

The last time shown on Fig. 4.2.3 is $t=1.7$. Shortly after this time, one of the turbulence normal stresses became negative, i.e., not realizable, and the glyph could not be drawn. The values of the c_1 and c_2 coefficients used for Figs. 4.2.2 and 4.2.3 are those for the SSG1991 model (Table 3.5.1). It is therefore useful to consider the combined effects, i.e., the entire slow pressure-strain and dissipation,

$$\frac{db_{ij}}{dt} = -\frac{1}{2}(c_1 - 2)\frac{\varepsilon}{k}b_{ij} + \frac{1}{2}c_2\frac{\varepsilon}{k}\left(b_{ik}b_{jk} - \frac{1}{3}\delta_{ij}II\right), \tag{4.2.9}$$

for the SSG1991 model. Figure 4.2.4 shows that the combined effect is a return to isotropy, but at a slightly slower pace than the c_1 term alone. With the combined effect, Eq. 4.2.9, $-b_{12}$ reduces from its initial value of 0.15 to 0.051 at $t=2$, compared with a reduction to 0.037 at $t=2$ when the c_2 term is omitted.

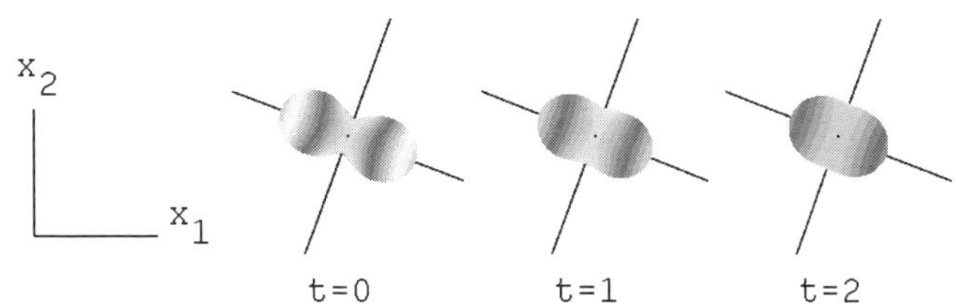

Fig. 4.2.4. Return to isotropy, solution of Eq. 4.2.9, using $\varepsilon/k=1$, and SSG1991 coefficients, $c_1 = 3.4$ and $c_2 = 4.2$.

For this shear flow, the c_3 term affects only the equation for b_{12},

$$\frac{db_{12}}{dt} = \frac{1}{2}\left(c_3 - \frac{4}{3}\right)S_{12} + \ldots\ldots \tag{4.2.10}$$

The "–4/3" in the equation came from the first term of the Reynolds stress production in Eq. 2.2.5. Since, in general, c_3 is less than 4/3, the effect of c_3 is simply to partially oppose that part of the production which stretches and contracts the Reynolds stresses along the axes of the strain rate tensor, as shown in Fig. 2.2.1.

Turbulence production in shear flow has no effect on the out-of-plane normal Reynolds stress, $\overline{u_3u_3}$ (see discussion of Eq. 2.2.8 and Fig. 2.2.3). Therefore positive turbulence production causes $\overline{u_3u_3}$ to be left behind and b_{33} decreases. This is reflected in the equation for b_{33},

$$\frac{db_{33}}{dt}=-\frac{1}{2}(c_4-2)S_{12}\frac{4}{3}b_{12}+\ldots\ldots\quad=\frac{1}{6}(c_4-2)\frac{P}{k}+\ldots\ldots, \tag{4.2.11}$$

by the "–2". The c_4 pressure-strain term opposes this effect of production. With positive production, if $c_4>2$, as it is for the isotropic limit, Eq. 3.3.4, an above-average proportion of the turbulence goes into $\overline{u_3u_3}$ and b_{33} increases. If $c_4<2$, as it is for most of the models in Table 3.5.1, then b_{33} declines.

In Fig. 2.2.4 it was shown that the vorticity part of the Reynolds stress production causes a rotation of the glyph in the direction opposite to the sense of the vorticity. This is reflected by the "–2" in the equations for $b_{11}-b_{22}$ and b_{12},

$$\frac{d}{dt}\begin{bmatrix}b_{11}-b_{22}\\ b_{12}\end{bmatrix}=\frac{1}{2}(c_5-2)W_{12}\begin{bmatrix}4b_{12}\\ b_{22}-b_{11}\end{bmatrix}+\ldots\ldots\ . \tag{4.2.12}$$

Positive values of c_5, between 0 and 2, as used in the models in Table 3.5.1, oppose part of the rotation.

The effect of the c_7 term, which is proportional to the strain rate,

$$\frac{d}{dt}\begin{bmatrix}b_{11}-b_{22}\\ b_{12}\end{bmatrix}=\frac{1}{2}c_7S_{12}(b_{22}-b_{11})\begin{bmatrix}4b_{12}\\ b_{22}-b_{11}\end{bmatrix}+\ldots\ldots, \tag{4.2.13}$$

is also a rotation of the glyph. As shown in Fig. 4.2.5a, with the initial stress tensor having $b_{11}>b_{22}$, the glyph rotates in a counter-clockwise direction, asymptotically approaching a position where $b_{11}=b_{22}$ and the principal directions of the stress and strain rate are aligned. If the initial stress tensor has $b_{22}>b_{11}$, Fig. 4.2.5b, the glyph rotates clockwise towards the same final state. Note that this state is 90 degrees from the Boussinesq position, and corresponds to a position of negative turbulence production. With negative values of c_7, this pressure-strain term would tend to move the glyph towards the Boussinesq position with maximum positive production.

The SL1985 model uses $c_7=0.8$, as demonstrated in Fig. 4.2.5. However, it also uses $c_8=c_7$. For shear flow without rotation, $W_{12}=S_{12}$, and the combined effects of the two terms are represented by the model equation,

Fig. 4.2.5. Effect of the c_7 term on turbulence. Solution of Eq. 4.2.13, with $c_7 = 0.8$ and S_{12}=3.

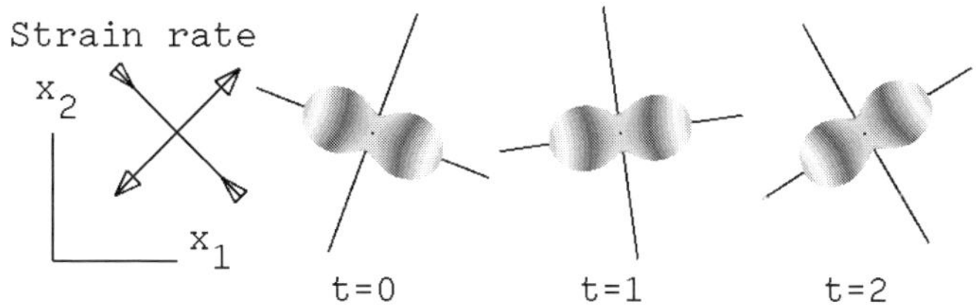

Fig. 4.2.5a. Glyph at t=0 has b_{11}>b_{22}.

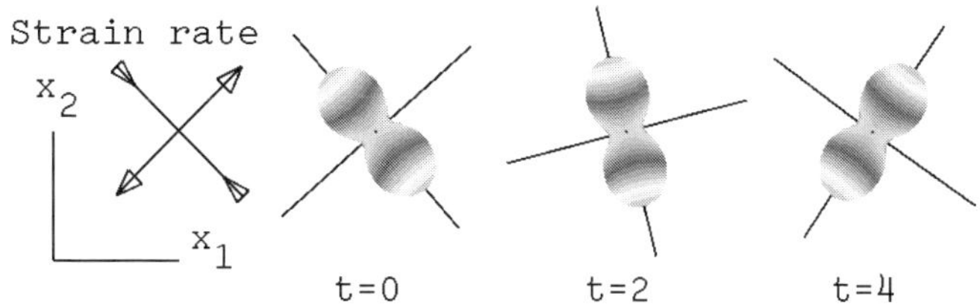

Fig. 4.2.5b. Glyph at t=0 has been rotated so that $b_{22} > b_{11}$.

$$\frac{d}{dt}\begin{bmatrix} b_{11}-b_{22} \\ b_{12} \end{bmatrix} = \frac{1}{2}\left(c_7 S_{12}(b_{22}-b_{11}) - c_8 W_{12} b_{33}\right)\begin{bmatrix} 4b_{12} \\ b_{22}-b_{11} \end{bmatrix} + \ldots\ldots$$
$$= c_7 S_{12} b_{22}\begin{bmatrix} 4b_{12} \\ b_{22}-b_{11} \end{bmatrix} + \ldots\ldots \quad . \tag{4.2.14}$$

The solution to this model equation is similar to that in Fig. 4.2.5, except that the glyph approaches the state $b_{22} = 0$, and the sign of b_{22}, instead of the sign of $b_{22} - b_{11}$, determines the direction of rotation.

Exercise 4.2.

Show that $$\frac{d}{dt}\begin{bmatrix} b_{11}-b_{22} \\ b_{33} \\ b_{12} \end{bmatrix} = A\begin{bmatrix} 4b_{12} \\ 0 \\ b_{22}-b_{11} \end{bmatrix}$$

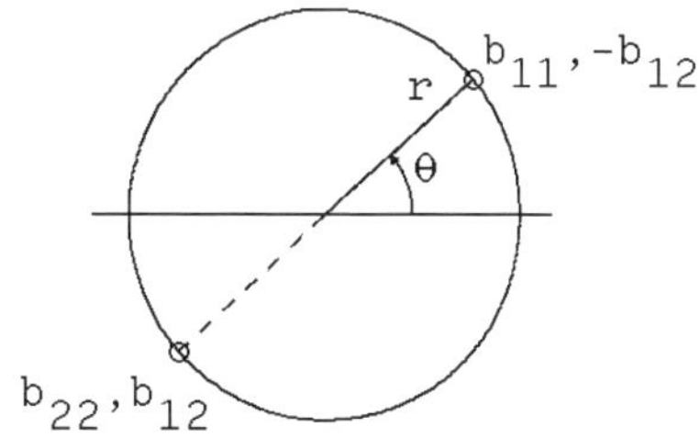

represents a pure rotation of the glyph about the x_3 axis at an angular rate of $-A$. Using the Mohr's circle representation of the stresses at right, show that $dr^2/dt = 0$ and $d\theta/dt = -2A$.

4.3 Effects of Rotation

Bertoglio (1982) analyzed homogeneous shear flow in a rotating reference frame using rapid distortion theory (RDT). The rotation vector is normal to the plane of the 2-d shear flow, as shown in Fig. 4.2.1. He started with isotropic turbulence and considered the development in time of the turbulence for various rotation rates. He showed that when the rotation rate, Ω_3, equaled $(1/4)\partial U_1/\partial x_2$, the principal axes of the stress and strain rate tensors remained closely aligned throughout the time of his calculation (his Fig. 3b), both being at 45 degrees to the x_1 and x_2 directions. Correspondingly, for this rotation rate his solution showed $b_{11} \approx b_{22}$. The large eddy simulation of Bardina (1983), with an initial Sk/ε of 2.0, also gave $b_{11} \approx b_{22}$ for homogeneous shear flow with $\Omega_3 = (1/4)\partial U_1/\partial x_2$.

Substituting $\Omega_3 = (1/4)\partial U_1/\partial x_2$, $W_{12} = (1/2)\partial U_1/\partial x_2 - \Omega_3 = (1/4)\partial U_1/\partial x_2$, and $b_{11} = b_{22}$, into the equation for $d(b_{11} - b_{22})/dt$, Eq. 4.2.5, yields,

$$0 = kb_{12}\frac{\partial U_1}{\partial x_2}\left(c_5 - c_8 b_{33} + c_{11}(b_{11}b_{22} - b_{12}^2)\right) . \tag{4.3.1}$$

This is satisfied by $c_5 = c_8 = c_{11} = 0$, which is consistent with Donaldson's model which took the rapid pressure-strain as zero (D1971, Table 3.5.1).

Vortex Flow

Donaldson (1973) applied his model to a classical line vortex, and determined regions where turbulence will be sustained and grow, and a core region where turbulence will decay.

The relationship between this vortical flow and shear flow with rotation is shown in Fig. 4.3.1. Figure 4.3.1a shows the mean tangential velocity used by Donaldson for the line vortex,

$$V_\theta = \frac{A}{2\pi r}\left(1 - e^{-1.256(r/r_m)^2}\right) , \tag{4.3.2}$$

where r_m is the radius at which the velocity is maximum. The flow at radius r_a will be stationary in the relative frame if the coordinates are rotated about the x_3 axis at a rate $\Omega_3 = V_\theta(r_a)/r_a$. The solid-body rotation of the coordinates is shown in Fig. 4.3.1b. The relative tangential velocity is the difference,

$$U_\theta(r) = V_\theta(r) - \Omega_3 r = V_\theta(r) - \frac{V_\theta(r_a)}{r_a}r , \tag{4.3.3}$$

and is shown in Fig. 4.3.1c. The same flow may be considered in Cartesian coordinates rotating at the same rate, as shown in Fig. 4.3.1d. For the point, $x_1 = 0$, $x_2 = r_a$, the only non-zero component of the mean velocity gradient tensor is (Appendix EQ.4.3.4),

$$\frac{\partial U_1}{\partial x_2} = -\frac{\partial U_\theta}{\partial r} . \tag{4.3.4}$$

Therefore, the flow at this point is equivalent to a shear flow with rotation $\Omega_3 = V_\theta(r_a)/r_a$.

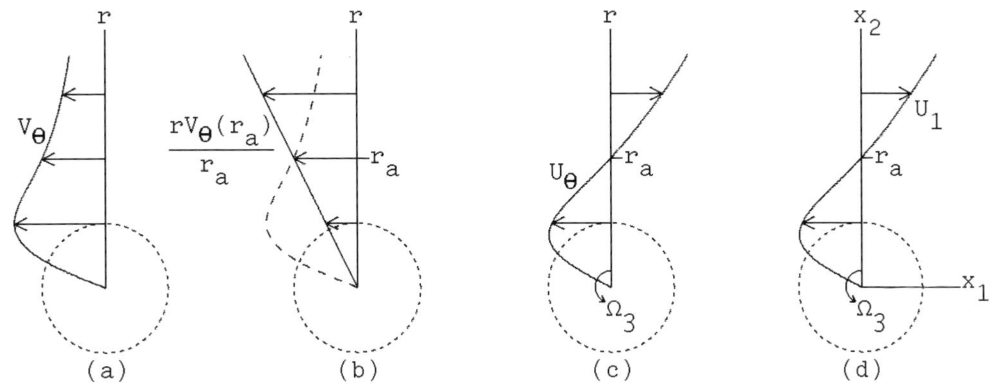

Fig. 4.3.1. Coordinates and mean velocity for vortex analysis.

The strain rate and absolute vorticity are independent of the chosen rotation rate for viewing the vortex. For cylindrical coordinates in the stationary frame,

$$S_{\theta r} = \frac{1}{2}\left(\frac{\partial V_\theta}{\partial r} - \frac{V_\theta}{r}\right), \quad \text{and} \quad W_{\theta r} = \frac{1}{2}\left(\frac{\partial V_\theta}{\partial r} + \frac{V_\theta}{r}\right). \tag{4.3.5}$$

From Fig. 4.3.1d, for the rotating Cartesian coordinates, at point r_a, the θ direction corresponds to the $-x_1$ direction, with the r and x_2 directions coincident. Therefore,

$$\text{for } x_1 = 0, \quad S_{12} = -S_{\theta r} = -\frac{1}{2}\left(\frac{\partial V_\theta}{\partial r} - \frac{V_\theta}{r}\right), \quad \text{and} \quad W_{12} = -W_{\theta r} = -\frac{1}{2}\left(\frac{\partial V_\theta}{\partial r} + \frac{V_\theta}{r}\right). \tag{4.3.6}$$

Donaldson analyzed the line vortex in what he referred to as a super-equilibrium analysis. This is consistent with assuming that the equivalent rotating shear flow at any point in the vortex is homogeneous and fully developed with $db_{ij}/dt = 0$. These assumptions give the algebraic stress limit of the Reynolds stress model. From Eq. 4.2.5, with only the c_1 pressure-strain term, as for model D1971,

$$\frac{\varepsilon}{k}\left((c_1 - 2) + 2\frac{P}{\varepsilon}\right)\begin{bmatrix} b_{11} - b_{22} \\ b_{33} \\ b_{12} \end{bmatrix} = -\frac{4}{3}S_{12}\begin{bmatrix} 0 \\ 0 \\ 1 \end{bmatrix} + 2S_{12}\begin{bmatrix} 0 \\ \frac{4}{3}b_{12} \\ b_{33} \end{bmatrix} + 2(\Omega_3 - W_{12})\begin{bmatrix} 4b_{12} \\ 0 \\ b_{22} - b_{11} \end{bmatrix}. \tag{4.3.7}$$

Noting that $W_{12} = S_{12} - \Omega_3$, this may be rewritten as

$$\frac{\varepsilon}{2S_{12}k}\left((c_1-2)+2\frac{P}{\varepsilon}\right)\begin{bmatrix} b_{11}-b_{22} \\ b_{33} \\ b_{12} \end{bmatrix} = -\frac{2}{3}\begin{bmatrix} 0 \\ 0 \\ 1 \end{bmatrix} + \begin{bmatrix} 0 \\ \frac{4}{3}b_{12} \\ b_{33} \end{bmatrix} + (\frac{2\Omega_3}{S_{12}}-1)\begin{bmatrix} 4b_{12} \\ 0 \\ b_{22}-b_{11} \end{bmatrix}. \quad (4.3.8)$$

Bifurcation of Shear with Rotation

Turbulence will be sustained or increased in solutions of Eq. 4.3.8 for which $P/\varepsilon \geq 1$, and the turbulence will decline when $P/\varepsilon < 1$. Therefore solutions for $P/\varepsilon = 1$ form a bifurcation line marking when turbulence is sustained. Since $P/\varepsilon = -2b_{12}(2S_{12}k/\varepsilon)$, the solutions are a function of two parameters, $\varepsilon/(2S_{12}k)$ and Ω_3/S_{12}, and the pressure-strain model.

The bifurcation line for the D1971 model at high R_t is shown in Fig. 4.3.2 in terms of the classic parameters, $1/s$, and the rotation parameter, R. For shear flow, the dimensionless strain rate is

$$s = \frac{k}{\varepsilon}\sqrt{2S_{ij}S_{ij}} = \frac{2k\,|S_{12}|}{\varepsilon}. \quad (4.3.9)$$

The rotation parameter can be expressed in terms of either the rotating Cartesian coordinates or the stationary cylindrical coordinates,

$$R = R_R = \frac{\Omega_3}{-S_{12}} \qquad \text{or} \qquad R = R_C = \frac{\dfrac{V_\theta}{r}}{\dfrac{1}{2}\left(\dfrac{\partial V_\theta}{\partial r} - \dfrac{V_\theta}{r}\right)}. \quad (4.3.10)$$

The bifurcation line peaks, and is symmetrical about $R = -0.5$. As discussed above, this is the rotation rate ($\Omega_3 = (1/4)\partial U_1/\partial x_2$) at which Bertoglio obtained $b_{11} \approx b_{22}$.

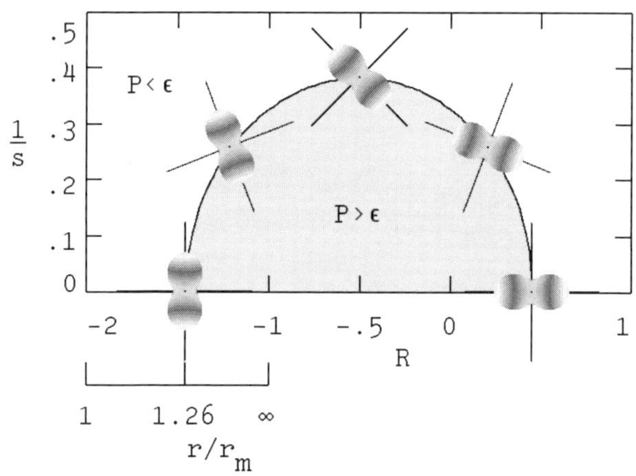

Fig. 4.3.2. Bifurcation plot for the D1971 model at high R_t, with an extra scale r/r_m for the line vortex. Glyph orientations correspond to the coordinate directions in Fig. 4.3.1d.

Donaldson used a stability parameter, N, where $N = R/2$, and noted that no locally sustained turbulent flow is possible for $N < -0.729$ ($R < -1.458$). From Fig. 4.3.2, this corresponds to the core of the line vortex where $r < 1.26 r_m$. The tendency of vortices to have laminar cores is consistent with the smoke visualization of Poppleton (1971, see Bradshaw, 1973, Fig. 30) for a wing tip vortex. This showed a "nearly laminar" viscous core with higher turbulence in the surrounding atmosphere (see Bilanin et al., BTDS1977, Fig. 4).

Figure 4.3.3 shows the effect of turbulence Reynolds number on the bifurcation line for the D1971 model. As R_t goes to zero, c_1 goes to 2, and the range of R for which turbulence can be sustained shrinks to $-1 \le R \le 0$.

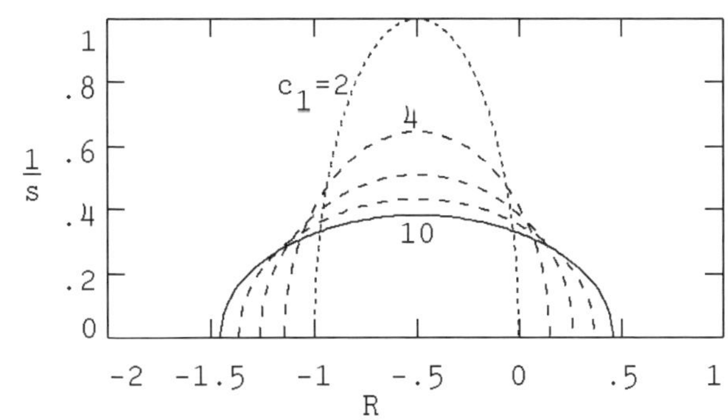

Fig. 4.3.3. Bifurcation lines for the D1971 model with $c_1 = 2$, 4, 6, 8, and 10, spanning $R_t = 0$ to ∞.

In general, for pressure-strain models which include only c_1, c_3, c_4, c_5 and c_6 terms, the range in R for which turbulence may be sustained is (Appendix EQ.4.3.11)

$$\frac{c_5 - 2 - C}{4 - c_5} \le R \le \frac{c_5 - 2 + C}{4 - c_5}, \quad \text{where} \quad C = \sqrt{\frac{(c_4 - 2)^2}{3} + \left(c_1 + \frac{c_6}{2}\right)\left(\frac{4}{3} - c_3\right)} . \qquad (4.3.11)$$

The mid-point of the sustainable range, where $1/s$ is maximum, is determined by c_5 alone when the coefficients are constants,

$$R_{mid} = -(2 - c_5)/(4 - c_5). \qquad (4.3.12)$$

As shown in Fig. 4.3.4, the GL1978 model has a very small range, $-0.81 < R < 0.24$, due to the low value of c_1 and the high value of c_3, and the mid-point is comparatively far to the right, $R_{mid} = -0.286$, due to the high value of c_5. This model, with a lower limit of $R = -0.81$, would predict, not just a laminar core, but laminarization of the entire classical line vortex (Eq. 4.3.2, Fig. 4.3.2).

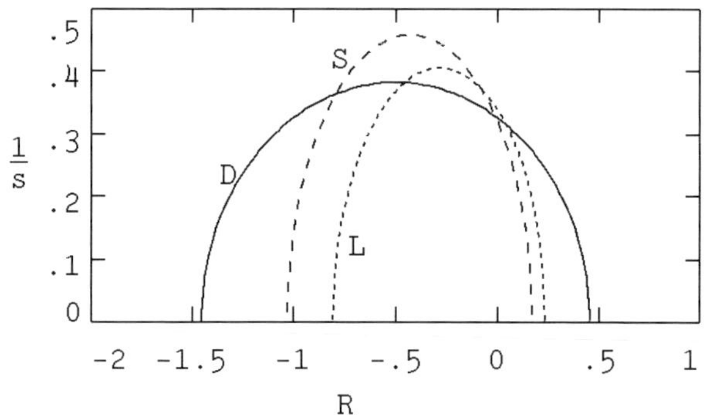

Fig. 4.3.4. Comparison of bifurcation lines for three models: D, D1971 at high R_t; S, SSG1991; L, GL1978.

Figure 4.3.4 also shows the bifurcation line for the SSG1991 model. This model includes a c_2 term, and c_3 depends on II, so that the mid-point and range vary slightly from Eqs. 4.3.11 and 4.3.12. The mid-point is at $R_{mid} = -0.43$, only slightly to the right of the D1971 model. The lower limit at $R = -1.028$ means that turbulence is sustainable in the vortex only for $r/r_m > 2.05$.

Exercise 4.3. Show that for constant pressure-strain coefficients, including only c_1, c_3, c_4, c_5 and c_6 terms, the bifurcation lines are ellipses in the R, $1/s$ plane.

Curved and Rotating Boundary Layers

Boundary layers with curvature and rotation may also be shear flow with rotation. For the analogy to be valid, the streamlines should be circular arcs with constant velocity, U_θ, along the streamline. The rotation parameter for curved boundary layers, in the rotating cylindrical coordinates of Fig. 4.3.5c, is

$$R = \frac{\dfrac{U_\theta}{r} + \Omega_3}{\dfrac{1}{2}\left(\dfrac{\partial U_\theta}{\partial r} - \dfrac{U_\theta}{r}\right)} = \frac{\text{effective rotation rate}}{\text{mean strain rate}} . \tag{4.3.13}$$

Note, that R is the ratio of the effective rotation rate, V_θ / r, to the mean strain rate, $S_{\theta r}$.

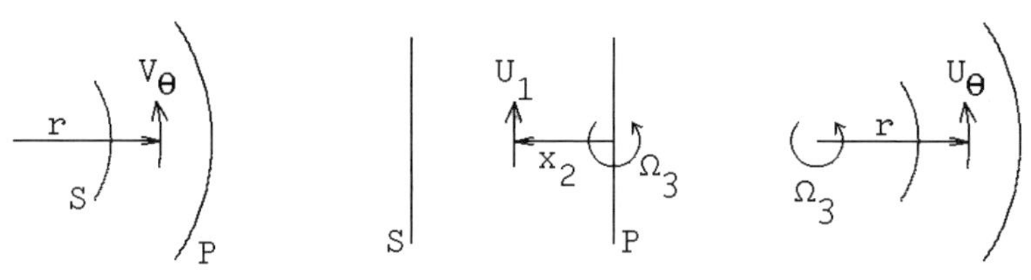

Curved Channel $R = R_C$, Eq. 4.3.10 | Rotating Plane Channel $R = R_R$, Eq. 4.3.10 | Rotating Curved Channel R, Eq. 4.3.13

Fig. 4.3.5. Coordinates for curved and rotating channels. S, suction side; P, pressure side.

The super-equilibrium analysis, used for the bifurcation plots, applies if the turbulence may be considered homogeneous, with negligible laminar and turbulent diffusion, and if the anisotropies are fully developed.

These algebraic stress limits may be used to formulate rotation/curvature modifications to mixing length models. A parameter which is symmetric about $R = -0.5$, as is the bifurcation line for the D1971 model, is the Richardson number,

$$R_i = R(R+1) = \frac{\left(\frac{U_\theta}{r} + \Omega_3\right) W_{\theta r}}{\left(S_{\theta r}\right)^2} = \frac{\frac{V_\theta}{r} W_{\theta r}}{\left(S_{\theta r}\right)^2}, \tag{4.3.14}$$

which may be physically interpreted as,

$$R_i = \frac{(\text{effective rotation rate}) \times (\text{absolute vorticity})}{(\text{mean strain rate})^2} \quad . \tag{4.3.15}$$

So (1975) proposed a rotation/curvature modification as a function of the Richardson number,

$$-\overline{u_1 u_2} = L^2 \left(1 - \frac{1}{2}\beta_{so} R_i\right)^{3/2} \left(2S_{12}\right)^2 . \tag{4.3.16}$$

This reduces to the standard mixing length model when $R_i = 0$. By setting $c_5 = 0$, the results of the super-equilibrium analysis, with only c_1, c_3, c_4, and c_6 pressure-strain terms, may be put in the form of Eq. 4.3.16 using

$$L = c_L \frac{k^{3/2}}{\varepsilon} , \tag{4.3.17}$$

(Appendix EQ.4.3.18). Taking $P = \varepsilon$, and with only a c_1 term in the pressure-strain model (like D1971),

$$\frac{1}{2}\beta_{so} = \frac{12}{c_1 - 2} \qquad \text{and} \qquad c_L = \left(\frac{4}{3c_1} - \frac{8}{3c_1^2}\right)^{3/4} \quad . \tag{4.3.18}$$

Donaldson's model with $c_1 = 10$ corresponds to $\beta_{so} = 3$. So and Mellor (1972) and So (1975) recommended values between 4 and 6, which correspond to c_1 of 8 and 6, respectively. (Donaldson and Bilanin, 1975, used $c_1 = 8$.)

Planar boundary layers without rotation correspond to $R = 0$, and $R_i = 0$. In boundary layers with convex curvature, or correspondingly on the suction side of a rotating passage, R is greater than zero, R_i is positive, and there is a tendency for lower turbulence levels. In boundary layers with concave curvature, or on the pressure side of a rotating passage, $R < 0$, and there is a tendency for turbulence levels to be higher throughout the region where $R_i < 0$. From Eq. 4.3.16, the most unstable condition is at $R = -0.5$, where R_i has its minimum value of -0.25.

Regions of enhanced and suppressed turbulence in curved and rotating channels, together with values of relevant parameters, are described in Table 4.3.1. Figure 4.3.6 shows a corresponding bifurcation plot with multiple scales. These representations are based on the work of Eskinazi et al. (EY1956, EE1969), Bradshaw (1969), Johnston et al. (JHL1972), Donaldson, and So. Note that the absolute-vorticity and mean-strain-rate parameters, W and S, are defined as magnitudes,

$$W = \sqrt{2W_{ij}W_{ij}} \qquad \text{and} \qquad S = \sqrt{2S_{ij}S_{ij}} \ , \tag{4.3.19}$$

so that W, S and W/S are always positive.

The symmetry of the shear stresses with Richardson number, in Eq. 4.3.16, corresponds to the symmetry of the bifurcation line. Note however that, unlike the shear stresses, the normal stresses do not exhibit Richardson number similarity. The glyphs on Fig. 4.3.2 show $\overline{u_1u_1} > \overline{u_2u_2}$, for $R > -0.5$, and $\overline{u_1u_1} < \overline{u_2u_2}$ for $R < -0.5$.

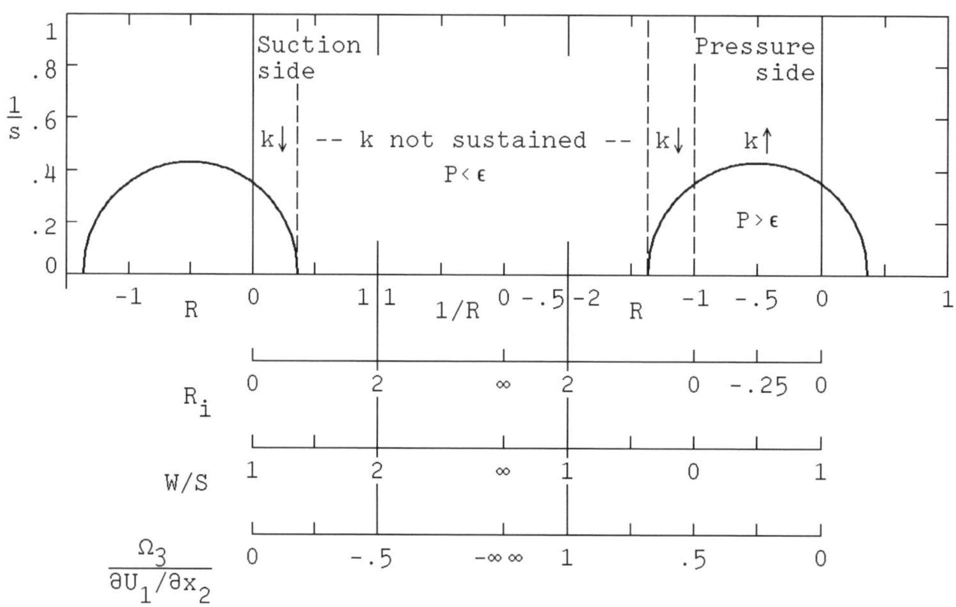

Fig. 4.3.6. Wrapped bifurcation plot with multiple scales, showing continuous variation between the suction and pressure sides of curved or rotating channels. Bifurcation line for D1971 model, but with c_1=8.

Table 4.3.1. Richardson number in regions of curved and rotating channel flows. Cutoff limits determined using D1971 model, but with $c_1 = 8$.

Curved Channel Flow Fig. 4.3.5a	------- Common Parameters -------					Rotating Plane Channel Flow Fig. 4.3.5b
$\frac{V_\theta}{r} \geq 0$	R	R_i	Turbulence	$\frac{W}{S}$	$\frac{\Omega_3}{\partial U_1/\partial x_2}$	$\Omega_3 > 0$
Suction side wall	$\to 0^*$	$\to 0^*$	Reduced	$\to 1^*$	$\to 0^*$	Suction side wall
Near wall			$P \geq \varepsilon$			Near wall
$\partial V_\theta/\partial r > V_\theta/r$	+ ve	+ ve	Stable		− ve	$\partial U_1/\partial x_2 < 0$
Cutoff $P = \varepsilon$ at $1/s = 0$	+0.366	+0.5	$P < \varepsilon$	1.366	−0.183	Cutoff $P = \varepsilon$ at $1/s = 0$
$-\overline{u_r u_\theta} = 0$, $P = 0$						$-\overline{u_1 u_2} = 0$, $P = 0$
$P/\varepsilon < 0^{\#}$			not sustained			$P/\varepsilon < 0^{\#}$
$\partial V_\theta/\partial r = V_\theta/r$ strain rate = 0	+∞ −∞	+∞ +∞	$P = 0$	∞ ∞	−∞ +∞	$\partial U_1/\partial x_2 = 0$ strain rate = 0
$\partial V_\theta/\partial r = 0$	− 2	+ 2	not	1	1	
Cutoff $P = \varepsilon$ at $1/s = 0$	−1.366	+ 0.5	sustained $P < \varepsilon$	0.366	+ 0.683	Cutoff $P = \varepsilon$ at $1/s = 0$
$\partial V_\theta/\partial r = -V_\theta/r$ vorticity = 0	− 1	0	Neutral	0	+ 0.5	$\partial U_1/\partial x_2 = 2\Omega_3$ vorticity = 0
$\partial V_\theta/\partial r = -3V_\theta/r$	−0.5	−0.25	Most unstable	0.5	+0.25	$\partial U_1/\partial x_2 = 4\Omega_3$
$\partial V_\theta/\partial r < -V_\theta/r$	− ve	− ve	Unstable		+ ve	$\partial U_1/\partial x_2 > 0$
Near wall			$P \geq \varepsilon$			Near wall
Pressure side wall	$\to 0^*$	$\to 0^*$	Enhanced	$\to 1^*$	$\to 0^*$	Pressure side wall

* The wall values in the curved channel without rotation are at the limits shown since $V_\theta = 0$ at the wall. However, in the rotating channel, they only approach the limits as $R_D \to \infty$, or $\Omega_3 \to 0$.

\# $P < 0$, from experimental observations in a curved channel (Eskinazi and Yeh, 1956), and in a rotating channel (Johnston, Halleen and Lezius, 1972), are not consistent with the assumptions used in the super-equilibrium analysis.

Chapter 5

PRESSURE-STRAIN MODEL BASED ON HOMOGENEOUS SHEAR

Launder et al. (LRR1975) used the nearly fully-developed homogeneous shear data of Champagne et al. (CHC1970) in determining the coefficients in their pressure-strain model. Speziale et al. (SSG1991) chose to match the nearly fully-developed homogeneous shear data of Tavoularis and Corrsin (TC1981). Here, we consider a large group of homogeneous shear data, both experimental and DNS. Of particular interest is the strain-rate dependence of the turbulence anisotropy, and the turbulence Reynolds number dependence of the slow pressure-strain.

Constant model coefficients are chosen to represent the bulk of the data at high turbulence Reynolds number. Then the coefficients are made functions of the two-dimensionality parameter, g, to satisfy isotropic and realizability limits. The new models are compared with the shear data, and then tested, using a two-equation analysis, to make sure it will be possible to extend them to near-wall flow.

5.1 Homogeneous Shear Data

Strain Rate Dependence

Craft et al. (CLS1993) considered (non-rotating, $\Omega_3 = 0$) homogeneous fully-developed shear flow and plotted the turbulence stress anisotropies as a function of the dimensionless strain rate, s. They used measurements and DNS data to develop an explicit algebraic Reynolds stress model. The dimensionless strain rate, for shear flow with Cartesian coordinates as shown in Fig. 4.2.1, is

$$s = \frac{k}{\varepsilon}\frac{\partial U_1}{\partial x_2} \quad . \tag{5.1.1}$$

Figure 5.1.1 shows a compilation of data for shear flow which is essentially homogeneous, and for which the turbulence anisotropy, b_{ij}, is nearly fully developed, i.e., $db_{ij}/dt \approx 0$. There is some time dependence in the data and this is shown by dotted lines, with the final data point in each sequence marked with a symbol. Of primary interest is the shear component, b_{12}, since this controls turbulence production, P. For shear flow,

$$P/\varepsilon = -2b_{12}s \quad . \tag{5.1.2}$$

The data shows b_{12} to have a minimum value of –0.15 to –0.17, near $s = 3$ to 4. At large s, the DNS data shows the magnitude of b_{12} becoming smaller.

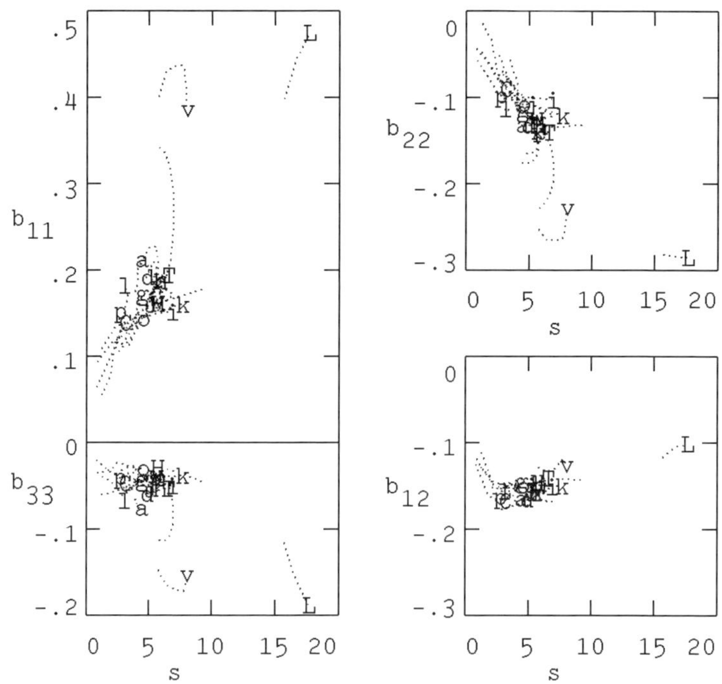

Fig. 5.1.1. Turbulence anisotropies in nearly fully-developed homogeneous shear flow. Measurements (final data points): C–Champagne et al. (CHC1970); H–Harris et al. (HGC1977); T–Tavoularis and Corrsin (TC1981); a,,p–Tavoularis and Karnik (TK1989). DNS: r,,x–Rogers et al. (RMR1986); L–Lee et al. (LKM1990). ········ time dependent data sets. (See Appendices CHC1970 etc.)

This reduction in the magnitude of the shear component at high s is significant for the modeling of near-wall boundary layer flows. The fully developed DNS channel calculations of Kim et al. (KMM1987), Fig. 5.1.2, show that s increases near the wall to levels well above the log-law region value of about 3.3. Therefore at least part of the reduction in magnitude of b_{12} observed near walls (Fig. 1.3.3) is due to the effects of high strain rates.

Some of the pressure-strain models in Table 3.5.1 have been evaluated in their fully developed or algebraic stress limit for homogeneous shear, i.e. when $db_{ij}/dt = 0$. The results, as a function of strain rate, s, are shown in Fig. 5.1.3. The D1971 and SL1985 models have been evaluated at a turbulence Reynolds number, $R_t = 1000$. Again, concentrating on b_{12}, the SL1985 model shows the best qualitative agreement. The model shows a minimum b_{12} at about $s = 5$ and then the magnitude decreases significantly at high s. The D1971 model shows some of this tendency. For the other models, the magnitude of b_{12} increases continually and approaches an asymptotic value at high s.

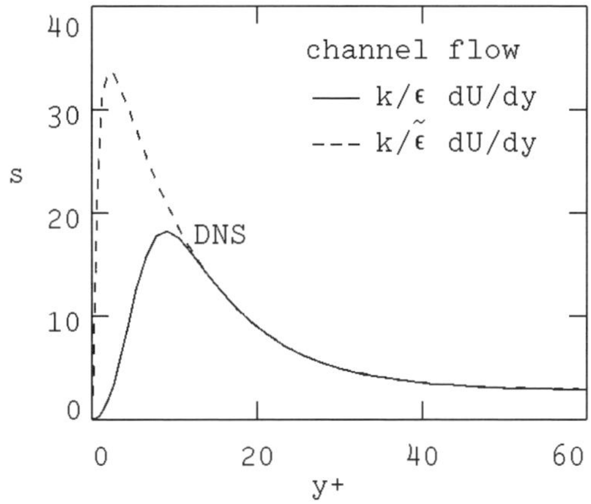

Fig. 5.1.2. Near-wall variation of dimensionless strain rate in fully developed channel flow. DNS data of Kim et al. (KMM1987), $R_\tau = 180$.

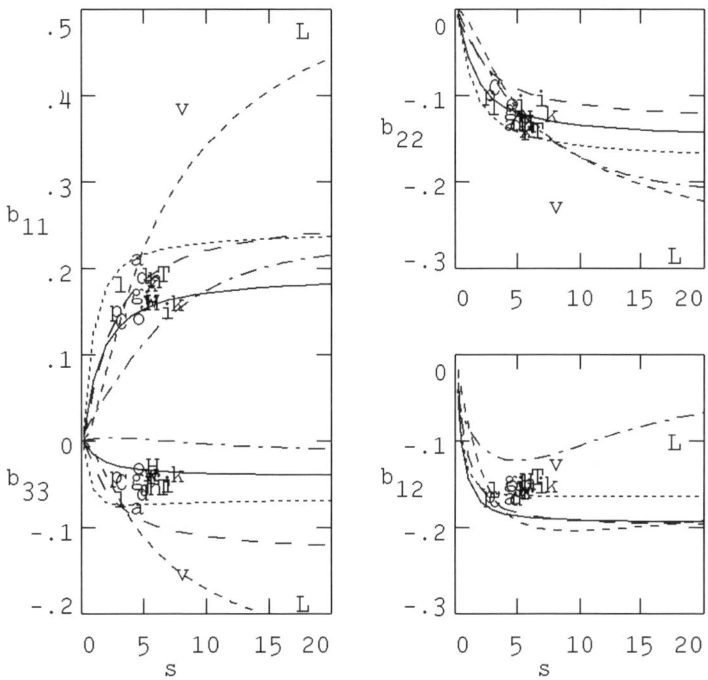

Fig. 5.1.3. Turbulence anisotropies in nearly fully-developed homogeneous shear flow. Data as for Fig 5.1.1. Models in Table 3.5.1:
-------- SSG1991;
— — — D1971;
——— GL1978;
———— LRR1975;
— - — - SL1985.

In terms of the normal stresses, only the D1971 model, which omitted rapid pressure-strain totally, shows the strong approach to two-dimensionality shown by the data at high s.

Another interesting feature of the D1971 and SL1985 models is the Reynolds number dependence of c_1 (Eqs. 3.5.2 and 3.5.5). In Fig. 5.1.4, b_{12} versus s is shown at several Reynolds numbers for the SL1985 model. Lowering the Reynolds number appears to compact the s axis so that lower magnitudes of b_{12} are reached at lower values of s. This suggests that with a correctly formulated pressure-strain model, where the magnitude of b_{12} decreases at high s, it may be possible to model near-wall flows with simply the Reynolds number dependence of c_1.

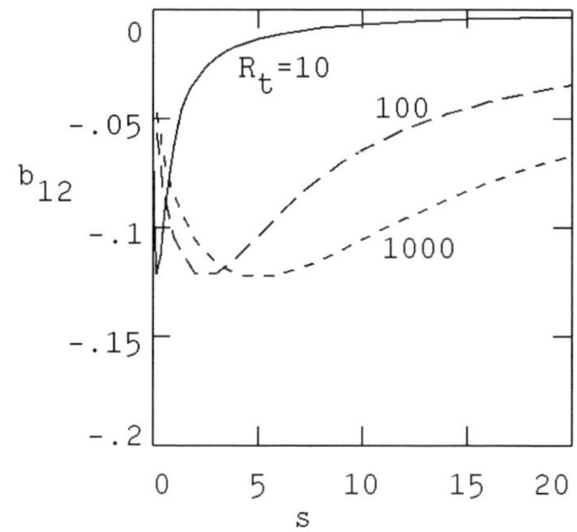

Fig. 5.1.4. Turbulence Reynolds number dependence of the shear anisotropy for fully-developed homogeneous shear flow using the SL1985 model.

Reynolds Number Dependence of Slow Pressure-Strain

The homogeneous shear DNS calculations of Rogers et al. (Appendix RMR1986) provide some information on the components of the slow pressure-strain. In particular, Rogers et al. give the principal angles of the pressure-strain, and of the slow and rapid pressure-strains, separately. From the data, Π_{12}^S and $\Pi_{11}^S - \Pi_{22}^S$ can be determined, as can the anisotropy of the dissipation.

The c_1 coefficient provides the simplest model of the effective slow pressure-strain,

$$\phi_{ij}^S = \Pi_{ij}^S - (\varepsilon_{ij} - \frac{2}{3}\varepsilon\delta_{ij}) = -c_1\varepsilon b_{ij} \ . \qquad (5.1.3)$$

Using Eq. 5.1.3 separately with the data for ϕ_{12}^S and $\phi_{11}^S - \phi_{22}^S$, values for c_1 have been obtained. These are shown in Fig. 5.1.5 as a function of turbulence Reynolds number. The data from the four DNS calculations have been sorted by the two-dimensionality parameter, g. For all values of g, c_1 shows significant increases with increasing R_t. Values of c_1 range from just above 2, with R_t near 10, to nearly 8 with R_t approaching 1000.

The differences between the c_1 values determined from ϕ_{12}^S and those determined from $\phi_{11}^S - \phi_{22}^S$ are due to the limitations of an algebraic model which is tensorially only a function of the turbulence anisotropy, b_{ij}. For the DNS data, there was an angle difference between the principal directions of the slow pressure-strain tensor and the Reynolds stress tensor of up to 17 degrees. Therefore, any slow pressure-strain model based on this data will need to be a compromise.

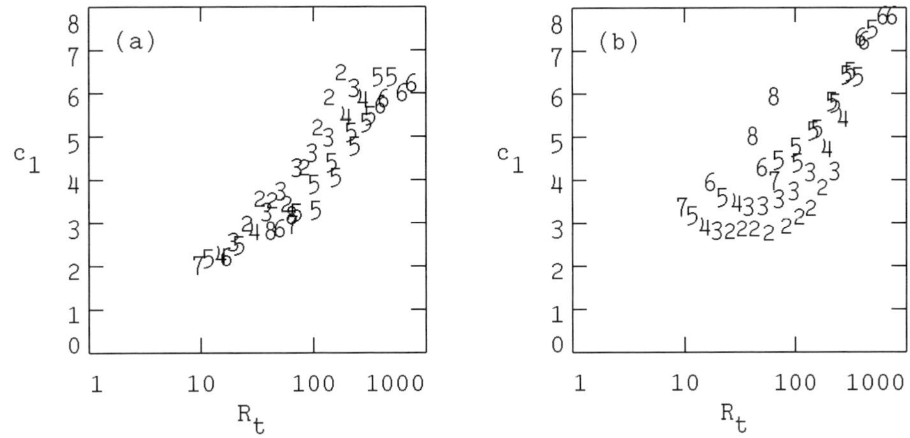

Fig. 5.1.5. Turbulence Reynolds number dependence of c_1 based on the DNS data of Rogers et al. (RMR1986). c_1 determined from: (a) ϕ_{12}^S; (b) $\phi_{11}^S - \phi_{22}^S$. Symbols indicate value of g: 2, $g = 0.2 \pm 0.05$; to 8, $g = 0.8 \pm 0.05$.

One approach is to use the data and modeled values (Eq. 5.1.3) of the radius of the Mohr's circle for ϕ_{ij}^S,

$$r_\phi^2 = \left(\phi_{12}^S\right)^2 + \left(\frac{\phi_{11}^S - \phi_{22}^S}{2}\right)^2 = c_1^2 \varepsilon^2 \left(b_{12}^2 + \left(\frac{b_{11} - b_{22}}{2}\right)^2\right), \tag{5.1.4}$$

which yields

$$c_1 = \frac{1}{\varepsilon} \frac{\sqrt{\left(2\phi_{12}^S\right)^2 + \left(\phi_{11}^S - \phi_{22}^S\right)^2}}{\sqrt{\left(2b_{12}\right)^2 + \left(b_{11} - b_{22}\right)^2}} . \tag{5.1.5}$$

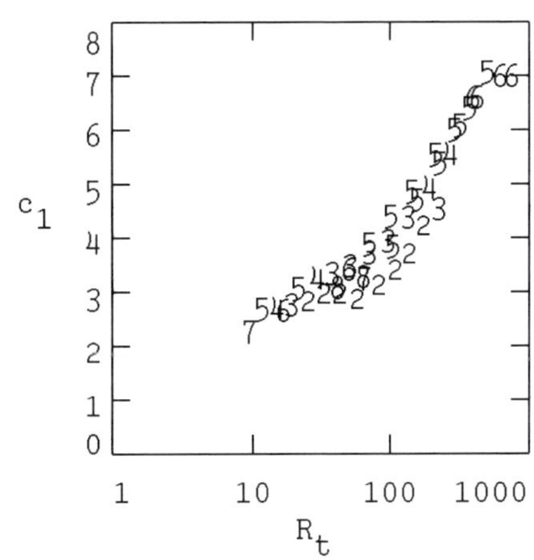

Fig. 5.1.6. c_1 determined from the DNS data of Rogers et al. (RMR1986) using Eq. 5.1.5. Symbols indicate value of g: 2, $g = 0.2 \pm 0.05$; to 8, $g = 0.8 \pm 0.05$.

As shown in Fig. 5.1.6, this brings coherence to the data, showing clearly the turbulence Reynolds number dependence of c_1, and some reduction of c_1 at low g.

5.2 *Choosing Pressure-Strain Model Coefficients*

Fitting Homogeneous Shear Data

The homogeneous shear data gives three independent anisotropies, e.g., $b_{11}-b_{22}$, b_{33} and b_{12}, which may be used to set three of the pressure-strain coefficients, c_1 to c_{12}. Considering only the pressure-strain terms which are tensorially linear in b_{ij}, i.e. taking all but c_1, c_3, c_4 and c_5 as zero, three of the coefficients may be calculated from the anisotropies of the shear flow when the fourth coefficient is set. In particular, if c_5 is set, c_1 may be calculated directly from the equation for $b_{11}-b_{22}$, which for shear flow without rotation (from Eq. 4.2.5) is

$$2k\frac{d(b_{11}-b_{22})}{dt}=-\left((c_1-2)\varepsilon+2P\right)(b_{11}-b_{22})+4(c_5-2)kW_{12}b_{12}\,. \tag{5.2.1}$$

Having found c_1, c_4 may be determined from the equation for b_{33},

$$2k\frac{db_{33}}{dt}=-\left((c_1-2)\varepsilon+2P\right)b_{33}-\frac{4}{3}(c_4-2)kS_{12}b_{12}\,, \tag{5.2.2}$$

and then c_3 from the equation for b_{12},

$$\begin{aligned}2k\frac{db_{12}}{dt}&=-\left((c_1-2)\varepsilon+2P\right)b_{12}+\left(c_3-\frac{4}{3}\right)kS_{12}-(c_4-2)kS_{12}b_{33}\\&\quad-(c_5-2)kW_{12}(b_{11}-b_{22})\,.\end{aligned} \tag{5.2.3}$$

Values of c_1, c_3 and c_4, obtained from homogeneous shear data using $c_5=0$ and $c_5=1.2$, are shown in Fig. 5.2.1. The last pair of points for each data set were used to evaluate the terms in the equations. The results are plotted as a function of g, so that the limits of two-dimensional turbulence, $g=0$, and isotropic turbulence, $g=1$, may be considered. However, except for two of the DNS points (L and v) all the g values lie between 0.5 and 0.7.

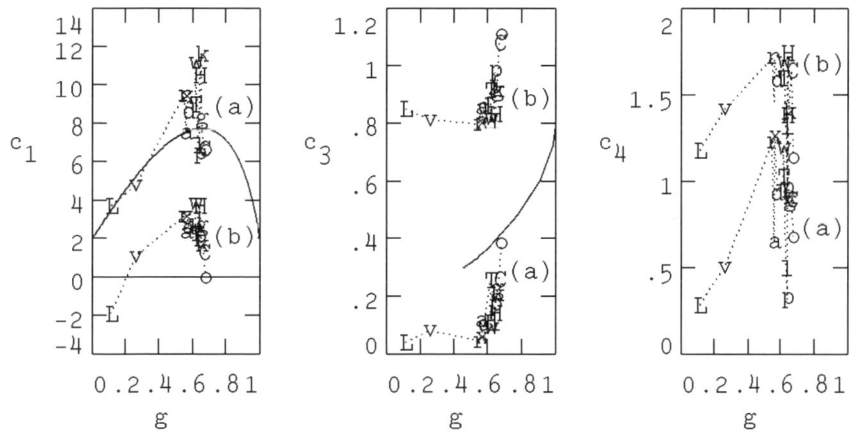

Fig. 5.2.1. Determination of c_1, c_3 and c_4 from homogeneous shear data using (a) $c_5 = 0$, and (b) $c_5 = 1.2$. Data symbols as for Fig. 5.1.1. —— c_1 for axisymmetric contraction at high R_t from Lumley (1978), Eq. 3.5.5. —— c_3 for shear flow from SSG1991 model. Experimental data, $850<R_t<10000$; DNS data, $200<R_t<700$.

Figure 5.2.1 can be used to see the consistency in the choice of constants for the models in Table 3.5.1. For example, the GL1978 and LRR1975 models use $c_5 = 1.2$ and 1.309, respectively. Consistent with data points (b), they use $c_3 = 0.8$, c_4 between 1.2 and 1.75, and low values of c_1, 3 to 3.6. The D1971 model with $c_5 = 0$ is consistent with data points (a) using $c_1 = 10$ at high R_t. Data points (a) give low values of c_3, near the zero value used by D1971, and lower values for c_4 than data points (b). The SSG1991 model falls between these limits using $c_5 = 0.4$. Correspondingly, the c_3 values for the SSG1991 model, as shown on Fig. 5.2.1, lie between data points (a) and (b).

Choosing c_5 for a New Model

Lumley's c_1, evaluated at high R_t, is also shown on Fig. 5.2.1. His model was based on data for turbulence returning to isotropy following axisymmetric contraction. It gives peak values of c_1 between 7 and 8, near $g = 0.6$. These high values of c_1 are consistent with the values of c_1 of up to 8 for R_t approaching 1000, determined from the DNS shear flow data (Fig. 5.1.5). From Fig. 5.2.1, the high values of c_1 are consistent with $c_5 = 0$. The RDT analysis of Bertoglio for rotating shear flow, as discussed in Section 4.3, is also consistent with $c_5 = 0$, as is the isotropic limit, Eq. 3.3.4. Therefore we take $c_5 = 0$.

The MC model at High Reynolds Number

Using $c_5 = 0$, the c_1, c_3 and c_4 coefficients, determined from the shear data in two ways, are shown in Fig. 5.2.2. Points marked '+' use experimental values of db_{ij}/dt, and are the same as shown in Fig. 5.2.1. Points marked 'o' evaluate the coefficients from the data using $db_{ij}/dt = 0$. The results for the two methods mostly overlap, confirming that the data represents b_{ij}'s which are essentially fully developed. From Fig. 5.2.2, constant coefficients for a pressure-strain model, based on the bulk of the data for $g = 0.5$ to 0.7, can be determined. In particular, for the MC model at high R_t, we take,

$$\text{MC model:} \qquad c_1 = 9, \qquad c_3 = 0.15, \qquad \text{and} \qquad c_4 = 1 \,. \tag{5.2.4}$$

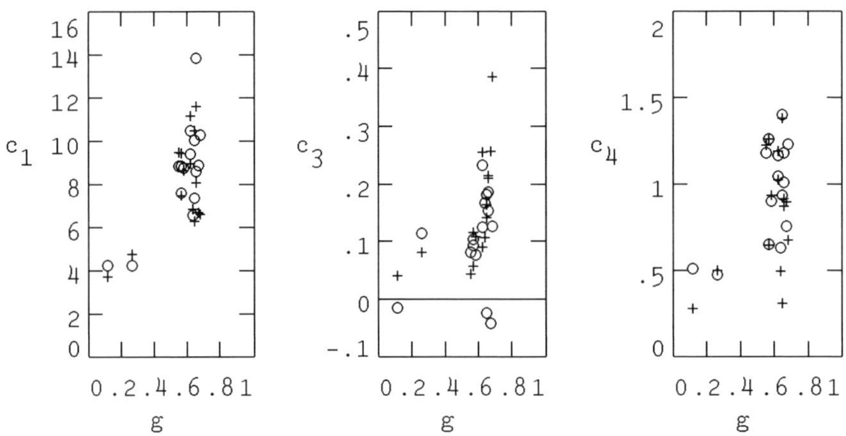

Fig. 5.2.2. Determination of c_1, c_3 and c_4 from homogeneous shear data using $c_5 = 0$. Coefficients determined using: +, experimental values of db_{ij}/dt; o, $db_{ij}/dt=0$.

The results from the MC model for homogeneous shear in the fully developed limit are shown in Fig. 5.2.3. The shear anisotropy has the desired character, with a minimum value of b_{12} of -0.154 at $s = 5.4$, and b_{12} returning towards zero at high s, so that it will be possible to extend the model to near-wall flow by adding a turbulence Reynolds number dependence to c_1. For the normal anisotropies, the model goes through the bulk of the data from $s = 3$ to 8, but shows less anisotropy than the DNS points v and L.

The MC model does not satisfy the isotropic limits, $c_3 = 0.8$ and $c_4 = 24/7$, Eqs. 3.3.2 and 3.3.4. It also does not satisfy the two-dimensional limits, Eqs. 3.4.5 - 3.4.7. The MC model does satisfy two of the conditions for realizability, Eqs. 3.4.10 and 3.4.11, by taking $c_1 > 2$, and $c_6 = 0 > -2c_4$, but not the third, $c_3 = 2c_4/3$, Eq. 3.4.6.

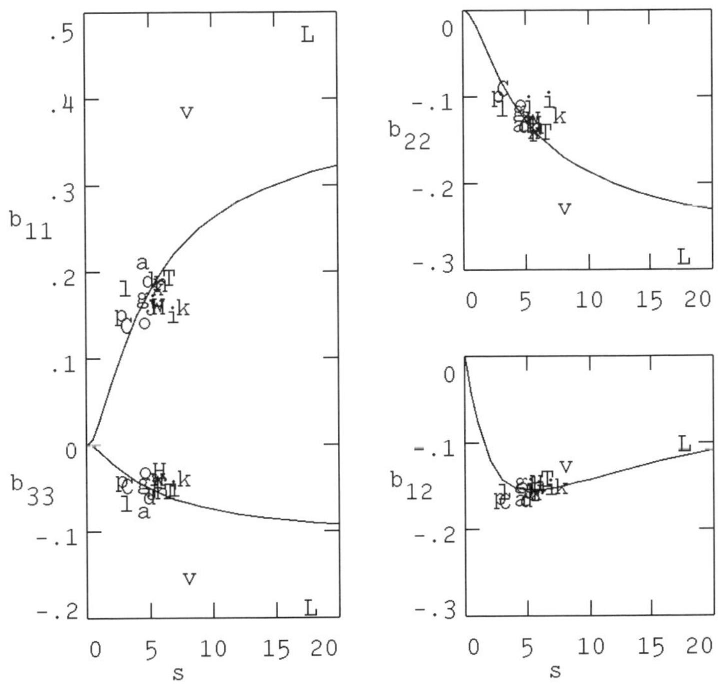

Fig. 5.2.3. Turbulence anisotropies in nearly fully-developed homogeneous shear flow. Data as for Fig 5.1.1. ——— MC model.

<u>**Add Isotropic and Two-Dimensional Limits**</u>

The isotropic and two-dimensional turbulence limits can be easily satisfied if the c_1, c_3 and c_4 coefficients are taken as functions of g. The shear data sheds insufficient light to say how the coefficients vary with g. Therefore we will start with the MC values for $g = 0.5$ to 0.7. The isotropic limits for $g = 1$ can be satisfied by adding a term which is quadratic in g above $g = 0.7$. In particular, defining

$$f_{g1}(g,g_L) = \left(\max\left(0, \frac{g - g_L}{1 - g_L}\right)\right)^2, \tag{5.2.5}$$

the isotropic limits are satisfied by

$$c_3 = 0.15 + (0.8 - 0.15) f_{g1}(g,0.7) \quad \text{and} \quad c_4 = 1 + \left(\frac{24}{7} - 1\right) f_{g1}(g,0.7). \tag{5.2.6}$$

Continuing to omit the c_6 coefficient, the two-dimensional turbulence limits for $g = 0$ reduce to, $c_4 = 0$, $c_1 = 2$ and $c_3 = 0$. Defining a function which is quadratic in g below $g = g_L$,

$$f_{g0}(g,g_L) = \left(\max\left(0, \frac{g_L - g}{g_L}\right)\right)^2 , \qquad (5.2.7)$$

we can form the MA.1 model which satisfies both the isotropic and two-dimensional limits,

MA.1 model:
$$\begin{aligned} c_1 &= 9 + (2-9)f_{g0}(g,0.5) + (3.4-9)f_{g1}(g,0.7) \\ c_3 &= 0.15 + (0-.15)f_{g0}(g,0.5) + (0.8-0.15)f_{g1}(g,0.7) \\ c_4 &= 1 + (0-1)f_{g0}(g,0.5) + \left(\frac{24}{7}-1\right)f_{g1}(g,0.7) . \end{aligned} \qquad (5.2.8)$$

The coefficients for the MA.1 model are shown in Fig. 5.2.4. It should be considered an intermediate or temporary model. Although the coefficients satisfy the desired limits, it is not known whether or not they approach them in a reasonable fashion. In addition there is no theoretical isotropic limit for c_1. The value of 3.4 used is consistent with the SSG1991 model, which was based on data near the isotropic limit.

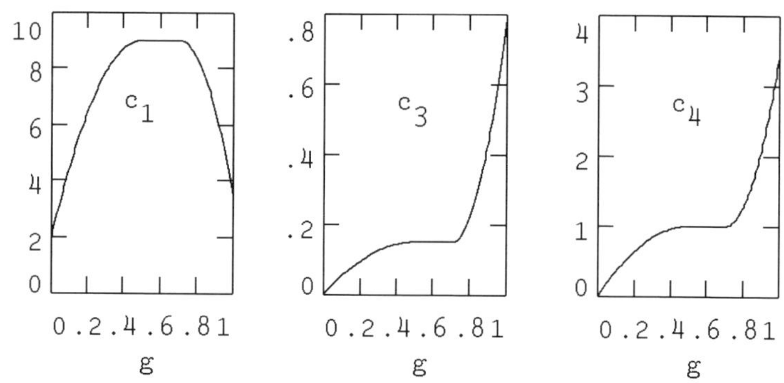

Fig. 5.2.4. Pressure-strain model coefficients for the MA.1 model, Eq. 5.2.8.

5.3 Model Behavior at High Strain Rate

Two-Equation Analysis

The limiting values obtained for the anisotropies at high strain rate for shear flow can be understood through a two-equation analysis of the models. Consider the equations for homogeneous shear, Eq. 4.2.5, in the limit of fully developed flow with $db_{ij}/dt = 0$, and with only a limited set of pressure-strain coefficients, c_1, and c_3 through c_6. For the two-equation analysis, we define the intermediate variable, D,

$$D = \left(c_1 - 2 + (2 + \frac{1}{2}c_6)\frac{P}{\varepsilon}\right)\frac{\varepsilon}{2kS_{12}} = (c_1 - 2)\frac{\varepsilon}{2kS_{12}} - b_{12}(4 + c_6) . \qquad (5.3.1)$$

Taking S_{12} as positive, the dimensionless strain rate is $s = 2kS_{12}/\varepsilon$, so that

$$D = \frac{(c_1 - 2)}{s} - b_{12}(4 + c_6). \tag{5.3.2}$$

In terms of D, the equations for the turbulence anisotropies (Eq. 4.2.5) in the fully developed limit are (Appendix EQ.5.3.1-6),

$$b_{11} - b_{22} = \left\{(4 - c_5)\frac{\Omega_3}{S_{12}} + c_5 - 2\right\}\frac{2b_{12}}{D}, \tag{5.3.3}$$

$$b_{33} = -(c_4 - 2)\frac{2b_{12}}{3D} , \tag{5.3.4}$$

and

$$2Db_{12} = \left(c_3 - \frac{4}{3}\right) - (c_4 - 2)b_{33} - \left\{(4 - c_5)\frac{\Omega_3}{S_{12}} + c_5 - 2\right\}(b_{11} - b_{22}). \tag{5.3.5}$$

Combining Eqs. 5.3.3-5, yields

$$b_{12} = \frac{\frac{D}{2}\left(c_3 - \frac{4}{3}\right)}{D^2 - \frac{1}{3}(c_4 - 2)^2 + \left\{(4 - c_5)\frac{\Omega_3}{S_{12}} + c_5 - 2\right\}^2} , \tag{5.3.6}$$

which for zero rotation reduces to

$$b_{12} = \frac{\frac{D}{2}\left(c_3 - \frac{4}{3}\right)}{D^2 - \frac{1}{3}(c_4 - 2)^2 + (c_5 - 2)^2} . \tag{5.3.7}$$

With the coefficients c_1 to c_6, and the dimensionless strain rate, s, taken as parameters, Eqs. 5.3.2 and 5.3.7 each give b_{12} as a function of D. The solution for fully developed flow is then the simultaneous solution of these equations, or the point of intersection of the lines represented by Eqs. 5.3.2 and 5.3.7.

Lines for the GL1978 model are shown in Fig. 5.3.1. Lines representing Eq. 5.3.2 are straight lines with a slope of –1/4, and move to the right as s is lowered. The line representing Eq. 5.3.7 has a steeper (more negative) slope at small D, and shows a minimum b_{12} before it intersects the Eq. 5.3.2 line for s of infinity. Therefore, the GL1978 model shows no decrease in $-b_{12}$ with high strain rate.

Fig. 5.3.1. Two-equation analysis lines for the GL1978 model.
——— Eq. 5.3.7.
- - - - Eq. 5.3.2, for various s.

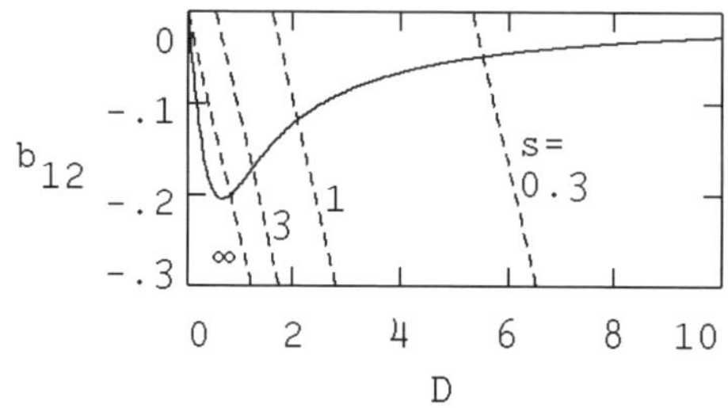

Corresponding lines for the MC model at high turbulence Reynolds numbers are shown on Fig. 5.3.2. Lines representing Eq. 5.3.2 are again straight lines with a slope of –1/4. The slope of the Eq. 5.3.7 line is shallower than –1/4 at small D. Therefore as s increases to large values, b_{12} goes to zero.

Fig. 5.3.2. Two-equation analysis lines for the MC model at high R_t.
——— Eq. 5.3.7.
- - - - Eq. 5.3.2, for various s.

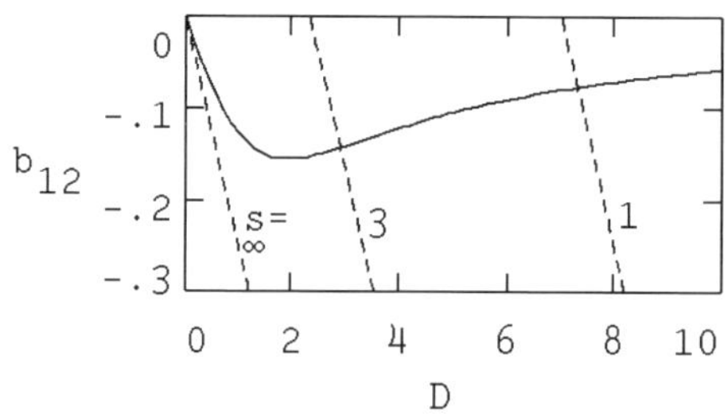

From Eq. 5.3.2, D, for $c_1 = 2$, is the same as for $s = \infty$. Therefore with c_1 taken as a function of R_t and reducing to 2 as R_t approaches zero, the MC model will tend to give the desired result that b_{12} goes to zero as a wall is approached, Fig. 1.3.3.

Overall, for models with constant rapid pressure-strain coefficients, b_{12} will approach zero with increasing strain rate, s, when the slope of Eq. 5.3.7 at $D = 0$, is greater than the slope of Eq. 5.3.2,

$$\left(\frac{db_{12}}{dD}\right)_{D=0,Eq.5.3.7} = \frac{-\frac{1}{2}\left(\frac{4}{3} - c_3\right)}{(c_5 - 2)^2 - \frac{1}{3}(c_4 - 2)^2} \geq -\frac{1}{4 + c_6} = \left(\frac{db_{12}}{dD}\right)_{Eq.5.3.2} . \quad (5.3.8)$$

Consider models which take $c_6 = 0$, so that the slope from Eq. 5.3.2 is –0.25. The slope from Eq. 5.3.7 for the MC model is –0.16, satisfying Eq. 5.3.8 as an inequality. The D1971 model satisfies Eq. 5.3.8 as an equality, both slopes being –0.25. However, the slopes from Eq. 5.3.7 for the S1973, LRR1975, GL1978 and Y1984 models (Table 3.5.1) all give values below –0.25, so that Eq. 5.3.8 is not satisfied and b_{12} does not approach zero at high strain rate.

Exercise 5.3. Consider the linearized version of the SSG model given by Gatski and Speziale (1993) with
GS1993: $c_1 = 6.8$, $c_2 = 0$, $c_3 = 0.36$, $c_4 = 1.25$, $c_5 = 0.4$, and $c_6 = 0$,
and a partially-linearized version with
SSG-mod: $c_1 = 3.4$, $c_2 = 0$, $c_3 = 0.36$, $c_4 = 1.25$, $c_5 = 0.4$, and $c_6 = 3.6$.
Show that b_{12} approaches zero at high strain rate for the linearized version, GS1993, but does not for the partially-linearized version, SSG-mod.

Analysis with Variable Coefficients

For the MA.1 model (Eq. 5.2.8) and similar models where the coefficients are functions of g, Eqs. 5.3.3-5 give (implicitly, since $g = g(b_{11}, b_{22}, b_{12})$) the turbulence anisotropies b_{11}, b_{22} and b_{12}, and the parameter D, as functions of g. Equation 5.3.2 can then be turned around to show at what strain rate, s, they occur.

The results for the MA.1 model and, for comparison, the MC model are shown in Fig. 5.3.3a. The MC model is well behaved with solutions in the range $g = 0.28$ to 1. For each strain rate, s, there is one and only one solution. The MA.1 model is poorly behaved. Figure 5.3.3a shows dual solutions for this model for strain rates from 1 to 50, and no fully developed solutions for $s > 50$.

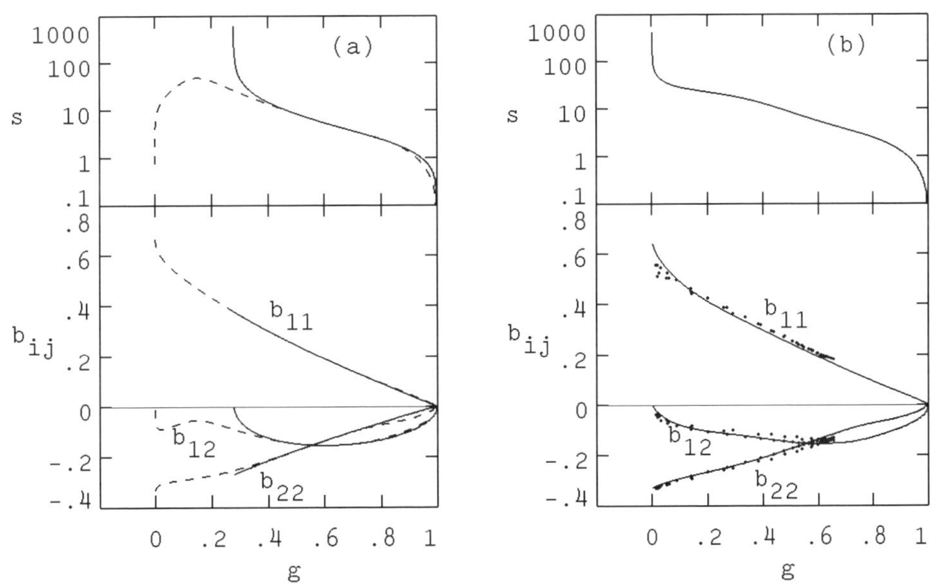

Fig. 5.3.3. Fully developed solutions for shear flow as a function of g. (a) ——— MC model at high R_t; ----- MA.1 model. (b) ——— MA.2 model; ······ Fully developed channel DNS data of Moser et al. (MKM1999) from $y^+ = 5$ to $y/\delta = 0.5$, for $R_\tau = 180$, 395 and 590.

Improving the MA Model at Low g

Figure 5.3.3a also shows that the MA.1 model is poorly behaved in that it has two locations in g for which b_{12} has a minimum value, the expected one, near $g = 0.6$, and an inappropriate one, near $g = 0$. The latter is due to the way in which the MA.1 model reduces both c_3 and c_4 to zero as g approaches zero.

Shown on Fig. 5.3.3b are anisotropies in fully developed channel flow from the DNS data of Moser et al. (MKM1999). Although this data does not represent homogeneous shear, it provides a guide as to how b_{12} should vary with g. This was used to determine modifications to the ways in which c_3 and c_4 approach zero as g goes to zero. The resulting MA.2 model, also shown on Fig. 5.3.3b, is well behaved, with only a single minimum in $b_{12}(g)$, and with b_{12} approaching zero as s approaches infinity, allowing the model to be extended to near-wall flow.

The revisions to the c_3 and c_4 coefficients were for $g < 0.5$ and are shown in Fig. 5.3.4.

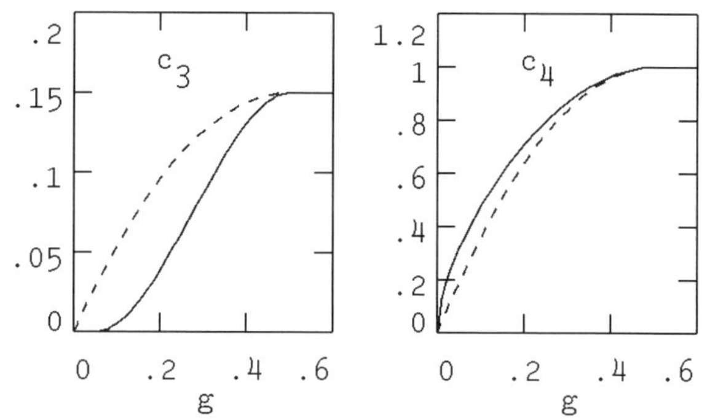

Fig. 5.3.4. Modification of the c_3 and c_4 coefficients.
----- Model MA.1.
—— Model MA.2.

For the MA.2 model, the c_3 coefficient is zero from $g = 0$ to $g = 0.05$, and then has a cubic profile ending at $g = 0.5$. Defining a generic cubic function, f_{g3}, going from 0 for $g \le g_L$ to 1 for $g \ge g_H$,

$$f_{g3}(g,g_L,g_H) = 1 - 3(f_{lin}(g,g_L,g_H))^2 + 2(f_{lin}(g,g_L,g_H))^3, \qquad (5.3.9)$$

where

$$f_{lin}(g,g_L,g_H) = \frac{\max(0,\, g_H - g_L - \max(0,\, g - g_L))}{g_H - g_L}, \qquad (5.3.10)$$

then the revised c_3 coefficient is,

MA.2 model: $$c_3 = 0.15 f_{g3}(g,0.05,0.5) + (0.8 - 0.15) f_{g1}(g,0.7)\ . \qquad (5.3.11)$$

The model results at high strain rate and low g are also quite sensitive to the distribution of c_4 near $g = 0$. To obtain the good results shown in Fig. 5.3.3b, model MA.2 uses a $\sqrt{g}$ variation in c_4 near $g = 0$,

$$\text{MA.2 model:}\quad \begin{aligned} c_4 &= 0.414 - 0.414\left(\max(0, 1-\sqrt{2g})\right)^2 + 0.086 f_{g3}(g,0,0.5) \\ &+0.5 - 0.5 f_{g0}(g,0.5) + \left(\frac{24}{7} - 1\right) f_{g1}(g,0.7) \quad . \end{aligned} \tag{5.3.12}$$

The results for the MA.2 model, in the form $b_{ij}(s)$, are compared with the homogeneous shear data in Fig. 5.3.5. The results go through the bulk of the data from $s = 3$ to 8, similar to the results from the MC model (Fig. 5.2.3).

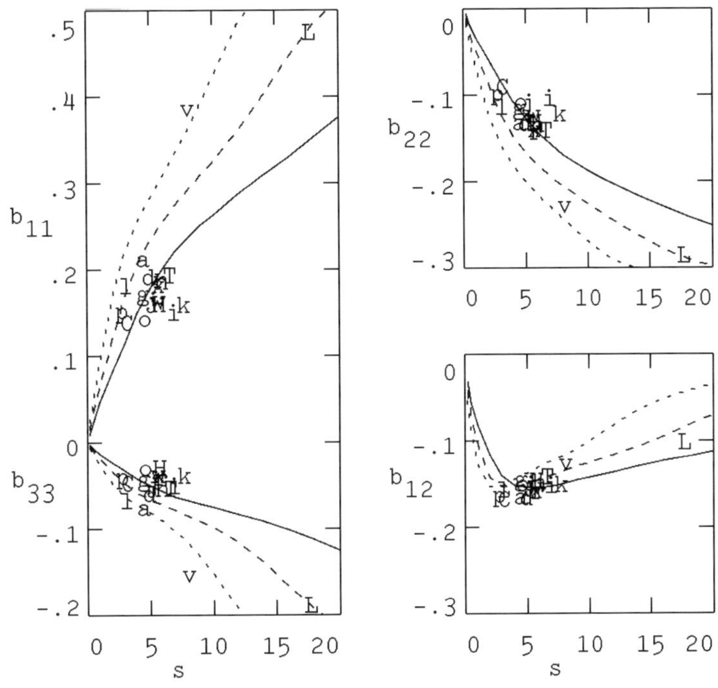

Fig. 5.3.5. Turbulence anisotropies in nearly fully-developed homogeneous shear flow for the MA.2 model and with modified c_1, Eq. 5.3.13.
——— MA.2, $c_{1,max} = 9$;
– – – – – $c_{1,max} = 6.5$;
- - - - - - $c_{1,max} = 5$.
Data as for Fig 5.1.1.

In addition to the results for the full c_1, which has a maximum value $c_{1,max} = 9$, results for reduced values of c_1 are also shown. With

$$c_1 = 2 + \frac{c_{1,max} - 2}{9-2}\left((9-2)\left(1 - f_{g0}(g,0.5)\right) + (3.4-9) f_{g1}(g,0.7)\right) , \tag{5.3.13}$$

the DNS data point L, at a turbulence Reynolds number, $R_t = 500$, is approximately fit by the model with $c_{1,\max} = 6.5$. With $c_{1,\max} = 5$, the model comes close to the DNS data point v, which had $R_t = 225$. These reductions are consistent with the turbulence Reynolds number dependence of c_1 shown in Fig. 5.1.5.

Chapter 6

DISSIPATION OF TURBULENCE

To convert an algebraic model for pressure-strain into a full Reynolds stress model, additional models are needed for the dissipation rate of turbulence kinetic energy, and turbulent diffusion. In this chapter we investigate the modeling of turbulence dissipation, including theoretical considerations, and data which may be used to determine models for the production and decay of dissipation. Turbulent diffusion will be considered in the next chapter.

6.1 Transport Equation for Dissipation

A transport equation for dissipation can be derived from the fluctuating momentum equation. The dissipation rate, which needs to be modeled, is (Eq. 2.1.18),

$$\varepsilon = \nu \overline{\frac{\partial u_i}{\partial x_k}\frac{\partial u_i}{\partial x_k}} = \nu \overline{\left(\frac{\partial u_i}{\partial x_k}\right)^2} \quad , \tag{6.1.1}$$

so that (replacing subscript k with n)

$$\frac{\partial \varepsilon}{\partial t} = 2\nu \overline{\frac{\partial u_i}{\partial x_n}\frac{\partial}{\partial t}\frac{\partial u_i}{\partial x_n}} = 2\nu \overline{\frac{\partial u_i}{\partial x_n}\frac{\partial}{\partial x_n}\frac{\partial u_i}{\partial t}} \quad . \tag{6.1.2}$$

Therefore, taking $\partial/\partial x_n$ of the fluctuating momentum equation for u_i (Eq. 2.1.11), multiplying by $2\nu\partial u_i/\partial x_n$, and ensemble averaging will yield an equation for $\partial\varepsilon/\partial t$ (Appendix EQ.6.1.3),

$$\begin{aligned}\frac{\partial \varepsilon}{\partial t} + U_k\frac{\partial \varepsilon}{\partial x_k} &= \frac{\partial}{\partial x_k}\nu\frac{\partial \varepsilon}{\partial x_k} - 2\nu\left(\overline{\frac{\partial u_k}{\partial x_i}\frac{\partial u_n}{\partial x_i}} + \overline{\frac{\partial u_i}{\partial x_n}\frac{\partial u_i}{\partial x_k}}\right)S_{kn} - 2\nu^2\overline{\left(\frac{\partial^2 u_i}{\partial x_k \partial x_n}\right)^2} \\ &-\nu\frac{\partial}{\partial x_k}\left(\overline{u_k\frac{\partial u_i}{\partial x_n}\frac{\partial u_i}{\partial x_n}} + \overline{\frac{2}{\rho}\frac{\partial u_k}{\partial x_n}\frac{\partial p}{\partial x_n}}\right) - 2\nu\overline{\frac{\partial u_i}{\partial x_n}\frac{\partial u_k}{\partial x_n}\frac{\partial u_i}{\partial x_k}} - 2\nu \overline{u_k\frac{\partial u_i}{\partial x_n}}\frac{\partial (S_{ik}+W_{ik})}{\partial x_n}\end{aligned} \quad . \tag{6.1.3}$$

The convection and laminar diffusion terms are easily recognized and need no modeling. The term proportional to S_{kn} has the form of a production term. The next term is the viscous decay term. The first term on the second line of Eq. 6.1.3 has the form of turbulent diffusion. The second term on the second line is considered to be the main production term, at high turbulence Reynolds numbers, due to the stretching of vortex filaments (Hanjalić and Launder, 1976, and Tennekes and Lumley, 1972). The last term may contribute to either production or destruction of dissipation.

Models in the Literature

Because so much of the dissipation equation needs to be modeled, in practice a generic transport equation, which contains the basic elements of convection, laminar diffusion, turbulent diffusion, production and decay, is applied either to the turbulence dissipation or to some alternative property from which the dissipation may be algebraically obtained. For property ψ,

$$\frac{\partial \psi}{\partial t} + U_k \frac{\partial \psi}{\partial x_k} = \frac{\partial}{\partial x_k} \nu \frac{\partial \psi}{\partial x_k} + T_\psi + P_\psi - \varepsilon_\psi \, . \tag{6.1.4}$$

Two such alternative properties, used in low Reynolds number near-wall modeling, were noted in the Introduction, $\tilde{\varepsilon} = \varepsilon - 2\nu\left(\partial\sqrt{k}/\partial y\right)^2$ (Eq. 1.3.15), and $\omega = \tilde{\varepsilon}/k$ (Eq. 1.3.18).

A variety of models are compared in Table 6.1.1. Column 2 lists the dissipation variable which is transported and diffused. For purposes of comparison, the source term is given in a common form, as the source term for dissipation

$$S_\varepsilon = P_\varepsilon - \varepsilon_\varepsilon = c_{\varepsilon 1} P \frac{\varepsilon}{k} - c_{\varepsilon 2} \tilde{\varepsilon} \frac{\varepsilon}{k} + E_\varepsilon \, . \tag{6.1.5}$$

For models which use alternative variables, i.e. for $\psi = k^a \varepsilon^b$, or $\psi = k^a \tilde{\varepsilon}^b$,

$$S_\psi = P_\psi - \varepsilon_\psi = (b c_{\varepsilon 1} + a) P \frac{\psi}{k} - (b c_{\varepsilon 2} + a) \tilde{\varepsilon} \frac{\psi}{k} + b E_\varepsilon \frac{\psi}{\varepsilon} \, . \tag{6.1.6}$$

Constant or base levels of $c_{\varepsilon 1}$, range from 1 to 1.555. The CH1985 model has an effective $c_{\varepsilon 1}$ of 1.61, since the E_ε term is $0.055 P\varepsilon/k$ when P is evaluated without the van Driest modification to ν_t. The overall effect is a relative increase in $c_{\varepsilon 1}$ near the wall. The added terms, E_ε, in the JL1972, DM1981, S1973 and HL1976 models also increase the dissipation in the near-wall region. Jones and Launder noted that there is no theoretical basis for the added term in the JL1972 model. However, it does have some similarity to the last term in Eq. 6.1.3. The DM1981 model decreases $c_{\varepsilon 1}$ based on turbulence Reynolds number.

Table 6.1.1. Some models for the production and decay of dissipation.
(* Models which do not distinguish between ε and $\tilde{\varepsilon}$.
Uses ε instead of $\tilde{\varepsilon}$ in Eq. 6.1.6.)

Model	Variable	$c_{\varepsilon 1}$	$c_{\varepsilon 2}$	E_ε
JL1972	$\tilde{\varepsilon}$	1.55	$2-0.6e^{-\tilde{R}_t^2}$	$2\nu\nu_t\left(\frac{\partial^2 U}{\partial y^2}\right)^2$
DM1981#	$\tilde{\varepsilon}$	$1.35-0.054e^{-(\tilde{R}_t/50)^2}$	$2-0.6e^{-(\tilde{R}_t/50)^2}$	$\left(\frac{13.5}{4}P+1.6\varepsilon\right)\frac{\nu\tilde{\varepsilon}}{\varepsilon y^2}$
CH1985	ω	1.555	1.833	$0.055(0.09k)\left(\frac{\partial U_i}{\partial x_j}\right)^2$
S1973*	L	1	2	$\frac{a}{(1+cy/\delta)^2}\frac{k^2}{y^2}$
LRR1975*	ε	1.44	1.9	0
HL1976	ε	1.275	$1.8-0.4e^{-(R_t/6)^2}$	$8\nu\frac{\overline{u_1u_2}^2}{\varepsilon}\left(\frac{\partial^2 U}{\partial y^2}\right)^2$
SSG1991*	ε	1.44	1.83	0
BFR1985*	ε	–	$\frac{11}{6}$	$-0.15\varepsilon\sqrt{\frac{W_{ij}W_{ij}}{2}}$

From 2-eq. models: JL1972, Jones and Launder; DM1981, Dutoya and Michard (see Patel et al, PRS1985); CH1985, Coakley and Hsieh. From Reynolds stress models: S1973, Shir; LRR1975, Launder, Reece and Rodi; HL1976, Hanjalić and Launder; SSG1991, Speziale, Sarkar and Gatski. Isotropic turbulence: BFR1985, Bardina, Ferziger and Rogallo.

Constant or base levels of $c_{\varepsilon 2}$ tend to range between 1.8 and 2. Several of the models reviewed by Patel et al. (PRS1985) reduce $c_{\varepsilon 2}$ to 1.4 (or about 1.4) as R_t goes to zero, similar to the JL1972, HL1976 and DM1981 models.

The BFR1985 model was based solely on the decay of isotropic turbulence subjected to fluid rotation. It effectively increases $c_{\varepsilon 2}$ linearly with rotation rate and was based on the observation that turbulence decays less rapidly (therefore dissipation decays more rapidly) when subjected to high rotation rates. This model, and rotating isotropic turbulence in general, is discussed further in Section 18.3.

Following the presentation of dissipation rates for DNS calculations of fully developed channel flow, there was a flurry of activity in modifying dissipation equations, which was reviewed by Hanjalić (1994). The modifications included making $c_{\varepsilon 1}$ a function of P/ε, and making both $c_{\varepsilon 1}$ and $c_{\varepsilon 2}$ functions of turbulence anisotropy, as well as wall-distance modifications.

6.2 Decay of Dissipation

In the absence of turbulence production, the decay rate of dissipation may be determined from the decay of turbulence kinetic energy. Consider homogeneous flow in the absence of turbulence production,

$$dk/dt = -\varepsilon \ . \tag{6.2.1}$$

Modeling the decay of dissipation in the form

$$d\varepsilon/dt = -c_{\varepsilon 2}\varepsilon^2/k \ , \tag{6.2.2}$$

gives

$$c_{\varepsilon 2} = k\frac{d^2k}{dt^2}\Big/\left(\frac{dk}{dt}\right)^2 \ . \tag{6.2.3}$$

Theoretical Considerations

Batchelor and Townsend (1948) presented theoretical analyses showing that in the final period of decay, i.e. as R_t goes to zero, turbulence kinetic energy varies as $t^{-5/2}$, independent of the initial turbulence anisotropy. Therefore, from Eq. 6.2.3,

$$\text{For } R_t = 0, \qquad c_{\varepsilon 2} = \frac{(-5/2)(-7/2)}{(-5/2)(-5/2)} = 1.4 \ . \tag{6.2.4}$$

Aupoix (1987) showed how $c_{\varepsilon 2}$ could be obtained by integrating turbulence energy spectra. Following Comte-Bellot and Corrsin (1966) and Reynolds (1974), he used the assumption that the low wave-number part of the 3-d turbulence spectra is unchanged as turbulence decays,

$$E_{low}(K) = cK^b, \tag{6.2.5}$$

where K is the magnitude of the wave number and c and b are constants, c being determined by the initial spectrum. The high wave-number part of the spectra, in the inertial sub-range at high turbulence Reynolds numbers, depends on the dissipation and takes the form

$$E_{high}(K) = \alpha\varepsilon^{2/3}K^{-5/3}\ , \tag{6.2.6}$$

where α is the Kolmogorov constant. He assumed that at any wave number, K, the energy is the smaller of relations 6.2.5 and 6.2.6, so that

$$k = \int_0^\infty \min(E_{low}, E_{high})dK\ . \tag{6.2.7}$$

Using Eqs. 6.2.1 and 6.2.2, he obtained (Appendix A1987)

$$c_{\varepsilon 2} = \frac{3b+5}{2(b+1)}\ . \tag{6.2.8}$$

The exponent b (Eq. 6.2.5) is generally considered to be in the range 1 to 4, giving $c_{\varepsilon 2}$ at high R_t in the range 2 to 1.7, consistent with the models in Table 6.1.1. Reynolds (1974) argued that $b = 2$ for the decay of isotropic turbulence, giving $c_{\varepsilon 2} = 11/6 \cong 1.83$.

An analysis with more complex spectra may be used to obtain $c_{\varepsilon 2} = c_{\varepsilon 2}(R_t)$. Aupoix suggested a formula which extends Eq. 6.2.6 to higher wave numbers (beyond the inertial sub-range),

$$E_{high}(K) = \alpha\varepsilon^{2/3}K^{-5/3}\exp\{-3.5n^2[1-\exp(6n+1.2-\sqrt{196n^2-33.6n+1.4532})]\}, \tag{6.2.9}$$

where n is the dimensionless wave number, $n = K\varepsilon^{-1/4}\nu^{3/4}$. Equations 6.2.5 and 6.2.9 may be smoothly combined using Chapman's (1979) method,

$$E(K) = \frac{E_{high}E_{low}}{E_{high}+E_{low}}. \tag{6.2.10}$$

Integrating the spectra, using the assumption that b and c (in Eq. 6.2.5) remain unchanged as turbulence decays, and using $\alpha = 1.7$ (Pao, 1965) and $b = 2$, $c_{\varepsilon 2}$ is approximated by (Appendix A1987)

$$c_{\varepsilon 2} = 1.83 - .43\frac{1+10\sqrt{R_t}}{1+15\sqrt{R_t}+3R_t+0.01R_t^2}\ . \tag{6.2.11}$$

Experimental Data

Comte-Bellot and Corrsin (Appendix CC1966) considered measurements of turbulence downstream of a grid. They present the results for both their own data and other data in the literature in the form

$$\frac{U_o^2}{\overline{u_1 u_1}} = A_1\left(\frac{U_o t}{M} - \frac{U_o t_1}{M}\right)^{n_1} \quad \text{and} \quad \frac{U_o^2}{\overline{u_2 u_2}} = A_2\left(\frac{U_o t}{M} - \frac{U_o t_2}{M}\right)^{n_2}, \tag{6.2.12}$$

where U_o is the flow velocity in the x_1 direction, M is the turbulence grid mesh size, and, A_1, n_1, t_1, A_2, n_2 and t_2 are parameters determined by graphical analysis of each data set. By symmetry, $\overline{u_2 u_2}(t) = \overline{u_3 u_3}(t)$. Using Eq. 6.2.3, $c_{\varepsilon 2}$, can be determined from the correlations of the data. The data is in the range $60 < R_t < 39000$, as shown in Fig. 6.2.1.

Batchelor and Townsend (Appendix BT1948) and Bennett and Corrsin (Appendix BC1978) measured decaying turbulence far downstream of turbulence grids in the range $1 < R_t < 20$. Fitting each data set, $k = k(t)$, to the equation

$$k = A(t+B)^{-n} \qquad \text{or} \qquad \ln(k) = \ln(A) - n\ln(t+B), \tag{6.2.13}$$

and determining A, B and n to minimize the errors in $\ln(k)$, then from Eq. 6.2.3,

$$c_{\varepsilon 2} = \frac{n+1}{n}. \tag{6.2.14}$$

Since the values of $c_{\varepsilon 2}$ depend on the second derivative of the measured quantity, k, errors can be quite high and an error analysis is important. Using a finite difference analysis (see Appendix BT1948), the error is estimated from the root-mean-square error in $\ln(k)$ by

$$c_{\varepsilon 2,error} \approx \frac{\ln(k)_{error}}{\ln(k_{\max}) - \ln(k_{\min})}\left(\frac{16}{\ln(k_{\max}) - \ln(k_{\min})} + 4(c_{\varepsilon 2} - 1)\right). \tag{6.2.15}$$

The results for $c_{\varepsilon 2}$ are shown as boxes on Fig. 6.2.1, indicating the error band on $c_{\varepsilon 2}$, and the range of R_t for each data set.

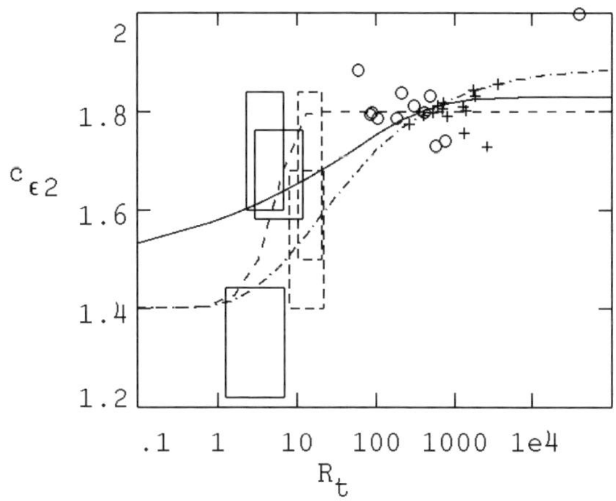

Fig. 6.2.1. The decay-of-dissipation coefficient, $c_{\varepsilon 2}$.
Analyzed data: +, CC1966 – Table 1, data of Comte-Bellot and Corrsin; o, CC1966 – Table 2, data from earlier sources; □, Batchelor and Townsend (BT1948); I⌶I, Bennett and Corrsin (BC1978).
Models: —— Eq. 6.2.11 (Aupoix method); - - - - HL1976 (Table 6.1.1); – - – - Lumley and Newman (1977), Eq. 6.2.16.

Three models for $c_{\varepsilon 2}$ are also shown on Fig. 6.2.1. To within the scatter of the data, all three models are reasonable. Hanjalić and Launder (HL1976) based their model on Batchelor and Townsend's data, so that $c_{\varepsilon 2}$ falls sharply between $R_t = 10$ and 1, Lumley and Newman (1977) fit Comte-Bellot and Corrsin's data (+ on Fig. 6.2.1) with a form that could be extrapolated to high and low Reynolds numbers. Since all the data on Fig. 6.2.1 is nearly isotropic ($g > 0.9$, except for one point), the isotropic limit for their model is shown,

$$c_{\varepsilon 2} = 1.4 + 0.49\exp\left(-4.245/\sqrt{R_t}\right) . \tag{6.2.16}$$

This model also gives $c_{\varepsilon 2}$ close to 1.4 at $R_t = 1$, while the model obtained using Aupoix's method falls more gradually at low R_t.

__Exercise 6.2__. Show that, using Eq. 6.2.13, the turbulence Reynolds number is given by $R_t = \dfrac{A}{\nu n}(t+B)^{1-n}$. For what values of n, and correspondingly $c_{\varepsilon 2}$, will the turbulence Reynolds number decay with time?

6.3 Production of Dissipation

Most two-equation turbulence models take $c_{\varepsilon 1}$ as a constant with values between 1.35 and 1.55. If $c_{\varepsilon 2}$ is taken as known, then $c_{\varepsilon 1}$ can be determined from the time dependence of the turbulence kinetic energy for homogeneous shear flow. The equations are

$$dk/dt = P - \varepsilon , \tag{6.3.1}$$

and

$$d\varepsilon/dt = (c_{\varepsilon 1}P - c_{\varepsilon 2}\varepsilon)\varepsilon/k \; . \qquad (6.3.2)$$

Chen and Jaw (1998), using the Champagne et al. (CHC1970) data, inferred that $dk/dt = 0$, and concluded that $d\varepsilon/dt$ is also approximately zero, so that $c_{\varepsilon 1} \approx c_{\varepsilon 2}$. Speziale et al. (SSG1991) concluded that the Tavoularis and Corrsin (TC1981) data was consistent with a fully developed flow limit with $d\omega/dt = 0$. From the equation for ω,

$$\frac{d\omega}{dt} = \left((c_{\varepsilon 1} - 1)P - (c_{\varepsilon 2} - 1)\varepsilon\right)\frac{\omega}{k} - \left((c_{\varepsilon 1} - 1)\frac{P}{\varepsilon} - (c_{\varepsilon 2} - 1)\right)\omega^2 , \qquad (6.3.3)$$

it follows that for $d\omega/dt = 0$,

$$(c_{\varepsilon 1} - 1)\frac{P}{\varepsilon} = c_{\varepsilon 2} - 1 \; . \qquad (6.3.4)$$

Speziale et al. used $c_{\varepsilon 1} = 1.44$ with $c_{\varepsilon 2} = 1.83$, which corresponds to $P/\varepsilon = 1.89$ in this fully developed limit.

Figure 6.3.1 shows the results of analyzing a large group of homogeneous shear data for $c_{\varepsilon 1}$. For all the results, $c_{\varepsilon 2}$ was set using Eq. 6.2.11 (Aupoix's method). For the DNS data, ω was available, and $c_{\varepsilon 1}$ was found by evaluating Eq. 6.3.3 between each pair of data points. For the experimental data, Eqs 6.3.1 and 6.3.2 were integrated to obtain $k(t)$. A value of $c_{\varepsilon 1}$ for each data set was determined by minimizing the mean square errors in $\ln(k)$, with respect to the choices for $c_{\varepsilon 1}$, and initial values for k and ε (see Appendix CHC1970). It was found that including data points less than 50 mesh spacings downstream of either the turbulence grid, or the shear generator, gave spurious results with $c_{\varepsilon 1}$ varying from –2 to 1.67 (Appendix TK1989). Therefore, the analysis results shown on Fig. 6.3.1 use only data greater than 50 mesh spacings from the grid or shear generator. The errors in $c_{\varepsilon 1}$ were estimated from the root mean square errors in $\ln(k)$ from (see Appendix CHC1970),

$$c_{\varepsilon 1,error} = \pm \frac{\ln(k)_{error}}{\delta(St)}\left(\frac{8s}{-b_{12}\delta(St)} + 2s\left|c_{\varepsilon 1} - 1\right| + \frac{2(c_{\varepsilon 2} - 1)}{-b_{12}}\right) \; . \qquad (6.3.5)$$

The data which most closely satisfies the condition $d\omega/dt = 0$ over a long period of time is Case (a) from Tavoularis and Karnik, which yields $c_{\varepsilon 1} = 1.59 \pm 0.03$; and the corresponding data fit gave P/ε between 1.37 and 1.4 over a dimensionless time, $\delta(St) = 16$.

Overall, the data shows large scatter with $c_{\varepsilon 1}$ varying from 1 to 1.6, and with no obvious functionality in R_t, P/ε or g. Therefore a constant value of $c_{\varepsilon 1}$ in the range,

$$c_{\varepsilon 1} = 1.5 \pm 0.1 \; , \qquad (6.3.6)$$

which is consistent with earlier models, appears to be appropriate.

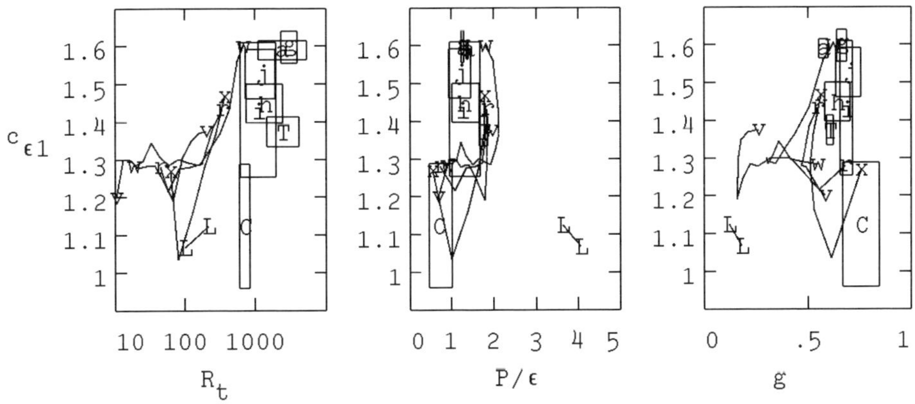

Fig. 6.3.1. The production-of-dissipation coefficient, $c_{\varepsilon 1}$. Analyzed experimental data: C, Champagne et al. (CHC1970); T, Tavoularis and Corrsin (TC1981); a-j, Tavoularis and Karnik (TK1989). DNS data: L, Lee et al. (LKM1990); r-x, Rogers et al. (RMR1986)

6.4 Near-Wall Considerations

An important consideration, in choosing the primary variable for the dissipation, length, or time scale, is how sensitive the results will be to grid resolution and discretization of the differential transport equation. Other considerations are boundary conditions and the interdependence of the equations.

From the definition of the dissipation, Eq. 2.1.18, an inhomogeneous part which is significant near walls, can be determined by considering separately the inhomogeneity of the turbulence velocity scale q. In particular,

$$\varepsilon = \nu \overline{\frac{\partial u_i}{\partial x_k} \frac{\partial u_i}{\partial x_k}} = \nu \overline{\frac{\partial q(u_i/q)}{\partial x_k} \frac{\partial q(u_i/q)}{\partial x_k}} = \nu \overline{\left(\frac{u_i}{q} \frac{\partial q}{\partial x_k} + q \frac{\partial (u_i/q)}{\partial x_k} \right)^2} \tag{6.4.1}$$

$$= \nu \left(\frac{\overline{u_i u_i}}{q^2} \left(\frac{\partial q}{\partial x_k} \right)^2 + \frac{1}{2} \frac{\partial q^2}{\partial x_k} \frac{\overline{\partial (u_i/q)^2}}{\partial x_k} + q^2 \overline{\left(\frac{\partial (u_i/q)}{\partial x_k} \right)^2} \right),$$

so that defining

$$\tilde{\varepsilon} = \nu q^2 \overline{\left(\frac{\partial (u_i / q)}{\partial x_k} \right)^2} , \tag{6.4.2}$$

gives

$$\varepsilon = 2\nu \left(\frac{\partial q}{\partial x_k} \right)^2 + \tilde{\varepsilon} \; . \tag{6.4.3}$$

Many of the two-equation models in Table 6.1.1 use $\tilde{\varepsilon}$ as the transported variable. The remaining term, $2\nu(\partial q / \partial x_k)^2$, is evaluated algebraically. The equation for the turbulence kinetic energy (2.1.17) then becomes

$$\frac{\partial k}{\partial t} + U_k \frac{\partial k}{\partial x_k} = -\overline{u_i u_k} \frac{\partial U_i}{\partial x_k} - \frac{\partial}{\partial x_k} \left(\frac{\overline{u_i u_i u_k}}{2} + \frac{\overline{p u_k}}{\rho} \right) + \frac{\partial}{\partial x_k} \nu \frac{\partial k}{\partial x_k} - 2\nu \left(\frac{\partial q}{\partial x_k} \right)^2 - \tilde{\varepsilon} \; . \tag{6.4.4}$$

The equation for the turbulence velocity scale, $q = \sqrt{k}$, is formed by taking $1/(2q)$ times the equation for k,

$$\frac{\partial q}{\partial t} + U_k \frac{\partial q}{\partial x_k} = -\frac{\overline{u_i u_k}}{2q} \frac{\partial U_i}{\partial x_k} - \frac{1}{2q} \frac{\partial}{\partial x_k} \left(\frac{\overline{u_i u_i u_k}}{2} + \frac{\overline{p u_k}}{\rho} \right) + \frac{\partial}{\partial x_k} \nu \frac{\partial q}{\partial x_k} - \frac{\tilde{\varepsilon}}{2q} . \tag{6.4.5}$$

Note that the algebraic dissipation term does not appear in the equation because it has been absorbed into the laminar diffusion term.

Choice of Primary Variables

Figure 6.4.1 shows sample near-wall profiles of k, q, and several possible choices for a dissipation variable. The dissipation, ε, has a complex shape near the wall, while ε / q and ε / k are infinite at the wall. Either $\tilde{\varepsilon}$, $\tilde{\varepsilon}/q$, or $\tilde{\varepsilon}/k = \omega$, are therefore better choices for the dissipation variable.

Regardless of whether k, q, or $\overline{u_i u_j}$ equations are solved for the turbulence, choosing ω as the dissipation variable reduces the interdependence of the equations. The source term for ω depends only indirectly on the turbulence kinetic energy,

$$\frac{D\omega}{Dt} = (c_{\varepsilon 1} - 1)(-2b_{ij}S_{ij})\omega - (c_{\varepsilon 2} - 1)\omega^2 + \ldots\ldots \; . \tag{6.4.6}$$

With the dissipation of turbulence taken proportional to k, when ω is used, there is less interdependence between k and ω equations, than between k and $\tilde{\varepsilon}$ equations,

$$\frac{Dk}{Dt} = (-2b_{ij}S_{ij})k - \omega k + \ldots\ldots \quad = (-2b_{ij}S_{ij})k - \tilde{\varepsilon} + \ldots\ldots \; . \tag{6.4.7}$$

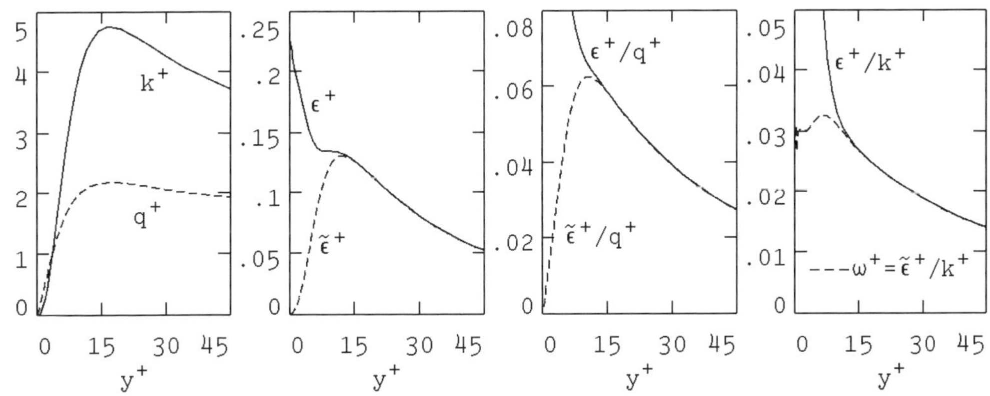

Fig. 6.4.1. Near-wall profiles of turbulence and dissipation properties for fully developed channel flow, DNS data of Moser et al. (MKM1999), R_τ=590.

Correspondingly, there is less interdependence between q and ω equations, than between equations for q and $\tilde{\varepsilon}/q$,

$$\frac{Dq}{Dt} = (-b_{ij}S_{ij})q - \frac{1}{2}\omega q + ... \quad = (-b_{ij}S_{ij})q - \frac{1}{2}\left(\frac{\tilde{\varepsilon}}{q}\right) + ... \quad . \tag{6.4.8}$$

Since q varies linearly near the wall, while k varies as y^2, q is a better choice than k for the turbulence variable. To set the Reynolds stresses, the natural complement to a $q-\omega$ model is differential equations for the turbulence anisotropies, b_{ij}. Equations for five independent b_{ij}'s are needed. The sixth being redundant, since $b_{ii} = 0$.

Wall Boundary Conditions

The breakdown of the Reynolds stresses, in terms of primary variables q and b_{ij},

$$\overline{u_i u_j} = 2(b_{ij} + \frac{1}{3}\delta_{ij})q^2, \tag{6.4.9}$$

has the advantage, that with $q = 0$ being the obvious wall boundary condition for q, the Reynolds stresses all vary as y^2 or higher powers of y near the wall. Further discussion of the b_{ij} equations near the wall is given in Chapter 12. The boundary conditions for b_{ij} are set in Sections 8.2 and 8.3.

From Fig. 6.4.1, it can be seen that the bulk of the dissipation near the wall comes from the algebraic term, $2\nu(\partial q/\partial x_k)^2$, so that $\omega << \varepsilon/k$. For the case shown, ω reaches 10 percent of ε/k at about $y^+ = 2.5$. Therefore the wall boundary condition on ω may not be critical. Possible choices include

$$\frac{\partial\omega}{\partial y} = 0, \qquad \omega^+ = c, \qquad \text{or} \qquad \omega = c\frac{\partial q}{\partial y}, \tag{6.4.10}$$

where c is a constant.

Figure 6.4.2 shows the specific dissipation, ω, near the wall from direct numerical simulations of fully developed channel flow, Table 6.4.1. Below $y^+ = 2$, the large spread of values is due to the evaluation error, since ω is obtained as the difference between two large numbers. An appropriate wall boundary condition may be obtained by considering the values at $y^+ = 2$. Here, $\omega^+ = 0.027 \pm 0.005$. There is less variation when ω is normalized with the wall value of $\partial q/\partial y$,

$$\frac{\omega}{(\partial q/\partial y)_{wall}} = 0.086 \pm 0.006 , \tag{6.4.11}$$

suggesting that this, or $\partial\omega/\partial y = 0$, are the best choices for the boundary condition on ω.

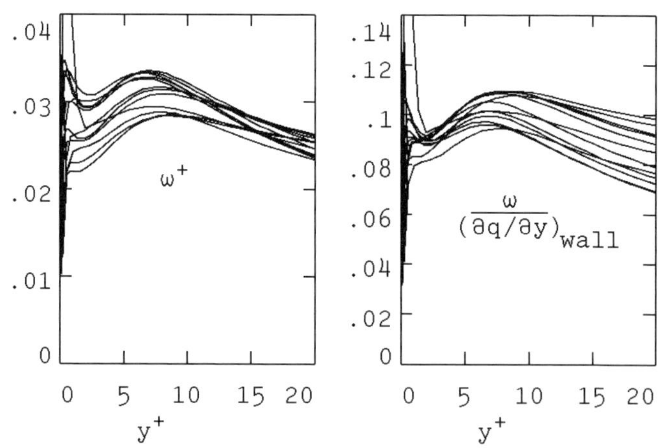

Fig. 6.4.2. The specific dissipation, ω, near the wall from a large selection of fully developed channel flow DNS cases, $R_\tau = 100$ to 643, Table 6.4.1.

Table 6.4.1. DNS Channel Data. Direct numerical simulations of fully developed two-dimensional channel flow without rotation.

Author	R_τ	R_D	$y^+_{\text{near wall}}$
Moser et al. (MKM1999)	178	5585	0.0536
	392	13763	0.0295
	587	21907	0.0442
From data base by Kasagi et al. (KHMMN1990)[1]			
Kuroda and Kasagi	100	2890	0.0535
	150	4560	0.0357
Horiuti	180	5602	0.0542
Gilbert and Kleiser	211	6666	0.0063
Iwamoto, Suzuki and Kasagi	109	3220	0.0330
	150	4586	0.0201
	298	10039	0.0100
	396	13925	0.0075
	643	24272	0.0215

[1] Additional references: Kuroda et al. (KKH1990); Kuroda (1990); Gilbert and Kleiser (1991); Gilbert (1993); Iwamoto (2002); and Iwamoto et al. (ISK2002).

6.5 Summary

The Reynolds stress models developed in this book will include a differential equation for the specific dissipation, $\omega = \tilde{\varepsilon}/k$. The equation will take the form,

$$\frac{\partial \omega}{\partial t} + U_k \frac{\partial \omega}{\partial x_k} = \frac{\partial}{\partial x_k} \nu \frac{\partial \omega}{\partial x_k} + T_\omega + (c_{\varepsilon 1} - 1) P \frac{\omega}{k} - (c_{\varepsilon 2} - 1)\omega^2 + E_\omega \ , \qquad (6.5.1)$$

where $c_{\varepsilon 1}$ is most likely a constant between 1.4 and 1.6, and $c_{\varepsilon 2}$ is a function of at least $\tilde{R}_t = k/(\omega\nu)$, with $c_{\varepsilon 2} = 1.4$ when $\tilde{R}_t = 0$. The form of the turbulent diffusion, T_ω, and possibly any extra terms, E_ω, are as yet to be determined. For the wall boundary condition, either $\partial\omega/\partial y = 0$, or $\omega = c\,\partial q/\partial y$ will be used.

Chapter 7

TURBULENT DIFFUSION

In this chapter we investigate possible choices for turbulent diffusion terms in the modeled Reynolds stress and dissipation, or alternative, equations. The asymptotic near-wall behavior of the triple-product correlation term in the Reynolds stress equations suggests that a model between the standard isotropic, and Daly-Harlow models is appropriate. A new model based on the square root of the Reynolds stress tensor is proposed and shown to be consistent with DNS channel data.

7.1 Models in the Literature

Three different models for turbulent diffusion are commonly associated with Reynolds stress modeling. Referring to Eq. 2.2.1, the turbulent diffusion term models both the diffusion due to the triple velocity correlation, T_{ij}, and the non-homogeneous part of the pressure-velocity correlation, D_{ij},

$$T_{ij} = -\frac{\partial \overline{u_i u_j u_k}}{\partial x_k}, \qquad D_{ij} = -\frac{1}{\rho}\left(\frac{\partial \overline{p u_j}}{\partial x_i} + \frac{\partial \overline{p u_i}}{\partial x_j}\right). \tag{7.1.1}$$

The simplest model is an isotropic diffusion model, with the same form as used in two-equation turbulence models. With $c_{\mu q}$ a constant,

$$T_{ij} + D_{ij} = \frac{\partial}{\partial x_n} \nu_{tq} \frac{\partial \overline{u_i u_j}}{\partial x_n}, \quad \text{with} \quad \nu_{tq} = c_{\mu q}\frac{k^2}{\varepsilon}, \quad \text{or} \quad \nu_{tq} = c_{\mu q}\frac{k^2}{\tilde{\varepsilon}}. \tag{7.1.2}$$

The most popular model is probably the Daly and Harlow (1970) model. They suggested a directional diffusion proportional to the Reynolds stresses. With c_q a constant,

$$T_{ij} = \frac{\partial}{\partial x_k} c_q \frac{k}{\varepsilon} \overline{u_k u_n} \frac{\partial \overline{u_i u_j}}{\partial x_n}. \tag{7.1.3}$$

This is sometimes applied as a model for $T_{ij} + D_{ij}$, and $\tilde{\varepsilon}$ may be used instead of ε.

Hanjalić and Launder (1972), based on the symmetry of u_i, u_j and u_k in T_{ij}, suggested

$$T_{ij} = \frac{\partial}{\partial x_k} c_s \frac{k}{\varepsilon}\left(\overline{u_i u_n} \frac{\partial \overline{u_j u_k}}{\partial x_n} + \overline{u_j u_n} \frac{\partial \overline{u_i u_k}}{\partial x_n} + \overline{u_k u_n} \frac{\partial \overline{u_i u_j}}{\partial x_n} \right), \tag{7.1.4}$$

where c_s is a constant. This is generally used with the assumption that $D_{ij} = 0$. However, Donaldson and Bilanin (1975) used it together with a model for D_{ij}, which was suggested by Daly and Harlow (1970),

$$D_{ij} = c_D \left(\frac{\partial}{\partial x_j} \frac{k^2}{\varepsilon} \frac{\partial \overline{u_i u_n}}{\partial x_n} + \frac{\partial}{\partial x_i} \frac{k^2}{\varepsilon} \frac{\partial \overline{u_j u_n}}{\partial x_n} \right), \qquad c_D = \text{constant}, \tag{7.1.5}$$

based on an earlier model by Donaldson (1969).

Realizability and Two-Dimensional Limits

At a point in a flow for which a component of the fluctuating velocity, for example, u_2, is zero for all time, then $\overline{u_2 u_2} = 0$, and the turbulence is two-dimensional. Since u_2 is continuous, and $u_2^2 \geq 0$ in the surrounding flow, then $\partial u_2^2 / \partial x_k = 0$. From Eq. 7.1.1, the two-dimensional limit for T_{22} is then

$$T_{22} = 0 \qquad \text{when} \qquad \overline{u_2 u_2} = 0 \, . \tag{7.1.6}$$

The two-dimensional limit for D_{22} is obtained by observing that D_{22} is the difference between the velocity pressure-gradient and the pressure-strain.

$$D_{22} = -\frac{2}{\rho} \frac{\partial \overline{p u_2}}{\partial x_2} = -\frac{2}{\rho} \overline{u_2 \frac{\partial p}{\partial x_2}} - \frac{2}{\rho} \overline{p \frac{\partial u_2}{\partial x_2}} = -\frac{2}{\rho} \overline{u_2 \frac{\partial p}{\partial x_2}} - \Pi_{22} \, . \tag{7.1.7}$$

If $\Pi_{22} = 0$ when $\overline{u_2 u_2} = 0$, then the two-dimensional limit for D_{22} is

$$D_{22} = 0 \qquad \text{when} \qquad \overline{u_2 u_2} = 0 \, . \tag{7.1.8}$$

None of the models, Eqs. 7.1.2-7.1.5, satisfy these two-dimensional limits (see, for example, Exercise 7.1).

For realizability, it is only necessary that the diffusion terms make positive or zero contributions to the Reynolds stress equations when the turbulence is two-dimensional,

$$\frac{\partial \overline{u_2 u_2}}{\partial t} = T_{22} + D_{22} + \ldots\ldots\ldots\ldots \geq 0 \qquad \text{when} \quad \overline{u_2 u_2} = 0 \ . \tag{7.1.9}$$

The models in Eqs. 7.1.2-7.1.4 satisfy realizability, but Eq. 7.1.5 does not (Exercise 7.1), therefore Eq. 7.1.5 will not be considered further.

Exercise 7.1. Consider the fluctuating velocities, $u_1 = a \sin \omega t$, $u_2 = (bx_1 + cx_3)\sin \omega t$, and $u_3 = d \sin \omega t$, where a, b, c, d and ω are constants. For $x_1 = x_3 = 0$, show that

Eq. 7.1.2 yields $\quad T_{22} + D_{22} = \nu_{tq}(b^2 + c^2) \geq 0$,

Eq. 7.1.3 yields $\quad T_{22} = \dfrac{c_q}{2}\dfrac{k}{\varepsilon}(ab + cd)^2 \geq 0,$

Eq. 7.1.4 yields $\quad T_{22} = c_s \dfrac{k}{\varepsilon}(ab + cd)^2 \geq 0,$ and

Eq. 7.1.5 yields $\quad D_{22} = c_D \dfrac{\partial (k^2/\varepsilon)}{\partial x_2}(ab + cd)$ which may have either sign.

Diffusion of Dissipation

Models which use isotropic diffusion for the turbulence, or turbulence stresses, also use isotropic diffusion of the dissipation variable. The turbulent diffusion of ω is then

$$T_\omega = \frac{\partial}{\partial x_n}\nu_{t\omega}\frac{\partial \omega}{\partial x_n} \ , \quad \text{with} \quad \nu_{t\omega} = c_{\mu\omega}\frac{k^2}{\tilde{\varepsilon}} = c_{\mu\omega}\frac{k}{\omega} \ , \tag{7.1.10}$$

where $c_{\mu\omega}$ is a constant which may be different from $c_{\mu q}$.

Reynolds stress models which use Eq. 7.1.3 or 7.1.4 for the diffusion of Reynolds stresses use the Daly and Harlow (1970) form for the diffusion of dissipation. Here we apply it to the diffusion of ω,

$$T_\omega = \frac{\partial}{\partial x_k} c_\omega \frac{k}{\omega}\frac{\overline{u_k u_n}}{k}\frac{\partial \omega}{\partial x_n} . \tag{7.1.11}$$

but unlike Daly and Harlow, this is frequently used with c_ω not equal to c_q.

An estimate of the diffusion rate for the dissipation variable is obtained by considering the log-law region of high Reynolds number fully-developed turbulent boundary layers. In this region, the turbulence is essentially uniform, so that the equation for k reduces to $P = \varepsilon$, and the mean velocity gradient is given by Eq. 1.3.4, $\partial U_1/\partial x_2 = \sqrt{-\overline{u_1 u_2}}/(\kappa y)$. Combining these gives

$$P = \varepsilon = -\overline{u_1 u_2}\frac{\partial U_1}{\partial x_2} = \frac{\left(-\overline{u_1 u_2}\right)^{3/2}}{\kappa y} \qquad \text{or} \qquad \omega = \frac{q}{\kappa y}(-2b_{12})^{3/2}, \tag{7.1.12}$$

so that both ε and ω vary as $1/y$. Substituting into the equation for ω (Eq. 6.5.1) using isotropic diffusion (laminar diffusion is negligible), and neglecting any extra terms, E_ω,

$$0 = \frac{\partial}{\partial x_k} c_{\mu\omega} \frac{k}{\omega}\frac{\partial \omega}{\partial x_k} + (c_{\varepsilon 1} - 1)P\frac{\omega}{k} - (c_{\varepsilon 2} - 1)\omega^2, \tag{7.1.13}$$

yields

$$c_{\mu\omega} = \frac{(c_{\varepsilon 2} - c_{\varepsilon 1})}{\kappa^2}(-2b_{12})^3 . \tag{7.1.14}$$

Correspondingly, using instead the Daly and Harlow diffusion model,

$$c_\omega = \frac{(c_{\varepsilon 2} - c_{\varepsilon 1})}{\kappa^2}\frac{(-2b_{12})^3}{(2b_{22} + 2/3)}, \tag{7.1.15}$$

where b_{12} and b_{22} are the anisotropies of the turbulence in the log-law region. A corollary to this analysis is that the coefficient of diffusion, $c_{\mu\omega}$ or c_ω, controls the slope of the velocity profile, $\partial U^+/\partial(\ln y^+)$, that the model gives.

If it is considered that $\tilde{\varepsilon}$ is the dissipation variable that diffuses, for example, using isotropic diffusion,

$$T_{\tilde{\varepsilon}} = \frac{\partial}{\partial x_n} \nu_{t\tilde{\varepsilon}} \frac{\partial \tilde{\varepsilon}}{\partial x_n} , \tag{7.1.16}$$

then using

$$\frac{\partial \omega}{\partial t} = \frac{1}{k}\frac{\partial \tilde{\varepsilon}}{\partial t} - \frac{\omega}{k}\frac{\partial k}{\partial t}, \tag{7.1.17}$$

the result for the ω equation (Eq. 6.5.1) is the diffusion of ω plus extra terms (Appendix EQ.7.1.18),

$$T_{\tilde{\varepsilon}} \;\Rightarrow\; T_\omega = \frac{\partial}{\partial x_n}\nu_{t\tilde{\varepsilon}}\frac{\partial \omega}{\partial x_n}, \quad E_\omega = \frac{2\nu_{t\tilde{\varepsilon}}}{k}\frac{\partial k}{\partial x_n}\frac{\partial \omega}{\partial x_n} + \frac{\omega}{k}\frac{\partial}{\partial x_n}(\nu_{t\tilde{\varepsilon}} - \nu_{tq})\frac{\partial k}{\partial x_n} . \tag{7.1.18}$$

Menter (1994) proposed a model in which an extra term was phased out near the wall, so that essentially ε/k diffused near the wall, and ε diffused in the outer parts of boundary layers and beyond.

7.2 Diffusion of Turbulence

The direct numerical simulations of fully developed channel flow provide the opportunity to evaluate the models for the diffusion of turbulence, and their importance. For these flows, the turbulence kinetic energy equation (Eq. 6.4.4) reduces to

$$0 = -\overline{u_1 u_2}\frac{\partial U_1}{\partial x_2} \underbrace{-\frac{\partial}{\partial x_2}\left(\frac{\overline{u_i u_i u_2}}{2}\right)}_{+T_{ii}/2} \underbrace{-\frac{\partial}{\partial x_2}\left(\frac{\overline{p u_2}}{\rho}\right)}_{+D_{ii}/2} + \frac{\partial}{\partial x_2}\nu\frac{\partial k}{\partial x_2} - 2\nu\left(\frac{\partial q}{\partial x_2}\right)^2 - \tilde{\varepsilon}\ , \qquad (7.2.1)$$

where x_1 is the flow direction, and x_2 is normal to the channel walls. Terms in this equation normalized by the total dissipation, ε, are shown in Fig. 7.2.1. With this normalization, the curves marked by P (production), L (laminar diffusion), and T (total turbulent diffusion) add up to one and are balanced by the dissipation (–1).

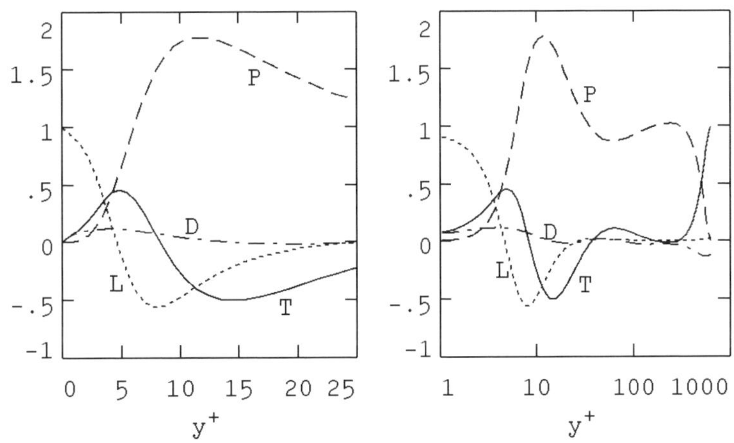

Fig. 7.2.1. Terms in the turbulence kinetic energy equation for the DNS channel data of Iwamoto et al. (ISK2002) at R_τ=643. P, P/ε; L, $L_{ii}/(2\varepsilon)$; D, $D_{ii}/(2\varepsilon)$; T, $(T_{ii}+D_{ii})/(2\varepsilon)$.

The turbulent diffusion makes significant contributions to the balance of the equation at three locations. Near $y^+ = 5$, the turbulent diffusion tends to increase the turbulence, and balances approximately half of the dissipation. Near $y^+ = 15$, the turbulent diffusion term is negative, balancing approximately one third of the turbulence production. On the channel centerline, production is zero, and turbulent diffusion balances dissipation. The turbulent diffusion due to the velocity-pressure correlation, line D on Fig. 7.2.1, makes only a small contribution to the overall balance of the equation.

For fully developed channel flow, the models for turbulent diffusion, Eq. 7.1.2-4, become,

$$\text{Isotropic:} \quad \frac{T_{ii} + D_{ii}}{2} = \frac{\partial}{\partial x_2} c_{\mu q} \frac{k^2}{\tilde{\varepsilon}} \frac{\partial k}{\partial x_2}, \tag{7.2.2}$$

$$\text{Daly and Harlow:} \quad \frac{T_{ii}}{2} = \frac{\partial}{\partial x_2} c_q \frac{k}{\tilde{\varepsilon}} \overline{u_2 u_2} \frac{\partial k}{\partial x_2}, \tag{7.2.3}$$

$$\text{Hanjalić and Launder:} \quad \frac{T_{ii}}{2} = \frac{\partial}{\partial x_2} c_s \frac{k}{\tilde{\varepsilon}} \left(\overline{u_1 u_2} \frac{\partial \overline{u_1 u_2}}{\partial x_2} + u_2 u_2 \frac{\partial \overline{u_2 u_2}}{\partial x_2} + \overline{u_2 u_2} \frac{\partial k}{\partial x_2} \right), \tag{7.2.4}$$

when $\tilde{\varepsilon}$, instead of ε, is used for the dissipation scale.

These models were evaluated using the DNS channel data of Iwamoto et al. (ISK2002) for $R_\tau = 643$ (the same data set as shown in Fig. 7.2.1). The coefficients, $c_{\omega q}$, c_q and c_s were set so that the minimum value of $(T_{ii} + D_{ii})/(2\varepsilon)$ matched the minimum value in the data, as shown in Fig. 7.2.2.

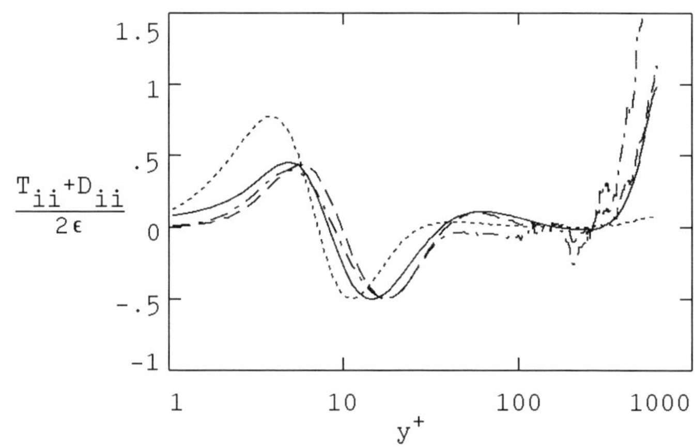

Fig. 7.2.2. Diffusion of turbulence from the DNS channel data of Iwamoto et al. (ISK2002) for R_τ=643. —— data; ---- Eq. 7.2.2 with $c_{\mu q}$=0.012; — — Eq. 7.2.3 with c_q=0.29; — - — Eq. 7.2.4 with c_s=0.58.

The Hanjalić and Launder model, Eq. 7.2.4, required $c_s = 0.58$ to match the minimum value of the data, and gave a very high value, $(T_{ii} + D_{ii})/(2\varepsilon) = 2.4$, on the centerline. The model's authors recommended a much lower coefficient, $c_s = 0.08$. Overall, it appears to be a poor fit to the data.

The Daly and Harlow model, Eq. 7.2.3, appears to be the best fit to the data, matching well the magnitude of the peaks and valleys. However, in the buffer layer (approx. $5 < y^+ < 40$), the model results are shifted to the right relative to the data. The isotropic diffusion model, Eq. 7.2.2, shows just the opposite trend, with the model results to the left of the data in the buffer region.

Asymptotic Analysis

For fully developed channel flow, the term we are primarily trying to model from Eq. 7.2.1 is

$$\frac{T_{ii}}{2} = \frac{1}{2}\frac{\partial}{\partial x_2}\overline{u_2 u_i u_i} = \frac{1}{2}\frac{\partial}{\partial y}\overline{u_2 u_i u_i} \ . \tag{7.2.5}$$

Near the wall $u_i u_i$ varies as y^2, as does the turbulence kinetic energy, k. The wall normal component of the fluctuating velocity, u_2, also varies as y^2, since from continuity, $\partial u_2 / \partial y = -\partial u_1 / \partial x_1 - \partial u_3 / \partial x_3 = 0$ at the wall. Therefore, from Eq. 7.2.5, $T_{ii}/2$ varies as y^3. With $k/\tilde{\varepsilon}$ $(= 1/\omega)$ having a value at the wall (Eq. 6.4.10), the models, given by Eqs. 7.2.2 and 7.2.3, vary as y^2 and y^4, respectively.

A gradient diffusion model between the isotropic diffusion model and the Daly-Harlow model, e.g.

$$\frac{T_{ii}}{2} = \frac{\partial}{\partial x_2} c_{vq} \sqrt{\overline{u_2 u_2}} \frac{q}{\omega}\frac{\partial k}{\partial x_2} \ , \tag{7.2.6}$$

which is asymptotically correct,

$$y^3 = \frac{1}{y} \qquad y^2 \ \frac{y}{1} \frac{y^2}{y} \ ,$$

would appear to be a reasonable representation for the diffusion. The problem is to generalize the model so that it is tensorially consistent.

The Square-Root of the Reynolds Stress Tensor

Since the Reynolds stress tensor is symmetric with positive normal stresses, we may define a turbulence velocity scale tensor, v_{ij} , so that

$$v_{ik} v_{kj} = \overline{u_i u_j} \ . \tag{7.2.7}$$

From the Cayley-Hamilton theorem (Appendix C), the principal directions of $v_{ik} v_{kj}$ are the same as for v_{ij}. In addition, when considered in principal coordinates, the normal stresses of $v_{ik} v_{kj}$ are the square of the normal components of v_{ij}. Therefore, in the principal coordinates for the Reynolds stress tensor,

$$v_{P_iP_j} = \begin{bmatrix} \sqrt{\overline{u_{P_1}u_{P_1}}} & 0 & 0 \\ 0 & \sqrt{\overline{u_{P_2}u_{P_2}}} & 0 \\ 0 & 0 & \sqrt{\overline{u_{P_3}u_{P_3}}} \end{bmatrix} , \tag{7.2.8}$$

and v_{ij} may be found from $\overline{u_iu_j}$. As a shorthand one may write,

$$v_{ij} = \left(\overline{u_iu_j}\right)^{1/2} . \tag{7.2.9}$$

New Diffusion Model

A tensorially consistent new model for the turbulent diffusion of Reynolds stresses, based on the square-root of the Reynolds stress tensor, is

$$T_{ij} = \frac{\partial}{\partial x_k} c_{vq} \frac{q}{\omega} v_{kn} \frac{\partial \overline{u_iu_j}}{\partial x_n} . \tag{7.2.10}$$

This will be referred to as the v_{ij} diffusion model. When this diffusion is considered in principal stress coordinates, it may be interpreted as being based on a turbulent viscosity composed of a length times a velocity, in particular, an isotropic length scale, $c_{vq}q/\omega$, times the r.m.s. value of the fluctuating velocity in the direction of the diffusion,

$$T_{ij} = \frac{\partial}{\partial x_{P_1}} c_{vq} \frac{q}{\omega} \sqrt{u_{P_1}^2} \frac{\partial \overline{u_iu_j}}{\partial x_{P_1}} + \frac{\partial}{\partial x_{P_2}} c_{vq} \frac{q}{\omega} \sqrt{u_{P_2}^2} \frac{\partial \overline{u_iu_j}}{\partial x_{P_2}} + \frac{\partial}{\partial x_{P_3}} c_{vq} \frac{q}{\omega} \sqrt{u_{P_3}^2} \frac{\partial \overline{u_iu_j}}{\partial x_{P_3}} . \tag{7.2.11}$$

For the diffusion of k, in the fully developed channel, Eq. 7.2.10 reduces to,

$$\frac{T_{ii}}{2} = \frac{\partial}{\partial x_2} c_{vq} \frac{q}{\omega} v_{22} \frac{\partial k}{\partial x_2} . \tag{7.2.12}$$

As shown in Fig. 7.2.3, the location of the minimum value for $T_{ii}/2$ coincides quite well with the data.

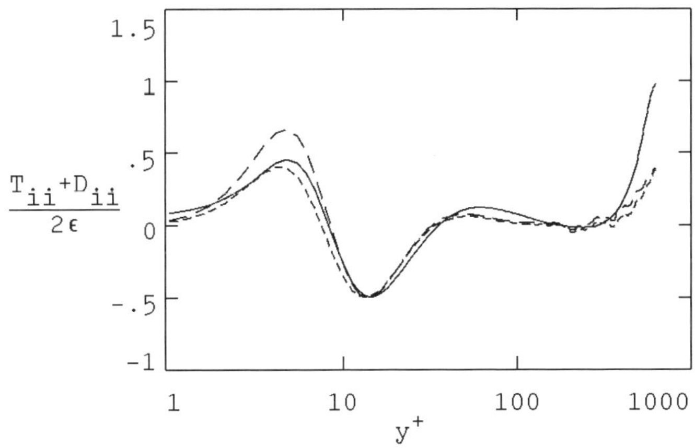

Fig. 7.2.3. Diffusion of turbulence from the DNS channel data of Iwamoto et al. (ISK2002) for R_τ=643. —— data; – – Eq. 7.2.12 with c_{vq} = 0.076; ---- Eq. 7.2.14 with c_{vq} = 0.074.

Also shown on Fig. 7.2.3, is the q-equation version of the new v_{ij} model. The equation for q (Eq. 6.4.5) in the fully developed channel may be written as,

$$0 = -\frac{\overline{u_1 u_2}}{2q}\frac{\partial U_1}{\partial x_2} \underbrace{- \frac{1}{2q}\frac{\partial}{\partial x_2}\left(\frac{\overline{u_i u_i u_2}}{2}\right)}_{+T_{ii}/(4q)} \underbrace{- \frac{1}{2q}\frac{\partial}{\partial x_2}\left(\frac{\overline{p u_2}}{\rho}\right)}_{+D_{ii}/(4q)} + \frac{\partial}{\partial x_2}\nu\frac{\partial q}{\partial x_2} - \frac{\tilde{\varepsilon}}{2q}. \tag{7.2.13}$$

Modeling $T_{ii}/(4q)$ as a turbulent diffusion of q,

$$\frac{T_{ii}}{4q} = \frac{\partial}{\partial x_k} c_{vq}\frac{q}{\omega} v_{kn}\frac{\partial q}{\partial x_n} , \tag{7.2.14}$$

gives quite good agreement with the data through the buffer layer.

The diffusion of q and the diffusion of k are related. Expanding Eq. 7.2.12 gives

$$\frac{T_{ii}}{2} = \frac{\partial}{\partial x_2} c_{vq}\frac{q}{\omega} v_{22}\frac{\partial k}{\partial x_2} = 2q\frac{\partial}{\partial x_2} c_{vq}\frac{q}{\omega} v_{22}\frac{\partial q}{\partial x_2} + 2c_{vq}\frac{q}{\omega} v_{22}\left(\frac{\partial q}{\partial x_2}\right)^2 . \tag{7.2.15}$$

Therefore, using Eq. 7.2.14 for T_{ii}, includes the assumption that the last term in Eq. 7.2.15 may be neglected, or that there is an algebraic contribution to the dissipation, which cancels this term in the q-equation.

Chapter 8

COMPLETE THE MC AND MA MODELS

In this chapter the MC and MA Reynolds stress models are completed using fully developed pipe and channel flows, without rotation. This involves choosing $c_{\varepsilon 1}$, $c_{\varepsilon 2}$ and the turbulent diffusion models for q, ω and b_{ij}, as well as the turbulence Reynolds number dependence of c_1. The objectives are to obtain good agreement with:

a) Skin friction as a function of Reynolds number, R_D,
b) Mean flow velocity profiles, which require good shear stress distributions,
c) Turbulence kinetic energy,
d) Turbulence dissipation, and
e) Turbulence anisotropy.

Variations on the choices in model coefficients are investigated within the framework of the MC model. The final MC model, in keeping with the simplicity of its pressure-strain model, uses isotropic diffusion for the turbulence properties.

Unlike the MC model, the more complex MA model has the potential to give very good agreement with the turbulence anisotropies near the wall. To fulfill this potential over a range of Reynolds numbers, it was found that more functionality than $c_1 = c_1(\tilde{R}_t, g)$ was needed.

8.1 Transport Equations for the Reynolds Stress Models

In Section 6.4, based on near-wall considerations, it was determined that the best choice of primary turbulence variables for the Reynolds stress models, is q, ω and five independent b_{ij}'s. The transport equations take the form,

$$\frac{\partial \psi}{\partial t} + U_k \frac{\partial \psi}{\partial x_k} = \frac{\partial}{\partial x_k} \nu \frac{\partial \psi}{\partial x_k} + T_\psi + S_\psi , \qquad \text{for} \qquad \psi = q, \omega, b_{ij} , \tag{8.1.1}$$

where T_ψ is the turbulent diffusion and S_ψ is the source term.

The model form chosen for turbulent diffusion,

$$T_\psi = \frac{\partial}{\partial x_k} c_{t\psi} \frac{q^2}{\omega} \left(\frac{\overline{u_k u_n}}{k} \right)^{n_t} \frac{\partial \psi}{\partial x_n} , \tag{8.1.2}$$

includes the choice of: isotropic diffusion with $n_t = 0$, and $(\overline{u_k u_n}/k)^0 = \delta_{kn}$, Eq. 7.1.2; the new v_{ij} diffusion model, $n_t = 0.5$, Eq. 7.2.10; or the Daly and Harlow diffusion model, $n_t = 1$, Eq. 7.1.3. While it is certainly possible to let n_t have different values for different properties, and there may be some advantages to isotropic diffusion for ω, regardless of the choice for other properties, this option was not chosen for any of the models, and n_t has only one value for each Reynolds stress model. The option of different coefficients of diffusion, $c_{t\psi}$, for different properties was retained.

The source term for the q equation neglects any extra term due to the difference between the turbulent diffusion of q and k (Eq. 7.2.15), and so consists only of production and dissipation terms (Eq. 6.4.8),

$$S_q = \frac{P}{2q} - \frac{\tilde{\varepsilon}}{2q} = (-b_{ik}S_{ik})q - \frac{\omega q}{2} \quad . \tag{8.1.3}$$

Extra source terms in the ω equation, based on whether it is more appropriate to assume that $q\omega$ or $\tilde{\varepsilon}$ diffuses, instead of ω, were considered but did not prove useful, therefore the source term for ω consists of the standard production and dissipation terms (Eq. 6.4.6),

$$S_\omega = (c_{\varepsilon 1} - 1)(-2b_{ij}S_{ij})\omega - (c_{\varepsilon 2} - 1)\omega^2 \ . \tag{8.1.4}$$

The source terms for the b_{ij} equations come from the production, dissipation, pressure-strain and Coriolis terms of the Reynolds stress equations. Writing the source terms for pressure-strain models with only c_1, and c_3 through c_6, terms, and taking the slow pressure-strain proportional to $\tilde{\varepsilon}$, instead of ε, from Eq. 4.1.6,

$$\begin{aligned} S_{bij} &= -\frac{1}{2}b_{ij}\left((c_1 - 2)\omega + (2 + \frac{1}{2}c_6)\frac{P}{k}\right) + \frac{1}{2}\left(c_3 - \frac{4}{3}\right)S_{ij} \\ &+ \frac{1}{2}(c_4 - 2)(b_{ik}S_{jk} + b_{jk}S_{ik} - \frac{2}{3}\delta_{ij}b_{mn}S_{mn}) \\ &+ \frac{1}{2}(c_5 - 2)(b_{ik}W_{jk} + b_{jk}W_{ik}) - \Omega_k\left(\varepsilon_{ikm}b_{jm} + \varepsilon_{jkm}b_{im}\right) . \end{aligned} \tag{8.1.5}$$

Again extra terms, due to the transport equation having b_{ij} diffuse, instead of $\overline{u_i u_j}$ diffusing, have been omitted. As discussed in Section 12.3, it may be considered that these extra terms are balanced by inhomogeneous contributions to pressure-strain, so that they do not appear in S_{bij}.

Equations for Fully Developed Channel Flow

The coordinates for two-dimensional channel flow are shown in Fig. 8.1.1. The flow is in the x_1 direction and x_2 is normal to the wall. For fully developed flow, the momentum equation (Eq. 2.1.10), without rotation, reducces to

$$0 = -\frac{\partial \overline{u_1 u_2}}{\partial x_2} - \frac{1}{\rho}\frac{\partial \overline{P}}{\partial x_1} + \frac{\partial}{\partial x_2}\nu\frac{\partial U_1}{\partial x_2}\,. \tag{8.1.6}$$

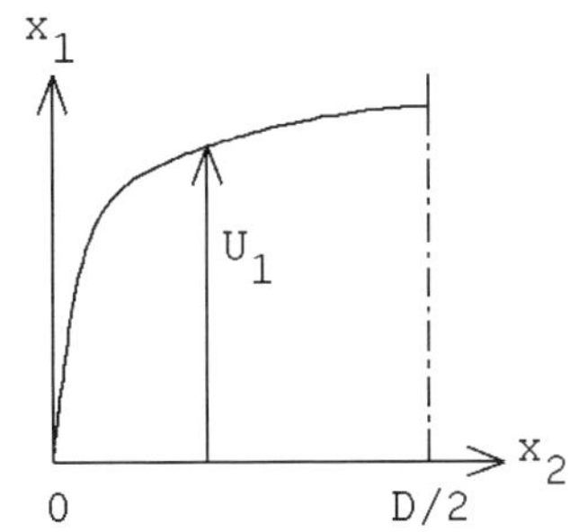

Fig. 8.1.1. Coordinates for half of the two-dimensional channel.

The continuity equation is not needed since the flow is fully developed, so that $U_2 = U_3 = 0$ and $\partial U_1/\partial x_1 = 0$. The desired overall flow rate is obtained by adjusting $\partial \overline{P}/\partial x_1$. The channel Reynolds number is defined using the area-mean velocity,

$$U_{mean} = \frac{1}{D}\int_0^D U_1 dx_2\,, \tag{8.1.7}$$

so that

$$R_D = U_{mean} D/\nu\,. \tag{8.1.8}$$

The skin friction coefficient is defined in terms of the wall shear stress,

$$c_f = \frac{\tau_w}{\frac{1}{2}\rho U_{mean}^2} = 2\left(\frac{U_\tau}{U_{mean}}\right)^2 = \frac{2}{\left(U_{mean}^+\right)^2}\,, \tag{8.1.9}$$

where

$$\tau_w = \mu\left(\frac{\partial U_1}{\partial x_2}\right)_{wall}\,. \tag{8.1.10}$$

The skin friction is then inversely proportional to the square of the average value of U^+, so that if U^+ values are high, c_f will be low.

The equations for the primary turbulence variables (Eq. 8.1.1) reduce to

$$0 = \frac{\partial}{\partial x_2} \nu \frac{\partial \psi}{\partial x_2} + \frac{\partial}{\partial x_2} C_{22} \frac{\partial \psi}{\partial x_2} + S_\psi , \qquad (8.1.11)$$

where, using Eq. 8.1.2, C_{22} is the $i = 2,\ j = 2$ component of

$$C_{ij} = c_{t\psi} \frac{q^2}{\omega} \left(\frac{\overline{u_i u_j}}{k} \right)^{n_t} . \qquad (8.1.12)$$

By symmetry, $b_{13} = b_{23} = 0$, so that only 5 of the 7 equations are needed, for example, equations for q, ω, b_{11}, $b_{22} - b_{33}$, and b_{12}.

Equations for Fully Developed Pipe Flow

Fully developed pipe flow is analyzed in cylindrical, r,θ,z, coordinates, Fig. 8.1.2. The area-mean velocity and the pipe Reynolds number are given by

$$U_{mean} = \int_0^{D/2} U_z r dr \Big/ \int_0^{D/2} r dr \quad \text{and} \quad R_D = U_{mean} D/\nu . \qquad (8.1.13)$$

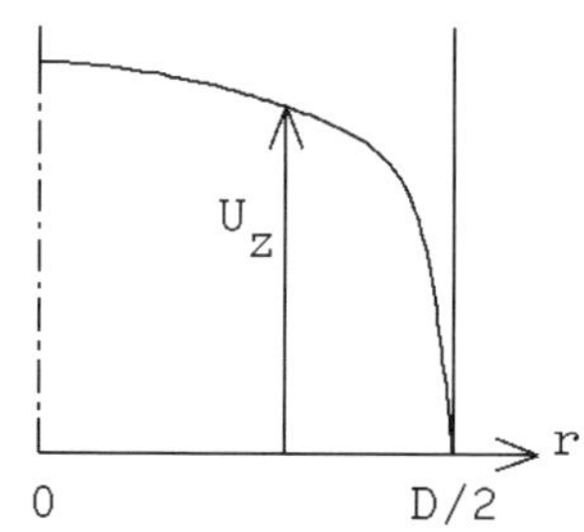

Fig. 8.1.2. Cylindrical coordinates for a pipe.

The conversion of the differential equations to cylindrical coordinates is given in Appendix D. The axial momentum equation reduces to

$$0 = -\frac{1}{r} \frac{\partial r \overline{u_z u_r}}{\partial r} - \frac{1}{\rho} \frac{\partial \overline{P}}{\partial z} + \frac{1}{r} \frac{\partial}{\partial r} \nu r \frac{\partial U_z}{\partial r} . \qquad (8.1.14)$$

The equations for the scalar turbulence properties, q and ω, are

$$0 = \frac{1}{r} \frac{\partial}{\partial r} r \nu \frac{\partial \psi}{\partial r} + \frac{1}{r} \frac{\partial}{\partial r} r \left(C_{rr} \frac{\partial \psi}{\partial r} \right) + S_\psi , \qquad (8.1.15)$$

where C_{rr} is the rr component of C_{ij}, Eq. 8.1.12.

By symmetry, $b_{r\theta} = b_{\theta z} = 0$. On the centerline of the pipe, $b_{rr} = b_{\theta\theta}$, so that it is convenient to solve equations for $b_{rr} - b_{\theta\theta}$, b_{zz} and b_{rz},

$$0 = \frac{1}{r}\frac{\partial}{\partial r} r\left((\nu + C_{rr})\frac{\partial(b_{rr} - b_{\theta\theta})}{\partial r}\right) + S_{brr} - S_{b\theta\theta} - \frac{4(\nu + C_{\theta\theta})}{r^2}(b_{rr} - b_{\theta\theta}) , \quad (8.1.16)$$

$$0 = \frac{1}{r}\frac{\partial}{\partial r} r\left((\nu + C_{rr})\frac{\partial b_{zz}}{\partial r}\right) + S_{bzz} , \quad (8.1.17)$$

$$0 = \frac{1}{r}\frac{\partial}{\partial r} r\left((\nu + C_{rr})\frac{\partial b_{rz}}{\partial r}\right) + S_{brz} - \frac{(\nu + C_{\theta\theta})}{r^2} b_{rz} . \quad (8.1.18)$$

The centerline boundary conditions are

$$b_{rr} - b_{\theta\theta} = 0, \ \frac{\partial b_{zz}}{\partial r} = 0, \ b_{rz} = 0, \ \frac{\partial q}{\partial r} = 0, \ \frac{\partial \omega}{\partial r} = 0, \text{ and } \frac{\partial U_z}{\partial r} = 0. \quad (8.1.19)$$

8.2 Completing the MC Model

The coefficients for the pressure-strain part of the MC model, at high R_t, are given by Eq. 5.2.4,

$$c_1 = 9, \qquad c_3 = 0.15, \qquad \text{and} \qquad c_4 = 1 . \quad (8.2.1)$$

The pressure-strain model is extended to low Reynolds numbers by taking

$$c_1 = 2 + (c_{1\infty} - 2)f_{c1} , \quad \text{where} \quad c_{1\infty} = 9 \quad \text{and} \quad f_{c1} = 1 - \exp(-c_{1a}\tilde{R}_t^{\,c_{1b}}) . \quad (8.2.2)$$

To complete the source term for the ω equation, $c_{\varepsilon1}$ is taken as a constant between 1.4 and 1.6 (Eq. 6.3.6). Variations in the turbulence Reynolds number dependence of $c_{\varepsilon2}$ will be considered in the form

$$c_{\varepsilon2} = 1.4 + (1.83 - 1.4)\left(1 - \exp(-c_{\varepsilon2a}\tilde{R}_t^{\,c_{\varepsilon2b}})\right) . \quad (8.2.3)$$

This builds in the limits of 1.4 and 1.83 at $\tilde{R}_t = 0$ and infinity, respectively, consistent with Eq. 6.2.11.

The MC model is then completed by choosing the constants: c_{1a} and c_{1b} for slow pressure-strain; $c_{\varepsilon1}$, $c_{\varepsilon2a}$ and $c_{\varepsilon2b}$ for the ω equation; and c_{tq}, $c_{t\omega}$, c_{tbij} and n_t for turbulent diffusion. For the MC model and variations of it considered here, all the diffusion coefficients were taken to be the same,

$$c_{tq} = c_{t\omega} = c_{tbij} \ . \tag{8.2.4}$$

Model Completion Procedure

Begin with the calculation of fully developed pipe flow for a Reynolds number, $R_D = 10^6$. Use the wall boundary conditions,

$$q = 0, \qquad d\omega / dy = 0, \quad \text{and} \quad db_{ij} / dy = 0 \ . \tag{8.2.5}$$

After choosing the other coefficients:

a) Adjust $c_{t\omega}$ to give a reasonable slope for $dU^+ / d\ln y^+$ in the log-law region of the boundary layer. Increasing $c_{t\omega}$ increases $dU^+ / d\ln y^+$ and decreases the skin friction coefficient, c_f.

b) Adjust either c_{1a} or $c_{\varepsilon 2a}$ to obtain $c_f = 0.00291 \pm 0.00005$. (This is consistent with experimental data compiled by Schlichting, 1979, but slightly lower than recent data, 0.00296 for R_D=1,024,000, McKeon et al., MSZDS2004). Lowering c_{1a} lowers the slow pressure-strain near the wall, increasing the size of the highly anisotropic turbulence region near the wall, and lowering c_f. Lowering $c_{\varepsilon 2a}$, lowers the decay of dissipation near the wall, so that ω^+ is increased; c_f is lowered.

If the velocity profile is reasonable, use the results of the calculation to change the wall boundary condition on ω to Eq. 6.4.10,

$$\omega = c_{\omega w} \frac{dq}{dy} \qquad c_{\omega w} = \left(\frac{\omega}{dq / dy} \right)_{\text{Wall, Pipe } R_D = 10^6} \ . \tag{8.2.6}$$

This boundary condition improved the convergence rate of the fully developed flow calculation procedure, and slightly improved the results at low R_D.

Proceed to the calculation of fully developed channel flow over a range of Reynolds numbers. Check $c_f(R_D)$ and turbulence properties against DNS data.

MC Model and Results

The constants to complete the final MC model are given in the first line in Table 8.2.1. The values for $c_{\varepsilon 2a}$ and $c_{\varepsilon 2b}$ roughly approximate Eq. 6.2.11 (Aupoix's method), as shown in Fig. 8.2.1a. $n_t = 0$ gives isotropic diffusion. The remaining constants were chosen based on results. The value for $c_{\varepsilon 1}$, 1.55, is towards the high end of the acceptable range, 1.4 to 1.6. The turbulence Reynolds number dependence of c_1 is consistent with DNS data, as shown in Fig. 8.2.1b. The value of $c_{t\omega} = 0.05$ is slightly above the value of 0.043 expected from Eq. 7.1.14 at high R_t (using $\kappa = 0.418$ and $b_{12} = -0.15$).

Table 8.2.1. Constants to complete the final MC model, and test variations.

Model	c_{1a} (Eq. 8.2.2)	c_{1b} (Eq. 8.2.2)	$c_{\varepsilon 1}$ (Eq. 8.1.4)	$c_{\varepsilon 2a}$ (Eq. 8.2.3)	$c_{\varepsilon 2b}$ (Eq. 8.2.3)	$c_{t\omega}$ (Eq. 8.1.2)	n_t (Eq. 8.1.2)	$c_{\omega w}$ (Eq. 8.2.6)
MC – final	0.0138	0.75	1.55	0.6	0.25	0.05	0	0.07
Test variations								
MC.test.b	0.00435	1	1.55	0.6	0.25	0.05	0	0.065
MC.test.c	0.0432	0.5	1.55	0.6	0.25	0.05	0	0.08
MC.test.d	0.0138	0.75	1.5	0.6	0.25	0.0603	0	0.048
MC.test.e	0.0138	0.75	1.55	0.28	0.4	0.05	0	0.076
MC.test.f	0.0122	0.75	1.55	0.08	1	0.05	0	0.053
MC.test.g	0.037	0.5	1.55	0.6	0.25	0.0943	0.5	0.066
MC.test.h	0.0249	0.5	1.55	0.6	0.25	0.1545	1	0.07

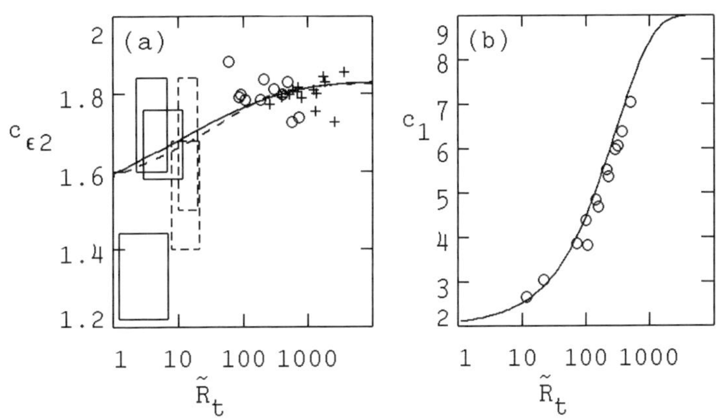

Fig. 8.2.1. Reynolds number dependence of $c_{\varepsilon 2}$ and c_1.
——— MC model;
– – – Eq. 6.2.11.
(a) with data as for Fig. 6.2.1.
(b) o, data from Fig. 5.1.6 for $g = 0.5 \pm 0.05$.

Table 8.2.1 also contains the constants for seven test variations of the model. Tests b and c vary $c_1(R_t)$. Test d lowers $c_{\varepsilon 1}$, increasing $c_{t\omega}$ to compensate. Test e varies $c_{\varepsilon 2}(R_t)$. Test f effectively removed the R_t dependence of $c_{\varepsilon 2}$ in terms of model development choices, since all values $c_{\varepsilon 2a} > 0.08$ gave essentially the same results for the pipe and channel flows considered; c_{1a} was lowered to compensate. Tests g and h vary the diffusion model, modifying $c_1(R_t)$ to compensate.

The model completion procedure required that the mean velocity profiles for fully developed pipe flow at $R_D = 10^6$ should be reasonable. As shown in Fig. 8.2.2, the profiles for the MC model (black line) and all the test variations (grey lines) are nearly indistinguishable, and in reasonable agreement with experimental data.

Fig. 8.2.2. Mean velocity for fully developed pipe flow, $R_D = 10^6$.
- - - $U^+ = 5.45 + \ln(y^+)/0.418$; black line, MC model; grey lines, models MC.test.b-h; o, data of Zagarola (1996).

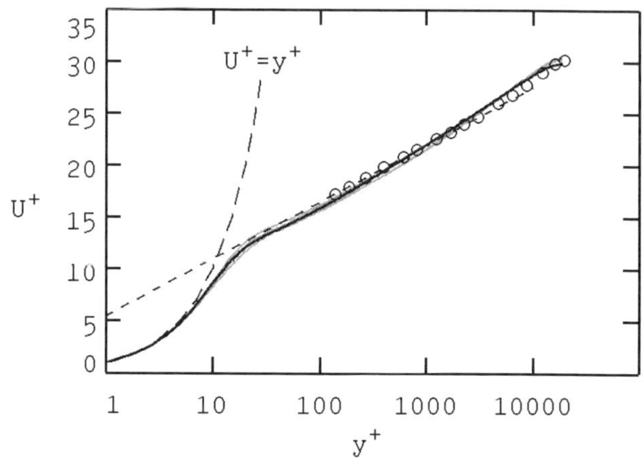

The skin friction coefficients for both fully developed pipe and channel flows are shown in Fig. 8.2.3. The MC model agrees well with the data, except at low R_D. Here, the model gives the laminar to turbulent transition at lower R_D than the data, especially for the pipe. Note that the computations were biased to obtain turbulent flow by starting with turbulent flow and then slowly lowering R_D.

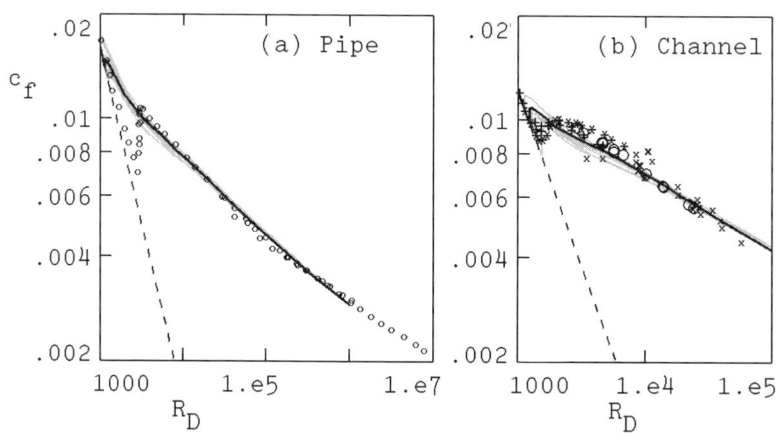

Fig. 8.2.3. Skin friction coefficients for fully developed pipe and channel flows. - - - laminar limits; black line, MC model; grey lines, MC.test.b-h. (a) Pipe data: o, McKeon et al. (MSZDS2004). (b) Channel data: +, Davies and White (DW1928); *, Patel and Head (1969); x, compilation from Dean (D1978, Fig. 2) for long channels, $x_1/D > 100$; o, DNS, see Table 6.4.1.

The skin friction coefficient at a channel Reynolds number, $R_D = 5600$, was one of the criteria used in choosing the final model. The test models gave values from $c_f = 0.0074$ to 0.0080, as shown in Table 8.2.2. The MC model with $c_f = 0.0079$ gives results close to the DNS value of 0.0082.

Comparison with Fully Developed Channel Flow DNS

The turbulence velocity scale profiles from the MC model, are compared with DNS data in Fig. 8.2.4 for three channel Reynolds numbers, $R_D = 5600$, 13,800, and 21,900. The q profiles agree well with the DNS data except near the wall where the peak values are 10 to 15 percent lower than the data. This may be attributed to the choice of isotropic diffusion with $c_{tq} = c_{t\omega} = 0.05$. This is much higher than the value $c_{tq} = 0.012$ shown to give reasonable diffusion of k near the wall in Fig. 7.2.2. Table 8.2.2 shows the maximum value of q^+ increasing from 1.80 for the MC model, to 1.87 for MC.test.g which used the v_{ij} diffusion model, to 2.07 for MC.test.h which used the Daly and Harlow diffusion model.

Table 8.2.2. Wall and near-wall results for fully developed channel flow with $R_D = 5600$ ($R_\tau = 180$).

Model	c_f	$\left(q^+\right)_{max}$	$\left(\dfrac{dq^+}{dy^+}\right)_{wall}$	$\tilde{s}_{wall} = \dfrac{1}{\omega^+_{wall}}$
MC – final	0.00791	1.795	0.396	36.1
MC.test.b	0.00783	1.789	0.394	39.0
MC.test.c	0.00793	1.803	0.393	31.8
MC.test.d	0.00770	1.804	0.498	41.8
MC.test.e	0.00782	1.791	0.380	34.6
MC.test.f	0.00749	1.833	0.442	42.7
MC.test.g	0.00796	1.871	0.377	40.2
MC.test.h	0.00743	2.073	0.362	39.5
DNS data				(at $y^+ = 2.5$)
H1992	0.00826	2.030	0.290	36.6
MKM1999	0.00814	2.026	0.292	36.7

H1992, Horiuti, in data base by Kasagi et al. (KHMMN1990); MKM1999, Moser et al.

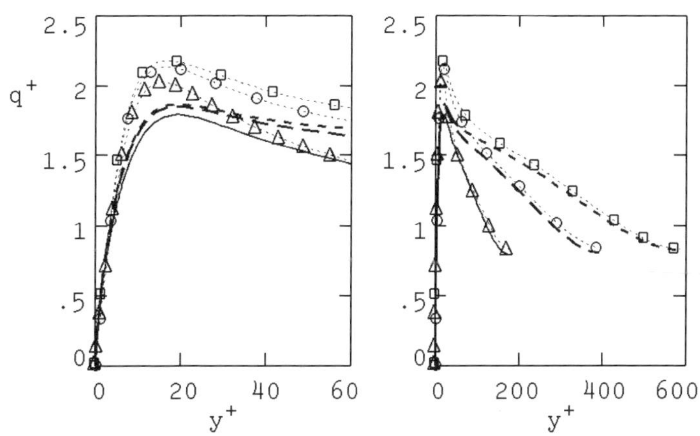

Fig. 8.2.4. Turbulence velocity scale for fully developed channel flow. Comparison of MC model with DNS data of Moser et al. (MKM1999).

R_D	model	data
5,600	———	△···
13,800	— — —	○···
21,900	- - - -	□···

The results from the different diffusion models are also compared in Fig. 8.2.5. Model MC.test.h, using Daly and Harlow diffusion, shows higher values of U^+ which is indicative of the lower skin friction coefficient. The higher values of b_{12} are due to the lowering of $c_1(R_t)$. Model MC.test.g with the v_{ij} diffusion model gives overall the best results. However, it was not adopted for Model MC, because the improvements were small compared to the added complexity.

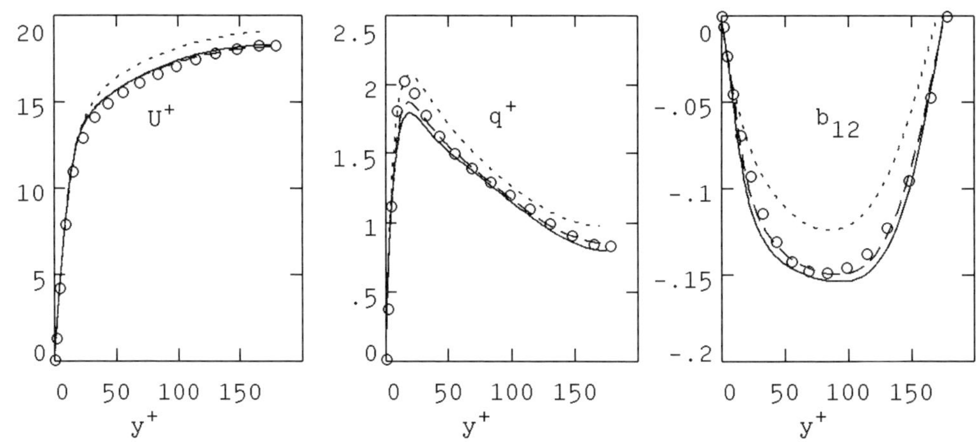

Fig. 8.2.5. Comparison of three diffusion models in fully developed channel flow at $R_D = 5600$. o, DNS data of Moser et al. (MKM1999). Models: ——— MC (isotropic); – – – MC.test.g (v_{ij}); - - - - MC.test.h (Daly and Harlow).

The profiles of specific dissipation rate, ω, for the MC model are shown in Fig. 8.2.6, and are in good agreement with the DNS data up to about $y/\delta = 0.5$. The coefficient for the wall boundary condition, $c_{\omega w} = 0.07$ (Eq. 8.2.6), being below the value 0.086 in Eq. 6.4.11, compensates for the high value of dq^+/dy^+, Table 8.2.2.

In the laminar sublayer, the dimensionless strain rate, $\tilde{s}$, is related to ω^+ by

$$\tilde{s} = \frac{1}{\omega^+}\frac{dU^+}{dy^+} = \frac{1}{\omega^+} , \tag{8.2.7}$$

so that $\tilde{s}$ characterizes ω^+ near the wall. As shown in Table 8.2.2, wall values of $\tilde{s}$ were affected by all variations to the model. Increasing the steepness of $c_1(R_t)$, MC.test.b, increased $\tilde{s}$, as did lowering $c_{\varepsilon 1}$, MC.test.d. Increasing the steepness of $c_{\varepsilon 2}(R_t)$, MC.test.e, lowered $\tilde{s}$.

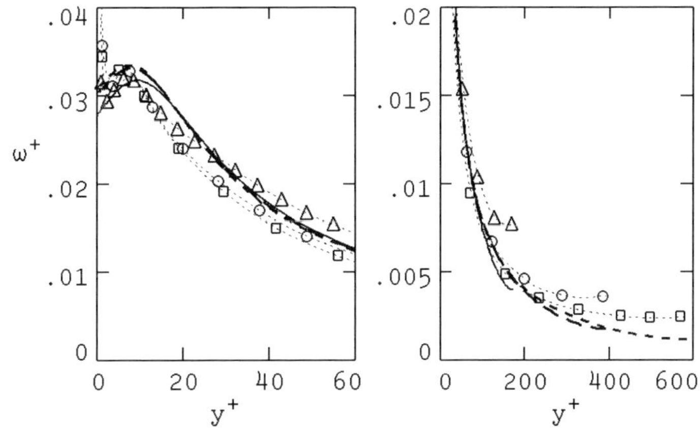

Fig. 8.2.6. Specific dissipation rate, ω, in fully developed channel flow. Comparison of MC model with DNS data of Moser et al. (MKM1999).

R_D	model	data
5,600	———	△···
13,800	— — —	○···
21,900	- - - -	□···

In the central portion of the channel, $y/\delta > 0.5$, the MC model values of ω^+ continue to fall, following approximately a profile $w^+(y^+)$ which is independent of R_D. The DNS data does not show this universality, with the result that the model values are about a factor of 2 lower than the DNS data values on the centerline of the channel. All the test variation models gave similar results.

The turbulence anisotropies for the MC model are compared with the DNS data in Fig. 8.2.7. The shear component, b_{12}, is in overall good agreement with data, the MC model giving a minimum value, $b_{12} = -0.155$. Note that in the range $y^+ = 10$ to 40, the lower modeled values of b_{12} compensate for the low modeled values of q, so that overall the shear stress, $\overline{u_1 u_2} = 2b_{12}q^2$, is well modeled. This is shown indirectly in Fig. 8.2.8, by the good agreement of the velocity profiles with the DNS data.

The anisotropy of the normal stresses near the wall, Fig. 8.2.7, is underpredicted. This is the fault of the constant values of the c_3 and c_4 pressure-strain coefficients which limit the normal stresses at high strain rate. The fully developed limiting values may be obtained from Eq. 5.3.7 (which yields $(b_{12}/D)_{D=0} = -0.1614$), and Eqs. 5.3.3 and 5.3.4, which yield

$$b_{11} = 0.377, \qquad b_{22} = -0.269, \qquad b_{33} = -0.108. \tag{8.2.8}$$

The wall values in Fig. 8.2.7 are consistent with this fully developed analysis.

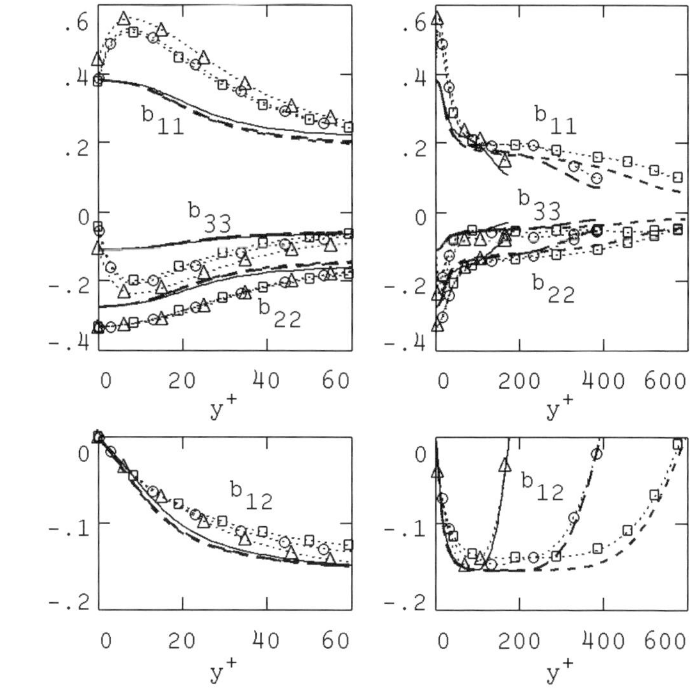

Fig. 8.2.7. Turbulence anisotropies for fully developed channel flow. Comparison of MC model with DNS data of Moser et al. (MKM1999).

R_D	model	data
5,600	———	△···
13,800	— — —	○···
21,900	- - - -	□···

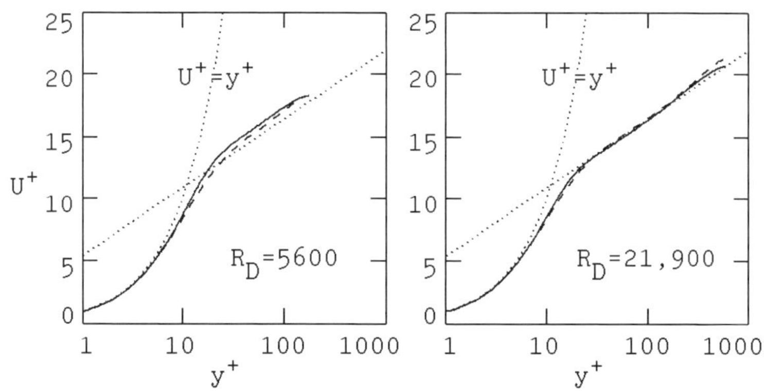

Fig. 8.2.8. Velocity profiles for fully developed channel flow. ——— MC model; - - - - DNS data of Moser et al. (MKM1999); ····· U^+=5.45+ln(y^+)/0.418.

8.3 Completing the MA Model

The pressure-strain coefficients for the MA.2 model are given in Eqs. 5.2.8, 5.3.11 and 5.3.12, with auxiliary relations, Eqs. 5.2.5, 5.2.7, 5.3.9 and 5.3.10. Some modifications to the g functionality were made for the final MA model. These, shown in Fig. 8.3.1 as model MA.3, are

Model MA.3:

$$c_1 = 8 + (2-8) f_{g0}(g, 0.5) + (2.5-8) f_{g1}(g, 0.7) \ ,$$
$$c_3 = 0.15 f_{g3}(g, 0.05, 0.5) + 0.65 f_{g3}(g, 0.7, 1) \ , \qquad (8.3.1)$$
$$c_4 = 0.414 - 0.414\left(\max(0, 1-\sqrt{2g})\right)^2 + 0.086 f_{g3}(g, 0, 0.5)$$
$$+0.5 - 0.5 f_{g0}(g, 0.5) + \left(\frac{24}{7} - 1\right) f_{g1}(g, 0.85) \quad .$$

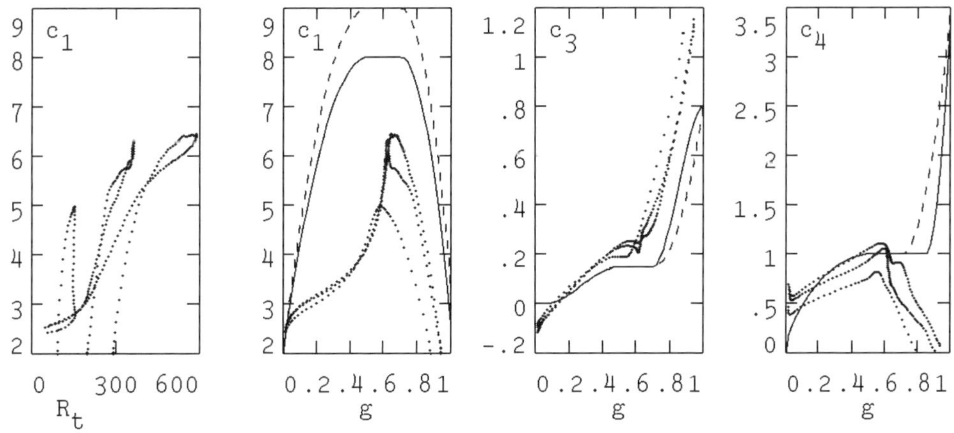

Fig. 8.3.1. Pressure-strain model coefficients. Models: - - - - MA.2; ——— MA.3. ······· Algebraic model limit values determined from the fully-developed channel flow DNS data of Moser et al. (MKM1999) for $y^+ > 5$, for $R_\tau = 180$, 395 and 590.

The maximum value of c_1 was reduced from 9 to 8. This modifies the anisotropies that the model will give in equilibrium when $P = \varepsilon$,

MA.2: $b_{11} = 0.130$, $b_{22} = -0.093$, $b_{33} = -0.037$ and $b_{12} = -0.147$ ($g = 0.72$)
MA.3: $b_{11} = 0.146$, $b_{22} = -0.104$, $b_{33} = -0.042$ and $b_{12} = -0.151$ ($g = 0.69$),

which affects the results obtained in the log-law regions of high Reynolds number turbulent boundary layers. In addition, the value for c_1 at $g = 1$ was reduced to 2.5. The basis for this is a reanalysis of data given in Chapter 15 (see Fig. 15.3.1).

Pressure-strain coefficients determined from the data of Moser et al., (MKM1999), assuming the algebraic limit applies (homogeneous, and $db_{ij}/dt = 0$), are also shown on Fig. 8.3.1. The data shows c_3 increasing above $g \sim 0.6$, while c_4 decreases. The ways in which c_3 and c_4 approach the isotropic limits were slightly modified based on this. The computational results showed only minor changes.

Modifications to Slow Pressure-Strain

As shown in Fig. 8.3.1, there is a large discrepancy between the c_1 coefficients determined from the data and model MA.3, below $g \sim 0.6$. It is clear that significant reductions in c_1 are needed to complete the model so that it will be able to calculate channel flow. Figure 8.3.1 suggests both a modification to the g dependence and a reduction based on turbulence Reynolds number. This functionality, however, was found to be inadequate.

The MA model was successfully completed by taking c_1 as a function of g, R_t and a dimensionless parameter characterizing the inhomogeneity of the anisotropy,

$$G = \frac{q^2}{\omega^2} \frac{\partial g}{\partial x_n} \frac{\partial g}{\partial x_n}, \tag{8.3.2}$$

so that $c_1 = c_1(g, R_t, G)$. As shown in Fig. 8.3.2, G peaks at about $y^+ = 30$, where the slope of g is largest, in the buffer layer, between the laminar sublayer and the log-law region. For the MA model, the inhomogeneity parameter is used to set an effective g,

$$g_{eff} = \frac{g^2}{g + 0.3G^{1.5}}, \tag{8.3.3}$$

which is then used to evaluate $c_1 = c_1(g_{eff}, \tilde{R}_t)$,

$$c_1 = 2 + \left(c_{1\infty}(g_{eff}) - 2\right) f_{c1}(R_t), \tag{8.3.4}$$

$$f_{c1} = 1 - \exp(-c_{1a} \tilde{R}_t^{\,c_{1b}} - c_{1c} \tilde{R}_t^{\,c_{1d}}). \tag{8.3.5}$$

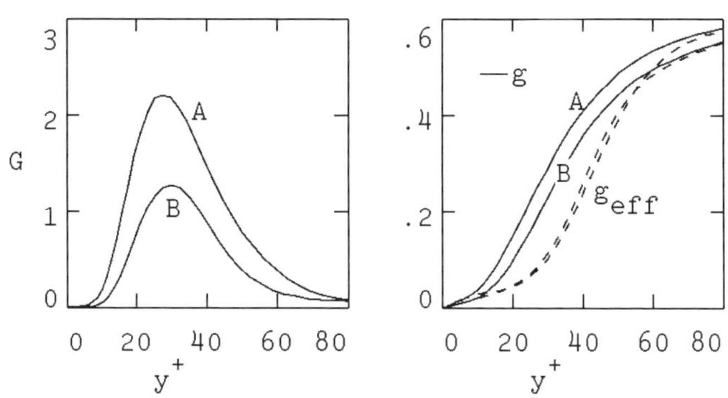

Fig. 8.3.2. Anisotropy parameters for slow pressure-strain model from the data of Moser et al. (MKM1999): A, $R_\tau = 590$; B, $R_\tau = 180$. (a) G. (b) —— g; – – g_{eff} for the MA model.

The net result is to lower c_1 where $\partial g/\partial y$ is large. Very little is known about the effect of inhomogeneity on pressure-strain. If there is any physical basis for the model, it could be argued that just as c_1 is lowered as the turbulence approaches a two-dimensional state due to the poor correlation of spectral components of the fluctuating velocity (see Chapter 9, Eq. 9.2.8), so too does inhomogeneity have the same effect when the turbulence has significant anisotropy.

For the final MA model, $c_{1\infty}$ is the same as for the MA.3 model,

MA: $$c_{1\infty}(g_{eff}) = 8 + (2-8)f_{g0}(g_{eff},0.5) + (2.5-8)f_{g1}(g_{eff},0.7)\ . \tag{8.3.6}$$

Two test variations of the MA model which take $g_{eff} = g$ are also considered. As shown in Fig. 8.3.3, model MA.test.b uses the same $\tilde{R}_t$ dependence for c_1 as the MA model but reduces $c_{1\infty}(g)$,

MA.test.b: $$c_{1\infty} = 3 - \left(\max(0, 1-\sqrt{10g})\right)^2 + 5f_{g3}(g,0.215,0.6) - 5.5f_{g1}(g,0.7)\ , \tag{8.3.7}$$

so it falls rapidly below $g = 0.6$. The MA.test.c model uses the same g dependence as the MA model but reduces $f_{c1}(\tilde{R}_t)$ so that c_1 falls more rapidly as $\tilde{R}_t$ is lowered. The parameters for the slow pressure-strain for the MA model and test variations are given in Table 8.3.1.

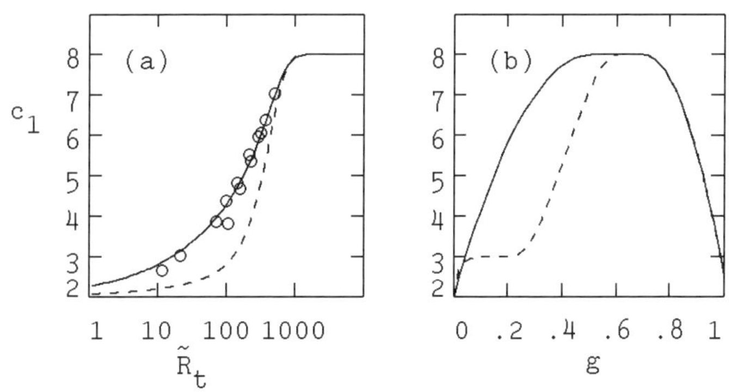

Fig. 8.3.3. Dependence of slow pressure-strain coefficient, c_1, on R_t and g. —— MA model. (a) models at $g = 0.7$. – – MA.test.c; o, data from Fig. 5.1.6 for $g = 0.5 \pm 0.05$. (b) $c_{1\infty}(g)$. – – MA.test.b.

Table 8.3.1. Summary of slow pressure-strain for the MA model and variations.

Model	g_{eff}	$\tilde{R}_t$ dependence, Eq. 8.3.5				$c_{1\infty}$
		c_{1a}	c_{1b}	c_{1c}	c_{1d}	
MA – final	Eq. 8.3.3	3.0e-6	2	0.045	0.5	Eq. 8.3.6
MA.test.b	$g_{eff} = g$	3.0e-6	2	0.045	0.5	Eq. 8.3.7
MA.test.c	$g_{eff} = g$	4.5e-6	2	0.01	0.5	Eq. 8.3.6

Diffusion and Dissipation Choices

As for the MC model, to complete the MA model, choices also need to be made for turbulent diffusion, and for $c_{\varepsilon 1}$ and $c_{\varepsilon 2}$. The MA model and the test variations all used the v_{ij} diffusion model with the same parameters (Eq. 8.1.2),

$$\text{MA:} \qquad n_t = 0.5, \qquad c_{tq} = c_{tbij} = 0.0735 \quad \text{and} \quad c_{t\omega} = 0.10 \ . \tag{8.3.8}$$

The value of c_{tq} is consistent with that determined from the data, Fig. 7.2.3.

For the MA model and test variations, the decay-of-dissipation coefficient, $c_{\varepsilon 2}$, is made a function of g as well as $\tilde{R}_t$. It should be noted that the basic data for $c_{\varepsilon 2}$, Fig. 6.2.1, are for nearly isotropic turbulence, so that any anisotropy functionality is essentially unknown. For the MA model and variations, factors are used to reduce $c_{\varepsilon 2}$ based on g and $\tilde{R}_t$,

$$c_{\varepsilon 2} = 1.4 + (1.83 - 1.4) f_{\varepsilon 2g} f_{\varepsilon 2Rt} \ , \tag{8.3.9}$$

with

$$f_{\varepsilon 2Rt} = 1 - \exp(-0.1\tilde{R}_t) \ , \tag{8.3.10}$$

and

$$\text{MA:} \qquad f_{\varepsilon 2g} = 1 - 0.65 f_{g0}(g, 0.7) \ , \tag{8.3.11}$$

$$\text{MA.test.b:} \qquad f_{\varepsilon 2g} = 0.87 + (0.35 - 0.87) f_{g0}(g, 0.5) + 0.13 f_{g3}(g, 0, 0.6) \ , \tag{8.3.12}$$

$$\text{MA.test.c:} \qquad f_{\varepsilon 2g} = 0.35 + 0.65 f_{g3}(g, 0, 0.6) \ . \tag{8.3.13}$$

$c_{\varepsilon 2}$ for the MA model is shown in Fig. 8.3.4. $c_{\varepsilon 2}$ begins to fall to its lower limit below $R_t = 100$. Thus, like model MC.test.f (Table 8.2.1), the R_t dependence is essentially immaterial to the completion of the model; all the significant near-wall reductions occur due to the g dependence. Attempts to obtain reasonable ω levels near the wall, using only R_t dependence, resulted in unrealistically low values for $c_{\varepsilon 2}$ at $R_t = 100$, compared to the data. The value for $c_{\varepsilon 2}$ at $g = 0$ and high R_t, 1.5505, is slightly higher than the value used for $c_{\varepsilon 1}$ for the MA model and test variations,

$$c_{\varepsilon 1} = 1.55 \ . \tag{8.3.14}$$

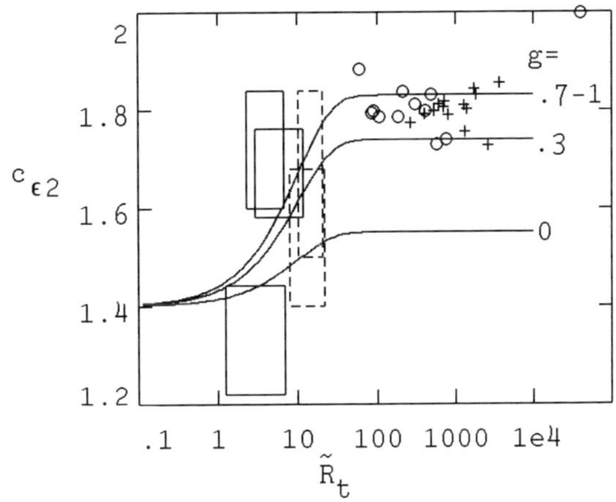

Fig. 8.3.4. $c_{\varepsilon 2}$ for the MA model at g =0, 0.3 and 0.7 to 1. Data for nearly isotropic turbulence, as in Fig. 6.2.1.

Boundary Conditions

The boundary conditions for q and ω are the same as those used for the MC model, Eqs. 8.2.5, and 8.2.6. From the DNS channel data, however, on Fig. 8.2.7, it is clear that to capture the details of the b_{ij} profiles near the wall, the boundary condition, $\partial b_{ij}/\partial y = 0$, will be inadequate. For the MA model, the limiting anisotropies in the algebraic limit for shear flow when $\tilde{R}_t$ goes to zero ($c_1 = 2$) are

$$b_{11} = 2/3, \quad b_{22} = -1/3, \quad b_{33} = -1/3, \; b_{12} = 0, \; b_{13} = 0, \text{ and } b_{23} = 0, \qquad (8.3.15)$$

where x_1 is in the limiting direction of the flow parallel to the wall, and x_2 is normal to the wall. For the fully developed channel flow data b_{13} and b_{23} are zero throughout the flow, and at the wall, $b_{22} = -1/3$ and $b_{12} = 0$, consistent with Eq. 8.3.15. However, the data on Fig. 8.2.7 shows that the b_{11} and b_{33} normal anisotropies become more equal in the laminar sublayer, so that the wall values are better approximated by

$$b_{33} = -0.11 \quad \text{and} \quad b_{11} = 1/3 + 0.11 \sim 0.44. \qquad (8.3.16)$$

If there is no flow parallel to the wall, it will not be possible to determine an x_1 direction, so that for the coordinate directions parallel to the wall, by symmetry,

$$b_{11} = b_{33} = 1/6 \; . \qquad (8.3.17)$$

These equations, 8.3.16 and 8.3.17, may be combined into a general wall boundary condition by taking b_{33} as a function of the dimensionless strain rate at the wall. The MA model (and test variations) uses

$$b_{33} = \frac{1}{6} - \left(0.2 + \frac{1}{6}\right)\left(1 - \exp\left(-0.0384\frac{1}{\omega}\left|\frac{\partial U_1}{\partial x_2}\right|_{\text{wall}}\right)\right), \tag{8.3.18}$$

together with the obvious choices

$$b_{22} = -1/3,\ b_{11} = -b_{22} - b_{33},\ b_{12} = 0,\ b_{13} = 0, \text{ and } b_{23} = 0, \tag{8.3.19}$$

as the wall boundary conditions for the b_{ij}'s. At the wall, b_{33} then varies from 1/6 when $\tilde{s}_{\text{wall}} = 0$, to –0.11 when $\tilde{s}_{\text{wall}} = 36.7$, consistent with the channel flow data (Table 8.2.2), to –0.2 when $\tilde{s}_{\text{wall}} = \infty$.

Results for the MA Model and Test Variations

The model completion procedure for the MC model was also used for the MA model. The initial test was fully developed pipe flow at $R_D = 10^6$, and all the models were anchored to $c_f = 0.00291$ using $d\omega/dy = 0$ at the wall. The constants for the ω boundary condition were then set using the results,

$$\text{MA: } c_{\omega w} = 0.089; \quad \text{MA.test.b: } c_{\omega w} = 0.086; \text{ and MA.test.c: } c_{\omega w} = 0.081. \tag{8.3.20}$$

This gave values consistent with Fig. 6.4.2.

The velocity profiles from the MA model and the test variations are nearly indistinguishable and, as required, are in reasonable agreement with the data, as shown in Fig. 8.3.5.

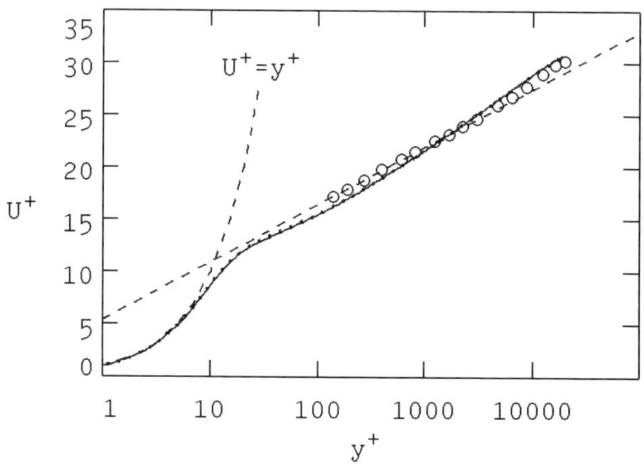

Fig. 8.3.5. Mean velocity for fully developed pipe flow, R_D=10^6.
- - - - U^+=5.45+ln(y^+)/0.418;
—— MA model;
······ models MA.test.b, and MA.test.c;
o, data of Zagarola (1996)

However, as shown in Fig. 8.3.6, the MA, MA.test.b and MA.test.c models give quite different results for the skin friction coefficient at low pipe and channel Reynolds numbers. A selection of results are also tabulated in Table 8.3.2. The MA model gives similar results to the MC model for pipe flow, and for channel flow down to R_D of about 5000, below which the channel flow results give a more gradual transition towards laminar flow. The MA.test.b model, which reduced c_1 using a different g dependence (Fig. 8.3.3b), shows a sharp reduction of c_f at about $R_D = 10^4$, and below this a turbulent flow mode with significantly lower skin friction coefficients. The MA.test.c model, which reduced c_1 using a different $\tilde{R}_t$ dependence (Fig. 8.3.3a), shows a decline in c_f as R_D is lowered followed by a abrupt transition to laminar flow. It was this inability to get $c_1 = c_1(g,\tilde{R}_t)$ to give reasonable skin friction coefficients at low R_D, which led to the expansion of the functionality of c_1 used in the MA model.

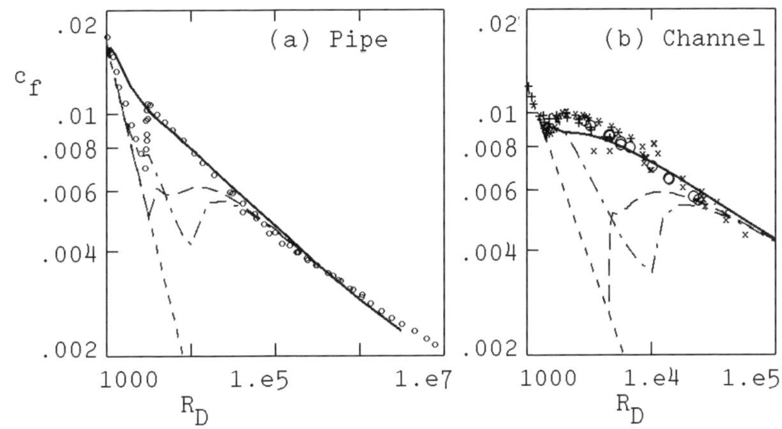

Fig. 8.3.6. Skin friction coefficients for fully developed pipe and channel flows.
- - - laminar flow;
——— MA model;
– – – - MA.test.b;
— — — MA.test.c.
o,+,x,*, data as for Fig. 8.2.3.

Table 8.3.2. Wall and near-wall results for fully developed channel and pipe flows.

Model	U^+ at $y^+ = 25$	c_f	$(q^+)_{max}$	$\left(\frac{dq^+}{dy^+}\right)_{wall}$	$\tilde{s}_{wall} = \frac{1}{\omega^+_{wall}}$
Channel, $R_D = 5600$:					
MA - final	13.7	0.00803	1.990	0.259	43.4
MA.test.b	18.3	0.00443	2.105	0.189	61.5
MA.test.c	17.9	0.00514	2.051	0.179	69.1
Data:H1992	13.4	0.00826	2.030	0.290	36.6 (at $y^+ = 2.5$)
Data:MKM1999	13.2	0.00814	2.026	0.292	36.7 (at $y^+ = 2.5$)
Pipe, $R_D = 10^6$					
MA - final	12.5	0.00291	2.095	0.313	35.9
MA.test.b	12.6	0.00291	2.094	0.311	37.5
MA.test.c	12.6	0.00291	2.102	0.302	40.6

H1992, Horiuti, data base by Kasagi et al. (KHMMN1990); MKM1999, Moser et al.

Results of the MA Model for Channel Flow

Comparisons of the MA model with fully developed channel flow DNS are given in Figs. 8.3.7 to 8.3.10, similar to Figs. 8.2.4 and 8.2.6 to 8.2.8 for the MC model.

Figure 8.3.7 shows that the turbulence velocity scale is well modeled for all three channel flows. The peak values of q^+ near the wall are much nearer the data values than were obtained with the MC model (Fig. 8.2.4). Modeled profiles of specific dissipation rate, Fig. 8.3.8, are in good agreement at the higher R_D flows from $y^+ = 15$ to $y/\delta = 0.5$. Like the MC model, the centerline values are about half what they should be. Unlike the data, there is little distinction below $y^+ = 100$ between ω levels at high and low R_D.

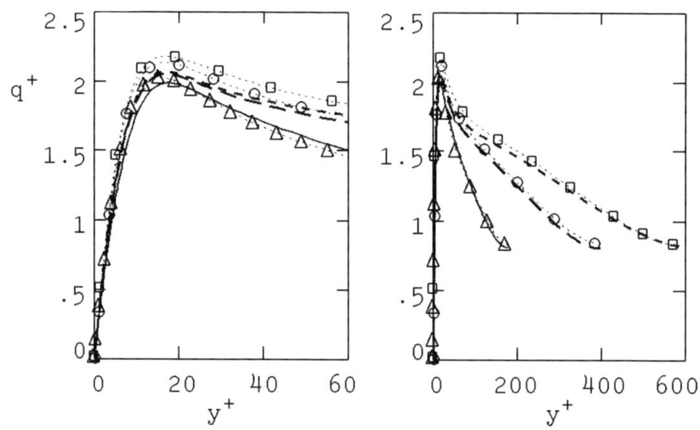

Fig. 8.3.7. Turbulence velocity scale for fully developed channel flow. Comparison of MA model with DNS data of Moser et al. (MKM1999).

R_D	model	data
5,600	———	△···
13,800	— — —	○···
21,900	- - - -	□···

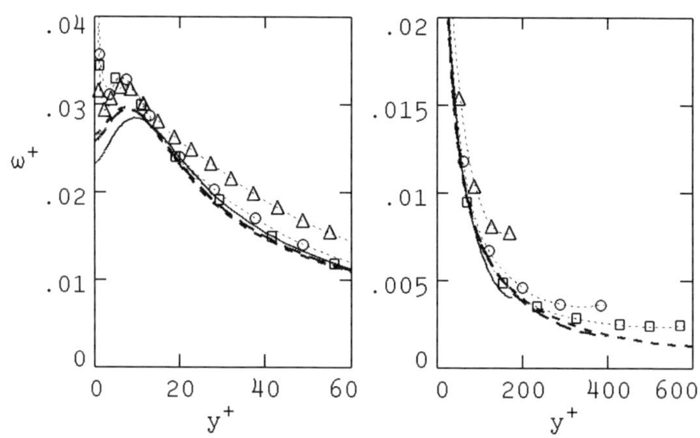

Fig. 8.3.8. Specific dissipation rate, ω, in fully developed channel flow. Comparison of MA model with DNS data of Moser et al. (MKM1999).

R_D	model	data
5,600	———	△···
13,800	— — —	○···
21,900	- - - -	□···

With the more complex pressure-strain model, the near-wall profiles of b_{ij}, Fig. 8.3.9, are in good agreement with the data. As the wall is approached, the anisotropies head towards the algebraic two-dimensional limit, Eq. 8.3.15, but b_{11} and b_{33} become more equal near the wall due to the boundary conditions (Eqs. 8.3.18 and 8.3.19), and laminar and turbulent diffusion.

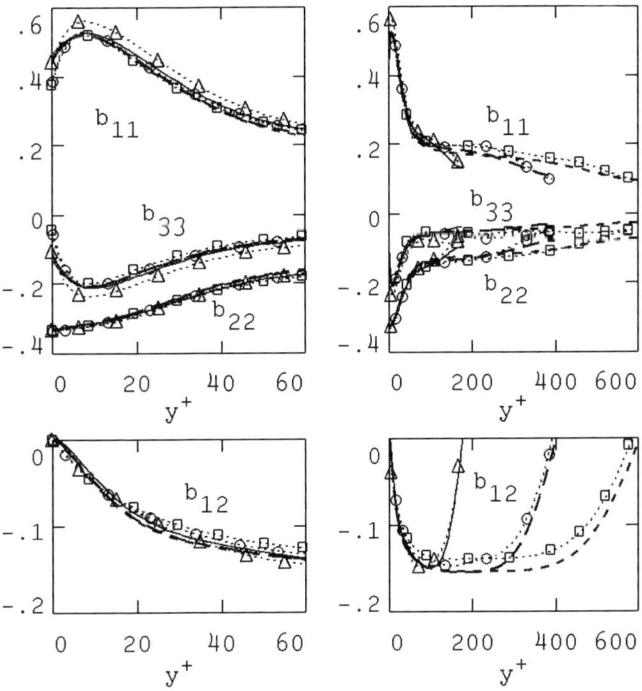

Fig. 8.3.9. Turbulence anisotropies for fully developed channel flow. Comparison of MA model with DNS data of Moser et al. (MKM1999).

R_D	model	data
5,600	———	△···
13,800	— — —	○···
21,900	- - - -	□···

The mean velocity profiles from the MA model, Fig. 8.3.10, are in good agreement with the data, as expected, since both q and b_{12} are well modeled.

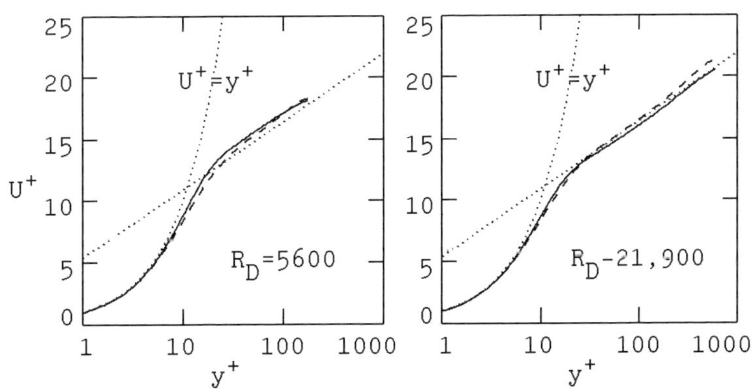

Fig. 8.3.10. Velocity profiles for fully developed channel flow. ——— MA model; - - - - DNS data of Moser et al. (MKM1999); ····· $U^+=5.45+\ln(y^+)/0.418$.

8.4 Developing Homogeneous Shear Flow

Both the MC and MA models are based on shear flow where the turbulence anisotropies are nearly fully developed. In this section the new models are tested on homogeneous shear flow with turbulence anisotropy developing from an initially isotropic state.

In particular we consider two DNS test cases, Case RR128 from Rogers et al. (RMR1986) which had a dimensionless strain rate, s, which varied from 1.1 to 5.8, and a case from Lee et al. (LKM1990) with a high strain rate, $s = 14$ to 18. The uncertainties in modeling of the turbulence dissipation rate are removed by using the DNS values for both dimensionless strain rate and turbulence Reynolds number. The calculations are then a pure test of the ability of the pressure-strain models to reproduce the development of the turbulence anisotropy as a function of the dimensionless time, St.

As shown in Fig. 8.4.1, the MC and MA models give similar results for case RR128, with some overprediction of the anisotropy of the turbulence. The results are worse for case LKM. Both models overpredict the rate at which the turbulence initially departs from isotropic, with the MA model, which anchored c_3, c_4 (and c_5) to isotropic limits at $g = 1$, giving the better results.

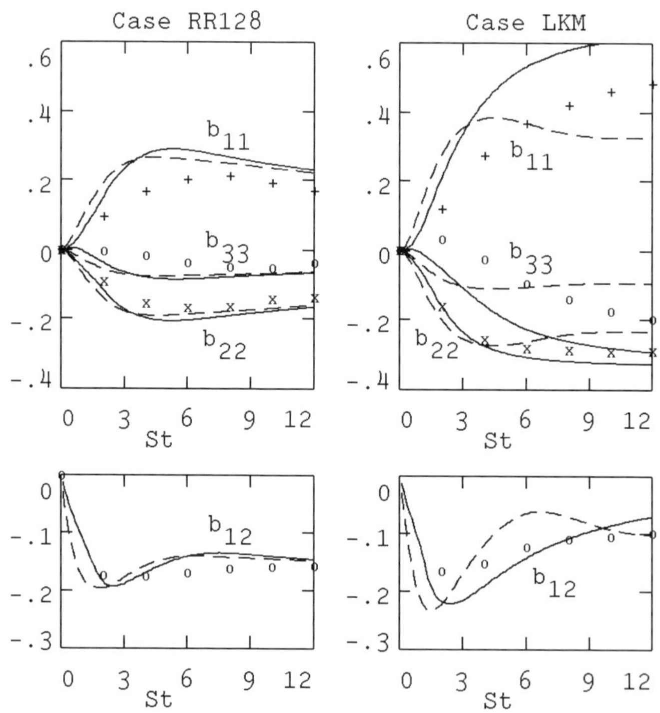

Fig. 8.4.1. Development of turbulence anisotropy due to homogeneous shear.
+, o, x, DNS data:
Case RR128, Rogers et al. (RMR1986), s= 1.1 to 5.8, R_t = 42 to 547;
Case LKM, Lee et al. (LKM1990), s= 14 to 18, R_t = 65 to 600.
– – – MC model;
——— MA model.

Lee et al. noted that their DNS results are similar to the results for rapid distortion theory (RDT), as shown in Fig. 8.4.2. As an exercise, coefficients for the pressure-strain model were determined which give good agreement with the early time development of the Lee et al. results,

$$c_1 = 2 + 5g, \quad c_3 = 0.8g, \quad c_4 = \frac{24}{7}g \ . \tag{8.4.1}$$

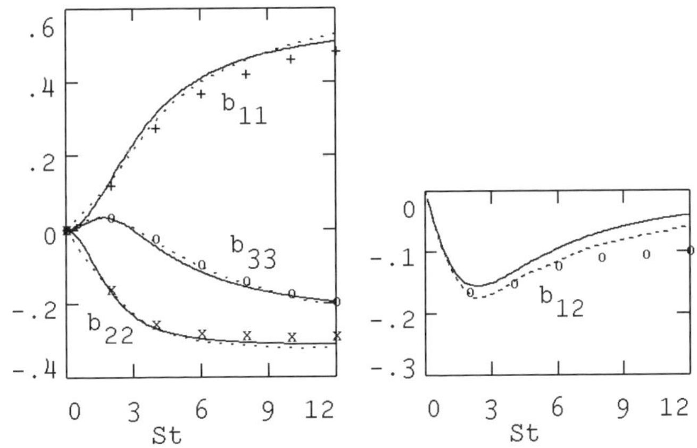

Fig. 8.4.2. Development of turbulence anisotropy due to homogeneous shear for Case LKM (Fig. 8.4.1). +, o, x, DNS data; ···· Rapid Distortion Theory results from Lee et al. (LKM1990) ; —— model, Eq. 8.4.1.

While this model, Eq. 8.4.1, does well for rapidly distorting shear flow, it gives unacceptable results for fully developed shear flow, as shown in Fig. 8.4.3. In particular, the lowest b_{12} value is –0.086, far higher that the expected minimum between –0.17 and –0.15. This is due to the higher values of c_3 and c_4, 0.48 and 2.057, respectively at $g = 0.6$, compared to the MC and MA models which use $c_3 = 0.15$ and $c_4 = 1$ at $g = 0.6$. An extension of the MA model, a new model, MARV, which has additional functionality to compute rapidly distorting flows as well as fully developed flows, is given in Chapter 10. But first, in Chapter 9, we introduce rapid distortion theory and consider its implications for Reynolds stress modeling.

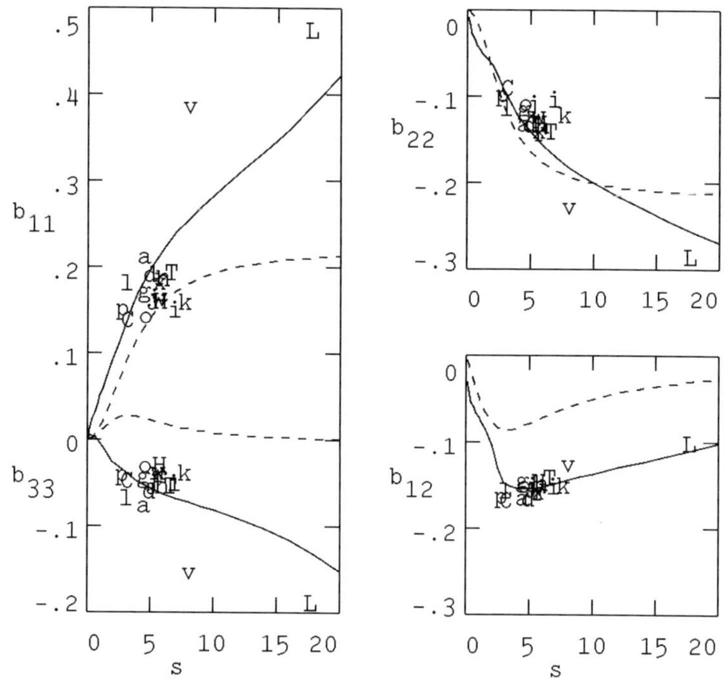

Fig. 8.4.3. Turbulence anisotropies for fully developed homogeneous shear flow.
Models: ——— MA model at high R_t;
– – – Eq. 8.4.1.
Data as for Fig 5.1.1.

Chapter 9

CONTRIBUTIONS OF RAPID DISTORTION THEORY TO REYNOLDS STRESS MODELING

Rapid distortion theory considers the spectral form of the Reynolds stresses. The analysis is valid for homogeneous turbulence where the gradients of the mean velocity are uniform and the flow is incompressible. It is useful to start as Townsend (1976) did and consider the Fourier transform of the equation for the fluctuating velocity. Townsend also presented the equations in a form which follows the development along a streamline of a single initial wave-number component and its movement in wave-number space.

The equation for the Fourier transform of the two-point double velocity correlation gives terms which may be integrated over wave-number space to give the terms in the Reynolds stress equations. In the limit of rapid distortion (high strain rate), the term which gives the cascade of energy, and the term which integrates to the slow pressure-strain, may be neglected, and a solvable set of equations which require no modeling, is obtained. Solutions to these equations, when integrated, give the development in time of both the Reynolds stresses and the rapid pressure-strain. These show the limitations of algebraic rapid pressure-strain modeling and suggest additional functionality to overcome at least part of the limitations.

9.1 Fluctuating Velocity Equation and Its Transform

The momentum equation for the fluctuating velocity was derived in Section 2.1 as Eq. 2.1.11. Noting that $\partial\overline{u_i u_k}/\partial x_k = 0$ for homogeneous turbulence, and that for incompressible flow, $\partial u_i u_k/\partial x_k = u_k \partial u_i/\partial x_k$, and with constant viscosity the last term in Eq. 2.1.11 is zero, the fluctuating momentum equation becomes

$$\frac{\partial u_i}{\partial t} + U_k \frac{\partial u_i}{\partial x_k} = -u_k \frac{\partial U_i}{\partial x_k} - 2\varepsilon_{ikm}\Omega_k u_m - \frac{1}{\rho}\frac{\partial p}{\partial x_i} - u_k \frac{\partial u_i}{\partial x_k} + \frac{\partial}{\partial x_k}\nu\frac{\partial u_i}{\partial x_k} \ . \tag{9.1.1}$$

Before taking the Fourier transform of this equation, the mean velocity needs to be expressed in terms of its constant gradients. With the origin of the coordinate system chosen to have $U_k(0,0,0) = 0$,

$$U_k = x_m \frac{\partial U_k}{\partial x_m} \ . \tag{9.1.2}$$

The Fourier transform of the fluctuating pressure term may be found by first forming a fluctuating pressure equation. Taking the divergence ($\partial/\partial x_i$) of Eq. 9.1.1 and using the incompressibility condition gives

$$\frac{1}{\rho}\frac{\partial^2 p}{\partial x_i^2} = -2\left(\frac{\partial U_i}{\partial x_m} + \varepsilon_{ikm}\Omega_k\right)\frac{\partial u_m}{\partial x_i} - \frac{\partial u_k}{\partial x_i}\frac{\partial u_i}{\partial x_k} \quad . \tag{9.1.3}$$

Defining a_i as the three-dimensional Fourier transform of u_i,

$$a_i = F(u_i) \quad , \tag{9.1.4}$$

then

$$u_i(x_1,x_2,x_3,t) = \iiint a_i(K_1,K_2,K_3,t)\exp(iK_n x_n)dK_1 dK_2 dK_3 \tag{9.1.5}$$

and

$$a_i(K_1,K_2,K_3,t) = \left(\frac{1}{2\pi}\right)^3 \iiint u_i(x_1,x_2,x_3,t)\exp(-iK_n x_n)dx_1 dx_2 dx_3 \quad , \tag{9.1.6}$$

(see, e.g., Batchelor, 1953, his Eqs. 2.4.2 and 2.4.3) where K_i, or $\mathbf{K}$, is the three-dimensional wave-number vector, and $i = \sqrt{-1}$, when i is not being used as a subscript. The integration in Eq. 9.1.5 is over the three-dimensional wave-number, or spectral, space, while the integration in Eq. 9.1.6 is over the physical space.

The Fourier transform of continuity for incompressible flow yields

$$K_i a_i = 0 \quad , \tag{9.1.7}$$

so that a_i may have no component in the direction of the wave vector K_i; the vector a_i may only have components normal to K_i.

The Fourier transform of the fluctuating momentum equation gives a differential equation for a_i (see Appendix EQ.9.1.8),

$$\begin{aligned}
&\underbrace{\frac{\partial a_i}{\partial t} - K_m \frac{\partial U_m}{\partial x_k}\frac{\partial a_i}{\partial K_k}}_{\text{convection}} = \underbrace{-a_m \frac{\partial U_i}{\partial x_m}}_{\text{production}} \underbrace{-\nu K^2 a_i}_{\text{dissipation}} \\
&\quad \underbrace{-iK_m \int a_m(\mathbf{K}-\mathbf{K}')a_i(\mathbf{K}')d\mathbf{K}'}_{\text{turbulent diffusion}} \underbrace{-2\varepsilon_{ikm}\Omega_k a_m}_{\text{Coriolis}} \\
&\quad \underbrace{+2\frac{K_i K_n}{K^2} a_m\left(\frac{\partial U_n}{\partial x_m} + \varepsilon_{nkm}\Omega_k\right)}_{\text{rapid pressure}-\text{strain}} \underbrace{+i\frac{K_i K_n K_m}{K^2}\int a_m(\mathbf{K}-\mathbf{K}')a_n(\mathbf{K}')d\mathbf{K}'}_{\text{slow pressure}-\text{strain}} \quad .
\end{aligned} \tag{9.1.8}$$

Here, the terms are given the labels of the corresponding terms in the Reynolds stress equations and the shorthand

$$\int d\mathbf{K} = \iiint dK_1 dK_2 dK_3 \tag{9.1.9}$$

is used to represent volume integrals. Note that the turbulent diffusion term and the slow pressure-strain term are integrals which give interactions between components in wave-number space.

Townsend (1976) presented the equation for a_i in terms of the substantial derivative, da_i/dt, so that when the turbulent diffusion and slow pressure-strain are neglected, the development of a single wave-number component may be followed independent of the other wave-number components. Using the chain rule,

$$\frac{da_i}{dt} = \frac{\partial a_i}{\partial t} + \frac{\partial a_i}{\partial K_k}\frac{dK_k}{dt}. \tag{9.1.10}$$

Comparing with the left hand side of Eq. 9.1.8, it becomes da_i/dt when

$$\frac{dK_k}{dt} = -K_m \frac{\partial U_m}{\partial x_k}. \tag{9.1.11}$$

Therefore, when following a single initial wave-number component, Eq. 9.1.11 describes the movement of the component in wave-number space under the influence of the mean velocity gradients. As Townsend (1976, p. 47) pointed out, “since the divergence of the ‘velocity’, dK_k/dt, in wave-number space is

$$\frac{\partial}{\partial K_k}\left(\frac{dK_k}{dt}\right) = -\frac{\partial U_m}{\partial x_m} = 0 \quad , \tag{9.1.12}$$

the motion of the allowed wave numbers in wave-number space is solenoidal and the originally constant number-density is preserved." It follows that the elemental volume does not change with time,

$$dK_1 dK_2 dK_3 = dK_{1,0} dK_{2,0} dK_{3,0} \ , \tag{9.1.13}$$

where $K_{n,0}$ is K_n at $t = 0$.

Figure 9.1.1 shows the movement of an initial circle of points in wave-number space under the influence of plane shear. With $\partial U_1 / \partial x_2 > 0$ as the only non-zero component of the mean velocity gradient, Eq. 9.1.11 yields movement in the K_2 direction, but not in the K_1 and K_3 directions. Figure 9.1.2 shows corresponding changes to the shape of elemental areas in the K_1-K_2 plane, while the magnitude of the elemental areas remains unchanged.

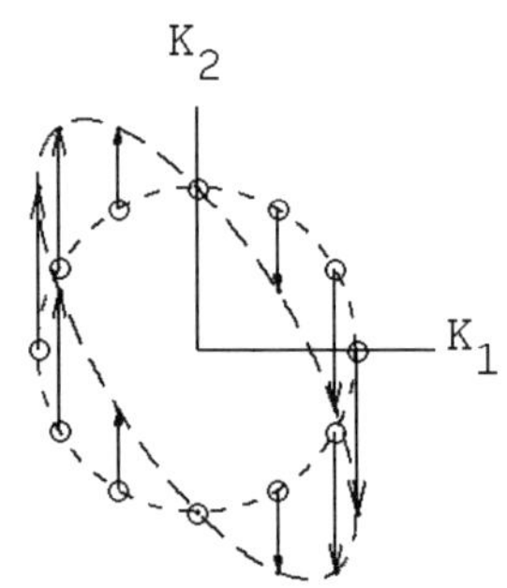

Fig. 9.1.1. Movement of points in wave-number space under the influence of shear, $\partial U_1 / \partial x_2 > 0$. o - - - o , initial location of points with K=1. ——, location at St=1.

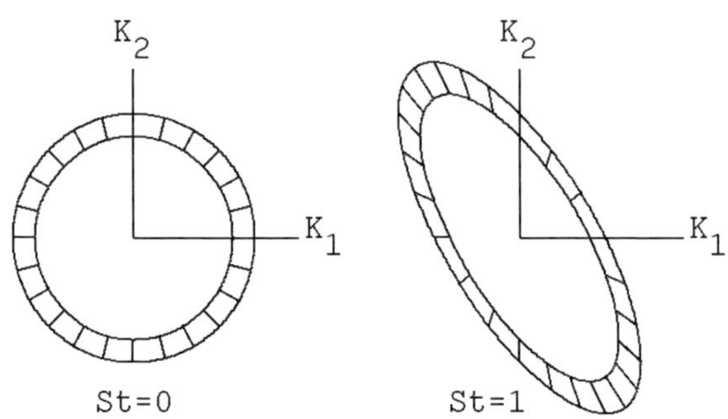

Fig. 9.1.2. Change in shape of elemental areas in wave-number space under the influence of shear, $\partial U_1 / \partial x_2 > 0$.

9.2 Two-Point Correlation

Rapid distortion theory approaches the Reynolds stresses by considering them as the limit at zero separation of two-point velocity correlations. Consider the fluctuating velocity at two independent points, $\mathbf{x}$ and $\mathbf{x}'$. For homogeneous turbulence the time-averaged correlation depends only on the difference in the locations, $\mathbf{r} = \mathbf{x}' - \mathbf{x}$, and not on $\mathbf{x}$ itself,

$$R_{ij}(\mathbf{r}) = \overline{u_i(\mathbf{x},t)u_j(\mathbf{x}',t)} . \tag{9.2.1}$$

Defining Φ_{ij} as the Fourier transform of R_{ij},

$$\Phi_{ij} = F(R_{ij}) , \tag{9.2.2}$$

then

$$R_{ij}(r_1,r_2,r_3) = \iiint \Phi_{ij}(K_1,K_2,K_3)\exp(iK_n r_n)dK_1 dK_2 dK_3 , \tag{9.2.3}$$

and the Reynolds stresses are

$$\overline{u_i u_j} = R_{ij}(0,0,0) = \iiint \Phi_{ij}(K_1,K_2,K_3)dK_1 dK_2 dK_3 . \tag{9.2.4}$$

Φ_{ij} is referred to as the spectrum tensor, as it gives the distribution in wave-number space of the contributions to the Reynolds stress tensor.

It can be shown (Appendix EQ.9.2.5) that Φ_{ij} can be obtained from the Fourier transform, a_i, of the fluctuating velocity. In particular,

$$\Phi_{ij} = \frac{(2\pi)^3}{V}\,\overline{a_i^* a_j} , \tag{9.2.5}$$

where $V/(2\pi)^3$ is the number density in wave-number space and

$$a_i^*(\mathbf{K},t) = a_i(-\mathbf{K},t) . \tag{9.2.6}$$

The symmetries of Φ_{ij} come from the turbulence being homogeneous, $R_{ij}(\mathbf{r}) = R_{ji}(\mathbf{r})$, and $R_{ij}(\mathbf{r}) = R_{ji}(-\mathbf{r})$, so that

$$\Phi_{ij}(\mathbf{K}) = \Phi_{ji}(\mathbf{K}) = \Phi_{ji}(-\mathbf{K}) . \tag{9.2.7}$$

The differential equation for Φ_{ij} may be written as (Appendix EQ.9.2.8)

$$\underbrace{\frac{\partial \Phi_{ij}}{\partial t} - K_m \frac{\partial U_m}{\partial x_k}\frac{\partial \Phi_{ij}}{\partial K_k}}_{\text{convection}} = \underbrace{-\left(\Phi_{mj}\frac{\partial U_i}{\partial x_m} + \Phi_{im}\frac{\partial U_j}{\partial x_m}\right)}_{\text{production}} \underbrace{- 2\nu K^2 \Phi_{ij}}_{\text{dissipation}}$$

$$+ \underbrace{\frac{(2\pi)^3}{V} iK_m \int \left(\overline{a_j(\mathbf{K}) a_m^*(\mathbf{K}+\mathbf{K}') a_i(\mathbf{K}')} - \overline{a_i^*(\mathbf{K}) a_m(\mathbf{K}-\mathbf{K}') a_j(\mathbf{K}')}\right) d\mathbf{K}'}_{\text{turbulent diffusion}}$$

$$\underbrace{-2\Omega_k\left(\varepsilon_{ikm}\Phi_{mj} + \varepsilon_{jkm}\Phi_{im}\right)}_{\text{Coriolis}} + \underbrace{2\left(\frac{\partial U_n}{\partial x_m} + \varepsilon_{nkm}\Omega_k\right)\left(\frac{K_i K_n}{K^2}\Phi_{mj} + \frac{K_j K_n}{K^2}\Phi_{im}\right)}_{\text{rapid pressure− strain}} \tag{9.2.8}$$

$$+ \underbrace{\frac{(2\pi)^3}{V}\frac{iK_n K_m}{K^2} \int \left(K_j \overline{a_i^*(\mathbf{K}) a_m(\mathbf{K}-\mathbf{K}') a_n(\mathbf{K}')} - K_i \overline{a_j(\mathbf{K}) a_m^*(\mathbf{K}+\mathbf{K}') a_n(\underline{\mathbf{K}'})}\right) d\mathbf{K}'}_{\text{slow pressure− strain}} \; .$$

Each term in this equation may be integrated over wave-number space to give the corresponding term in the differential equation for $\overline{u_i u_j}$ for homogeneous turbulence. For example, the dissipation is given by

$$\varepsilon_{ij} = 2\nu \iiint K^2 \Phi_{ij} dK_1 dK_2 dK_3 \; . \tag{9.2.9}$$

Contracting Eq. 9.2.4, the turbulence kinetic energy may be found from

$$2k = \iiint \Phi_{ii} dK_1 dK_2 dK_3 \; , \tag{9.2.10}$$

so that $\Phi_{ii}(\mathbf{K})$ represents (twice) the contribution to the turbulence kinetic energy from the wave number $\mathbf{K}$. From Eq. 9.2.5 and continuity (Eq. 9.1.7),

$$K_i \Phi_{mi} = 0, \text{ and } \quad K_i \Phi_{im} = 0, \tag{9.2.11}$$

and it follows that the rapid pressure-strain term makes no contribution to the equation for Φ_{ii} and thus to the equation for the turbulence kinetic energy. Similarly, by continuity, the slow pressure-strain term also makes no contribution to the equation for Φ_{ii} and correspondingly to the turbulence kinetic energy. Since both the rapid and slow pressure-strain terms make no contribution to the equation for Φ_{ii}, they may give only an interchange among the components of Φ_{ij} at the

same wave number. They do not give any net interchange in energy among different wave numbers.

The turbulent diffusion term, however, does not sum to zero for the equation for Φ_{ii}. For homogeneous turbulence it gives an interchange of energy in wave-number space and is responsible for the observed cascade of energy to higher wave numbers. (For inhomogeneous turbulence this term also gives an interchange of energy in physical space). When integrated over all wave numbers, this term gives no exchange of energy among the Reynolds stress components (see Appendix EQ.9.2.12),

$$\int K_m \int \left(\overline{a_j(\mathbf{K}) a_m^*(\mathbf{K}+\mathbf{K}') a_i(\mathbf{K}')} - \overline{a_i^*(\mathbf{K}) a_m(\mathbf{K}-\mathbf{K}') a_j(\mathbf{K}')} \right) d\mathbf{K}' d\mathbf{K} = 0 \ . \qquad (9.2.12)$$

9.3 Rapid Distortion Theory Equations

If the turbulence is undergoing rapid distortion, the terms in Eq. 9.2.8 which include the mean velocity gradients and rotation will dominate the development of Φ_{ij}, and the turbulent diffusion and slow pressure-strain terms containing triple products of a_i will be comparatively small and may be neglected. This leads to the solvable set of equations for Φ_{ij} used by Bertoglio (1982),

$$\underbrace{\frac{\partial \Phi_{ij}}{\partial t} - K_m \frac{\partial U_m}{\partial x_k} \frac{\partial \Phi_{ij}}{\partial K_k}}_{\text{convection}} = \underbrace{-\left(\Phi_{mj} \frac{\partial U_i}{\partial x_m} + \Phi_{im} \frac{\partial U_j}{\partial x_m} \right)}_{\text{production}} \underbrace{- 2\nu K^2 \Phi_{ij}}_{\text{dissipation}}$$
$$\underbrace{-2\Omega_k \left(\varepsilon_{ikm} \Phi_{mj} + \varepsilon_{jkm} \Phi_{im} \right)}_{\text{Coriolis}} + \underbrace{2 \left(\frac{\partial U_n}{\partial x_m} + \varepsilon_{nkm} \Omega_k \right) \left(\frac{K_i K_n}{K^2} \Phi_{mj} + \frac{K_j K_n}{K^2} \Phi_{im} \right)}_{\text{rapid pressure-strain}} . \qquad (9.3.1)$$

The resulting equations still include in full the rapid part of the pressure-strain so the results are useful for determining appropriate algebraic models for rapid pressure-strain for use in Reynolds stress modeling.

A more convenient form of Eq. 9.3.1 may be determined (again following Townsend, 1976) by putting the equation in terms of the substantial derivative,

$$\frac{d\Phi_{ij}}{dt} = \frac{\partial\Phi_{ij}}{\partial t} + \frac{\partial\Phi_{ij}}{\partial K_k}\frac{dK_k}{dt}. \tag{9.3.2}$$

Taking

$$\frac{dK_k}{dt} = -K_m\frac{\partial U_m}{\partial x_k}, \tag{9.3.3}$$

Eq. 9.3.1 becomes

$$\begin{aligned}\frac{d\Phi_{ij}}{dt} &= \underbrace{-\left(\Phi_{mj}\frac{\partial U_i}{\partial x_m} + \Phi_{im}\frac{\partial U_j}{\partial x_m}\right)}_{\text{production}} \underbrace{- 2\nu K^2\Phi_{ij}}_{\text{dissipation}} \\ &\underbrace{-2\Omega_k\left(\varepsilon_{ikm}\Phi_{mj} + \varepsilon_{jkm}\Phi_{im}\right)}_{\text{Coriolis}} + \underbrace{2\left(\frac{\partial U_n}{\partial x_m} + \varepsilon_{nkm}\Omega_k\right)\left(\frac{K_iK_n}{K^2}\Phi_{mj} + \frac{K_jK_n}{K^2}\Phi_{im}\right)}_{\text{rapid pressure–strain}}.\end{aligned} \tag{9.3.4}$$

Equations 9.3.3 and 9.3.4 may be used to follow the development in time of a single initial component of the turbulence spectrum. The wave numbers are time dependent, but Φ_{ij} is a function of time and the initial wave number which is independent of time, $\Phi_{ij}(K_{1,0},K_{2,0},K_{3,0},t)$. Since the elemental volume in wave-number space is unchanged (Eq. 9.1.13), the Reynolds stresses are obtained from

$$\overline{u_iu_j}(t) = \int_{-\infty}^{\infty}\int_{-\infty}^{\infty}\int_{-\infty}^{\infty}\Phi_{ij}(K_{1,0},K_{2,0},K_{3,0},t)dK_{1,0}dK_{2,0}dK_{3,0}\ . \tag{9.3.5}$$

In general the equation for Φ_{ij} is solved starting from isotropic turbulence. For isotropic turbulence, Batchelor (1953, his Eq. 3.4.12) relates Φ_{ij} to the turbulence energy spectrum function, E:

$$\Phi_{ij} = \frac{E(K)}{4\pi K^2}\left(\delta_{ij} - \frac{K_iK_j}{K^2}\right), \tag{9.3.6}$$

where K is the magnitude of the wave number.

When the dissipation term is neglected, the equation for Φ_{ij} is only dependent on the angular position in wave space, K_i/K, not on the magnitude of the wave number, K. This reduces the solution of Eq. 9.3.4 to a 2-d set of 1-d problems rather than a 3-d set. For many cases the set remains 2-d even when the dissipation term is included. This can be shown by putting Eq. 9.3.4 in the form

$$\frac{d\Phi_{ij}}{dt} = -2\nu K^2 \Phi_{ij} + L_{ijmn}\Phi_{mn} , \tag{9.3.7}$$

and by assuming a solution of the form

$$\Phi_{ij} = \exp(f)\Phi_{ij}(\nu = 0) . \tag{9.3.8}$$

Differentiating Eq. 9.3.8 and substituting into Eq. 9.3.7,

$$\begin{aligned} \frac{d\Phi_{ij}}{dt} &= \exp(f)\left(\Phi_{ij}(\nu = 0)\frac{df}{dt} + \frac{d\Phi_{ij}(\nu = 0)}{dt}\right) \\ &= \exp(f)\left(-2\nu K^2 \Phi_{ij}(\nu = 0) + L_{ijmn}\Phi_{mn}(\nu = 0)\right) \quad , \end{aligned} \tag{9.3.9}$$

which reduces to

$$\frac{df}{dt} = -2\nu K^2 = -2\nu\left(K_1^2 + K_2^2 + K_3^2\right) . \tag{9.3.10}$$

Both Townsend (1976) and Salhi and Cambon (1997) have given solutions of the form of Eq. 9.3.8 for homogeneous shear. In that case,

$$\frac{\partial U_m}{\partial x_k} = S\delta_{m1}\delta_{k2}, \tag{9.3.11}$$

so that from Eq. 9.3.3,

$$K_1 = K_{1,0}, \quad K_2 = K_{2,0} - K_{1,0}St, \quad K_3 = K_{3,0} \tag{9.3.12}$$

and

$$f = -2\nu\left(\left(K_{1,0}^2 + K_{2,0}^2 + K_{3,0}^2\right)t - K_{1,0}K_{2,0}St^2 + \frac{1}{3}K_{1,0}^2 S^2 t^3\right) . \tag{9.3.13}$$

Exercise 9.3. For isotropic turbulence, Eq. 9.3.6: (a) Show that Φ_{ij} satisfies continuity, $K_i\Phi_{ij} = 0$. (b) What are the components of the Φ_{ij} tensor at $K_1/K = 1$, $K_2 = K_3 = 0$ in wave-number space?

(c) Describe in general, for any turbulence, the limitations continuity imposes on the shape of the tensor Φ_{ij} at each point in wave-number space.

9.4 Rapid Pressure-Strain

Equation 9.3.4 may be integrated over wave-number space to give a corresponding Reynolds stress equation,

$$\frac{d\overline{u_i u_j}}{dt} = -\left(\overline{u_m u_j}\frac{\partial U_i}{\partial x_m} + \overline{u_i u_m}\frac{\partial U_j}{\partial x_m}\right) - 2\nu\int K^2 \Phi_{ij} d\mathbf{K}$$

production dissipation

$$-2\Omega_k\left(\varepsilon_{ikm}\overline{u_m u_j} + \varepsilon_{jkm}\overline{u_i u_m}\right) + 2\left(\frac{\partial U_n}{\partial x_m} + \varepsilon_{nkm}\Omega_k\right)\int\left(\frac{K_i K_n}{K^2}\Phi_{mj} + \frac{K_j K_n}{K^2}\Phi_{im}\right)d\mathbf{K}\,. \tag{9.4.1}$$

Coriolis rapid pressure—strain

The dissipation and the rapid pressure-strain are not simple functions of the Reynolds stresses, but weighted integrals of Φ_{ij} in wave-number space. The rapid pressure-strain is given by

$$\Pi_{ij}^{\mathrm{R}} = 2\left(\frac{\partial U_n}{\partial x_m} + \varepsilon_{nkm}\Omega_k\right)\int\left(\frac{K_i K_n}{K^2}\Phi_{mj} + \frac{K_j K_n}{K^2}\Phi_{im}\right)d\mathbf{K}\,. \tag{9.4.2}$$

As in the more general analysis in Section 3.1, this can be expressed in terms of the gradients of the absolute velocity, Eq. 3.1.2, to provide the basis for algebraic rapid pressure-strain modeling as described in Section 3.2. In particular,

$$\Pi_{ij}^{\mathrm{R}} = 2kM_{ijnm}\frac{\partial V_n}{\partial x_m} = 2kM_{ijnm}(S_{nm} + W_{nm})\,, \tag{9.4.3}$$

where

$$M_{ijnm} = \frac{1}{k}\iiint\left(\frac{K_i K_n}{K^2}\Phi_{mj} + \frac{K_j K_n}{K^2}\Phi_{im}\right)dK_1 dK_2 dK_3\,. \tag{9.4.4}$$

M_{ijnm} is symmetric in subscripts i and j, and from continuity, $M_{iinm} = 0$.

In Section 3.2, a general algebraic model was developed based on M_{ijnm} being a function of the anisotropy tensor b_{ij}. For example, with only the first four terms of Eq. 3.2.17,

$$\frac{\Pi_{ij}^{\mathrm{R}}}{k} = c_3 S_{ij} + c_4\left(b_{ik}S_{jk} + b_{jk}S_{ik} - \frac{2}{3}\delta_{ij}b_{mn}S_{mn}\right) + c_5\left(b_{ik}W_{jk} + b_{jk}W_{ik}\right) + c_6 b_{ij}b_{mn}S_{mn}\,. \tag{9.4.5}$$

This corresponds to

$$
\begin{aligned}
M_{ijnm} = & \frac{1}{4}c_3\left(\delta_{in}\delta_{jm} + \delta_{im}\delta_{jn} - \frac{2}{3}\delta_{nm}\delta_{ij}\right) \\
& + \frac{1}{4}c_4\left(\delta_{in}b_{jm} + \delta_{jn}b_{im} - \frac{2}{3}b_{nm}\delta_{ij} + \delta_{im}b_{jn} + \delta_{jm}b_{in} - \frac{2}{3}b_{mn}\delta_{ij}\right) \\
& + \frac{1}{4}c_5\left(\delta_{in}b_{jm} + \delta_{jn}b_{im} - \frac{2}{3}b_{nm}\delta_{ij} - \delta_{im}b_{jn} - \delta_{jm}b_{in} + \frac{2}{3}b_{mn}\delta_{ij}\right) \\
& + \frac{1}{2}c_6 b_{ij}b_{mn} \quad .
\end{aligned}
\tag{9.4.6}
$$

Taylor Series Analysis for Isotropic Limits

Rapid pressure-strain for isotropic turbulence may be evaluated using the isotropic spectral distribution for Φ_{ij}, Eq. 9.3.6, with Eqs. 9.4.4 and 9.4.3. This gives (Appendix EQ. 9.4.7)

$$
\Pi_{ij}^{\mathrm{R}} = 2k\frac{1}{5}\left(\delta_{in}\delta_{mj} + \delta_{im}\delta_{nj} - \frac{2}{3}\delta_{ij}\delta_{mn}\right)(S_{nm} + W_{nm}) \ , \tag{9.4.7}
$$

which reduces to the well known result from Crow (1968),

$$
\Pi_{ij}^{\mathrm{R}} = \frac{4}{5}kS_{ij} \ , \tag{9.4.8}
$$

or comparing with Eq. 9.4.5,

$$
c_3 = \frac{4}{5} \ , \tag{9.4.9}
$$

in the limit of isotropic turbulence.

A Taylor series analysis can be used to obtain the next higher order terms of the rapid pressure-strain in the limit of isotropic turbulence. Writing

$$
\Pi_{ij}^{\mathrm{R}} = \left(\Pi_{ij}^{\mathrm{R}}\right)_0 + \left(\frac{d}{dt}\Pi_{ij}^{\mathrm{R}}\right)_0 t + O(t^2) \quad , \tag{9.4.10}
$$

then

$$
\Pi_{ij}^{\mathrm{R}} = 2(S_{nm} + W_{nm})\left(\left(kM_{ijnm}\right)_0 + \left(\frac{dkM_{ijnm}}{dt}\right)_0 t\right) + O(t^2) \ . \tag{9.4.11}
$$

Differentiating Eq. 9.4.4 to obtain $d(kM_{ijnm})/dt$, yields two time derivatives, $d\Phi_{ij}/dt$, and dK_m/dt. Using the rapid distortion theory equation, Eq. 9.3.4, for $d\Phi_{ij}/dt$, and Eq. 9.3.3 for dK_m/dt, together with Eq. 9.3.6 for the initial spectrum, yields (Appendix EQ. 9.4.12)

$$\frac{\Pi_{ij}^{\mathrm{R}}}{k} = \frac{2}{5}(S_{nm} + W_{nm})\left(\delta_{in}\delta_{mj} + \delta_{im}\delta_{nj} - \frac{2}{3}\delta_{ij}\delta_{mn}\right)$$
$$- \frac{16}{35}(S_{nm} + W_{nm})\left(S_{im}\delta_{jn} + S_{in}\delta_{jm} + S_{jn}\delta_{im} + S_{jm}\delta_{in} - \frac{4}{3}S_{nm}\delta_{ij}\right)t + O(t^2), \tag{9.4.12}$$

or

$$\frac{\Pi_{ij}^{\mathrm{R}}}{k} = \frac{4}{5}S_{ij} - \frac{32}{35}S_{nm}\left(tS_{jn}\delta_{im} + tS_{in}\delta_{jm} - \frac{2}{3}tS_{nm}\delta_{ij}\right) + O(t^2). \tag{9.4.13}$$

For small departures, starting from isotropic turbulence, Eq. 9.4.1 yields (Appendix EQ.9.4.14)

$$2k\frac{db_{ij}}{dt} = -\frac{4}{3}kS_{ij} + \Pi_{ij}^{\mathrm{R}} + O(t), \tag{9.4.14}$$

so that substituting in Eq. 9.4.13,

$$b_{ij} = \frac{1}{2}\left(-\frac{4}{3} + \frac{4}{5}\right)S_{ij}t + O(t^2) \quad = -\frac{4}{15}S_{ij}t + O(t^2) \quad . \tag{9.4.15}$$

Using this to eliminate tS_{ij} in Eq. 9.4.13, yields

$$\frac{\Pi_{ij}^{\mathrm{R}}}{k} = \frac{4}{5}S_{ij} + \frac{24}{7}\left(b_{jn}S_{in} + b_{in}S_{jn} - \frac{2}{3}S_{nm}b_{nm}\delta_{ij}\right) + O(t^2). \tag{9.4.16}$$

Comparing with Eq. 9.4.5, it follows that for isotropic turbulence,

$$c_4 = \frac{24}{7} \qquad \text{and} \qquad c_5 = 0 \ . \tag{9.4.17}$$

These limits for c_3, c_4 and c_5 were found for non-rotating shear flow in Section 3.3. This analysis shows the results to be general and independent of the strain rate and vorticity. In the next section, flows with significant departures from the isotropic limit are considered.

9.5 Numerical Evaluation of Rapid Pressure-Strain

Expanding the equation for pressure-strain, Eq. 9.4.3, using the symmetries of the strain-rate and vorticity tensors, the coefficient of each component of the strain-rate and vorticity may be found,

$$\begin{aligned}\frac{\Pi_{ij}^{R}}{k} &= \left(2M_{ij11}\right)S_{11} + \left(2M_{ij22}\right)S_{22} + \left(2M_{ij33}\right)S_{33} \\ &+\left(2(M_{ij12} + M_{ij21})\right)S_{12} + \left(2(M_{ij13} + M_{ij31})\right)S_{13} + \left(2(M_{ij23} + M_{ij32})\right)S_{23} \\ &+\left(2(M_{ij12} - M_{ij21})\right)W_{12} + \left(2(M_{ij13} - M_{ij31})\right)W_{13} + \left(2(M_{ij23} - M_{ij32})\right)W_{23}\,.\end{aligned} \tag{9.5.1}$$

These coefficients have been evaluated numerically from rapid distortion theory calculations. The results are presented in this section and compared with possible pressure-strain models in which the coefficients are functions of the turbulence anisotropy.

In the limit of rapid strain rate, the turbulence production is much greater than the dissipation, and the dissipation term in Eq. 9.3.4 may be neglected. With this assumption, Eqs. 9.3.3 and 9.3.4 were solved to evaluate Φ_{ij} as a function of time, over a hemi-spherical grid of initial wave numbers. The equations were solved separately at each point on a 128x64 grid using a fifth-order Runge-Kutta solver. In terms of spherical coordinates, Fig. 9.5.1, the initial wave numbers were defined over a grid with uniform spacing in θ and $\cos\psi$, so that each point represented the same area on the surface. In Cartesian coordinates, the initial locations of the grid points are

$$K_{1,0} = K_0 \sin\psi\cos\theta,\; K_{2,0} = K_0 \sin\psi\sin\theta,\quad K_{3,0} = K_0 \cos\psi. \tag{9.5.2}$$

With an initial uniform thickness δK of the hemisphere, volume integrals are then simple sums of the values at the grid points – volumes associated with the grid points are uniform in size and do not change with time, Eq. 9.1.13. All the calculations were started with isotropic turbulence, Eq. 9.3.6, at $t = 0$.

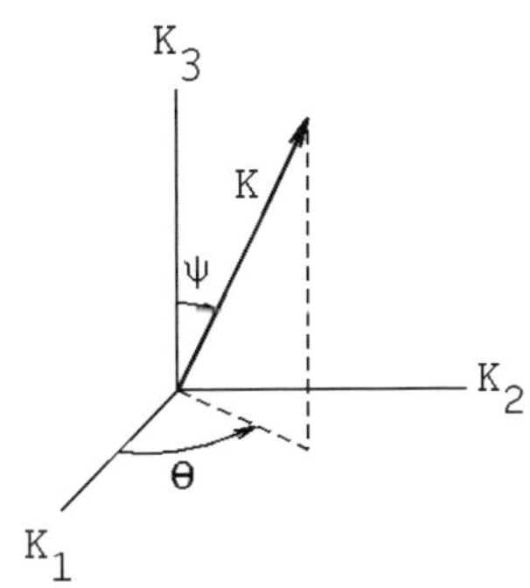

Fig. 9.5.1. Spherical coordinates in wave-number space.

Table 9.5.1 shows the coefficients of the strain rate and vorticity terms in Eq. 9.5.1 at $t = 0$. The corresponding algebraic pressure-strain model values evaluating M_{ijnm} from Eq. 9.4.6 using a unit value of c_3 are given in Table 9.5.2. As expected, the pressure-strain is precisely modeled with $c_3 = 0.8$, i.e., all the values in Table 9.5.1 are 0.8 times the values in Table 9.5.2.

Table 9.5.1. Coefficients of S_{nm} and W_{nm} in Π_{ij}^{R}/k for isotropic turbulence.

i	j	S_{11}	S_{22}	S_{33}	S_{12}	S_{13}	S_{23}	W_{12}	W_{13}	W_{23}
1	1	0.533	−0.267	−0.267	0	0	0	0	0	0
2	2	−0.267	0.533	−0.267	0	0	0	0	0	0
3	3	−0.267	−0.267	0.533	0	0	0	0	0	0
1	2	0	0	0	0.8	0	0	0	0	0
1	3	0	0	0	0	0.8	0	0	0	0
2	3	0	0	0	0	0	0.8	0	0	0

Table 9.5.2. Modeled coefficients of S_{nm} and W_{nm} in Π_{ij}^{R}/k for isotropic turbulence and a unit value for c_3.

i	j	S_{11}	S_{22}	S_{33}	S_{12}	S_{13}	S_{23}	W_{12}	W_{13}	W_{23}
1	1	0.667	−0.333	−0.333	0	0	0	0	0	0
2	2	−0.333	0.667	−0.333	0	0	0	0	0	0
3	3	−0.333	−0.333	0.667	0	0	0	0	0	0
1	2	0	0	0	1.0	0	0	0	0	0
1	3	0	0	0	0	1.0	0	0	0	0
2	3	0	0	0	0	0	1.0	0	0	0

Table 9.5.3 considers the case of plane shear with rotation but no vorticity,

$$\frac{\partial U_n}{\partial x_m} = S\delta_{n1}\delta_{m2}\,, \qquad \Omega_k = \frac{S}{2}\delta_{k3}\,. \tag{9.5.3}$$

The dimensionless time considered is sufficiently long, $St = 2$, that the flow is far from isotropic, $b_{11} = -0.16$, $b_{22} = 0.09$, $b_{12} = -0.13$. The corresponding model values for unit values of c_4 and c_5 are given in Table 9.5.4. The model values for a unit value of c_3 are independent of b_{ij}, so they are unchanged from the values given in Table 9.5.2. For this flow Table 9.5.3 shows there are no significant vorticity terms in the pressure-strain. A least-squares fit, minimizing the root-mean-square (rms) error in the modeled coefficients of S_{nm} and W_{nm} (Appendix EQ.9.5.1), yields $c_3 = 0.8$, $c_4 = 2.4$ and $c_5 = 0$. Using $c_3 = 0.8$ alone ($c_4 = 0$) reduced the rms error in the coefficients by a factor of 2.5 when compared with $c_3 = 0$. With both $c_3 = 0.8$ and $c_4 = 2.4$ the error is reduced by a factor of 8 compared with $c_3 = c_4 = 0$.

Table 9.5.3. Coefficients of S_{nm} and W_{nm} in Π_{ij}^{R}/k for plane rotating shear with no vorticity at $St = 2$.

i	j	S_{11}	S_{22}	S_{33}	S_{12}	S_{13}	S_{23}	W_{12}	W_{13}	W_{23}
1	1	0.186	–0.108	–0.078	–0.142	0	0	0.000	0	0
2	2	–0.108	0.743	–0.635	–0.415	0	0	0.000	0	0
3	3	–0.078	–0.635	0.712	0.558	0	0	0.000	0	0
1	2	–0.071	–0.208	0.279	0.572	0	0	0.000	0	0
1	3	0	0	0	0	0.436	–0.457	0	0.000	0.000
2	3	0	0	0	0	–0.457	1.350	0	0.000	0.001

Table 9.5.4. Modeled coefficients of S_{nm} and W_{nm} in Π_{ij}^{R}/k for plane rotating shear with no vorticity at $St = 2$, and unit values for c_4 and c_5.

i	j	S_{11}	S_{22}	S_{33}	S_{12}	S_{13}	S_{23}	W_{12}	W_{13}	W_{23}
1	1	–0.215	–0.062	–0.046	–0.085	0	0	–0.254	0	0
2	2	0.107	0.124	–0.046	–0.085	0	0	0.254	0	0
3	3	0.107	–0.062	0.091	0.169	0	0	0	0	0
1	2	–0.127	–0.127	0	–0.068	0	0	0.254	0	0
1	3	0	0	0	0	–0.093	–0.127	0	0.229	0.127
2	3	0	0	0	0	–0.127	0.161	0	0.127	–0.024

Tables 9.5.5 and 9.5.6 show the corresponding results for plane shear without rotation,

$$\frac{\partial U_n}{\partial x_m} = S\delta_{n1}\delta_{m2}\ , \qquad \Omega_k = 0\ , \tag{9.5.4}$$

at a dimensionless time of $St = 2$. The turbulence anisotropy, $b_{11} = 0.13$, $b_{22} = -0.17$ and $b_{12} = -0.16$ has a shear component similar to that in the log-law region of a turbulent boundary layer. Table 9.5.5 shows that vorticity now significantly influences the pressure-strain. The least-squares-fit values for the model coefficients are $c_3 = 0.8$, $c_4 = 1.4$ and $c_5 = 0.9$. Using c_3 alone reduces the error in the coefficients of S_{nm} by a factor of 2.9. With c_4 as well the reduction is a factor of 5.2. The c_5 value of 0.9 gives only a reduction factor of 1.4 in the error of the coefficients of W_{nm} when compared with $c_5 = 0$. This suggests that while vorticity clearly influences the pressure-strain, it does not do so in a manner consistent with standard algebraic models. While there is good correspondence in the zero coefficients, there is little correspondence between the non-zero coefficients of W_{nm} in Table 9.5.5 and those in Table 9.5.6. Note particularly that the coefficient of W_{12}

in Π^R_{33} is –0.291. This cannot be modeled by any algebraic model where M_{ijnm} is formed using only the anisotropy tensor, b_{ij}.

Table 9.5.5. Coefficients of S_{nm} and W_{nm} in Π^R_{ij}/k for plane shear at $St = 2$.

i	j	S_{11}	S_{22}	S_{33}	S_{12}	S_{13}	S_{23}	W_{12}	W_{13}	W_{23}
1	1	0.389	–0.038	–0.352	–0.405	0	0	0.068	0	0
2	2	–0.038	0.309	–0.271	–0.357	0	0	0.223	0	0
3	3	–0.352	–0.271	0.623	0.762	0	0	–0.291	0	0
1	2	–0.202	–0.178	0.381	0.838	0	0	–0.461	0	0
1	3	0	0	0	0	0.962	–0.428	0	–0.696	0.195
2	3	0	0	0	0	–0.428	0.879	0	–0.096	0.573

Table 9.5.6. Modeled coefficients of S_{nm} and W_{nm} in Π^R_{ij}/k for plane shear at $St = 2$, and unit values for c_4 and c_5.

i	j	S_{11}	S_{22}	S_{33}	S_{12}	S_{13}	S_{23}	W_{12}	W_{13}	W_{23}
1	1	0.178	0.114	–0.025	–0.110	0	0	–0.329	0	0
2	2	–0.089	–0.228	–0.025	–0.110	0	0	0.329	0	0
3	3	–0.089	0.114	0.050	0.219	0	0	0	0	0
1	2	–0.165	–0.165	0	–0.037	0	0	–0.304	0	0
1	3	0	0	0	0	0.171	–0.165	0	–0.096	0.165
2	3	0	0	0	0	–0.165	–0.134	0	0.165	0.208

The last example, in Tables 9.5.7 and 9.5.8, again considers plane shear but with the vorticity in a plane perpendicular to the strain,

$$\frac{\partial U_n}{\partial x_m} = S\delta_{n1}\delta_{m2}\,, \qquad \Omega_k = \frac{S}{2}\delta_{k2} + \frac{S}{2}\delta_{k3}\,. \tag{9.5.5}$$

At $St = 2$, all the components of the turbulence anisotropy have non-zero values, $b_{11} = -0.02$, $b_{22} = 0.05$, $b_{12} = -0.07$, $b_{13} = -0.11$, $b_{23} = -0.05$, although the magnitudes of the normal components are small. Correspondingly, there are no zero coefficients in Table 9.5.7 and few in Table 9.5.8. The least-squares fit gives $c_3 = 0.8$, $c_4 = 2.3$ and $c_5 = 0.4$. Again c_3 and c_4 are meaningful, with reduction factors of 3.3 for c_3 alone and 6.7 for them together. However, c_5 is meaningless with a reduction factor of only 1.03.

The terms in the full equation for Φ_{ij} (Eq. 9.2.8) which have been omitted in this analysis – the turbulent diffusion, the slow pressure-strain and the dissipation

– affect both the rapid pressure-strain and the turbulence anisotropy, b_{ij}. To estimate the effect, the calculations for the three examples were repeated including the dissipation term with a modest initial dimensionless strain rate, $(k/\varepsilon)\partial U_1/\partial x_2 = 5$. The results were similar except that the coefficients of W_{nm} for the zero vorticity case were now of the same order of magnitude as for the other cases. The c_5 term in the pressure-strain model still correlated poorly with the W_{nm} coefficients.

Table 9.5.7. Coefficients of S_{nm} and W_{nm} in Π^{R}_{ij}/k for S_{12} shear with W_{13} vorticity at $St = 2$.

i	j	S_{11}	S_{22}	S_{33}	S_{12}	S_{13}	S_{23}	W_{12}	W_{13}	W_{23}
1	1	0.295	–0.155	–0.140	–0.309	–0.098	0.252	0.289	–0.082	0.240
2	2	–0.155	0.613	–0.457	–0.154	0.302	–0.078	–0.338	–0.241	–0.275
3	3	–0.140	–0.457	0.597	0.463	–0.204	–0.173	0.049	0.323	0.035
1	2	–0.154	–0.077	0.231	0.815	–0.110	–0.163	–0.371	0.228	–0.348
1	3	–0.049	0.151	–0.102	–0.110	0.547	–0.346	–0.012	–0.180	0.045
2	3	0.126	–0.039	–0.087	–0.163	–0.346	1.133	0.107	0.094	–0.228

Table 9.5.8. Modeled coefficients of S_{nm} and W_{nm} in Π^{R}_{ij}/k for S_{12} shear with W_{13} vorticity at $St = 2$, and unit values for c_4 and c_5.

i	j	S_{11}	S_{22}	S_{33}	S_{12}	S_{13}	S_{23}	W_{12}	W_{13}	W_{23}
1	1	–0.027	–0.030	0.017	–0.047	–0.075	0.068	–0.141	–0.224	0
2	2	0.014	0.061	0.017	–0.047	0.149	–0.034	0.141	0	–0.102
3	3	0.014	–0.030	–0.033	0.094	–0.075	–0.034	0	0.224	0.102
1	2	–0.071	–0.071	0	0.025	–0.051	–0.112	0.066	–0.051	–0.112
1	3	–0.112	0	0.112	–0.051	–0.045	–0.071	–0.051	–0.005	0.071
2	3	0	–0.051	–0.051	–0.112	–0.071	0.020	0.112	0.071	–0.070

9.6 Algebraic Pressure-Strain Models Based on RDT Results

Section 9.5 shows that it is beyond the capabilities of the standard algebraic pressure-strain models to model all the coefficients of vorticity in the pressure-strain. In this section a more limited objective is sought. Assuming that strain rate and vorticity do not change with time, it is only necessary to model the net pressure-strain for each Reynolds stress component. For example if S_{12} is the only non-zero component of the strain rate, it is necessary to model only the coefficients of S_{12} in the pressure-strain. If $W_{12} = S_{12}$ are the only non-zero components, it is only necessary to model the sum of the S_{12} and W_{12} terms.

The standard vorticity term, which is linear in the turbulence anisotropy (c_5 term in Eq. 3.2.17), is zero for isotropic turbulence. Section 9.5 shows it correlates poorly with the vorticity coefficients for anisotropic turbulence. Therefore pressure-strain models will be sought which contain only strain terms.

For shear flow with rotation in the same plane as the strain, there are only three non-zero independent components of the turbulence anisotropy, b_{11}, b_{22} and b_{12}. The development of these can therefore be modeled with an appropriate three-term pressure-strain model. Here, we choose the c_3, c_4, and c_6 terms,

$$\Pi_{ij}^{R} = c_3 k S_{ij} + c_4 k(b_{ik}S_{jk} + b_{jk}S_{ik} - \frac{2}{3}\delta_{ij}b_{mn}S_{mn}) + c_6 b_{ij} k b_{mn} S_{mn} \ . \qquad (9.6.1)$$

The RDT results in the limit of high strain rate, described in Section 9.5, have been used to determine the c_3, c_4, and c_6 coefficients for two cases, shear with no vorticity (Eq. 9.5.3), and shear with no rotation (Eq. 9.5.4). The coefficients are functions of time but are shown in Fig. 9.6.1 as functions of the two-dimensionality parameter, g, which is suitable for use in algebraic pressure-strain models. The time range covered in both cases is $St = 0$ to 4. The two cases give the same results for c_3 and c_4 in the limit of isotropic turbulence, as expected, but quite different results away from this limit.

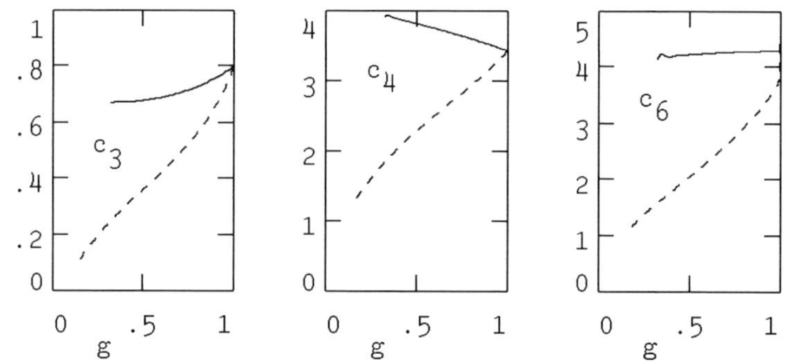

Fig. 9.6.1. Pressure-strain model coefficients, c_3, c_4 and c_6, determined from RDT calculations for shear flow with:
--- zero rotation;
—— zero vorticity.

Zero Vorticity Flows

For the zero vorticity case, the coefficients may be approximated by

Model Mrdt,W=0: $c_3 = 0.6 + 0.2g$,

$$c_4 = \frac{24}{7} + 0.8(1-g) + \left(0.1 - \frac{24}{7}\right)\left(\max(0, 1-\sqrt{4g})\right)^2 , \qquad (9.6.2)$$

$$c_6 = 4.2 - 6\left(\max(0, 1-\sqrt{4g})\right)^2 \ .$$

The last terms in the equations for c_4 and c_6 change the g dependence below $g = 0.25$ to satisfy two-dimensional turbulence limits, Eqs. 3.4.6-7, at $g = 0$. (The coefficients are plotted later in Fig. 9.6.4.)

Equation 9.6.2 captures the developing turbulence well for a large variety of zero vorticity cases, as shown in Fig. 9.6.2. The plots compare the model results with the corresponding RDT results. For these tests, both the model and the RDT calculations omit dissipation and slow pressure-strain terms. In the first three cases, axisymmetric contraction, axisymmetric expansion and plane strain, it is important to model the normal components of the turbulence anisotropy since they contribute to turbulence production. Correspondingly, for the last two cases where the strain rate contains only a shear component, the most important component of the turbulence anisotropy is the shear component, b_{12}.

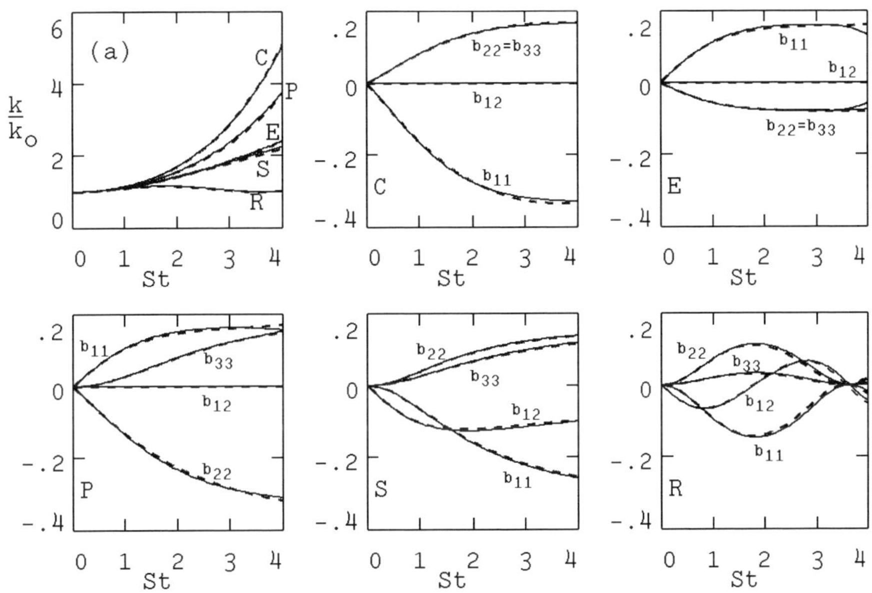

Fig. 9.6.2. Turbulence development for several zero vorticity test cases, without turbulence dissipation. ——— RDT; - - - model, Eq. 9.6.2. Velocity gradients and rotation for test cases:

	$\frac{1}{S}\frac{\partial U_1}{\partial x_1}$	$\frac{1}{S}\frac{\partial U_2}{\partial x_2}$	$\frac{1}{S}\frac{\partial U_3}{\partial x_3}$	$\frac{1}{S}\frac{\partial U_1}{\partial x_2}$	$\frac{1}{S}\frac{\partial U_2}{\partial x_1}$	$\frac{\Omega_3}{S}$
C: Axisymmetric contraction	0.5774	–0.2887	–0.2887	0	0	0
E: Axisymmetric expansion	–0.5774	0.2887	0.2887	0	0	0
P: Plane strain	–0.5	0.5	0	0	0	0
S: Zero-vorticity plane shear	0	0	0	1.0	0	0.5
R: High rotation	0	0	0	1.5	–0.5	1.0

Figure 9.6.2a shows the development of the turbulence kinetic energy. Axisymmetric contraction has the most rapid increase. The high rotation case shows a decrease in turbulence kinetic energy due to negative turbulence production.

Shear Flows

For shear with no rotation, the coefficients in Fig. 9.6.1 may be approximated as

Model Mrdt,W=Sa:

$$c_3 = 0.7g + 0.1f_{g1}(g,0.5)$$

$$c_4 = 1.2 - 1.2\left(\max(0,1-\sqrt{2g})\right)^2 + \left(\frac{24}{7} - 1.2\right)g \tag{9.6.3}$$

$$c_6 = 0.8 - 0.8f_{g0}(g,0.3) + 2.8g \quad .$$

This includes two-dimensional limits, $c_3 = c_4 = c_6 = 0$ at $g = 0$. (These coefficients are also plotted in Fig. 9.6.4.)

Figure 9.6.3a shows that this model compares well with the RDT results over a long dimensionless time period, $St = 0$ to 12.

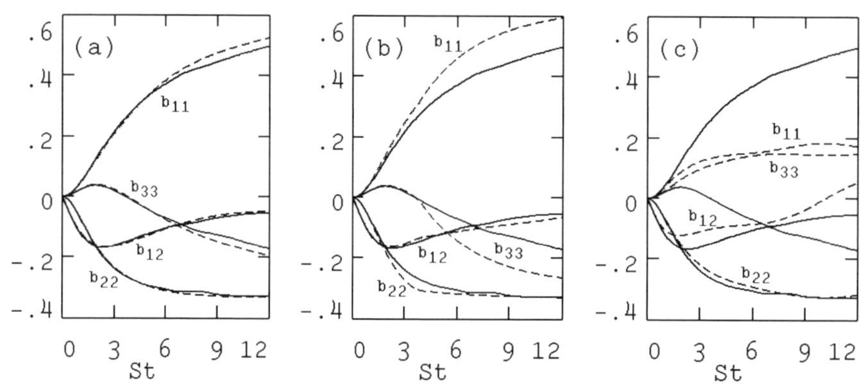

Fig. 9.6.3. Turbulence development for shear flow without rotation, in the RDT limit without dissipation. —— RDT results. – – – models: (a) Mrdt,W=Sa, Eq. 9.6.3; (b) Mrdt,W=Sb, Eq. 9.6.4; (c) Mrdt,W=0, Eq. 9.6.2.

Due to problems encountered when modeling flat plate transitional boundary layers (Chapter 11), an alternative model with $c_6 = 0$, was developed to approximate rapidly distorting shear flow,

Model Mrdt,W=Sb:

$$c_3 = 0.1 - 0.1\left(\max\left(0, 1 - \sqrt{\frac{g}{0.15}}\right)\right)^2 + 0.7g$$
$$c_4 = 1.9 - 1.4\left(\max\left(0, 1 - \sqrt{\frac{g}{0.15}}\right)\right)^2 - 0.5 f_{g0}(g, 0.5) + \left(\frac{24}{7} - 1.9\right)g^2 . \quad (9.6.4)$$

As shown in Fig. 9.6.3b, the model concentrated on maintaining good agreement with the shear anisotropy, b_{12}, and therefore the growth of turbulence kinetic energy, but is unable to do as well on the normal anisotropies over a long time period.

Figure 9.6.3c shows that the model developed for zero vorticity flows, Eq. 9.6.2, does not do well for shear flow without rotation beyond $St = 1$. The early minimum in b_{12} near $St = 2$, is under-predicted, which correspondingly gives lower turbulence kinetic energy. Because of the high values of c_4, the model gives high values for b_{33}, with $b_{33} \approx b_{11}$. The coefficients for the zero-vorticity and shear models are compared in Fig. 9.6.4.

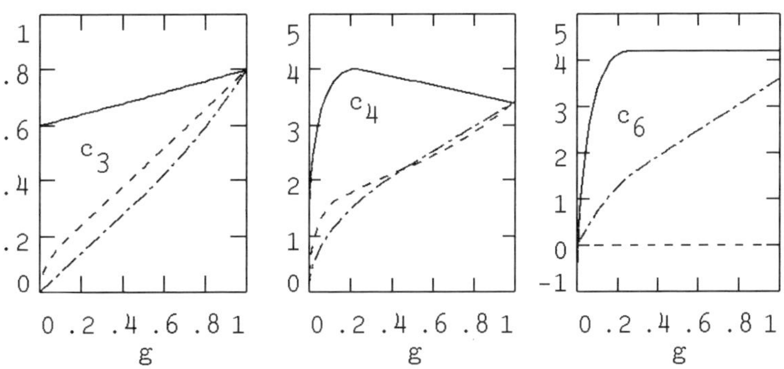

Fig. 9.6.4. Coefficients c_3, c_4 and c_6 for pressure-strain models: —— Mrdt,W=0, Eq. 9.6.2; –·–·- Mrdt,W=Sa, Eq. 9.6.3; – – – Mrdt,W=Sb, Eq. 9.6.4.

Adding a Vorticity Parameter

To model both the zero-rotation and zero-vorticity shear cases well, the model needs a more complex functionality. The difference in b_{33} for shear flow, with and without rotation, is dependent on c_4 and independent of the standard vorticity terms, c_5, c_8, and c_{11}, in Eq. 4.2.5. Since Eq. 9.6.2 worked well for all the zero vorticity test cases, here, we introduce a dimensionless invariant parameter to interpolate between zero vorticity flows and shear flow for which $W = S$.

Equations 9.6.2 and either 9.6.3 or 9.6.4, may be combined to give a model for rapidly distorting flows by taking

$$c_n = (1 - f_w)c_{n,W=0} + f_w c_{n,W=S} \quad \text{for} \quad n = 3,4,6 \; , \tag{9.6.5}$$

where $f_w = 0$ for zero vorticity, and $f_w = 1$ when the strain rate and vorticity are coplanar and equal. Two parameters on which f_w could be based were considered,

$$f_{w1} = \frac{2W_{ij}W_{ij}}{S^2} = \left(\frac{W}{S}\right)^2 \quad \text{and} \quad f_{w2} = \frac{8W_{ij}S_{jk}W_{kn}S_{ni}}{S^4} \; , \tag{9.6.6}$$

where $S^2 = 2S_{ij}S_{ij}$ and $W^2 = 2W_{ij}W_{ij}$. f_{w1} considers the relative magnitudes of the vorticity and the strain rate. f_{w2} considers only the component of the vorticity which is coplanar with the strain, giving the same results when vorticity is coplanar, $f_{w2} = (W/S)^2$, but a different result when the vorticity is not coplanar. For the three shear examples in Section 9.5,

	Ω_2	Ω_3	f_{w1}	f_{w2}	c_4 (rms fit)
Case 1. Zero vorticity, Eq. 9.5.3	0	S/2	0.	0.	2.4
Case 2. Zero rotation, Eq. 9.5.4	0	0	1.	1.	1.4
Case 3. 1-3 vorticity, Eq. 9.5.5	S/2	S/2	1.	0.	2.3

the rms values of c_4 suggest that cases 1 and 3 are most similar, so that f_{w2} should be used as the basis for f_w.

To determine $f_w(f_{w2})$, the case of shear with $\Omega_3 = S/4$ $(W/S = 0.5)$, half way between the zero-vorticity and zero-rotation cases, was used. Figure 9.6.5 shows the turbulence anisotropies using pressure-strain model coefficients which are (a) the simple average of models Mrdt,W=0 and Mrdt,W=Sa, and (b) the simple average of Mrdt,W=0 and Mrdt,W=Sb. Both give reasonable results.

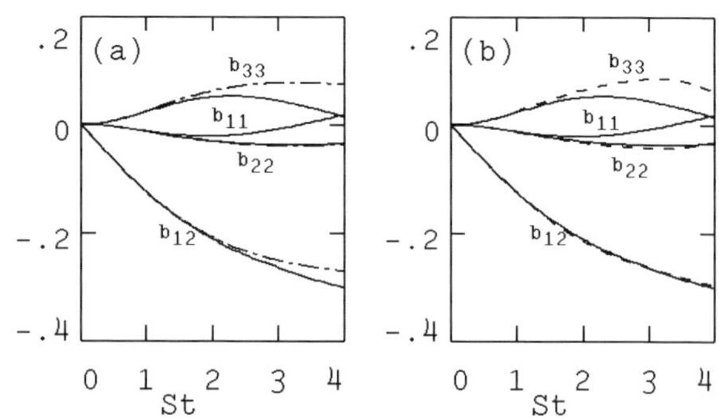

Fig. 9.6.5. Turbulence anisotropy for shear flow with $\Omega_3 = S/4$.
—— RDT results. Models:
(a) —·—·— average of Eqs. 9.6.2 and 9.6.3.
(b) — — — average of Eqs. 9.6.2 and 9.6.4.

Therefore, for $\Omega_3 = S/4$, we require $f_w = 0.5$. Since $W/S = 0.5$ for this case, for shear with coplanar rotation, we need $f_w = W/S = \sqrt{f_{w2}}$. With prudent bounding, we take

$$f_w = \left(\min\left(1,\ \max\left(0,\ \frac{8W_{ij}S_{jk}W_{kn}S_{ni}}{S^4}\right)\right)\right)^{1/2} . \tag{9.6.7}$$

A model for rapidly distorting turbulence may then be defined by Eqs. 9.6.5, 9.6.7, and

Mrdt: $c_{n,W=0}$ as for Eq. 9.6.2, and $c_{n,W=S}$ as for Eq. 9.6.4. (9.6.8)

Note that this is not the first attempt to use a more complex vorticity functionality than c_5, c_8, and c_{11} terms in pressure-strain models. Lumley (1975, see his Eq. 36), through a complicated expression for M_{ijnm}, dependent on not only b_{ij}, but also on the mean velocity gradient, involving terms like,

$$\frac{(S_{im} + W_{im})(S_{nk} + W_{nk})b_{kj}}{(S^2 + W^2)},\quad \frac{(S_{ik} + W_{ik})b_{km}(S_{nj} + W_{nj})}{(S^2 + W^2)},\ \text{etc.,} \tag{9.6.9}$$

set the scene for the pressure-strain coefficients to be different, in different strain/vorticity situations.

Shear Flow with Rotation

Sections 4.2 and 4.3 considered shear flow with coplanar rotation. Here, we continue consideration of these flows, but in the rapidly distorting limit without dissipation and slow pressure-strain.

The development of the turbulence kinetic energy for plane shear with several coplanar rotations is shown in Fig. 9.6.6a. The results from the Mrdt model (Eqs. 9.6.5-8) and from the RDT calculations are nearly indistinguishable for rotations, $\Omega_3 = 0$, $S/4$ and $S/2$. Large eddy simulations by Bardina (1983) for these three rotations, gave qualitatively the same results, with $\Omega_3 = S/4$ having the highest growth rate of turbulence, followed by $\Omega_3 = 0$, then $S/2$. Speziale et al. (SSG1991) used Bardina's results, and in particular the difference in the how k increased for $\Omega_3 = 0$ and $\Omega_3 = S/2$ to set $c_5 = 0.4$ for their model. Here, in model Mrdt, the difference is obtained by using different values for the strain term coefficients based on the vorticity parameter.

Outside the range $\Omega_3 = 0$ to $S/2$, both the model and RDT results give oscillating b_{12} profiles (Fig. 9.6.6b) with positive as well as negative values so that the turbulence production has regions where it is negative. Overall the results are reasonable in that the turbulence kinetic energy stays near its original value. (The modeled results for $\Omega_3 = -S/4$ and $3S/4$ are nearly indistinguishable in Fig. 9.6.6.)

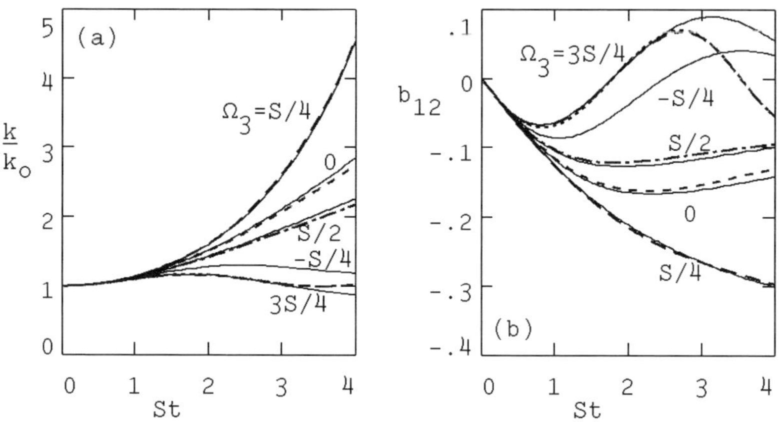

Fig. 9.6.6. Turbulence kinetic energy and shear anisotropy for shear flow with various rotations.
—— RDT results.

Ω_3/S	Model Mrdt
–0.25	- - - - -
0	— — —
0.25	—— —— ——
0.5	——-——
0.75	— — — —

Figure 9.6.7 replots the development of the turbulence kinetic energy as a function of the rotation parameter R, as used for the bifurcation plots in Section 4.3. Also shown are some two-dimensional outlines of Reynolds stress tensor glyphs. At $t = 0$ all the glyphs start out as spherical (not shown). The initial departure from isotropic turbulence is a stretching at 45 degrees from the x_1 and x_2 axes in the direction of compressive strain rate (Fig 2.2.1). Correspondingly, from $St = 0$ to $St = 1$, there is a small but fairly uniform increase in turbulence kinetic energy.

By $St = 1$, and later, the influence of vorticity and rotation on the glyphs can be seen. At $R = -0.5$ ($\Omega_3 = S/4$), the glyphs remain at or near 45 degrees and the turbulence increases rapidly. Either side of this rotation, the glyphs twist with time, misaligning the principal directions of the turbulence stress and the strain rate, which reduces turbulence production.

As the glyphs at $R = -1$ ($\Omega_3 = S/2$) develop, they appear smaller than the glyphs at $R = 0$, because more of the turbulence energy is in the $\overline{u_3 u_3}$ component. This also results in less turbulence being produced. The rotation parameter corresponding to the peak turbulence is slightly above $R = -0.5$, with the shift being more pronounced for the model.

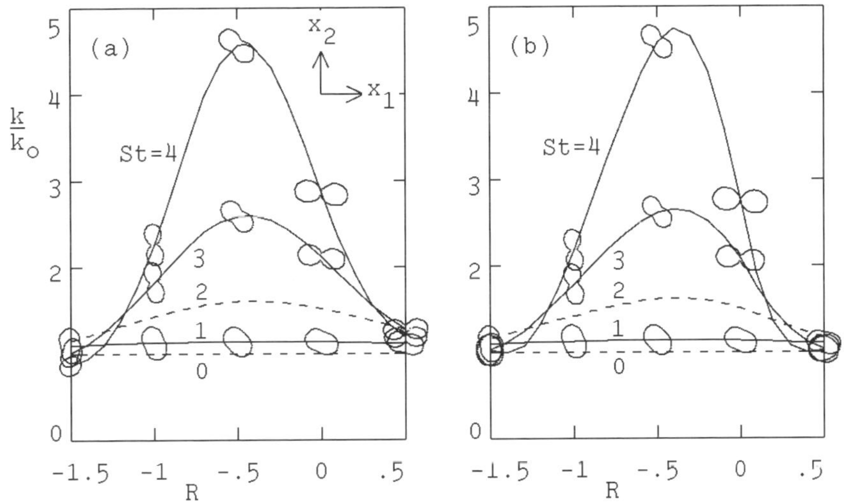

Fig. 9.6.7. Development of turbulence kinetic energy for shear flow as a function of rotation parameter, R. (a) RDT results. (b) Model Mrdt (Eqs. 9.6.5-8). Two-dimensional outlines of Reynolds stress glyphs of $\overline{u_i u_j}/k$ at $St = 1$, 3 and 4 show the turbulence anisotropy in the x_1, x_2 plane.

Chapter 10

MODELING RAPID DISTORTION AND VORTICITY EFFECTS – THE MARV MODEL

In the last chapter it was determined that vorticity effects for rapidly distorting flows cannot be modeled using the standard algebraic form, and an interpolation parameter was used to combine models with and without vorticity. In Section 8.4 it was shown that the MA model, which was based on fully developed shear flow, did not adequately address the problem of rapidly distorting shear flow. It is the task of this chapter to add rapid distortion and zero vorticity effects to the rapid pressure-strain MA model, forming the MARV model.

Experimental measurements and DNS calculations in plane strain shed light on zero vorticity flows at low Sk/ε. DNS, and rapid distortion theory calculations with dissipation, at modest Sk/ε, aid in determining a criterion to distinguish between the RDT limit without dissipation ($Sk/\varepsilon = \infty$), and fully developed flows.

10.1 Functionality of the Rapid Pressure-Strain Coefficients

For the MARV model, the rapid pressure-strain coefficients are taken as functions of:

the two-dimensionality parameter, g, which was used in the MA model;

the vorticity parameter, f_w, which was used for model Mrdt in Chapter 9;

and a new parameter, f_B, which distinguishes between rapidly distorting developing flow and nearly fully developed results,

$$c_n = c_n(g, f_w, f_B), \qquad \text{for} \qquad n = 3,\ 4,\ 6. \tag{10.1.1}$$

If we assume that f_B may be defined as an interpolation parameter, with $f_B = 0$ for fully developed flow, and $f_B = 1$ for the rapid distortion limit, then we may write

$$\begin{aligned} c_n(g, f_w, f_B) = {} & (1 - f_w)(1 - f_B)c_{n,fdfW=0} + f_w(1 - f_B)c_{n,fdfW=S} \\ & +(1 - f_w)f_B c_{n,rdtW=0} + f_w f_B c_{n,rdtW=S} \quad , \end{aligned} \tag{10.1.2}$$

and the MARV model is formed of four g-dependent sub-model limits for rapid pressure-strain, which are linearly combined using interpolation factors. The fully developed shear flow limit with $W = S$ is the MA model (Eq. 8.3.1),

$$c_{3,fdfW=S} = 0.15 f_{g3}(g,0.05,0.5) + 0.65 f_{g3}(g,0.7,1) \ , \tag{10.1.3}$$

$$\begin{aligned} c_{4,fdfW=S} = {} & 0.414 - 0.414\left(\max(0,1-\sqrt{2g})\right)^2 + 0.086 f_{g3}(g,0,0.5) \\ & +0.5 - 0.5 f_{g0}(g,0.5) + \left(\frac{24}{7} - 1\right) f_{g1}(g,0.85) \ , \end{aligned} \tag{10.1.4}$$

$$c_{6,fdfW=S} = 0 \ . \tag{10.1.5}$$

The rapid-distortion limits for $W = 0$ and $W = S$ are given in Chapter 9, Eqs. 9.6.2 and 9.6.4, respectively,

$$c_{3,rdtW=0} = 0.6 + 0.2g \, , \tag{10.1.6}$$

$$c_{4,rdtW=0} = \frac{24}{7} + 0.8(1-g) + \left(0.1 - \frac{24}{7}\right)\left(\max(0,1-\sqrt{4g})\right)^2 , \tag{10.1.7}$$

$$c_{6,rdtW=0} = 4.2 - 6\left(\max(0,1-\sqrt{4g})\right)^2 \ , \tag{10.1.8}$$

$$c_{3,rdtW=S} = 0.1 - 0.1\left(\max\left(0, 1 - \sqrt{\frac{g}{0.15}}\right)\right)^2 + 0.7g \ , \tag{10.1.9}$$

$$c_{4,rdtW=S} = 1.9 - 1.4\left(\max\left(0, 1 - \sqrt{\frac{g}{0.15}}\right)\right)^2 - 0.5 f_{g0}(g,0.5) + \left(\frac{24}{7} - 1.9\right) g^2 \ , \tag{10.1.10}$$

$$c_{6,rdtW=S} = 0 \ , \tag{10.1.11}$$

as is the vorticity interpolation parameter, Eq. 9.6.7,

$$f_w = \left(\min\left(1, \ \max\left(0, \ \frac{8 W_{ij} S_{jk} W_{kn} S_{ni}}{S^4}\right)\right)\right)^{1/2} \ . \tag{10.1.12}$$

The main tasks left to do are then to determine a fully developed $W = 0$ model, and to determine the interpolation parameter to be used between the fully-developed, and rapid-distortion, limits.

10.2 Fully Developed Turbulence Without Vorticity

In Section 8.4 it was shown that the MA model, based on fully developed shear flow, did a reasonable job of modeling the turbulence anisotropy in developing shear flow, provided the dimensionless strain rate was low. Therefore, as a basis for the fully developed limit for zero vorticity flow, we use the development of turbulence under the influence of plane strain at low dimensionless strain rates.

Tucker (T1970) presents measurements for seven zero-vorticity plane strain cases, varying the straining duct, the grid which generated the turbulence, and the turbulence Reynolds number. Sketches of the straining ducts are shown in Fig. 10.2.1. All the cases started with low dimensionless strain rate, $s = 0.5$ to 2, and most maintained low s throughout. The turbulence is essentially homogeneous near the duct centerlines, and the mean flow has constant strain rate, S, for the length of the straining duct,

$$\frac{\partial U_i}{\partial x_j} = \begin{bmatrix} S/2 & 0 & 0 \\ 0 & 0 & 0 \\ 0 & 0 & -S/2 \end{bmatrix}. \qquad (10.2.1)$$

Note that the coordinate directions have been chosen differently for the ducts, so that $\partial U_1/\partial x_1 = -\partial U_3/\partial x_3 > 0$ for both configurations.

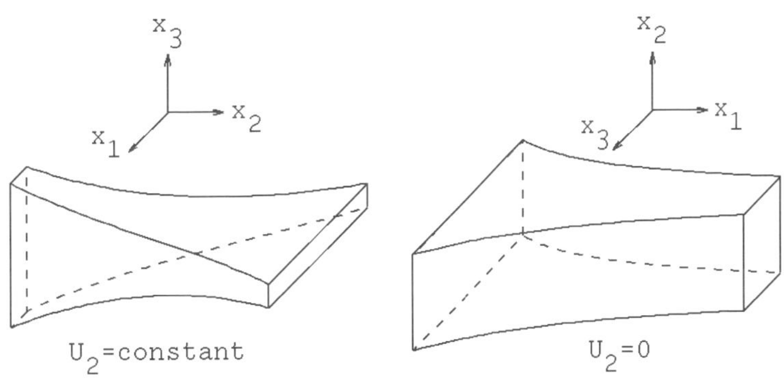

Fig. 10.2.1. Sketches of the straining ducts used by Tucker (T1970) for plane strain.

For this plane strain, and considering pressure-strain models with only c_1, c_3 and c_4 terms, Eq. 4.1.6, for the development of b_{ij}, yields two independent equations (Appendix EQ.10.2.2),

$$2\frac{d(b_{11} - b_{33})}{dt} = -(b_{11} - b_{33})\left((c_1 - 2)\omega + 2\frac{P}{k}\right) + \left(c_3 - \frac{4}{3}\right)S + (c_4 - 2)(b_{11} + b_{33})S, \quad (10.2.2)$$

$$2\frac{db_{22}}{dt} = -b_{22}\left((c_1-2)\omega + 2\frac{P}{k}\right) + (c_4-2)\frac{1}{3}\frac{P}{k} \ , \tag{10.2.3}$$

which may be used to set two of the three pressure-strain coefficients. Since c_1 is assumed to be independent of the strain and vorticity, we use c_1 from the MA model. For homogeneous turbulence, with $g_{eff} = g$ and $\tilde{R}_t = R_t$, Equations 8.3.4, 8.3.1 and 8.3.5, respectively, with the constants in Table 8.3.1, become

$$c_1 = 2 + (c_{1\infty} - 2) f_{c1}(R_t) \ , \tag{10.2.4}$$

where

$$c_{1\infty} = 8 + (2-8) f_{g0}(g, 0.5) + (2.5-8) f_{g1}(g, 0.7) \tag{10.2.5}$$

and

$$f_{c1} = 1 - \exp\left(-0.000003 R_t^2 - 0.045\sqrt{R_t}\right) \ . \tag{10.2.6}$$

Figure 10.2.2 shows the results, using Eq. 10.2.3 to obtain c_4 from Tucker's data and from DNS calculations for plane strain by Lee and Reynolds (LR1985, presented by Shabbir and Shih, 1993). Despite the scatter, the data is mostly well below the rapid distortion $W = 0$ model for c_4, and mostly well above the fully developed $W = S$ model. For the fully developed $W = 0$ model limit, we choose something in between, with $c_4 = 2.2$ over the bulk of the g range, and with $c_4 = 0$ and 24/7, at the two-dimensional and isotropic turbulence limits, respectively,

$$c_{4,fdfW=0} = 2.2 - 2.2\left(\max(0,\ 1-\sqrt{4g})\right)^2 + \left(\frac{24}{7} - 2.2\right) f_{g1}(g, 0.7) \ . \tag{10.2.7}$$

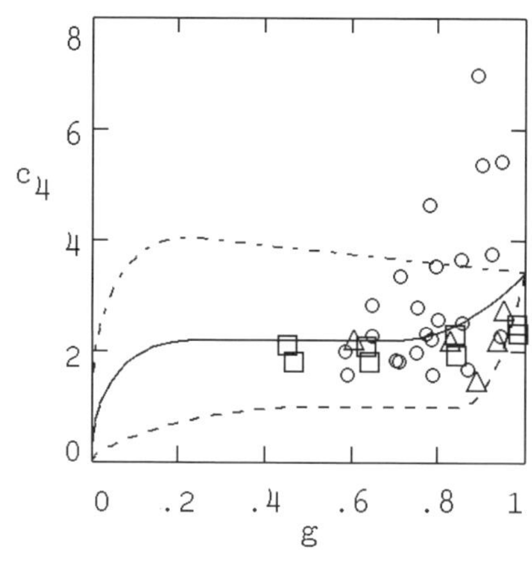

Fig. 10.2.2. c_4 from plane strain data for $s < 4$, using c_1 from the MA model. Smoothed data of Tucker (T1970): o, duct 1 with uniform mean velocity; Δ, ducts with accelerating mean velocity. □, DNS data of Lee and Reynolds (LR1985), cases PXA and PXB. Models:
- - - - $c_{4,rdtW=0}$, Eq. 10.1.7;
— — — $c_{4,fdfW=S}$, Eq. 10.1.4;
—— $c_{4,fdfW=0}$, Eq. 10.2.7.

Having chosen a model for c_4, values for c_3 may be obtained from Eq. 10.2.2, and the data. Figure 10.2.3 shows that, again, the data is mostly well below the rapid distortion $W=0$ model, and mostly well above the fully developed $W=S$ model. For the fully developed $W=0$ model limit, we choose

$$c_{3,fdfW=0} = 0.44 - 0.44\left(\max(0,\ 1-\sqrt{4g})\right)^2 + 0.36 f_{g3}(g,0.7,1) \;, \qquad (10.2.8)$$

which has $c_3 = 0.44$ over the bulk of the g range, and $c_3 = 0$ and 0.8, at the two-dimensional and isotropic turbulence limits, respectively.

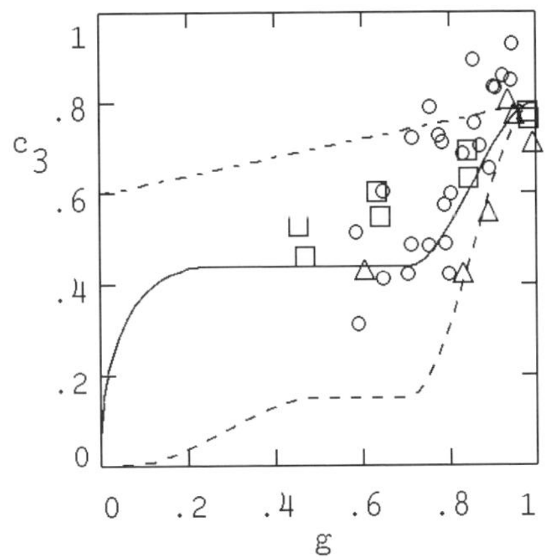

Fig. 10.2.3. c_3 from plane strain data for $s<4$, using c_1 from the MA model, and $c_4 = c_{4,fdfW=0}$, Eq. 10.2.7. Data as for Fig. 10.2.2. Models:
- - - - $c_{3,rdtW=0}$, Eq. 10.1.6;
— — — $c_{3,fdfW=S}$, Eq. 10.1.3;
—— $c_{3,fdfW=0}$, Eq. 10.2.8.

10.3 Rapid Distortion Theory with Dissipation

The models developed in Section 9.6 were based on the high-strain-rate limit of the rapid distortion theory equation (Eq. 9.3.4), i.e., the dissipation term was neglected, which is equivalent to setting $\nu = 0$. At modest dimensionless strain rates,

$$s = \frac{Sk}{\varepsilon} = \frac{S\int \frac{1}{2}\Phi_{ii} d\mathbf{K}}{\int \nu K^2 \Phi_{ii} d\mathbf{K}} \;, \qquad (10.3.1)$$

dissipation is no longer negligible.

The slow pressure-strain term may also not be negligible. Townsend (1976) suggested a combined model for the slow pressure-strain and dissipation which replaces the laminar viscosity, ν, in the dissipation term of the rapid distortion theory equation (Eq. 9.3.4), with an effective value, ν_t. While this is qualitatively incorrect, in that the slow pressure-strain model does not then sum to zero for the turbulence kinetic energy, it appears to give reasonable results for turbulence anisotropy (Salhi and Cambon, 1997). Therefore, RDT calculations with dissipation may indicate how the development of turbulence anisotropy is affected by dimensionless strain rate.

In this section, we test this concept by comparing RDT results with the DNS plane strain data of Lee and Reynolds (LR1985). The DNS calculations began with dimensionless strain rates, $s_o = 1.15$, 2.31, 4.62 and 44. The initial turbulence Reynolds number was low, $R_t = 11.6$, which reduces the importance of slow pressure-strain.

Corresponding RDT calculations were started with isotropic turbulence and the spectrum tensor, Φ_{ij}, specified over a hemi-spherical surface of grid points with uniform initial wave-number magnitude, $K = K_0$ (see Eqs. 9.3.6 and 9.5.2). At $t = 0$, Eq. 10.3.1 then reduces to

$$s_o = \frac{S}{2\nu K_0^2} , \tag{10.3.2}$$

and the dimensionless strain rate is initialized by setting νK_0^2. For each grid point the development of Φ_{ij} with time was obtained by solving Eq. 9.3.4, and the movement of the point in wave-number space was tracked solving Eq. 9.3.3. Because the points move in wave-number space, the weighted average value of K^2 changes; this in turn changes the dimensionless strain rate (Eq. 10.3.1). Dissipation of Φ_{ij}, which is proportional K^2, also becomes non-uniform.

As shown in Fig. 10.3.1, the first set of RDT calculations did not reproduce the development of dimensionless strain rate obtained in the DNS data. s tended to be significantly lower for the RDT results.

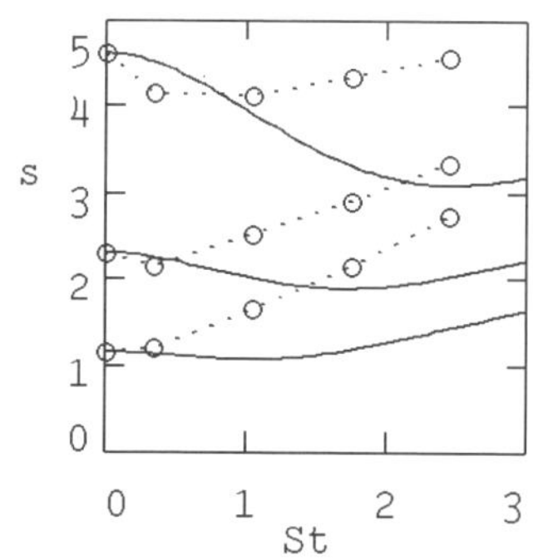

Fig. 10.3.1. Development of dimensionless strain rate, s, for plane strain test cases.
o · · o, DNS data of Lee and Reynolds (LR1985)
—— RDT calculations with constant ν.

Therefore, the RDT calculations were repeated, with the s development of the DNS data imposed on the calculations, by periodically changing the laminar viscosity, ν. These calculations gave remarkably good results for the development of the Reynolds stresses as shown in Fig. 10.3.2. Note that the coordinate convention used gives

$$\partial U_1 / \partial x_1 = -\partial U_3 / \partial x_3 > 0 , \tag{10.3.3}$$

consistent with Eq. 10.2.1. The calculations distinguish between the lowest s results, where $\overline{u_3u_3}/k$ increases more rapidly and $\overline{u_2u_2}/k$ falls slightly, and the highest s results, where $\overline{u_3u_3}/k$ increases more slowly and $\overline{u_2u_2}/k$ increases.

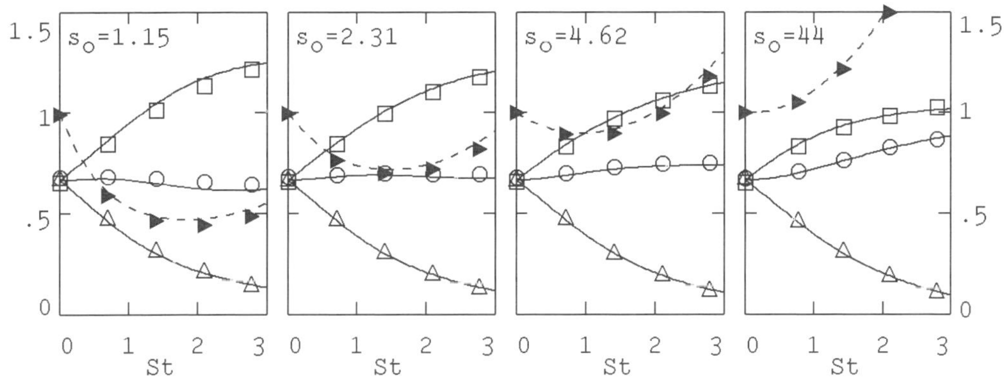

Fig. 10.3.2. Development of Reynolds stresses for plane strain test cases. DNS data of Lee and Reynolds (LR1985): △, $\overline{u_1u_1}/k$; ○, $\overline{u_2u_2}/k$; □, $\overline{u_3u_3}/k$; ▶, k/k_o. ——, - - - -, RDT calculations with variable ν to match the dimensionless strain rates of the DNS data.

This validates the concept that RDT calculations with dissipation can provide information on how dimensionless strain rate affects the development of turbulence anisotropy, at least at low turbulence Reynolds numbers.

10.4 A Rapid Distortion Parameter

In Fig. 10.3.2 we saw the effects of dimensionless strain rate on developing plane strain flows. Figure 8.4.1 showed that dimensionless strain rate significantly affected the time development of shear flow. In this section we combine the pressure-strain models developed for fully developed turbulence with those developed for rapidly distorting turbulence. While dimensionless strain rate has been an indicator in developing flows, it is not a suitable parameter on which to base the interpolation factor f_B (Eq. 10.1.2). Turbulence anisotropy can become fully developed at high, as well as low, dimensionless strain rate. We therefore seek a parameter, which is dimensionless, a scalar, and includes db_{ij}/dt.

One such parameter is

$$B = \frac{1}{2\omega II}\frac{dII}{dt} = \frac{b_{ij}}{\omega II}\frac{db_{ij}}{dt} \quad . \tag{10.4.1}$$

It has a value of zero in the fully developed limit when the turbulence anisotropy is unchanging. It has a value of infinity at the high-strain-rate or RDT limit of $\omega = 0$. It is also infinite at the isotropic limit ($II = 0$) so that it becomes optional whether or not the isotropic limits, $c_3 = 0.8$ and $c_4 = 24/7$, need to be included in the g functionality of the coefficients for fully developed turbulence. (In the end, the isotropic limits were retained for the *fdf* coefficients, Eqs. 10.1.3-4, 10.2.7-8.)

To avoid any ambiguity, db_{ij}/dt in the parameter B is a theoretical value of how fast b_{ij} is changing in the absolute frame in the absence of convection, laminar diffusion and turbulent diffusion. In other words for the purpose of evaluating B, we use Eq. 4.1.6, without rotation. With only c_1, c_3, c_4 and c_6 pressure-strain terms, this reduces to

$$\begin{aligned}2\frac{db_{ij}}{dt} &= -b_{ij}\left((c_1 - 2)\omega + (2 + \frac{1}{2}c_6)\frac{P}{k}\right) + \left(c_3 - \frac{4}{3}\right)S_{ij} \\ &+(c_4 - 2)(b_{ik}S_{jk} + b_{jk}S_{ik} - \frac{2}{3}\delta_{ij}b_{mn}S_{mn}) - 2(b_{ik}W_{jk} + b_{jk}W_{ik}),\end{aligned} \tag{10.4.2}$$

which can be evaluated locally at any point in a flow.

An interpolation function $f_B(B)$ is sought. The fully developed pipe and channel flows used to complete the MA model had B values from –0.8 to 0.5. If f_B is zero for $B < 0.5$, then the results of Section 8.3 will remain valid, as the MARV model will reduce to the MA model for these flows. f_B being zero for $B < 0$ will also mean that the *fdf* limits for rapid pressure-strain will be used in situations when slow pressure-strain is dominant, and b_{ij} is returning towards isotropy.

The DNS data of Lee and Reynolds has been supplemented by RDT calculations with dissipation (Section 10.3), at various dimensionless strain rates, to estimate $f_B(B)$. For the plane strain data, c_3 and c_4 coefficients are determined from the data, assuming that the same interpolation factor applies to c_6 as applies to c_3 or c_4. First obtaining data values for c_3 and c_4 with $c_6 = 0$ and $c_6 = c_{6,rdtW=0}$, the interpolation factor was determined (for $g < 0.7$) from

$$\left(c_{n,data}\right)_{c_6=0}(1 - f_B) + f_B\left(c_{n,data}\right)_{c_6=c_{6,rdtW=0}} = c_{n,fdfW=0}(1 - f_B) + f_B c_{n,rdtW=0}\ , \quad \text{for } n = 3,4. \tag{10.4.3}$$

RDT calculations with dissipation were made for shear flow with no rotation ($W = S$) and for $\Omega_3 = S/2$ so that $W = 0$. From these, c_1, c_3 and c_4 coefficients can be determined, and f_B was estimated using

$$c_{n,data} = c_{n,fdf}(1 - f_B) + f_B c_{n,rdt}, \quad \text{for} \quad n = 3,4. \tag{10.4.4}$$

Figure 10.4.1 shows the results of these data analyses for $f_B(B)$. The data presented is limited to $g < 0.7$ to remove any ambiguity near the isotropic limit. f_B falls sharply for all the data below $B = 3$. The results may be approximated by

$$f_B(B) = \frac{B - 0.7}{B + 1} \tag{10.4.5}$$

or

$$f_B(B) = \frac{\max(0, B - 0.7)}{\max(0, B - 0.7) + 1.7}, \tag{10.4.6}$$

which insures that the values of f_B are always between 0 and 1. This also satisfies the condition $f_B = 0$ for $B < 0.5$, needed for consistency with prior development of the model on fully developed pipe and channel flows.

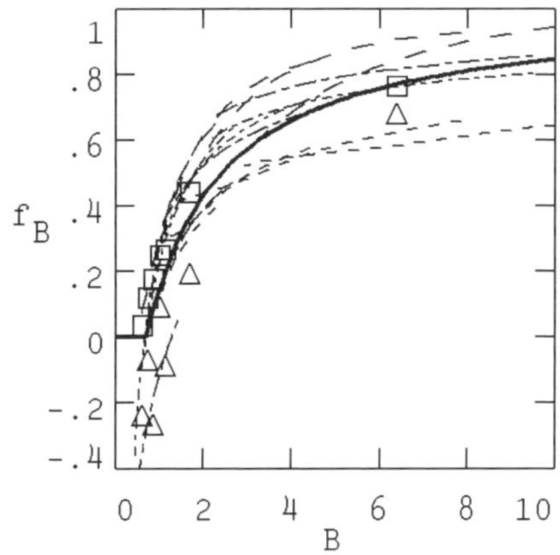

Fig. 10.4.1. Interpolation factor $f_B(B)$. Determined from c_3 and c_4 for data with g<0.7:
□, △, from Eq. 10.4.3 and Lee and Reynolds (LR1985) plane strain data, s = 1.15 to 44; from Eq. 10.4.4 and RDT calculations for shear flow with s = 5 to 40:
- - -, — — —, W=S;
— — —, — — — —, W=0.
———, Eq. 10.4.6.

10.5 Pictorial Summary

In this chapter, the functionality of the rapid pressure-strain model coefficients, c_3, c_4, and c_6, has been expanded to three parameters, the two-dimensionality parameter, g, a vorticity parameter, f_w, and a rapid distortion parameter, f_B. This three-dimensional functionality may be represented as a cube, as all three parameters vary between zero and one. Figure 10.5.1 shows the four sub-model edges of the cube. The entire domain is covered by interpolating between these edges.

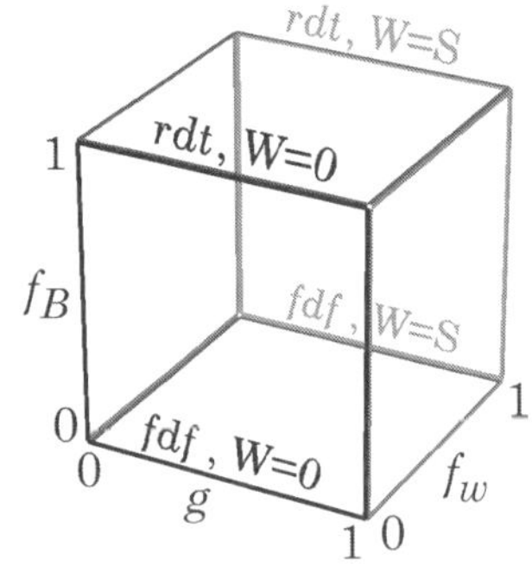

Fig. 10.5.1. Functionality cube for rapid pressure-strain model coefficients, showing the sub-model edges.

The four sub-models have significantly different coefficients, as shown in Fig. 10.5.2. Thus, for example, the pressure-strain, for turbulence encountering a rapidly distorting plane strain situation as it approaches the leading edge of a body (the rdt, W=0 limit, labeled r0), will be many times that which would be expected from using the fully developed boundary layer limit model (fdf, W=S, labeled fs).

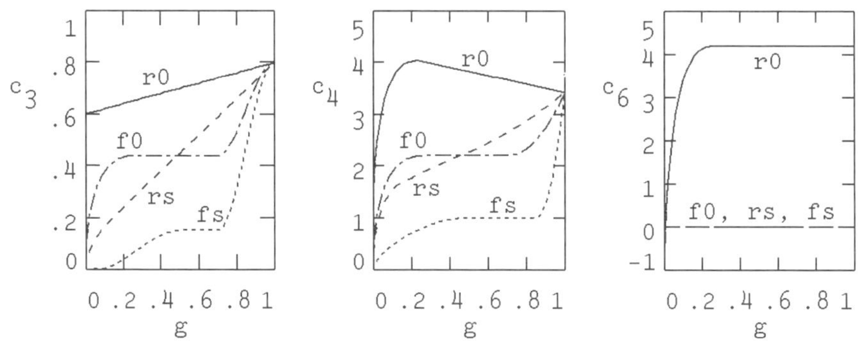

Fig. 10.5.2. The four sub-models for rapid pressure-strain in the MARV model.
—— rdt,W=0, Eqs. 10.1.6-8; –·– fdf,W=0, Eqs. 10.2.7-8; – – – rdt,W=S, Eqs. 10.1.9-11; ---- fdf,W=S, Eqs. 10.1.3-5.

Figure 10.5.3 shows the main elements of the data basis of the model on the surfaces of the functionality cube. The front left surface for zero vorticity flows is based primarily on plane strain, away from the two-dimensional limit, $g = 0$. The Tucker (T1970) data is mostly on or near the fully developed limit. The Lee and Reynolds (LR1985) data spreads across the surface. This was supplemented by RDT calculations at moderate s (not shown) for shear flow with rotation to give $W = 0$.

The model for the back right surface, $f_w = 1$, was based on shear flow without rotation. Along the bottom edge is the homogeneous shear flow data at mid to high g, and fully developed channel data (at almost all g, but not shown) including the important log-law boundary layer point. The Lee et al. (LKM1990) DNS data at high s cuts across the surface. This was supplemented by RDT calculations at moderate s (not shown).

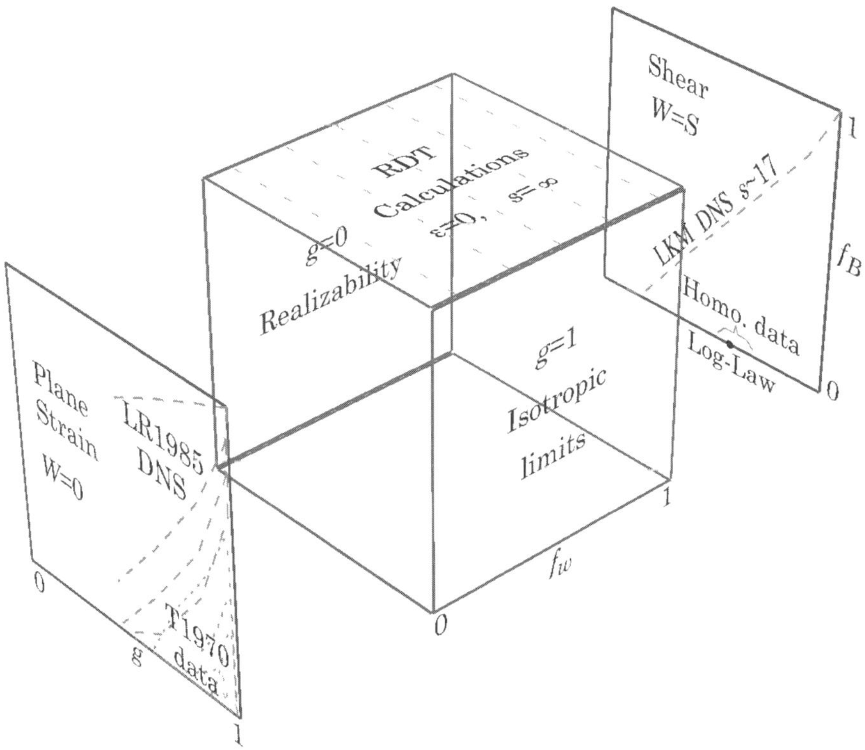

Fig. 10.5.3. Functionality cube, showing the primary data and limit bases for rapid pressure-strain model coefficients.

RDT calculations, without dissipation, for shear flow with rotation were used to cover the top of the cube. The model edge of this surface, corresponding to $W = 0$, was validated to model turbulence in numerous types of zero vorticity flows. Corresponding validation was not made for various flow situations with vorticity.

It should be noted that the f_w parameter has been min/max limited to values between zero and one (Eq. 10.1.2), and is constructed to consider only vorticity where there is also strain in the same plane. There will therefore be flows with vorticity which will use the $W = 0$ model face, and flows with high vorticity which will use the $W = S$ model face.

The remaining three sides of the cube are conspicuous, as is the interior, by the absence of data. Modeling of the $g = 1$ face is based on rapid pressure-strain model limits determined from rapid distortion theory, which are general for all types of strain and vorticity (but perhaps not all types of isotropic turbulence, see Exercise 18.2). Choices made for the $g = 0$ face ensure that the modeled Reynolds stresses will remain realizable. The limits chosen, however, provide only one of many ways to do this. DNS calculations of homogeneous turbulence could provide more data for the interior and bottom surfaces of the cube.

10.6 Simple Developing Turbulence Applications of the MARV Model

Several developing turbulence test cases have been used and mentioned in the development of the MARV model. In this section we complete the exercise of seeing how well the model does on these cases. For comparison some results for the MC and MA models will also be shown.

Plane Strain DNS

MARV model results for the plane strain DNS cases of Lee and Reynolds (LR1985) are shown in Fig. 10.6.1. The equations for b_{ij}, q and ω were solved in time, starting with the initial conditions for the DNS data. The anisotropy of the turbulence, shown as $\overline{u_i u_j}/k$, is well represented by the model results, as expected, since this DNS data contributed to the choice of the rapid pressure-strain model coefficients. The development of the turbulence kinetic energy is reasonably modeled, although slightly high for the lower s_o cases. Therefore, the choices made for dissipation equation coefficients, $c_{\varepsilon 1}$ and $c_{\varepsilon 2}$, are adequate for these low turbulence Reynolds number test cases.

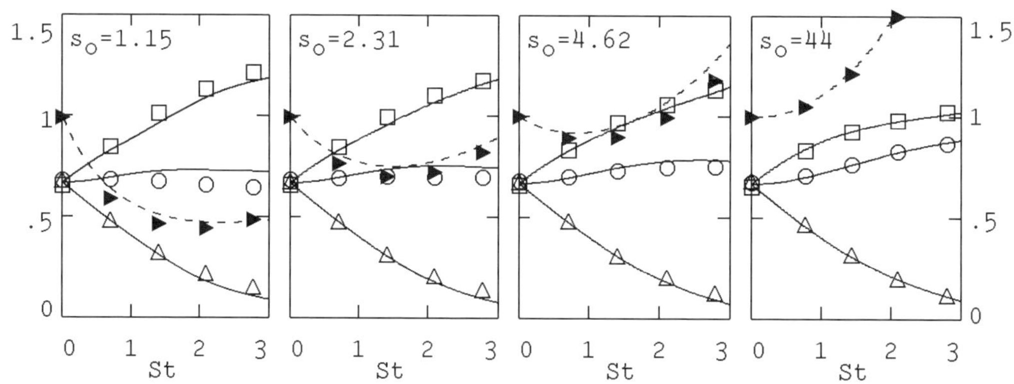

Fig. 10.6.1. Development of Reynolds stresses in plane strain. s_o as shown, $R_{t,0} = 11.6$. DNS data of Lee and Reynolds (LR1985): △, $\overline{u_1 u_1}/k$; ○, $\overline{u_2 u_2}/k$; □, $\overline{u_3 u_3}/k$; ▶, k/k_o. MARV model: ——, $\overline{u_i u_j}/k$; ----, k/k_o.

Some corresponding results using the MC and MA models are shown in Fig. 10.6.2. Shown are b_{11} and b_{33}, the difference of which governs turbulence production. $b_{33} - b_{11}$ increases most rapidly for the MC model which uses the lowest values for the c_3 and c_4 pressure-strain coefficients. The MA model, which includes the isotropic limits for c_3 and c_4, does well to about $St = 1$, after which, $b_{33} - b_{11}$ increases too rapidly.

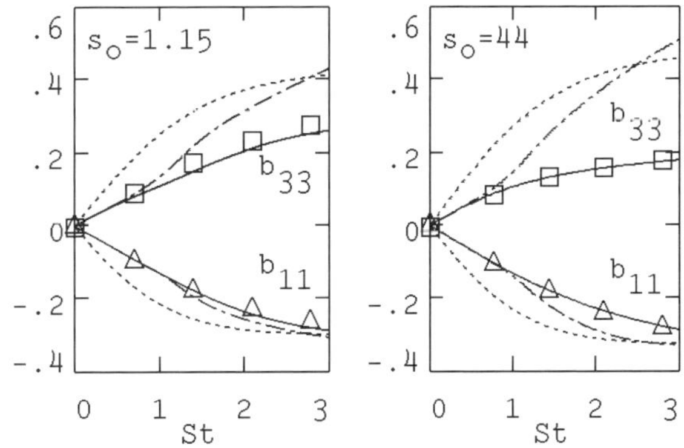

Fig. 10.6.2. Development of b_{11} and b_{33}, plane strain test cases of Lee and Reynolds (LR1985). △, □, DNS data. Models: ——, MARV; — - — - MA; - - - - MC.

Both the MC and MA models give larger values of $b_{33} - b_{11}$ for the high dimensionless strain rate case, $s_o = 44$, than for the low s case. This is consistent with the lower relative importance of slow pressure-strain when s is high, and correspondingly less return to isotropy from slow pressure-strain. The MARV model, consistent with the data, gives less of an increase in $b_{33} - b_{11}$ at high s, the opposite of the MC and MA models. This was achieved with the additional functionality in the rapid pressure-strain coefficients; in particular, the MARV model's use of a rapid distortion parameter, f_B. As shown in Fig. 10.6.3, f_B remains high for the $s_o = 44$ case, while f_B falls to zero by $St = 2$ for the $s_o = 1.15$ case.

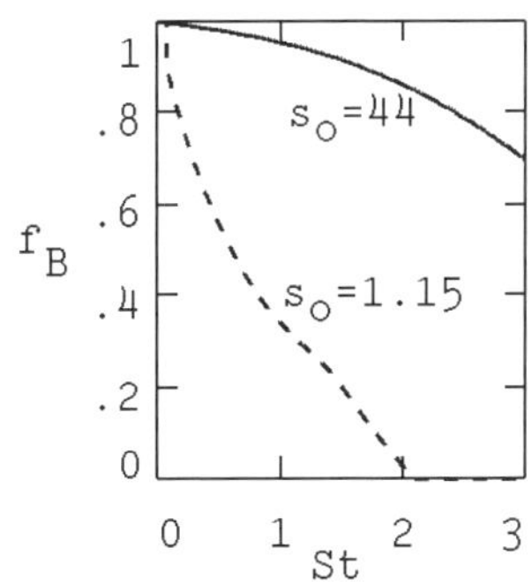

Fig. 10.6.3. f_B values from the MARV model, for the plane strain DNS test cases.

Plane Strain Experiments of Tucker

The plane strain data of Tucker (T1970) contributed to the $fdf, W = 0$ limit of the MARV model. The coordinate convention for the straining directions is given in Eq. 10.2.1. Case numbers for the data correspond to the figure numbers on which Tucker presented the data.

Three of the cases, 16, 19, and 31, had similar initial conditions and are shown together in Fig. 10.6.4. Calculations using the MC, MA and MARV models, and average initial conditions, were made. The flow is in the x_2 direction so that the grid-generated turbulence gives b_{22} higher than b_{11} and b_{33}, at the start of the distortion duct. The difference, between the results for the MARV model, and those for the MC and MA models, is primarily due to the difference between the $fdf, W = 0$ limit of the MARV model and the lower rapid pressure-strain coefficients used in the MA and MC models. Correspondingly, while the MARV model gives about the right decline in b_{22}, and the right spread in $b_{33} - b_{11}$, the MC and MA models over-predict both. All three models give reasonable results for the development of the turbulence kinetic energy, which supports the choices made for $c_{\varepsilon 1}$ and $c_{\varepsilon 2}$ at these moderate turbulence Reynolds numbers.

Fig. 10.6.4. Tucker (T1970) plane strain test cases, 16, 19 and 31, with s_o=0.62-0.634, $R_{t,0}$=178-228. Data: △, b_{11}; ○, b_{22}; □, b_{33}; ▽, k/k_o. Models: ——, MARV; – – – MA; - - - - MC.

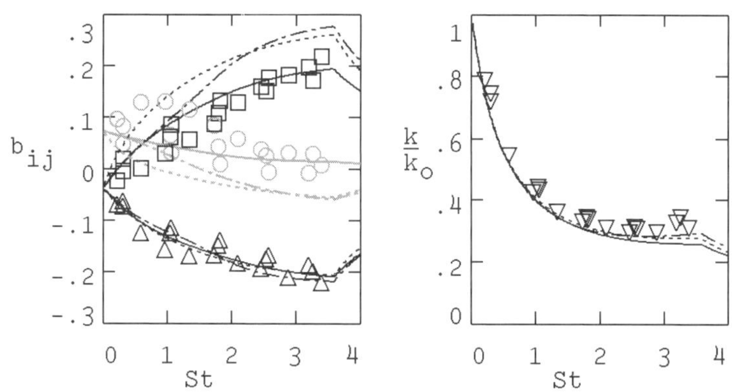

Figure 10.6.5 shows results for Tucker's cases 39 and 41. These cases differ from those shown in Fig. 10.6.4, in that the flow is in the x_1 direction, so that initially $b_{11} > b_{33}$, and turbulence production is negative. There is little difference in the development of b_{ij} in the two cases, even though s_o is higher for case 39, 1.9 compared to 0.61 for case 41. The MARV model gives good agreement for b_{ij} in both cases; the decreased importance of slow pressure-strain in the high s case, approximately cancels the effects of slightly higher values of f_B.

The marked difference between the cases comes in the development of the turbulence kinetic energy. For the high strain rate case, there is less dissipation and the turbulence increases. The MARV model under-predicts the development of turbulence in both cases. The cause could be either an inadequate model for dissipation, perhaps related to negative turbulence production, or an insufficient understanding of the experimental situation.

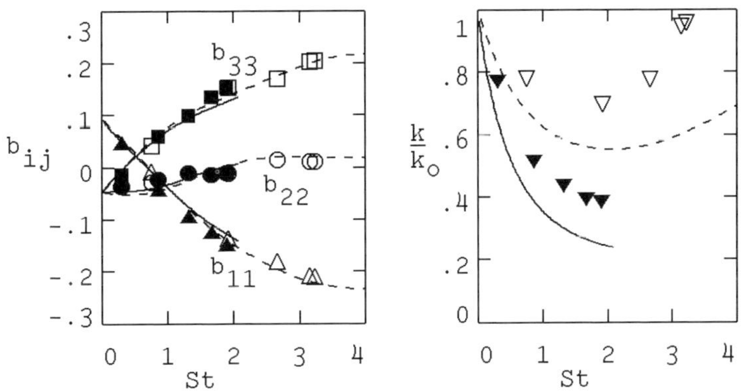

Fig. 10.6.5. Tucker (T1970) plane strain test cases. Case 39, s_o= 1.9, open symbols. Case 41, s_o=0.61, filled symbols. Both, $R_{t,0}$=202. MARV model: ---- case 39; ——, case 41.

Homogeneous Shear DNS

Calculations using the MARV model, for the developing homogeneous shear cases discussed in section 8.4, are shown in Fig. 10.6.6. As before, the DNS distributions of both dimensionless strain rate and turbulence Reynolds number were imposed to provide a pure test of the pressure-strain model. For case RR128, with low s, there is little difference between the MA and MARV model results. For the LKM case, f_B values are much higher as shown in Fig. 10.6.7, and the MARV model gives significantly improved results over the MA model.

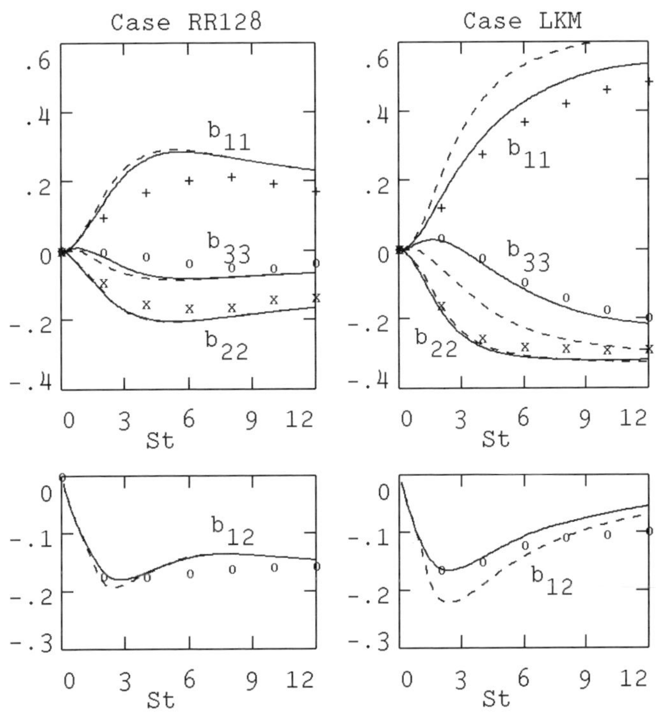

Fig. 10.6.6. Development of turbulence anisotropy due to homogeneous shear. +, o, x, DNS data: Case RR128, Rogers et al. (RMR1986), s=1.1 to 5.8, R_t= 42 to 547; Case LKM, Lee et al. (LKM1990), s=14 to 18, R_t= 65 to 600. — — — MA model; ———— MARV model.

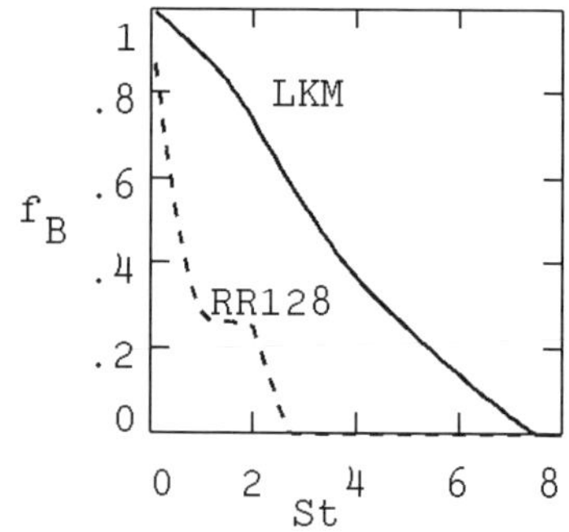

Fig. 10.6.7. f_B values from the MARV model, for shear flow DNS test cases, RR128 and LKM.

Figure 10.6.8 shows the difficulties encountered reproducing the turbulence Reynolds number when including the solution of the ω equation. Most of the DNS data, Fig. 10.6.8a, shows a sharp fall in R_t between $St = 0$ and 2 (all except cases O and L). This is followed by increasing R_t, with a slope which is fairly consistent for all the data. The MARV model, Fig. 10.6.8b, gives only a small decline in R_t, followed, in general, by a less steep rise in R_t. If $c_{\varepsilon 1}$ is lowered to 1.3 in the model, Fig. 10.6.8c, R_t increases at about the right rate. The model will not be changed on this basis. Instead it will be left as an unresolved issue, as to what more needs to be understood about modeling ω, and what more needs to be understood about DNS calculations.

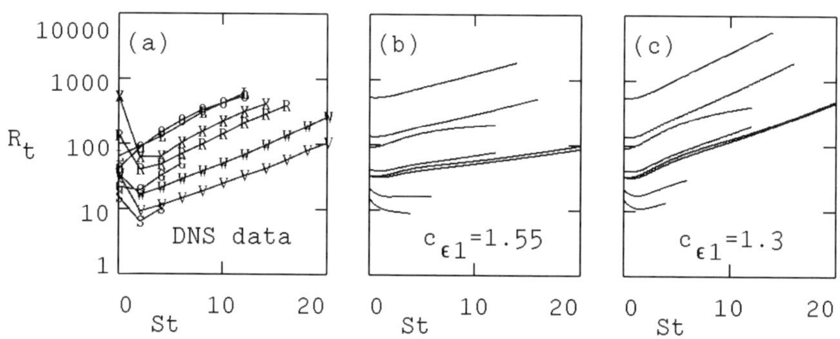

Fig. 10.6.8. Turbulence Reynolds number for homogeneous shear DNS test cases. (a) DNS data: L, case LKM; all other cases from Rogers et al. (RMR1986). (b) MARV model. (c) Modified MARV model with $c_{\varepsilon 1} = 1.3$.

Chapter 11

FLAT PLATE BOUNDARY LAYER TRANSITION

In this chapter we consider the laminar to turbulent transition of boundary layers developing along a flat plate. We limit our consideration to cases with zero pressure gradient, so that the free stream velocity is constant. Data from numerous sources for the start of transition as a function of free stream turbulence intensity is compared with results from the MARV model.

Important features of the MARV model for the calculation of transition are the inclusion of rapid distortion limits for the rapid pressure-strain coefficients, and the ω boundary condition. The MARV model results show both a turbulence intensity and a turbulence Reynolds number dependence for the start of transition, which are consistent with the data.

11.1 Transition Data

As an initially laminar boundary layer develops on a flat plate, it will grow and then become turbulent if the plate is sufficiently long and the boundary layer is perturbed, for example, by inherited and free stream turbulence. The laminar to turbulent transition is marked by a change in shape of the mean velocity profile, an increase in the wall skin friction, and an increase in turbulence in the boundary layer.

The basic geometry and coordinates are shown in Fig. 11.1.1. The x_1 coordinate is measured from the plate leading edge. It is made dimensionless as a Reynolds number using the constant free stream velocity, U_∞,

$$R_x = \frac{x_1 U_\infty}{\nu} \quad . \tag{11.1.1}$$

The x_2, or y, direction is normal to the plate.

Fig. 11.1.1. Coordinate directions for flat plate.

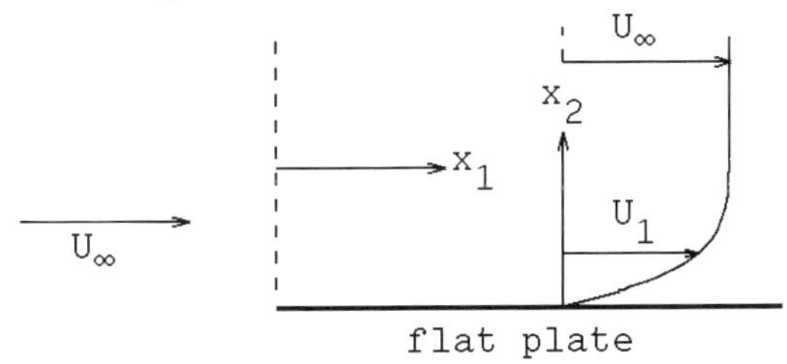

An example of the changes that occur due to transition is shown in Fig. 11.1.2. The experimental data used in the figure is the T3A case of Roach and Brierley (1990). The data for this and other cases has been made available to ERCOFTAC (European Research Community on Flow, Turbulence and Combustion) in tabulated form.

Upstream of transition, the turbulence in the boundary layer is low and the velocity profiles are consistent with the theoretical profiles of Blasius for laminar flow (see e.g. Schlichting, 1979). The boundary layer grows, and the skin friction develops as for laminar flow.

The skin friction coefficient, defined using the free stream velocity, is

$$c_f = \frac{\tau_w}{\frac{1}{2}\rho U_\infty^2} \quad . \tag{11.1.2}$$

For laminar flow (Schlichting's Eq. 7.32),

$$c_f = \frac{0.664}{\sqrt{R_x}} \quad . \tag{11.1.3}$$

Figure 11.1.2 shows c_f at first gradually departing from laminar flow and then rapidly increasing as the turbulence grows in the layer and the boundary layer becomes turbulent in nature. After transition, the skin friction data is consistent with correlations for developing turbulent boundary layers. Shown as the turbulent limit (using Schlichting's Eq. 21.11) is

$$c_f = 0.0592(R_x)^{-1/5} \quad . \tag{11.1.4}$$

The start of transition may be defined as the location where c_f has a minimum, and the end of transition, where c_f has a maximum. Between these locations, Fig. 11.1.2 shows a peak in $\overline{u_1 u_1}$, a rapid increase in the boundary layer thickness and a change in shape in the mean velocity profile.

The shape of the velocity profile may be characterized by the ratio of the displacement thickness, δ^*, and the momentum thickness, θ, which are defined by

$$\delta^* = \int_0^\infty \left(1 - \frac{U_1}{U_\infty}\right) dy \tag{11.1.5}$$

and

$$\theta = \int_0^\infty \frac{U_1}{U_\infty}\left(1 - \frac{U_1}{U_\infty}\right) dy \quad . \tag{11.1.6}$$

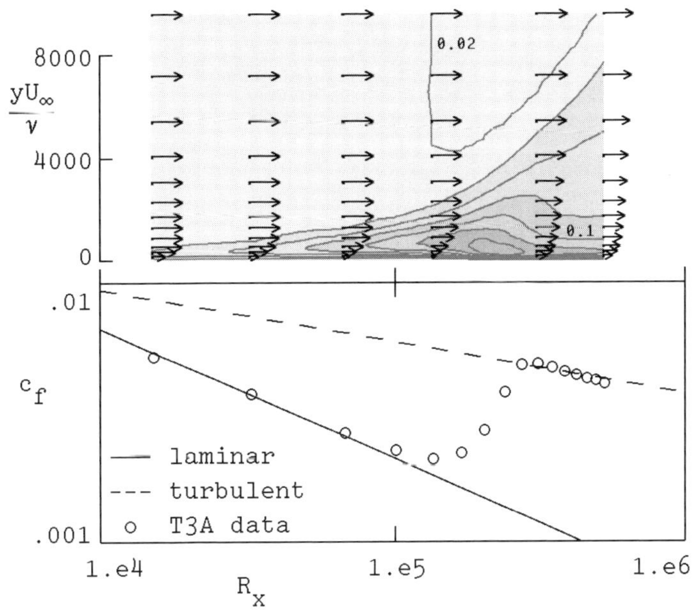

Fig. 11.1.2. Flat plate transition. Case T3A data of Roach and Brierley (1990). Bottom: skin friction coefficient as a function of R_x on a log scale. Top: mean velocity vectors and contours of $\left(\overline{u_1 u_1}\right)^{1/2} / U_\infty$, on the same log scale for R_x. Contour interval 0.02.

For a laminar boundary layer with a Blasius velocity profile, $\delta^* / \theta = 2.59$. After transition, when the boundary layer was fully turbulent, the T3A data had $\delta^* / \theta = 1.46$. Note that like R_x, θ increases monotonically along the plate. The location of the start of transition may therefore be defined in terms of a momentum thickness Reynolds number,

$$R_\theta = \frac{\theta U_\infty}{\nu} , \tag{11.1.7}$$

instead of R_x.

Dryden (1936) concluded that the location of the start of transition was primarily a function of the turbulence intensity of the free stream flow, with high turbulence resulting in significantly earlier transition. For data that measured only the velocity fluctuations in the x_1 direction, the turbulence intensity is defined by

$$T_u(\%) = 100 \frac{\sqrt{\overline{u_1 u_1}}}{U_\infty} . \tag{11.1.8}$$

Some authors, having more information available, define the turbulence intensity in terms of the turbulence kinetic energy,

$$T_u(\%) = 100 \frac{\sqrt{\frac{2}{3} k}}{U_\infty} = 100 \frac{\sqrt{\frac{\overline{u_1 u_1} + \overline{u_2 u_2} + \overline{u_3 u_3}}{3}}}{U_\infty} , \tag{11.1.9}$$

so that the definitions are consistent when the turbulence is isotropic. For grid generated turbulence with say $b_{11} = 0.1$, Eq. 11.1.8 gives turbulence intensities which are 14 percent higher than Eq. 11.1.9.

Figure 11.1.3 shows R_θ for the start of transition as a function of free stream turbulence intensity for data from a variety of sources. When only R_x was available, R_θ was calculated assuming a Blasius profile at the start of transition so that

$$R_\theta = 0.664\sqrt{R_x} \ . \qquad (11.1.10)$$

The trend suggested by Dryden is clearly present above a turbulence intensity of 0.1%, with R_θ at the start of transition decreasing from over 1000 to below 200 as the turbulence intensity increases from 0.1% to 5%.

Below $T_u = 0.1\%$, two sets of experiments suggest that R_θ is independent of turbulence intensity. The R_θ for the Wells (1967) data is higher than that for the Schubauer and Skramstad (1948) data, perhaps because Wells was better able to reduce externally generated disturbances to the flow, such as noise.

It is generally thought that there are two mechanisms for transition in zero pressure gradient flows: Tollmien-Schlichting waves which initiate transition at low free stream turbulence, and 'bypass' transition in which higher free stream turbulence preempts the initial stages of transition due to Tollmien-Schlichting waves (Morkovin, 1969). Observations and measurements of the detailed structures of turbulence during transition were also reviewed, more recently, by Mayle (1991). Such detail is considered only in an average sense by Reynolds stress models.

Hislop (1940) considered the influence of length scale as well as turbulence intensity on transition. He used three different mesh sizes in the grids which were used to generate the free stream turbulence upstream of the plate. At a turbulence intensity of 1.6 percent, R_x for the start of transition was delayed from 265,000 to 383,000 as the mesh spacing was increased from 0.5 to 2 inches. This corresponds to a 20 percent change in R_θ at the start of transition.

Turbulence length scale accounts for some of the scatter seen in Fig. 11.1.3. Other reasons for the scatter, besides experimental error, include variation in turbulence intensity along the plate and the location for which it is given, and the various methods used to determine the start of transition. From Fig. 11.1.2, it might be considered that transition starts as early as $R_x = 100{,}000$, based on the departure of c_f from the laminar line, or as late as $R_x = 200{,}000$ based on $\overline{u_1 u_1}$.

The filled squares and circles on Fig. 11.1.3 represent the T3A–, T3A and T3B cases available through ERCOFTAC, with the start of transition taken at the minimum c_f, but plotted in two different ways, to demonstrate the variations which can occur simply due to interpretation. For the squares, T_u is at the start of transition and Eq. 11.1.10 was used to obtain R_θ from R_x. The circles have higher values for both T_u and R_θ; these are the measured R_θ at the start of transition, and T_u midway between the plate leading edge and the start of transition.

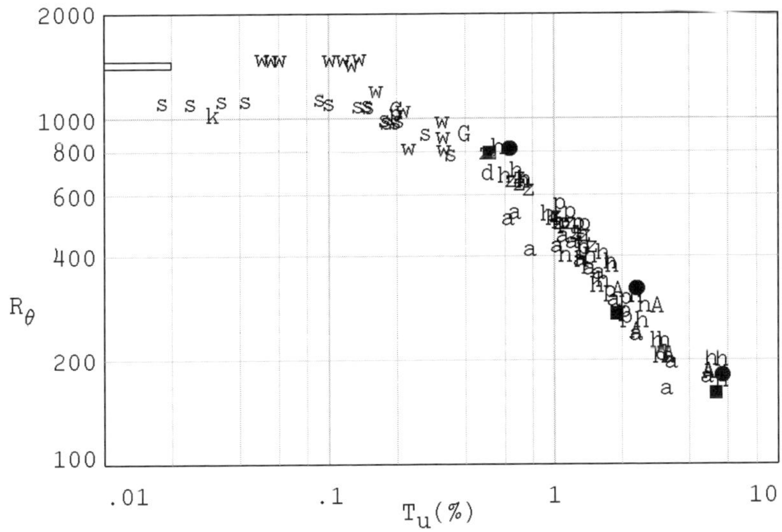

Fig. 11.1.3. Momentum thickness Reynolds number at start of transition as a function of free stream turbulence intensity. Data: a, A – Abu-Ghannam and Shaw (AS1980); □ – Boltz et al. (BKA1960); d – Dryden (D1936); k – Schubauer and Klebanoff (SK1956); s – Schubauer and Skramstad (SS1948); w – Wells (W1967); z –Zysina-Molozhen and Kuznetsova (ZK1969); ●, ■, – Roach and Brierley (1990), and Ryhming (1990). Data from Hall and Gibbings (HG1972): G – Hall and Gibbings; h – Hislop (1940); p – Hall and Hislop (1938). A, G, ● – measured R_θ; otherwise Eq. 11.1.10. a, A, ● – T_u midway between leading edge and start of transition; ■ – T_u at start of transition; otherwise unspecified.

11.2 Calculation Procedure

For the flat plate transition test cases, a marching calculation procedure is used to test the Reynolds stress models. While this necessitates some assumptions, it allows the economical use of fine grid spacing normal to the wall, and small step sizes in the mean flow direction, removing errors due to coarse grids.

The code used is a much modified version of the marching code pioneered by Patankar and Spalding (1968). The mean momentum and Reynolds stress equations are solved along a series of streamlines in the flow, as sketched in Fig. 11.2.1, with the stream variable, ξ, defined by

$$\xi = \int_0^y U_1 dy \quad . \tag{11.2.1}$$

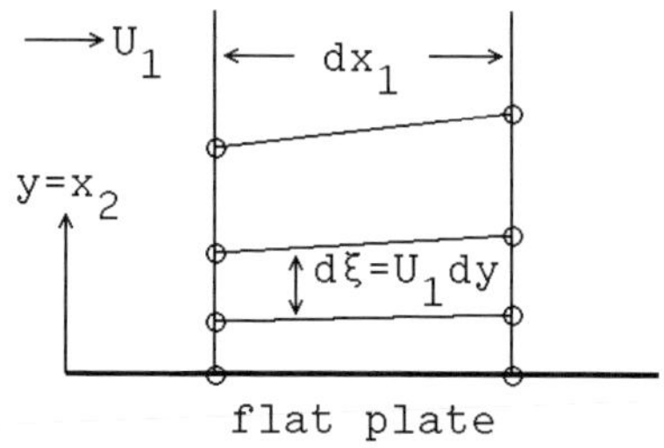

Fig. 11.2.1. x_1, ξ coordinates for one step in the marching calculation procedure.

The assumptions for the mean momentum equation are that diffusion, both laminar and due to turbulent stresses, is negligible in the x_1 direction. It is also assumed that there are no y variations in $\partial \overline{P}/\partial x_1$. For these zero pressure gradient test cases, $\partial \overline{P}/\partial x_1 = 0$ and the mean momentum equation (Eq. 2.1.10) reduces to

$$U_1 \frac{\partial U_1}{\partial x_1} + U_2 \frac{\partial U_1}{\partial x_2} = -\frac{\partial \overline{u_1 u_2}}{\partial y} + \frac{\partial}{\partial y} \nu \frac{\partial U_1}{\partial y} . \tag{11.2.2}$$

Changing to the von Mises, x_1,ξ, coordinates, where the derivative with respect to x_1 is taken at constant ξ,

$$\left(\frac{\partial U_1}{\partial x_1}\right)_\xi = \left(\frac{\partial U_1}{\partial x_1}\right)_{x_2} + \left(\frac{\partial U_1}{\partial x_2}\right)_{x_1} \left(\frac{\partial x_2}{\partial x_1}\right)_\xi = \left(\frac{\partial U_1}{\partial x_1}\right)_{x_2} + \left(\frac{\partial U_1}{\partial x_2}\right)_{x_1} \frac{U_2}{U_1} , \tag{11.2.3}$$

the equation to be discretized over the grid shown in Fig. 11.2.1 may be written as

$$\frac{\partial U_1}{\partial x_1} = -\frac{\partial \overline{u_1 u_2}}{\partial \xi} + \frac{\partial}{\partial \xi} \nu \frac{\partial U_1}{\partial y} . \tag{11.2.4}$$

Correspondingly, neglecting diffusion in the x_1 direction, the Reynolds stress equations (Eqs. 8.1.1-2) take the form

$$\frac{\partial \psi}{\partial x_1} = \frac{\partial}{\partial \xi} \nu \frac{\partial \psi}{\partial y} + \frac{\partial}{\partial \xi}\left(c_{t\psi} \frac{q^2}{\omega} \delta_{k2}\delta_{n2} \left(\frac{\overline{u_k u_n}}{k}\right)^{n_t} \frac{\partial \psi}{\partial y}\right) + \frac{S_\psi}{U_1} , \quad \text{for} \qquad \psi = q, \omega, b_{ij} . \tag{11.2.5}$$

The source terms, S_ψ, are evaluated assuming that the only mean velocity gradient of significance is $\partial U_1/\partial y$.

The equations are discretized using a mixture of upstream (known) and downstream (being calculated) values. For example, the gradients $\partial U_1/\partial y$ and

$\partial\psi/\partial y$ which appear in the viscous terms use downstream values of U_1 and ψ. The y location of the grid points at each x_1 is found by inverting Eq. 11.2.1,

$$y = \int_0^{\xi} \frac{1}{U_1} d\xi \quad . \tag{11.2.6}$$

All calculations were started at the plate leading edge with uniform properties at all grid points except the wall, i.e. a one-point boundary layer. The near-wall grid point was at $y^+ \approx 0.4$, and remained at this location throughout the calculation, because of the streamline grid, even though both y and U_1 changed. As the boundary layer grew it encompassed more grid points, so that at the start of transition there were about 50 grid points in the boundary layer. An adjustable step size, δx_1, was used to limit the change in properties over the step to a maximum of about 2 percent at each grid point. In general, the calculations used between 1500 and 20,000 steps to reach the start of transition.

11.3 Results from the MARV Model

To obtain the start of transition from the MARV model, several sets of calculations were made starting with isotropic turbulence and varying T_u and R_t. Results which started with 2 percent turbulence at the leading edge are shown in Fig. 11.3.1. For the calculations, the start of transition is taken as the location of minimum skin friction. The solid line gives the calculated R_θ and free stream turbulence intensity at the start of transition. The dashed lines represent the trajectories of the calculations with various starting turbulence Reynolds numbers. Lower R_t corresponds to higher dissipation so that the turbulence intensity falls more rapidly with distance along the plate, and correspondingly falls more rapidly with increased R_θ. Transition was obtained for values of R_t at the leading edge of 28 and above. Below this, the calculated boundary layer profiles remained laminar.

A more complete picture of transition from the MARV model is shown in Fig. 11.3.2. Here the data is overlaid with a grid showing the start of transition for lines of constant T_u and lines of constant R_t at the leading edge. The grid overlays all the data above $T_u = 0.2$ percent, so that the model predicts by-pass transition. It does not cover the data below 0.1 percent where the transition mechanism is Tollmien-Schlichting waves, and externally generated disturbances to the flow, such as noise, affect the R_θ of transition.

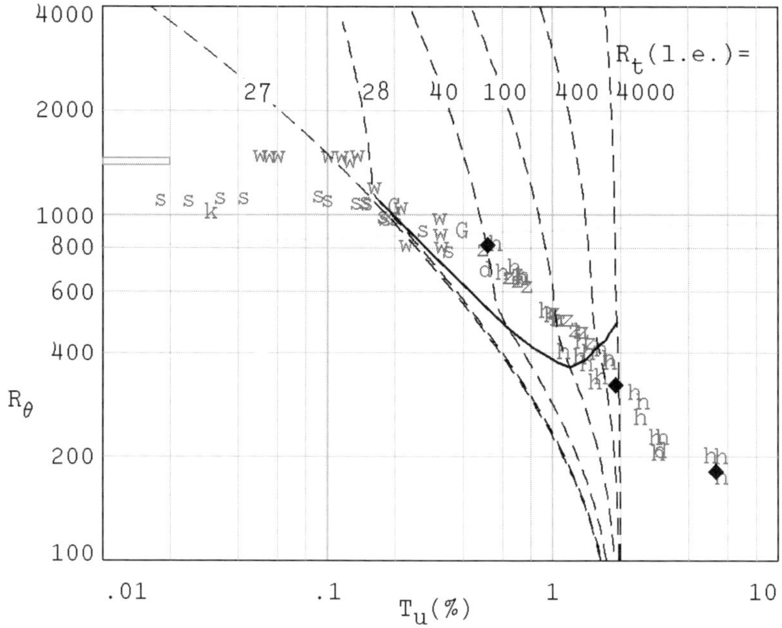

Fig. 11.3.1. Momentum thickness Reynolds number as a function of free stream turbulence intensity. MARV model calculations, $T_u(\%)=2$ and various R_t at leading edge: - - - free stream T_u versus R_θ; —— start of transition. Data for start of transition: ◆, ERCOFTAC cases T3A–, T3A and T3B; other data as for Fig. 11.1.3.

At the high T_u end, calculations above 6 percent are not included on the figure as the free stream turbulence levels are similar to maximum intensities expected in fully turbulent boundary layers – about 10 percent of the free stream velocity. Correspondingly, transition may be so gradual that there is no minimum c_f to mark the start.

Effect of Turbulence Length Scale

The results from the MARV model suggest that the effect of turbulence length scale, shown on Fig. 11.3.2 as an R_t effect, can be considerable. Most of Hislop's data (symbol h) falls between the calculations with $R_t = 100$ to 400. This may be consistent with the factor of four variation in the size of turbulence grids which he used. But the magnitude of the mean velocity and the turbulence intensity also affect R_t,

$$R_t = \frac{q^2}{\nu\omega} = \frac{1}{\nu} q \frac{q}{\omega} \sim \frac{1}{\nu} U_1 \, T_u \, Length \, . \tag{11.3.1}$$

For the ERCOFTAC test cases (symbol ◆ on Fig. 11.3.2), free stream turbulence kinetic energy data in the region of transition is given as a function of distance along the plate. From this, the turbulence intensity and Reynolds number at the plate leading edge may be estimated. These are shown in Table 11.3.1. Interpolating on the calculated grid in Fig. 11.3.2 for the location of transition of these points, they are in reasonable agreement with the data both in terms of T_u

and R_t. With the data having variations in R_t of nearly a factor of 15, this supports the R_t, or length scale, variation in the start of transition obtained from the MARV model.

The apparent cohesion of the data on Fig. 11.3.2, suggesting simply a T_u dependence for the start of transition, may reflect the similarity of the experimental setups in terms of overall dimensions, and velocities used for hot-wire measurements. What could be changed is the size of the upstream grids. Large grids were used to generate high turbulence intensities, while, for example, Wells (symbol w on Fig. 11.3.2) and Schubauer and Skramstad (symbol s) used very fine meshes to obtain low turbulence intensities. The fine meshes also resulted in small length scales and low turbulence Reynolds numbers.

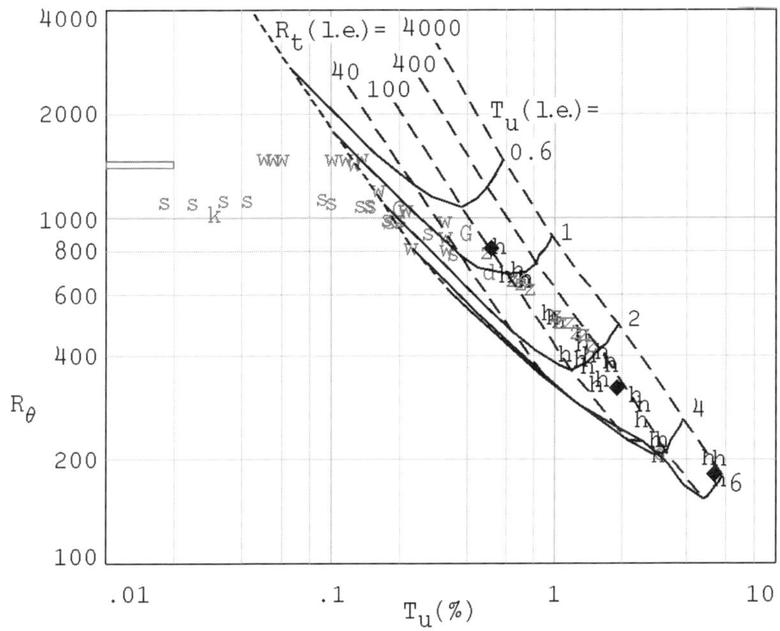

Fig. 11.3.2. Start of transition from the MARV model: —— lines with constant T_u at the leading edge; – – – lines with constant R_t at the leading edge; - - - lower limit for transition, R_t = 24 to 34. Data as for Fig. 11.3.1.

Table 11.3.1. Estimate of conditions at the plate leading for ERCOFTAC test cases (Appendix RB1990). T_u and R_t are based on least-square fits of free stream turbulence data. b_{ij} are from the free stream data at the first measurement location.

Case	T_u(%) based on k	R_t	b_{11}	b_{22}	b_{33}
T3A–	0.78	78	0.080	–0.088	0.008
T3A	2.94	127	0.066	–0.069	0.003
T3B	5.28	1133	0.072	–0.129	0.057

ERCOFTAC Test Cases

This section compares in more detail the MARV model results and data for the ERCOFTAC test cases. Calculations were made for both the anisotropic inlet conditions given in Table 11.3.1, and for isotropic turbulence, but with the same T_u and R_t. Figure 11.3.3 shows the development of skin friction as a function of both R_x and R_θ. The start of transition is well modeled for all three cases, although the calculations starting with anisotropic turbulence had slightly later starts for transition by about 10 percent on R_θ. The calculations also show the pre-transition departure of c_f from the laminar line, evident in the data at the higher free stream turbulence intensities.

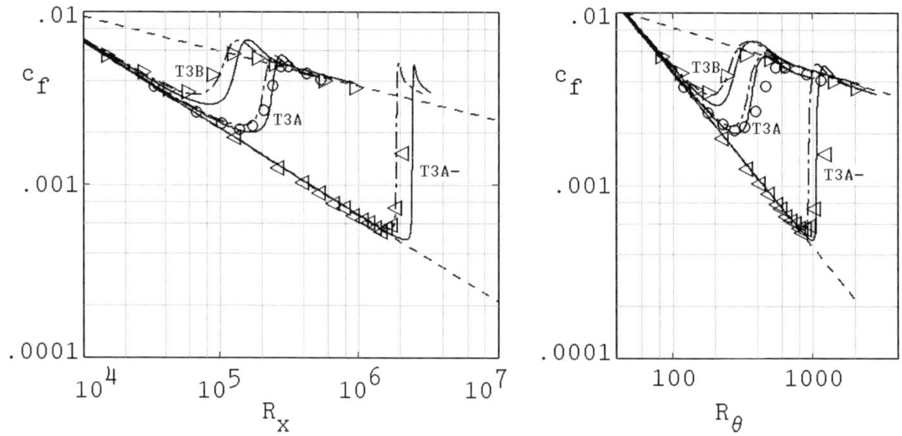

Fig. 11.3.3. Development of skin friction coefficient as a function of R_x and R_θ for ERCOFTAC cases. Measurements: ▷, T3B; ○, T3A; ◁, T3A–. MARV model: —— inlet conditions from Table 11.3.1; – – – b_{ij}=0 at inlet. - - - laminar and turbulent limits.

Figures 11.3.4 and 11.3.5 show additional comparisons between the model results and data for case T3A at three locations, x_1 = 0.095, 0.395 and 1.495 meters. The middle location is the start of transition, the minimum c_f in the data. The first location is the first full measurement station, a quarter of the distance to the start of transition, i.e. nominally in the laminar boundary layer region. The last location is the last measurement station, nearly four times the start location, where the boundary layer is fully turbulent.

Figure 11.3.4 shows the mean velocity, and the turbulence properties, q and b_{12}, which affect the mean velocity through the shear stress, $\overline{u_1 u_2} = 2q^2 b_{12}$. In agreement with the data the velocity profiles are laminar upstream and at the start of transition, and fully turbulent at the downstream location.

Both the data and the model show (Fig. 11.3.4b) increased levels of turbulence in the boundary layers, above free stream levels, at all locations. The model, however, under-predicts the level of pre-transitional turbulence in the boundary layer.

The turbulence production rate, $P = -2kb_{12}\partial U_1/\partial y$, is proportional to the shear anisotropy which is shown in Fig. 11.3.4c. Downstream, both the data and the model show b_{12} going to about –0.15 as expected in the log-law region of a fully turbulent boundary layer. In the pre-transitional boundary layers, both the data and the model show suppressed levels of $-b_{12}$, with the data, in general, showing more suppression than the model. The difference between the data and the modeled levels of b_{12} do not explain the higher pre-transitional q levels in the data.

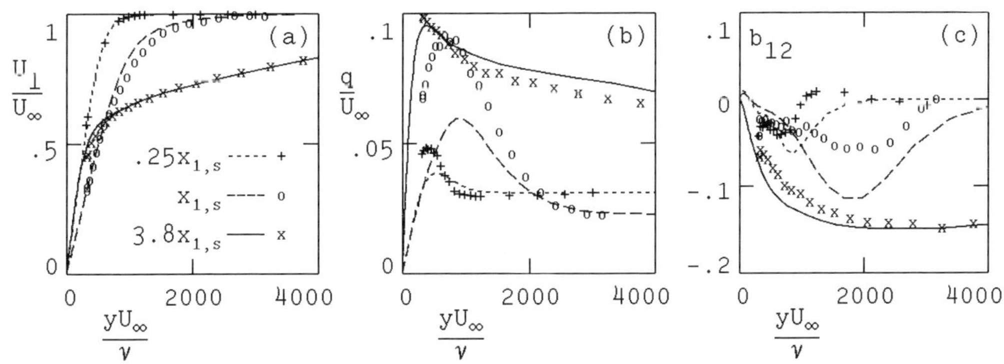

Fig. 11.3.4. Mean velocity, turbulence velocity and shear anisotropy for ERCOFTAC case T3A at three locations: +, ---, $x_{1,s}/4$; o, — —, $x_{1,s}$; x, ——, $3.8x_{1,s}$; where $x_{1,s}$ is the (data) location of the start of transition. Data: +, o, x; MARV model: ---, — —, ——.

The normal anisotropies are shown in Fig. 11.3.5. At the downstream location, Fig. 11.3.5c, the model is in good agreement with the data. At the pre-transitional locations, the model is in reasonable agreement with the data, but there are some interesting differences. At the first station, Fig. 11.3.5a, the data shows positive values for b_{33} at the edge of the boundary layer, and corresponding reduced values for b_{22}. This may be due to elliptic effects at the plate leading edge (which have not been included in the calculations) altering the near-wall anisotropy of the turbulence. The boundary layer thickness at this location is only 1.8 times the plate leading-edge diameter. The effect is less pronounced at the start of transition, Fig.11.3.5b.

Regardless of some of the shortcomings in the pre-transitional turbulence and anisotropy, it is clear that the MARV model captures the start of transition over a wide range of turbulence intensity and Reynolds number. In the next section we look at the features and model choices which made this possible.

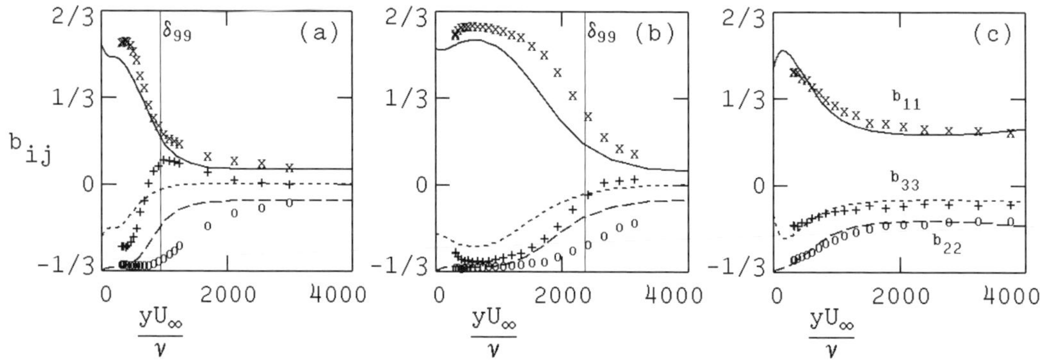

Fig. 11.3.5. Normal anisotropies for ERCOFTAC case T3A at three locations: (a) $x_{1,s}/4$, (b) $x_{1,s}$, (c) $3.8x_{1,s}$, where $x_{1,s}$ = (data) location of the start of transition. x, —, b_{11}; o, – –, b_{22}; +, - - -, b_{33}. Data: x, o, +; MARV model: —, – –, - - -.

11.4 Important Features of the MARV Model

The shear anisotropy and its pre-transitional behavior play an important role in controlling the onset of transition. In Fig. 11.4.1 various parameters are plotted to explain the calculated b_{12} profile for the T3A case at $x_{1,s}/4$, which was also shown earlier in Fig. 11.3.4c.

Reduction from Small Development Time

For the MARV model, the shear anisotropy at $x_{1,s}/4$ has a minimum value of about –0.05 just below $yU_\infty/\nu = 1000$. This is compared in Fig. 11.4.1a with the corresponding profile for b_{12} obtained using the algebraic (fdf) limit for the model, i.e. ignoring history, so that $b_{12} = b_{12}(R_t, S/\omega)$. This has a much lower minimum, about –0.15 between $yU_\infty/\nu = 1000$ and 1500, which corresponds to dimensionless strain rates, S/ω, of about 1 to 3, as shown on Fig. 11.4.1c. The main reason for the difference between the algebraic limit and the MARV solution beyond $yU_\infty/\nu = 1000$ is shown in Fig. 11.4.1b. The total strain, St, is so low that b_{12} has not had time to develop. Recall, Fig. 9.6.3 shows that for rapidly distorting turbulence, b_{12} has a minimum at $St = 2.4$, whereas in Fig. 11.4.1b, $St < 0.6$ for all $yU_\infty/\nu > 1000$. This reduction in $-b_{12}$ due to small development time is characteristic of all Reynolds stress models based on differential equations, but not on algebraic models (e.g. ARSM, k-ε).

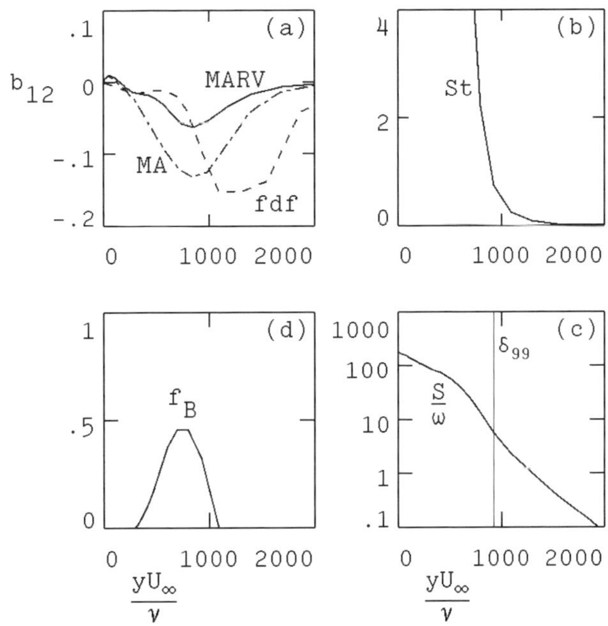

Fig. 11.4.1. Shear anisotropy and related parameters calculated for case T3A at $x_{1,s}/4$.
——, MARV model;
- - -, fully developed algebraic limit of MARV model;
—-—, MA model.
δ_{99}, boundary layer thickness where U_1/U_∞-0.99.

Reduction at High Strain Rate

Figure 11.4.1c shows very high dimensionless strain rates in the laminar boundary layer. All the Reynolds stress models developed in this book have the characteristic that $-b_{12}$ is reduced at high strain rate in the fully developed limit (Figs. 5.2.3 and 8.4.3). Correspondingly Fig. 11.4.1a shows near zero values of b_{12} for the fully developed limit below $yU_\infty/\nu = 700$, with the MARV model results, which are not yet fully developed, giving slightly more negative values. Note that, in two-equation turbulence models, some of which give reasonable results for transition (Savill, 1992), the reduction in shear stress near the wall is based on turbulence Reynolds number, independent of dimensionless strain rate.

Rapidly Distorting Turbulence

Figure 11.4.1d shows the rapid distortion parameter used in the MARV model to modify the pressure-strain coefficients, distinguishing between rapidly distorting flow ($f_B = 1$) and slowly changing or fully developed anisotropies ($f_B = 0$). This has the effect of reducing the magnitude of b_{12}, as shown on Fig. 11.4.1a, compared with results from a calculation using the MA model (equivalent to the MARV model but taking $f_B = 0$). Correspondingly, the MA model, which does not include rapid distortion effects, gives early transition as shown in Fig. 11.4.2.

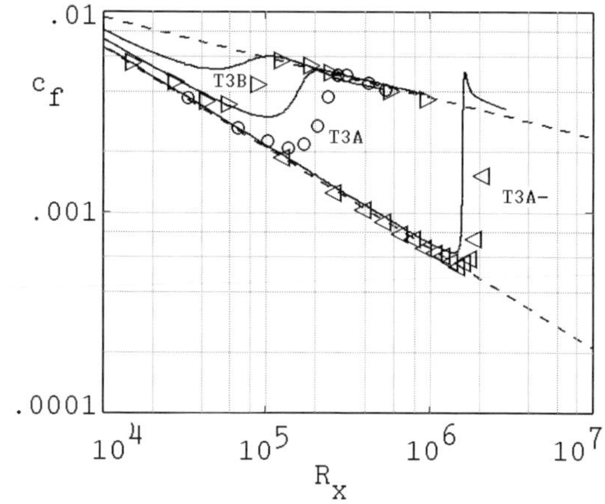

Fig. 11.4.2. Skin friction coefficients from the MA model for ERCOFTAC test cases. Measurements: ▷, T3B; o, T3A; ◁, T3A–. —— MA model. – – – laminar and turbulent limits.

Alternative rdt Limit

As noted in Chapter 9, the rdt,W=S limit for the pressure-strain coefficients in the MARV model does not include a c_6 term due to problems encountered modeling flat plate transition. Figure 11.4.3 compares results from the MARV model with results using the alternative limit including a c_6 term, Eq. 9.6.3. At $x_{1,s}/4$, the b_{12} values for the alternative model are lower than those for the MARV model, especially below $yU_\infty/\nu = 1000$ where the c_6 term acts to retard the time development, which is sending b_{12} towards zero at high strain rate. The net result in terms of skin friction is an early departure from the laminar line and early transition.

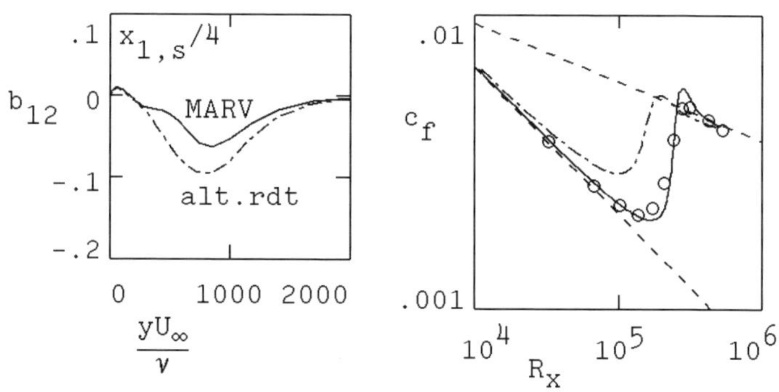

Fig. 11.4.3. Shear anisotropy at $x_{1,s}/4$, and skin friction coefficient for case T3A. ——, MARV model; – – –, model with alternative rdt,W=S limit, Eq. 9.6.3. o, data.

Boundary Condition for ω

In Chapter 6, three choices for the boundary condition on ω were considered (Eq. 6.4.10):

$$\text{(a)}\ \frac{\partial \omega}{\partial y} = 0, \qquad \text{(b)}\ \omega^+ = c, \quad \text{and} \qquad \text{(c)}\ \omega = c\frac{\partial q}{\partial y}. \tag{11.4.1}$$

Based on DNS data for fully developed channel flow, (c) appeared to be the best choice and was adopted (Sections 6.4 and 8.2). The influence of this choice on flat plate transition is shown in Fig. 11.4.4.

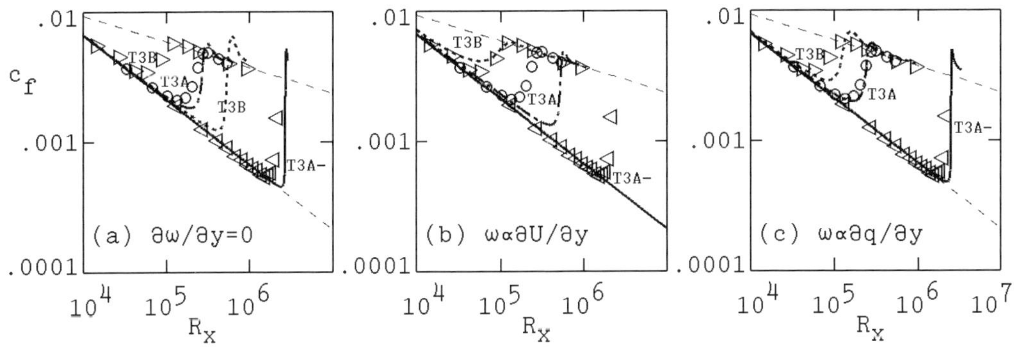

Fig. 11.4.4. Influence of different wall boundary conditions for ω on the development of skin friction coefficient for ERCOFTAC cases. Measurements: ▷, T3B; ○, T3A; ◁, T3A–. Calculations: --- T3B; — — — T3A; —— T3A–. (a) $\partial\omega/\partial y = 0$. (b) $\omega = 0.032\,\partial U_1/\partial y$. (c) $\omega = 0.089\,\partial q/\partial y$.

Figure 11.4.4a shows results for the MARV model modified to use $\partial\omega/\partial y = 0$ at the wall. Results are reasonable for cases T3A and T3A–, but transition is very late for case T3B. From Table 11.3.1, case T3B is different in that it begins with high turbulence Reynolds number, $R_t = 1133$. With the $\partial\omega/\partial y = 0$ boundary condition, ω remains low through the pre-transitional boundary layer. While this reduces turbulence dissipation, which should encourage transition, it also results in high S/ω, higher values of f_B (Eqs. 10.4.1 and 10.4.6) and high turbulent diffusion, all of which inhibit $-b_{12}$ and therefore inhibit transition.

The boundary condition of $\omega = 0.032\,\partial U_1/\partial y$, Fig. 11.4.4b, fixes S/ω at the wall; $S/\omega = 31$ independent of turbulence intensity or free stream conditions. For the three cases considered, this had the effect of increasing ω near the wall in the pre-transitional boundary layers, compared to the other wall boundary conditions. The T3B case had earlier transition than for the other two boundary conditions, while the T3A case gave later transition, and the T3A– case remained laminar. For the T3A and T3A– cases, transition was inhibited by high turbulence dissipation.

The boundary condition used in the MARV model (ω proportional to $\partial q/\partial y$ at the wall, Fig. 11.4.4c) gives a balance of the competing effects that ω has on the pre-transitional boundary layer, which results in reasonable locations for the start of transition over a wide range of turbulence intensity and initial turbulence Reynolds number.

11.5 Upgrade to the MC model – the MR model

In this section we add a rapid distortion functionality to the MC model, forming the MR model, to see how well transitional skin friction coefficients can be obtained within the framework of a relatively simple model.

For the fdf limit at $f_B = 0$, the MR model uses the MC model coefficients, $c_3 = 0.15$, and $c_4 = 1$, and for the rdt limit at $f_B = 1$, the MR model uses the isotropic limits, $c_3 = 0.8$ and $c_4 = 24/7$, so that

MR model:
$$c_3 = 0.15(1 - f_B) + 0.8 f_B \ ,$$
$$c_4 = (1 - f_B) + \frac{24}{7} f_B \ . \tag{11.5.1}$$

B, and therefore f_B, are slightly different than for the MARV model (Eqs. 10.4.1 and 10.4.6). In particular, ω is replaced by $0.9\tilde{\varepsilon}/k + 0.1\varepsilon/k$,

MR model:
$$B = \frac{k}{(0.9\tilde{\varepsilon} + 0.1\varepsilon)} \frac{b_{ij}}{II} \frac{db_{ij}}{dt}; \quad f_B(B) = \frac{\max(0, B - 0.7)}{\max(0, B - 0.7) + 1.7} \ . \tag{11.5.2}$$

This ensures that B and f_B go to zero at the wall, a condition compatible with the $\partial b_{ij}/\partial y = 0$ wall boundary condition.

Skin friction coefficients from the MC and MR models for the ERCOFTAC test cases are compared in Fig. 11.5.1. The addition of rapid distortion functionality has delayed the start of transition so that the MR model is in reasonable agreement with the data.

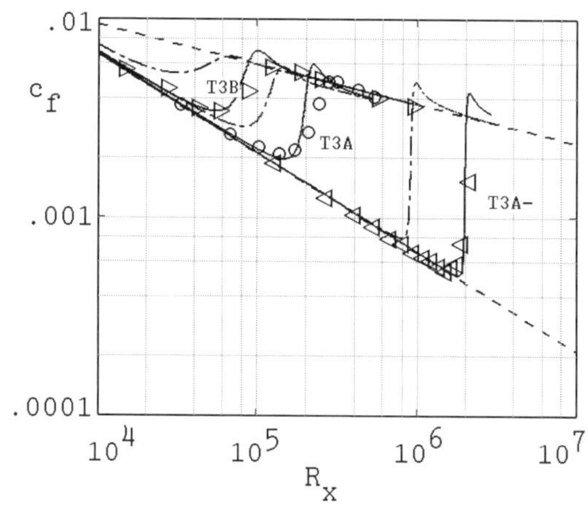

Fig. 11.5.1. Skin friction coefficients from the MC and MR models for ERCOFTAC test cases. Measurements: ▷, T3B; ○, T3A; ◁, T3A–. Models: ——, MR; – – –, MC. – – – laminar and turbulent limits.

Chapter 12

THE EFFECTS OF WALLS ON TURBULENCE

The purpose of this chapter is two-fold. The first is to look at the near-wall limiting behavior of Reynolds stresses and terms in the Reynolds stress equations, and to compare with results of the MARV model, for fully developed channel flow. The second is to determine what additions to the models are needed for the calculation of flows with rotating walls.

Taylor series analysis suggests the limiting behavior of Reynolds stresses, and therefore the turbulence anisotropy near walls. Limiting analyses may also be applied to dissipation and other terms in the Reynolds stress equations, and correspondingly the b_{ij} equations. These analyses result in regrouping of terms, defining inhomogeneous pressure-strain terms which are implied by the b_{ij} equation formulation.

Considering the near-wall turbulence to be approximately fully developed, together with the limiting behavior from the Taylor series analysis, suggests that the turbulence anisotropies near walls are the same, whether or not the wall is rotating. The MARV model is modified to make the near-wall turbulence anisotropies material frame indifferent by an addition to the pressure-strain model, which is essentially a wall reflection term. Additional discussion of wall reflection in developing flows, and of material frame indifference in high vorticity flows, is deferred to Chapters 16 and 18.

12.1 Wall Coordinates

It is convenient for the discussion of walls to define a coordinate system relative to the wall. It is also convenient to use names for the fluctuating velocities which do not have subscripts. Therefore, we use interchangeably,

$$u = u_1, \qquad v = u_2, \qquad \text{and} \qquad w = u_3. \tag{12.1.1}$$

Consider the flow near a smooth wall in a relative frame such that the wall appears stationary. Consider Cartesian coordinates, Fig. 12.1.1, such that at the point under consideration on the wall, the x_1 direction is parallel to the wall in the direction of the limiting mean streamline, and the x_2 or y direction is normal to the wall. At the wall, viscous effects force a no slip condition so that both the relative mean, and fluctuating components, of the velocity are zero.

$$U_1 = U_2 = U_3 = 0, \qquad u = v = w = 0. \tag{12.1.2}$$

It should also be noted that $U_2 = 0$ and $v = 0$ at the wall due to impermeability. This blockage condition is imposed through pressure gradients.

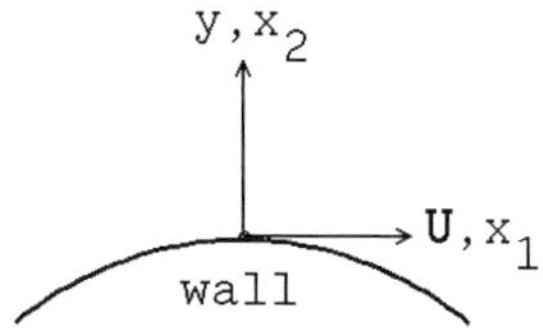

Fig. 12.1.1. Cartesian coordinates defined relative to a point on the wall.

Because the wall is smooth without sharp corners, the velocity gradients parallel to the no-slip wall are zero at the wall point.

$$\frac{\partial U_1}{\partial x_1} = \frac{\partial U_2}{\partial x_1} = \frac{\partial U_3}{\partial x_1} = \frac{\partial U_1}{\partial x_3} = \frac{\partial U_2}{\partial x_3} = \frac{\partial U_3}{\partial x_3} = 0 \ , \tag{12.1.3}$$

$$\frac{\partial u}{\partial x_1} = \frac{\partial v}{\partial x_1} = \frac{\partial w}{\partial x_1} = \frac{\partial u}{\partial x_3} = \frac{\partial v}{\partial x_3} = \frac{\partial w}{\partial x_3} = 0 \ . \tag{12.1.4}$$

For incompressible flow, continuity gives

$$\frac{\partial U_1}{\partial x_1} + \frac{\partial U_2}{\partial x_2} + \frac{\partial U_3}{\partial x_3} = 0 \qquad \text{and} \qquad \frac{\partial u}{\partial x_1} + \frac{\partial v}{\partial x_2} + \frac{\partial w}{\partial x_3} = 0 \ , \tag{12.1.5}$$

so that the combination of the no-slip wall, the smooth wall, and continuity give

$$\frac{\partial U_2}{\partial x_2} = 0 \qquad \text{and} \qquad \frac{\partial v}{\partial x_2} = 0, \qquad \text{at the wall.} \tag{12.1.6}$$

Since the x_1 direction was chosen as the direction of the limiting mean streamline, the gradient $\partial U_3 / \partial x_2$ is also zero, and the only non-zero component of the mean velocity gradient tensor, $\partial U_i / \partial x_j$, is $\partial U_1 / \partial x_2$. The non-zero components of the fluctuating velocity gradient tensor are $\partial u / \partial y$ and $\partial w / \partial y$.

12.2 Taylor Series Analysis for Near-Wall Reynolds Stresses

Near the wall the fluctuating pressure and fluctuating velocity can be written as Taylor series in y about the wall point. Since $u = v = w = 0$, and $\partial v / \partial y = 0$ at the wall, we may write

$$
\begin{aligned}
p &= p_0 + a_0 y + \ldots. \\
u &= \qquad a_1 y + b_1 y^2 + \ldots. \\
v &= \qquad\qquad b_2 y^2 + \ldots. \\
w &= \qquad a_3 y + b_3 y^2 + \ldots. \quad ,
\end{aligned}
\tag{12.2.1}
$$

where p_0, a_i and b_i are functions of t, x_1 and x_3. It follows that

$$
\begin{aligned}
&\overline{uu} = \overline{a_1 a_1}\ y^2 + \ldots, \quad &&\overline{vv} = \overline{b_2 b_2}\ y^4 + \ldots, \quad &&\overline{ww} = \overline{a_3 a_3}\ y^2 + \ldots\ , \\
&\overline{uv} = \overline{a_1 b_2}\ y^3 + \ldots, \quad &&\overline{uw} = \overline{a_1 a_3}\ y^2 + \ldots, \quad &&\overline{vw} = \overline{b_2 a_3}\ y^3 + \ldots\ .
\end{aligned}
\tag{12.2.2}
$$

A review of DNS data for fully developed channel flow suggests the extent of these near-wall dependencies. Figure 12.2.1 shows the data of Iwamoto et al. (ISK2002), for $R_\tau = 396$. This data set is typical, but was chosen because the near-wall point is close to the wall, below $y^+ = 0.01$. Figure 12.2.1 shows that $\overline{uu}$ is proportional to y^2 up to about $y^+ = 4$, $\overline{ww}$ is proportional to y^2 up to perhaps $y^+ = 1$, $\overline{uv}$ is proportional to y^3 to nearly $y^+ = 4$, and $\overline{vv}$ is proportional to y^4 to perhaps $y^+ = 2$. Since these limiting functionalities are valid only very near the wall, deep in the laminar sublayer, they may not be of any practical importance.

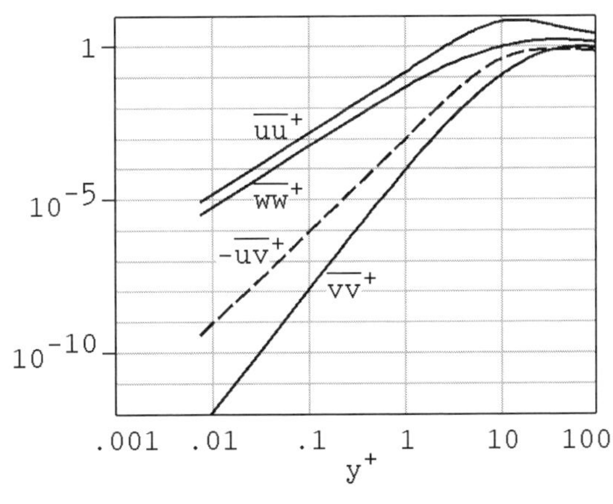

Fig. 12.2.1. Reynolds stresses in fully developed channel flow. DNS data of Iwamoto et al. (ISK2002) for $R_\tau = 396$.

The above Taylor series analysis shows the desired near-wall behavior of the variables q and b_{ij} used in the Reynolds stress models developed in this book. From Eq. 12.2.2,

$$
2q^2 = 2k = \left(\overline{a_1 a_1} + \overline{a_3 a_3}\right) y^2 + \ldots\ , \tag{12.2.3}
$$

and

$$b_{11} = \frac{\overline{a_1a_1}}{\overline{a_1a_1} + \overline{a_3a_3}} - \frac{1}{3} + ..., \quad b_{22} = -\frac{1}{3} + \frac{\overline{b_2b_2}}{\overline{a_1a_1} + \overline{a_3a_3}}y^2 + ..., \quad b_{33} = \frac{\overline{a_3a_3}}{\overline{a_1a_1} + \overline{a_3a_3}} - \frac{1}{3} + ... \tag{12.2.4}$$

$$b_{12} = \frac{\overline{a_1b_2}}{\overline{a_1a_1} + \overline{a_3a_3}}y + ..., \quad b_{13} = \frac{\overline{a_1a_3}}{\overline{a_1a_1} + \overline{a_3a_3}} + ..., \quad b_{23} = \frac{\overline{b_2a_3}}{\overline{a_1a_1} + \overline{a_3a_3}}y + ... \;.$$

Analysis of the MARV Model

For the MARV model, the wall boundary conditions for b_{ij} are given in Eqs. 8.3.18-19. Also, $q = 0$ at the wall. With these boundary conditions, the near-wall behavior takes the form

$$q = b_q y + ...$$

$$b_{11} = A_{11} + B_{11}y + ..., \quad b_{22} = -\frac{1}{3} + B_{22}y + C_{22}y^2 ...., \quad b_{33} = A_{33} + B_{33}y + ... \tag{12.2.5}$$

$$b_{12} = B_{12}y + ..., \quad b_{13} = B_{13}y + ..., \quad b_{23} = B_{23}y + ... \;.$$

The Reynolds stresses from the MARV model, $\overline{u_iu_j} = 2(b_{ij} + \delta_{ij}/3)q^2$, for the conditions of the DNS data in Fig. 12.2.1, are shown in Fig. 12.2.2. With $q = 0$ at the wall, all the Reynolds stresses go as at least y^2. $\overline{uu}$ and $\overline{ww}$ vary as y^2, and are consistent both qualitatively and quantitatively with the DNS data. With $b_{12} = 0$ and $b_{22} = -1/3$ at the wall, both $\overline{uv}$ and $\overline{vv}$ vary as y^3. This is qualitatively correct for $\overline{uv}$ but not $\overline{vv}$. Quantitatively, both $\overline{uv}$ and $\overline{vv}$ are in error by more than a factor of 2 below $y^+ = 3$. However, these errors have no significant effect on the mean velocity since $\overline{uv}$ contributes only 3 percent of the total shear stress at $y^+ = 3$, and 0.1 percent at $y^+ = 1$.

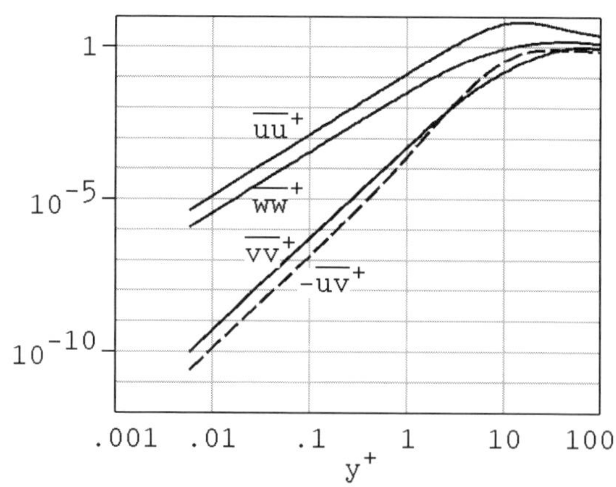

Fig. 12.2.2. Reynolds stresses for fully developed channel flow. MARV model calculations, for $R_\tau = 396$.

To obtain $\overline{vv}$ proportional to y^4 in the model would require the additional condition, $\partial b_{22}/\partial y = 0$ at the wall, which also follows from Eq. 12.2.4. This cannot be directly imposed within the framework of the MARV model. Durbin (1993), by solving additional differential equations for a normalized pressure-strain, Π_{ij}/k, and using appropriate boundary conditions on those, has managed to devise a model in which $\overline{vv}$ is proportional to y^4 in the laminar sublayer below $y^+ = 2$.

Analysis of Near-Wall Dissipation

Taylor series analysis has also been applied to the dissipation of the Reynolds stresses (Launder and Reynolds, 1983). Substituting the Taylor series, Eq. 12.2.1, into the equation for the dissipation (Eq. 2.2.21),

$$\varepsilon_{ij} = 2\nu \overline{\frac{\partial u_i}{\partial x_k}\frac{\partial u_j}{\partial x_k}}\,, \tag{12.2.6}$$

yields

$$\varepsilon_{11} = 2\nu\overline{a_1 a_1} + \ldots,\quad \varepsilon_{12} = 4\nu\overline{a_1 b_2}\,y + \ldots,\quad \varepsilon_{22} = 8\nu\overline{b_2 b_2}\,y^2 + \ldots,\quad \varepsilon_{33} = 2\nu\overline{a_3 a_3} + \ldots\,, \tag{12.2.7}$$

so that at the wall,

$$\frac{\varepsilon_{11}/\varepsilon}{\overline{u_1 u_1}/k} = 1,\quad \frac{\varepsilon_{12}/\varepsilon}{\overline{u_1 u_2}/k} = 2,\quad \frac{\varepsilon_{22}/\varepsilon}{\overline{u_2 u_2}/k} = 4,\quad \frac{\varepsilon_{33}/\varepsilon}{\overline{u_3 u_3}/k} = 1\,. \tag{12.2.8}$$

These ratios are shown for DNS data in Fig. 12.2.3. Again this limiting functionality is only valid very near the wall, deep in the laminar sublayer, although the dissipation of $\overline{vv}$ remains proportionally higher than that of the other normal stresses into the buffer region.

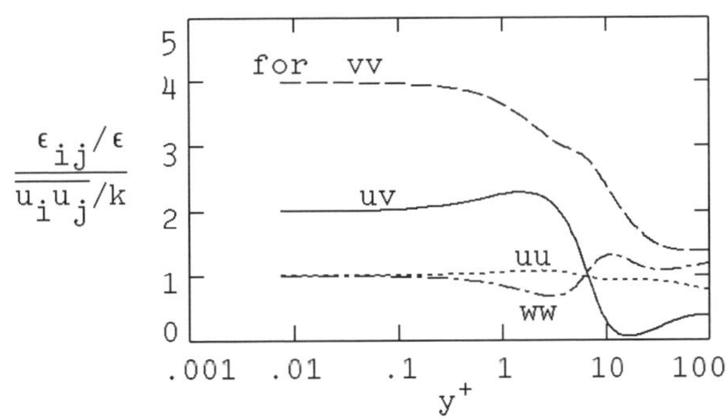

Fig. 12.2.3. Normalized components of dissipation for fully developed channel flow. DNS data of Iwamoto et al. (ISK2002) for $R_\tau = 396$.

12.3 Near-Wall Behavior of Terms in the Reynolds Stress Equations

Applying the Taylor series analysis near the wall to the terms in the $\overline{u_i u_j}$ equations, Eq. 2.1.13 or rewritten as Eq. 2.2.1,

$$\frac{\partial \overline{u_i u_j}}{\partial t} + U_k \frac{\partial \overline{u_i u_j}}{\partial x_k} = L_{ij} + P_{ij} + F_{ij} + T_{ij} + D_{ij} + \Pi_{ij} - \varepsilon_{ij} \,, \tag{12.3.1}$$

the terms vary from order one up to order y^5, as shown in Table 12.3.1.

The b_{ij} equations are formed from the $\overline{u_i u_j}$ equations. For $\overline{u_i u_j}$,

$$\overline{u_i u_j} = 2k(b_{ij} + \frac{1}{3}\delta_{ij}) = \overline{u_n u_n}(b_{ij} + \frac{1}{3}\delta_{ij}) \,, \tag{12.3.2}$$

the substantial derivative is

$$\frac{D\overline{u_i u_j}}{Dt} = \frac{\partial \overline{u_i u_j}}{\partial t} + U_k \frac{\partial \overline{u_i u_j}}{\partial x_k} = (b_{ij} + \frac{1}{3}\delta_{ij})\frac{D\overline{u_n u_n}}{Dt} + \overline{u_n u_n}\frac{Db_{ij}}{Dt} \,, \tag{12.3.3}$$

so that

$$\frac{Db_{ij}}{Dt} = \frac{1}{2k}\frac{D\overline{u_i u_j}}{Dt} - \frac{\overline{u_i u_j}}{2k}\frac{1}{2k}\frac{D\overline{u_n u_n}}{Dt} \,. \tag{12.3.4}$$

Noting that $F_{nn} = 0$, and $\Pi_{nn} = 0$, the exact equation for b_{ij} is

$$\frac{\partial b_{ij}}{\partial t} + U_k \frac{\partial b_{ij}}{\partial x_k} = \underbrace{\left(\frac{L_{ij}}{2k} - \frac{\overline{u_i u_j}}{2k}\frac{L_{nn}}{2k}\right)}_{\text{L term}} + \underbrace{\left(\frac{P_{ij}}{2k} - \frac{\overline{u_i u_j}}{2k}\frac{P_{nn}}{2k}\right)}_{\text{P term}} + \frac{F_{ij}}{2k} + \underbrace{\left(\frac{T_{ij}}{2k} - \frac{\overline{u_i u_j}}{2k}\frac{T_{nn}}{2k}\right)}_{\text{T term}}$$
$$+ \underbrace{\left(\frac{D_{ij}}{2k} - \frac{\overline{u_i u_j}}{2k}\frac{D_{nn}}{2k}\right)}_{\text{D term}} + \underbrace{\left(\frac{\Pi_{ij}}{2k}\right)}_{\Pi\text{ term}} - \underbrace{\left(\frac{\varepsilon_{ij}}{2k} - \frac{\overline{u_i u_j}}{2k}\frac{\varepsilon_{nn}}{2k}\right)}_{\text{E term}} \,. \tag{12.3.5}$$

Since k varies as y^2 near the wall, this would suggest that the terms in the b_{ij} equations are of order $1/y^2$ times the terms in the $\overline{u_i u_j}$ equations , or $1/y^2$ to y^3. The b_{ij} equations then appear to be ill posed, with the terms varying as $1/y$ and $1/y^2$ blowing up near the wall. However, as will be shown, rearrangement of the terms, together with some assumed algebraic contributions to the modeled terms due to inhomogeneity, consistent with DNS data, give well-posed b_{ij} equations. As shown in Table 12.3.1, the terms of concern are the ones containing L_{ij}, D_{ij}, Π_{ij} and ε_{ij}.

Table 12.3.1. Order of terms near the wall in the $\overline{u_i u_j}$ equations, Eq. 12.3.1.

Term	Equation for:	$\overline{uu}$	$\overline{vv}$	$\overline{ww}$	$\overline{uv}$	$\overline{uw}$	$\overline{vw}$
$\dfrac{\partial \overline{u_i u_j}}{\partial t}$		y^2	y^4	y^2	y^3	y^2	y^3
$U_k \dfrac{\partial \overline{u_i u_j}}{\partial x_k}$		y^3	y^5	y^3	y^4	y^3	y^4
$L_{ij} = \dfrac{\partial}{\partial x_k} \nu \dfrac{\partial \overline{u_i u_j}}{\partial x_k}$		1	y^2	1	y	1	y
$P_{ij} = -\overline{u_j u_k} \dfrac{\partial U_i}{\partial x_k} - \overline{u_i u_k} \dfrac{\partial U_j}{\partial x_k}$		y^3	y^5	y^3	y^4	y^3	y^4
$F_{ij} = -2\Omega_k \left(\varepsilon_{ikm} \overline{u_j u_m} + \varepsilon_{jkm} \overline{u_i u_m} \right)$		y^2	y^3	y^2	y^2	y^2	y^2
$T_{ij} = -\dfrac{\partial \overline{u_i u_j u_k}}{\partial x_k}$		y^3	y^5	y^3	y^4	y^3	y^4
$D_{ij} = -\dfrac{1}{\rho} \left(\dfrac{\partial \overline{p u_j}}{\partial x_i} + \dfrac{\partial \overline{p u_i}}{\partial x_j} \right)$		y	y	y	1	y	1
$\Pi_{ij} = \overline{\dfrac{p}{\rho} \left(\dfrac{\partial u_i}{\partial x_j} + \dfrac{\partial u_j}{\partial x_i} \right)}$		y	y	y	1	y	1
$\varepsilon_{ij} = 2\nu \overline{\dfrac{\partial u_i}{\partial x_k} \dfrac{\partial u_j}{\partial x_k}}$		1	y^2	1	y	1	y
$D_{ij} + \Pi_{ij} = -\dfrac{1}{\rho} \left(\overline{u_j \dfrac{\partial p}{\partial x_i}} + \overline{u_i \dfrac{\partial p}{\partial x_j}} \right)$		y	y^2	y	y	y	y

For the fully developed channel flow without rotation, the left hand side of Eq. 12.3.5 and F_{ij} are zero. The remaining terms in the b_{ij} equations are plotted in Fig. 12.3.1 with negative and positive values shown separately for the three independent b_{ij}'s, b_{11}, b_{22} and b_{12}, using the DNS data of Iwamoto et al. (ISK2002) for R_τ =396. In the b_{11} equation, the L, D, Π and E terms vary as $1/y$, with half of the positive L term balancing the E term. In the b_{22} equation, the D and Π terms vary as $1/y$, and balance out. In the b_{12} equation, the D and Π terms again appear to balance out, but vary as $1/y^2$. This suggests that the D and Π terms should be modeled together, not separately. This is confirmed by the Taylor series analysis of the combined velocity pressure-gradient term, $D_{ij}+\Pi_{ij}$ (last line in Table 12.3.1), which is higher order for the $\overline{vv}$ and $\overline{uv}$ equations, than the terms taken separately. In the b_{12} equation, half of the negative L term balances out the E term, with both varying as $1/y$.

The laminar diffusion term,

$$L_{ij}=\frac{\partial}{\partial x_k}\nu\frac{\partial\overline{u_iu_j}}{\partial x_k}\,, \tag{12.3.6}$$

may be expanded in terms of b_{ij},

$$L_{ij}=2k\frac{\partial}{\partial x_k}\nu\frac{\partial b_{ij}}{\partial x_k}+\frac{\overline{u_iu_j}}{2k}\frac{\partial}{\partial x_k}\nu\frac{\partial 2k}{\partial x_k}+2\nu\frac{\partial 2k}{\partial x_k}\frac{\partial b_{ij}}{\partial x_k}\,, \tag{12.3.7}$$

so that

$$\left(\frac{L_{ij}}{2k}-\frac{\overline{u_iu_j}}{2k}\frac{L_{nn}}{2k}\right)=\frac{\partial}{\partial x_k}\nu\frac{\partial b_{ij}}{\partial x_k}+\frac{2\nu}{k}\frac{\partial k}{\partial x_k}\frac{\partial b_{ij}}{\partial x_k}\,. \tag{12.3.8}$$

The last term may vary as $1/y$. This is also an inhomogeneous term, which depends on the spatial variation of mean turbulence properties. A rearrangement of the b_{ij} equation, combining half of this term with the pressure-velocity terms and half with the dissipation, gives

$$\begin{aligned}\frac{\partial b_{ij}}{\partial t}+U_k\frac{\partial b_{ij}}{\partial x_k}=&\underbrace{\left(\frac{\partial}{\partial x_k}\nu\frac{\partial b_{ij}}{\partial x_k}\right)}_{\text{L term}}+\underbrace{\left(\frac{P_{ij}}{2k}-\frac{\overline{u_iu_j}}{2k}\frac{P_{nn}}{2k}\right)}_{\text{P term}}+\frac{F_{ij}}{2k}+\underbrace{\left(\frac{T_{ij}}{2k}-\frac{\overline{u_iu_j}}{2k}\frac{T_{nn}}{2k}\right)}_{\text{T term}}\\&+\underbrace{\left(\frac{D_{ij}}{2k}-\frac{\overline{u_iu_j}}{2k}\frac{D_{nn}}{2k}+\frac{\Pi_{ij}}{2k}+\frac{\nu}{k}\frac{\partial k}{\partial x_k}\frac{\partial b_{ij}}{\partial x_k}\right)}_{\Pi\text{ term}}-\underbrace{\left(\frac{\varepsilon_{ij}}{2k}-\frac{\overline{u_iu_j}}{2k}\frac{\varepsilon_{nn}}{2k}-\frac{\nu}{k}\frac{\partial k}{\partial x_k}\frac{\partial b_{ij}}{\partial x_k}\right)}_{\text{E term}}.\end{aligned} \tag{12.3.9}$$

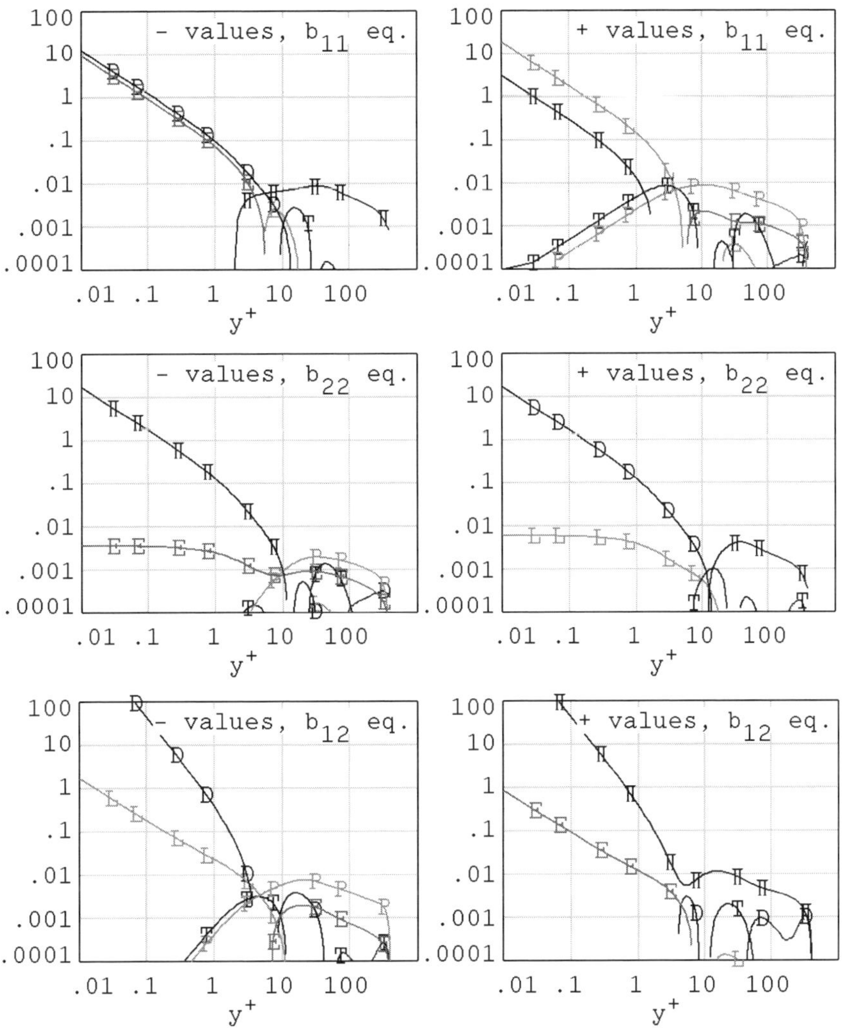

Fig. 12.3.1. Terms L, P, T, D, Π, and E, in Eq. 12.3.5 for b_{ij}, made dimensionless with U_τ and ν. Evaluated from the fully developed channel flow DNS data of Iwamoto et al. (ISK2002) for $R_\tau = 396$.

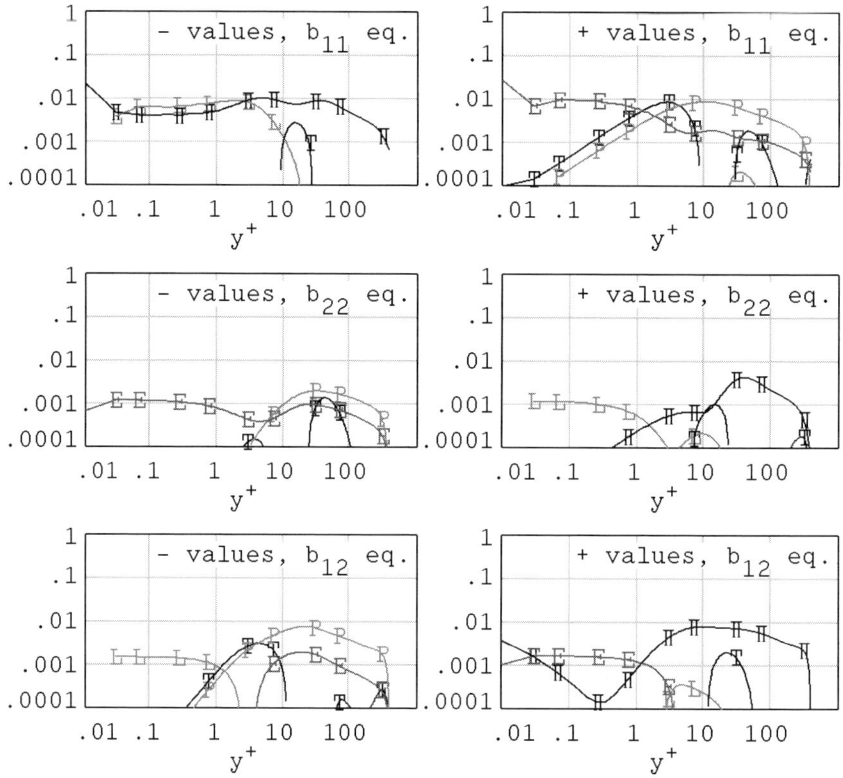

Fig. 12.3.2. Terms L, P, T, Π and E, in Eq. 12.3.9 for b_{ij}. Evaluated from the fully developed channel flow DNS data of Iwamoto et al. (ISK2002) for $R_\tau = 396$.

These rearranged terms are plotted in Fig. 12.3.2. None of the terms vary as $1/y$ or $1/y^2$.

The equations are then well posed, and terms containing the pressure-velocity correlation and dissipation may be modeled with a pressure-strain model of the standard form, as used for homogeneous turbulence. If we define a generalized effective pressure-strain, $\hat{\phi}_{ij}$, as

$$\frac{\hat{\phi}_{ij}}{2k} = \left(\frac{D_{ij}}{2k} - \frac{\overline{u_i u_j}}{2k}\frac{D_{nn}}{2k} + \frac{\Pi_{ij}}{2k} + \frac{\nu}{k}\frac{\partial k}{\partial x_k}\frac{\partial b_{ij}}{\partial x_k}\right) - \left(\frac{\varepsilon_{ij}}{2k} - \frac{\overline{u_i u_j}}{2k}\frac{\varepsilon_{nn}}{2k} - \frac{\nu}{k}\frac{\partial k}{\partial x_k}\frac{\partial b_{ij}}{\partial x_k}\right) - b_{ij}\frac{2\tilde{\varepsilon}}{2k}, \quad (12.3.10)$$

then for homogeneous turbulence, where spatial gradients are zero and $\varepsilon = \tilde{\varepsilon}$, $\hat{\phi}_{ij}$ reduces to the effective pressure-strain, ϕ_{ij}, as defined in Eq. 2.2.22,

$$\frac{\hat{\phi}_{ij}}{2k} = \frac{1}{2k}\left(\Pi_{ij} - \varepsilon_{ij} + \frac{\overline{u_i u_j}}{2k} 2\varepsilon\right) - b_{ij}\frac{2\varepsilon}{2k} = \frac{1}{2k}\left(\Pi_{ij} - \varepsilon_{ij} + \frac{2}{3}\varepsilon\delta_{ij}\right). \tag{12.3.11}$$

The equation for b_{ij} may then be written as

$$\begin{aligned} \frac{\partial b_{ij}}{\partial t} + U_k\frac{\partial b_{ij}}{\partial x_k} &= \frac{\partial}{\partial x_k}\nu\frac{\partial b_{ij}}{\partial x_k} + \left(\frac{P_{ij}}{2k} - \frac{\overline{u_i u_j}}{2k}\frac{P_{nn}}{2k}\right) + \frac{F_{ij}}{2k} \\ &+ \left(\frac{T_{ij}}{2k} - \frac{\overline{u_i u_j}}{2k}\frac{T_{nn}}{2k}\right) + \frac{\hat{\phi}_{ij}}{2k} + b_{ij}\frac{2\tilde{\varepsilon}}{2k} \quad , \end{aligned} \tag{12.3.12}$$

where items in the first line may be directly evaluated, and items in the second line require modeling.

Comparing this with the modeled equations for b_{ij}, Eqs. 8.1.1, 8.1.2 and 8.1.5 combined (Appendix EQ.12.3.13),

$$\begin{aligned} \frac{\partial b_{ij}}{\partial t} + U_k\frac{\partial b_{ij}}{\partial x_k} &= \frac{\partial}{\partial x_k}\nu\frac{\partial b_{ij}}{\partial x_k} + \left(\frac{P_{ij}}{2k} - \frac{\overline{u_i u_j}}{2k}\frac{P_{nn}}{2k}\right) + \frac{F_{ij}}{2k} \\ &+ \frac{\partial}{\partial x_k} c_{tbij}\frac{q^2}{\omega}\left(\frac{\overline{u_k u_n}}{k}\right)^{n_t}\frac{\partial b_{ij}}{\partial x_n} - \frac{1}{2}b_{ij}\left(c_1\omega + \frac{1}{2}c_6\frac{P}{k}\right) + \frac{1}{2}c_3 S_{ij} \\ &+ \frac{1}{2}c_4(b_{ik}S_{jk} + b_{jk}S_{ik} - \frac{2}{3}\delta_{ij}b_{mn}S_{mn}) + \frac{1}{2}c_5(b_{ik}W_{jk} + b_{jk}W_{ik}) + b_{ij}\frac{2\tilde{\varepsilon}}{2k}, \end{aligned} \tag{12.3.13}$$

then the modeled terms are

$$\begin{aligned} &\underbrace{\left\{\frac{T_{ij}}{2k} - \frac{\overline{u_i u_j}}{2k}\frac{T_{nn}}{2k}\right\}}_{\text{T term}} + \underbrace{\left\{\frac{\hat{\phi}_{ij}}{2k} + b_{ij}\frac{2\tilde{\varepsilon}}{2k}\right\}}_{\phi=\Pi+\text{E terms}} = \underbrace{\left\{\frac{\partial}{\partial x_k} c_{tbij}\frac{q^2}{\omega}\left(\frac{\overline{u_k u_n}}{k}\right)^{n_t}\frac{\partial b_{ij}}{\partial x_n}\right\}}_{\text{modeled T term}} \\ &+ \underbrace{\left\{ -\frac{1}{2}b_{ij}\left(c_1\omega + \frac{1}{2}c_6\frac{P}{k}\right) + \frac{1}{2}c_3 S_{ij} + \frac{1}{2}c_4(b_{ik}S_{jk} + b_{jk}S_{ik} - \frac{2}{3}\delta_{ij}b_{mn}S_{mn}) + \frac{1}{2}c_5(b_{ik}W_{jk} + b_{jk}W_{ik}) + b_{ij}\frac{2\tilde{\varepsilon}}{2k} \right\}}_{\phi=\text{ modeled }\Pi+\text{E terms}} . \end{aligned} \tag{12.3.14}$$

MARV model results (equivalent to the MA model for this case) for the terms in the b_{ij} equations are compared in Fig. 12.3.3 with the DNS data. Overall, for both the model and the data, b_{ij} is mostly determined by a balance between the production, term P, and the $\Pi + E$, or generalized effective pressure-strain term, marked as ϕ. These are in good agreement above $y^+ = 10$. In the model, the turbulent diffusion acts to smooth the results above about $y^+ = 10$, while the laminar diffusion smoothes and blends the results with the wall boundary conditions below about $y^+ = 10$.

Near the wall, below $y^+ = 3$, the data shows ϕ terms for the normal components, which have opposite signs than in the remainder of the flow. For b_{22} this results in $\partial b_{22} / \partial y = 0$ or $\overline{vv}$ proportional to y^4. There is no corresponding contribution to the ϕ term in the model.

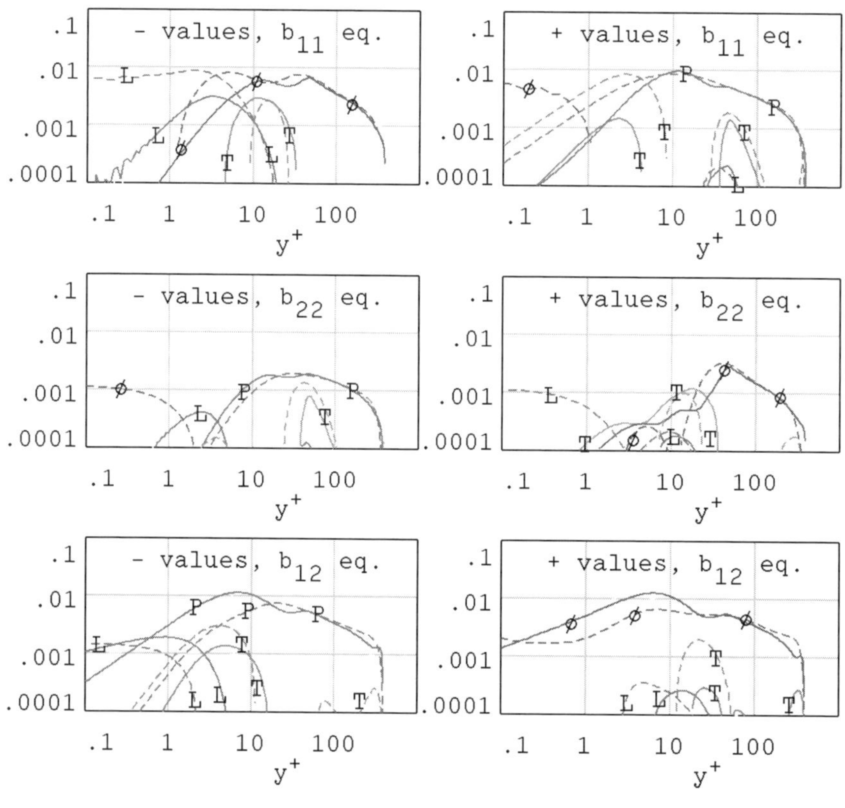

Fig. 12.3.3. Terms in the b_{ij} equations for fully developed channel flow, $R_\tau = 396$.
—— MARV model calculations; - - - evaluated from DNS data of Iwamoto et al. (ISK2002).
L, P, laminar diffusion and production of b_{ij}, Eq. 12.3.9. T, ϕ, turbulent diffusion and $\phi = \Pi + E$ terms, data and modeled, Eq. 12.3.14.

12.4 Algebraic Stress Analysis for Rotating Walls

In Section 8.2 it was noted that the b_{ij} values from the MC model near the wall in fully developed (non-rotating) channel flow were consistent with a fully developed homogeneous flow analysis of the b_{ij} equations at $R_t = 0$, Eq. 8.2.8. The MA model also shows this tendency, except that b_{11} and b_{33} were modified in the laminar sublayer through a specified b_{ij} boundary condition. Therefore, in this section we use a homogeneous fully developed flow analysis to shed light on pressure-strain near rotating walls. From the asymptotic analysis in Section 12.2, we seek modification (if necessary) of the pressure-strain model to give

$$b_{22} = -1/3, \qquad b_{12} = 0 \quad \text{and} \quad b_{23} = 0, \tag{12.4.1}$$

at the wall, regardless of the wall rotation.

Here we continue to consider pressure-strain models with c_1 and c_3 through c_6 terms,

$$\begin{aligned}\frac{\hat{\phi}_{ij}}{k} &= -c_1\omega b_{ij} + c_3 S_{ij} + c_4(b_{ik}S_{jk} + b_{jk}S_{ik} - \frac{2}{3}\delta_{ij}b_{mn}S_{mn}) \\ &+c_5(b_{ik}W_{jk} + b_{jk}W_{ik}) + c_6 b_{ij}b_{mn}S_{mn} \quad .\end{aligned} \tag{12.4.2}$$

In the homogeneous fully developed limit, the b_{ij} equations, Eq. 8.1.1, reduce to $S_{bij} = 0$, so that from Eq. 8.1.5,

$$\begin{aligned}&b_{ij}\left((c_1 - 2)\omega + (2 + \frac{1}{2}c_6)\frac{P}{k}\right) = \left(c_3 - \frac{4}{3}\right)S_{ij} \\ &+(c_4 - 2)(b_{ik}S_{jk} + b_{jk}S_{ik} - \frac{2}{3}\delta_{ij}b_{mn}S_{mn}) \\ &+(c_5 - 2)(b_{im}W_{jm} + b_{jm}W_{im}) - 2\Omega_k\left(\varepsilon_{ikm}b_{jm} + \varepsilon_{jkm}b_{im}\right).\end{aligned} \tag{12.4.3}$$

Rewriting this in terms of the relative vorticity using,

$$W_{im} = \omega_{im} + \varepsilon_{ikm}\Omega_k, \tag{12.4.4}$$

yields

$$b_{ij}\left((c_1-2)\omega+(2+\frac{1}{2}c_6)\frac{P}{k}\right)=\left(c_3-\frac{4}{3}\right)S_{ij}$$
$$+(c_4-2)(b_{ik}S_{jk}+b_{jk}S_{ik}-\frac{2}{3}\delta_{ij}b_{mn}S_{mn}) \tag{12.4.5}$$
$$+(c_5-2)(b_{im}\omega_{jm}+b_{jm}\omega_{im})+(c_5-4)\Omega_k\left(\varepsilon_{ikm}b_{jm}+\varepsilon_{jkm}b_{im}\right).$$

Noting that the wall coordinates have been defined so that $\partial U_1/\partial x_2$ is the only non-zero component of the mean velocity gradient tensor at the wall, and paralleling Section 5.3 by defining D, such that

$$D\frac{\partial U_1}{\partial y}=(c_1-2)\omega-b_{12}(4+c_6)\frac{\partial U_1}{\partial y}\ , \tag{12.4.6}$$

we can expand Eq. 12.4.5 to give (Appendix EQ.12.4.7-11)

$$b_{11}D\frac{\partial U_1}{\partial y}=b_{12}\frac{\partial U_1}{\partial y}\left(\frac{c_4-2}{3}+c_5-2\right)+2(c_5-4)\left(b_{13}\Omega_2-b_{12}\Omega_3\right)\ , \tag{12.4.7}$$

$$b_{22}D\frac{\partial U_1}{\partial y}=b_{12}\frac{\partial U_1}{\partial y}\left(\frac{c_4-2}{3}-(c_5-2)\right)+2(c_5-4)\left(b_{12}\Omega_3-b_{23}\Omega_1\right)\ , \tag{12.4.8}$$

$$b_{12}D\frac{\partial U_1}{\partial y}=\frac{1}{2}\frac{\partial U_1}{\partial y}\left(c_3-\frac{4}{3}+(c_4-2)(b_{11}+b_{22})+(c_5-2)(b_{22}-b_{11})\right)$$
$$+(c_5-4)\left(\Omega_3(b_{11}-b_{22})+\Omega_2b_{23}-\Omega_1b_{13}\right)\ , \tag{12.4.9}$$

$$b_{13}D\frac{\partial U_1}{\partial y}=\frac{1}{2}b_{23}\frac{\partial U_1}{\partial y}(c_4-2+c_5-2)$$
$$+(c_5-4)\left(\Omega_1b_{12}-\Omega_2(b_{22}+2b_{11})-\Omega_3b_{23}\right)\ , \tag{12.4.10}$$

and

$$b_{23}D\frac{\partial U_1}{\partial y}=\frac{1}{2}b_{13}\frac{\partial U_1}{\partial y}(c_4-2-(c_5-2))$$
$$+(c_5-4)\left(\Omega_1(2b_{22}+b_{11})-\Omega_2b_{12}+\Omega_3b_{13}\right)\ . \tag{12.4.11}$$

Substituting in the desired wall values, Eq. 12.4.1, and noting that $c_1\to2$ at the wall, since $R_t\to0$, so that $D\to0$, then Eqs. 12.4.7-11 reduce to

$$0=2(c_5-4)b_{13}\Omega_2\ , \tag{12.4.12}$$

$$0 = 0\ , \tag{12.4.13}$$

$$0 = \frac{1}{2}\frac{\partial U_1}{\partial y}\left(c_3 - \frac{1}{3}(c_4 + c_5) + b_{11}(c_4 - c_5)\right) + (c_5 - 4)\left(\Omega_3\left(b_{11} + \frac{1}{3}\right) - \Omega_1 b_{13}\right)\ , \tag{12.4.14}$$

$$0 = (c_5 - 4)\Omega_2\left(\frac{1}{3} - 2b_{11}\right)\ , \tag{12.4.15}$$

and

$$0 = \frac{1}{2}b_{13}\frac{\partial U_1}{\partial y}(c_4 - c_5) + (c_5 - 4)\left(\Omega_1\left(b_{11} - \frac{2}{3}\right) + \Omega_3 b_{13}\right)\ . \tag{12.4.16}$$

For a general rotation with $\Omega_2 \neq 0$, from Eq. 12.4.12, either $c_5 = 4$, or $b_{13} = 0$. Trying $c_5 \neq 4$ and $b_{13} = 0$, Eqs. 12.4.15 and 12.4.16 reduce to $b_{11} - 1/6$, and $b_{11} = 2/3$, which is inconsistent. Therefore,

$$c_5 = 4 \tag{12.4.17}$$

is the only solution for which c_5 is independent of the rotation vector Ω_n. From Eq. 12.4.5, this is the material frame indifferent limit, since then the equations and solution for the b_{ij}'s at the wall are independent of the wall rotation.

With $c_5 = 4$, Eqs. 12.4.7-9 reduce to

$$b_{11} = \frac{b_{12}}{D}\left(\frac{c_4 - 2}{3} + 2\right) \tag{12.4.18}$$

$$b_{22} = \frac{b_{12}}{D}\left(\frac{c_4 - 2}{3} - 2\right) \tag{12.4.19}$$

$$b_{12}D = \frac{1}{2}\left(c_3 - \frac{4}{3} + (c_4 - 2)(b_{11} + b_{22}) + 2(b_{22} - b_{11})\right)\ , \tag{12.4.20}$$

and paralleling the analysis in Section 5.3, these may be combined to give

$$\left(\frac{b_{12}}{D}\right)_{D=0} = \frac{\frac{1}{2}\left(c_3 - \frac{4}{3}\right)}{4 - \frac{1}{3}(c_4 - 2)^2} \quad \text{and} \quad (b_{22})_{D=0} = \frac{\frac{1}{2}\left(c_3 - \frac{4}{3}\right)}{4 - \frac{1}{3}(c_4 - 2)^2}\left(\frac{c_4 - 2}{3} - 2\right)\ . \tag{12.4.21}$$

Requiring that the coefficients are consistent with the general two-dimensional turbulence limit for arbitrary strain so that $c_3 = 2c_4/3$ and $c_6 = -2c_4$ (see Eqs. 3.4.6-

7), then Eq. 12.4.21 yields $b_{22} = -1/3$ when $c_4 = 4$, and the complete set of coefficients is

$$c_3 = 8/3, \qquad c_4 = 4, \qquad c_5 = 4 \qquad \text{and} \qquad c_6 = -8. \tag{12.4.22}$$

The fully developed solution at a rotating or stationary wall is then one-dimensional turbulence,

$$b_{11} = 2/3, \quad b_{22} = -1/3, \quad b_{33} = -1/3, \; b_{12} = 0, \; b_{13} = 0, \text{ and } b_{23} = 0. \tag{12.4.23}$$

Note that this is the same solution as was obtained for a stationary wall using the algebraic limit of the MA model (see Eq. 8.3.15).

In the vicinity of the wall, the coefficients in Eq. 12.4.22 may be combined with Eqs. 12.4.2 and 12.4.3 to give

$$\frac{\hat{\phi}_{ij}}{k} = -c_1\omega b_{ij} + 2b_{ij}(c_1 - 2)\omega + 4\Omega_k\left(\varepsilon_{ikm}b_{jm} + \varepsilon_{jkm}b_{im}\right) \tag{12.4.24}$$

or

$$\frac{\hat{\phi}_{ij}}{k} = -c_1\omega b_{ij} + 2b_{ij}(c_1 - 2)\omega - \frac{F_{ij}}{k} \ . \tag{12.4.25}$$

With $c_1 = 2$ at the wall, this reduces to

$$\frac{\hat{\phi}_{ij}}{2k} + \omega b_{ij} = -\frac{F_{ij}}{2k} \ . \tag{12.4.26}$$

Comparing with Eq. 12.3.12, the combined effective pressure-strain and anisotropy of dissipation to be modeled, exactly cancels the rotation force term in the b_{ij} equations. Therefore, at a rotating wall, the fully developed flow analysis suggests that wall turbulence is material frame indifferent. Note that this contribution is not part of the homogeneous pressure-strain, which has its basis in the volume integral in Eq. 3.1.4, and which is expected to be a function of absolute vorticity (independent of rotation alone). Instead, it is an inhomogeneous wall effect, which has its basis in the surface integral of Eq. 3.1.4.

To get a feel for the effect this extra term has, we can evaluate it using the b_{ij}'s at the wall, Eq. 12.4.23,

$$\begin{aligned} &-\frac{F_{11}}{2k} = -\frac{F_{22}}{2k} = -\frac{F_{33}}{2k} = 0, \qquad && -\frac{F_{12}}{2k} = 2(b_{11} - b_{22})\Omega_3 = 2\Omega_3 \ , \\ &-\frac{F_{13}}{2k} = -2(2b_{11} + b_{22})\Omega_2 = -2\Omega_2 \ , && -\frac{F_{23}}{2k} = 2(b_{11} + 2b_{22})\Omega_1 = 0 \ . \end{aligned} \tag{12.4.27}$$

The only non-zero components affect the shear anisotropies associated with the direction (x_1) of the limiting mean streamline.

12.5 Modifying the MARV Model for Material Frame Indifferent Walls

So far, the pressure-strain part of the MARV model has been developed without any direct consideration of the wall. The model gives good agreement with the stationary wall DNS data, including the highly anisotropic and streaky near-wall turbulence, without including any pressure-strain terms which depend on wall normal directions or distance. This has been possible because non-rotating shear flow at high strain rate and low turbulence Reynolds number tends to the same limits as wall flow with $b_{22} = -1/3$, and $b_{12} = 0$. The model does include a reduced slow pressure-strain due to the turbulence inhomogeneity.

The current model takes the form

$$\frac{\hat{\phi}_{ij}}{k} = -c_1 \omega b_{ij} + c_3 S_{ij} + c_4 (b_{ik} S_{jk} + b_{jk} S_{ik} - \frac{2}{3} \delta_{ij} b_{mn} S_{mn}) + c_6 b_{ij} b_{mn} S_{mn} \, . \tag{12.5.1}$$

At the wall, where $g = 0$, the model reduces to

$$\frac{\hat{\phi}_{ij}}{k} = -c_1 \omega b_{ij} \, , \tag{12.5.2}$$

for both the fully developed zero vorticity (W=0) and plane shear (W=S) limits. This is consistent with Eq. 12.4.25 for a stationary wall where $c_1 = 2$ and $F_{ij}/k = 0$. Therefore, to extend the model to rotating walls, it is sufficient to add a term which cancels out the rotation term at the wall, in particular $-(F_{ij}/k) f_R$, where f_R varies from 1 at the wall, to zero sufficiently far from the wall that the wall no longer directly influences the turbulence,

$$\frac{\hat{\phi}_{ij}}{k} = -c_1 \omega b_{ij} + c_3 S_{ij} + c_4 (b_{ik} S_{jk} + b_{jk} S_{ik} - \frac{2}{3} \delta_{ij} b_{mn} S_{mn}) + c_6 b_{ij} b_{mn} S_{mn} - \frac{F_{ij}}{k} f_R \, . \tag{12.5.3}$$

To combine together pressure-strain for homogeneous turbulence and wall turbulence, Durbin (1993) solved a relaxation equation

$$L^2 \frac{\partial^2 \left(\frac{\phi_{ij}}{k} + 2\omega b_{ij} \right)}{\partial x_k^2} - \left(\frac{\phi_{ij}}{k} + 2\omega b_{ij} \right) = -\left(\frac{\phi_{ij}}{k} + 2\omega b_{ij} \right)_{\text{homogeneous model}} \quad , \qquad (12.5.4)$$

(where L is a turbulence length scale) with wall boundary conditions for $\left(\frac{\phi_{ij}}{k} + 2\omega b_{ij} \right)$. Applying an equation of this form to the factor f_R,

$$L^2 \frac{\partial^2 f_R}{\partial y^2} - f_R = 0 \ , \qquad (12.5.5)$$

with the boundary condition $f_R = 1$ at the wall, and assuming that L is constant (which it most definitely is not) yields the solution $f_R = \exp(-y/L)$, i.e., an exponential decay with distance from the wall. Therefore to complete the wall model we considered factors of the form

$$f_R = \exp\left(-a \left(\frac{y\omega}{q} \tilde{R}_t^b \right)^c \right) . \qquad (12.5.6)$$

F_{ij} in Eq. 12.5.3 will be evaluated using local values of b_{ij} (not wall values as in the example, Eq. 12.4.27), so that the model will remain realizable and independent of the wall normal direction. The model will depend on the distance to the nearest wall, y, and the rotation rate vector of the nearest wall.

Values for a, b and c in Eq. 12.5.6 will be determined in Chapter 13 based on the DNS data for fully developed rotating channels. The estimated evaluation, Eq. 12.4.27, suggests that b_{12} and therefore the wall shear stresses will be sensitive to the extra term.

Chapter 13

ROTATING CHANNELS

The modification of turbulence in the boundary layers on the pressure and suction sides of rotating passages was discussed qualitatively in Section 4.3. In this chapter we continue this theme by considering, quantitatively, turbulence in fully developed two-dimensional rotating channel flows. These test cases form an important part of verifying Reynolds stress models, since turbulence modification due to rotation is a prime reason to use a Reynolds stress model, rather than a simpler two equation model.

The DNS data of Kristoffersen and Andersson (1993) provides a basis for setting the constants in the rotating-wall term for the MARV model, Eq. 12.5.6. DNS also provides data for detailed comparisons of the turbulence properties, q and b_{ij}. Experimental data, on the centerline of high aspect ratio rotating channels, provides information on wall shear stresses at higher Reynolds numbers, and on laminarization of the velocity profiles on the suction side.

The importance of various features of the MARV model for the computation of rotating channel flow is tested. Results from the simple MC model are also given.

13.1 Geometry and Nomenclature

The coordinate convention that is used in this chapter for fully developed two-dimensional rotating channel flow is the same as used for rotating boundary layers in Chapter 4. As shown in Fig. 13.1.1, the mean flow is in the x_1 direction; x_2 is normal to the walls and measured from the pressure surface, and the rotation is positive about the x_3 axis, $\Omega = \Omega_3$.

Fig. 13.1.1. Coordinates for rotating channels. S, suction side; P, pressure side.

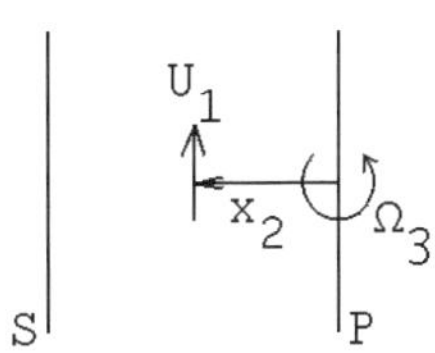

A rotation number, R_o, is defined using the channel width, D, and the mean velocity,

$$R_o = \frac{\Omega D}{U_{mean}} \quad . \tag{13.1.1}$$

Wall friction-velocity Reynolds numbers are defined for each side of the channel,

$$R_{\tau,s} = \frac{U_{\tau,s}}{\nu}\frac{D}{2}, \quad R_{\tau,p} = \frac{U_{\tau,p}}{\nu}\frac{D}{2} . \tag{13.1.2}$$

When the flow is fully developed, the x_1 component of the mean momentum equation, Eq. 2.1.10, reduces to

$$0 = \frac{\partial}{\partial x_1}\left(\frac{\Omega^2 r^2}{2}\right) - \frac{1}{\rho}\frac{\partial \overline{P}}{\partial x_1} + \frac{\partial}{\partial x_2}\nu\frac{\partial U_1}{\partial x_2} - \frac{\partial \overline{u_1 u_2}}{\partial x_2} . \tag{13.1.3}$$

The centrifugal and pressure-gradient terms may be combined by defining a reduced static pressure

$$\overline{P}_r = \overline{P} - \frac{1}{2}\rho\Omega^2 r^2 = \overline{P} - \frac{1}{2}\rho\Omega^2\left(\left(x_1 - x_{1,c}\right)^2 + \left(x_2 - x_{2,c}\right)^2\right) , \tag{13.1.4}$$

where $x_{1,c}, x_{2,c}$ is on the axis of rotation.

In the fully developed flow, the reduced mean pressure gradient, $\partial\overline{P}_r/\partial x_1$, is uniform and constant. Integrating the mean momentum equation across the channel yields

$$-\frac{D}{\rho}\frac{d\overline{P}_r}{dx_1} = U_{\tau,p}^2 + U_{\tau,s}^2 . \tag{13.1.5}$$

An average friction velocity,

$$U_{\tau,a} = \sqrt{\frac{1}{2}\left(U_{\tau,p}^2 + U_{\tau,s}^2\right)} , \tag{13.1.6}$$

may then be used to give a channel, or average friction-velocity, Reynolds number, R_τ, extending the definition used for non-rotating channels,

$$R_\tau = \frac{U_{\tau,a}}{\nu}\frac{D}{2} = \frac{D}{2\nu}\sqrt{-\frac{D}{2\rho}\frac{d\overline{P}_r}{dx_1}} . \tag{13.1.7}$$

13.2 Rotating Channel Data

The effect of rotation on channel flow is shown in Fig. 13.2.1. As rotation is increased, the magnitude of the turbulent shear stresses on the pressure-side of the channel ($x_2 = 0$) increases, while that on the suction side ($x_2/D = 1$) decreases. The total shear stress,

$$\frac{\tau}{\rho} = \nu \frac{\partial U_1}{\partial x_2} - \overline{u_1 u_2}, \qquad (13.2.1)$$

remains linear in the fully developed flow, but there is now an imbalance in the magnitude of the wall shear stresses, so that the location of zero total shear stress is towards the suction side. The shape of the mean velocity profile also changes, with the peak velocity being near, but not coincident with, the zero total shear stress location. At the high rotation case in Fig. 13.2.1, the turbulent shear stresses on the suction side have become small compared with the laminar shear stresses and the velocity profile is correspondingly laminar.

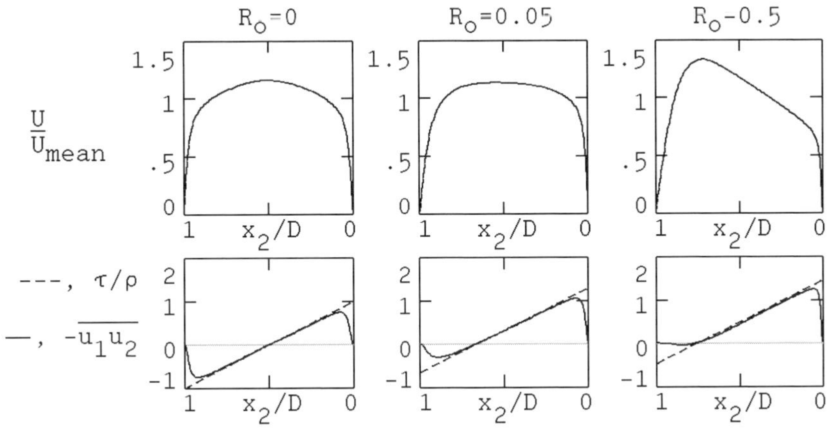

Fig. 13.2.1. Profiles of mean velocity and shear stresses from the DNS data of Kristoffersen and Andersson (KA1993). The total and turbulent shear stresses have been made dimensionless with the average wall shear stress, $U_{\tau,a}^2$.

Overall conditions and parameters for the Kristoffersen and Andersson (KA1993) DNS data, shown in Fig. 13.2.1, and some other fully-developed rotating channel DNS data are given in Table 13.2.1. Kristoffersen and Andersson set a fixed reduced pressure gradient corresponding to $R_\tau = 194$, while increasing the rotation number from zero to $R_o = 0.5$. By contrast, Wu and Kasagi (WK2004) kept the flow rate constant, $R_D = 4560$, while increasing the rotation from zero to $R_o = 1$. For the KA1993 data sets, R_D varied, and for the WK2004 data sets, R_τ varied.

In Fig. 13.2.2, wall friction velocities from the data sets listed in Table 13.2.1 are compared on a common basis, by normalizing $R_{\tau,p}$ and $R_{\tau,s}$ (Eq. 13.1.2) with the r.m.s average, R_τ (Eq. 13.1.7). On the suction side of the channel, $R_{\tau,s}/R_\tau$ falls rapidly until the layer becomes laminar, or nearly laminar. The data at $R_D \approx 6000$

for $R_{\tau,s}/R_\tau$ achieves lower values than the data at $R_D \approx 4600$, since $R_{\tau,s}$ may fall proportionally farther to reach a laminar state at higher Reynolds number, Fig. 13.2.3.

Table 13.2.1. Rotating Channel DNS Data.

Author	R_D	R_τ	R_o	$y^+_{\text{near wall}}$	Box, δx_1
KA1993	~6000	194	0 – 0.5	0.5	$2\pi D$
WK2004	4560	~150	0 –1		$(2.5\pi D)$
NK1996+	4750	150	0.159	0.04	$2.5\pi D$
EK2003+	4531	121	1.0	0.02	$2.5\pi D$
WK2004b+	10,756	295.5	0.137	0.02	$1.5\pi D$
AJ1999*	~6500	180	0.43, 0.77		$2\pi D - 4\pi D$

+ From data base by Kasagi et al, KHMMN1990. Additional references: NK1996, Nishimura and Kasagi, 1996; EK2003, El-Samni and Kasagi, 2000, 2001; WK2004b, Wu and Kasagi, 2004.
* AJ1999, Alvelius and Johansson, 1999, from Wallin and Johansson, 2002.

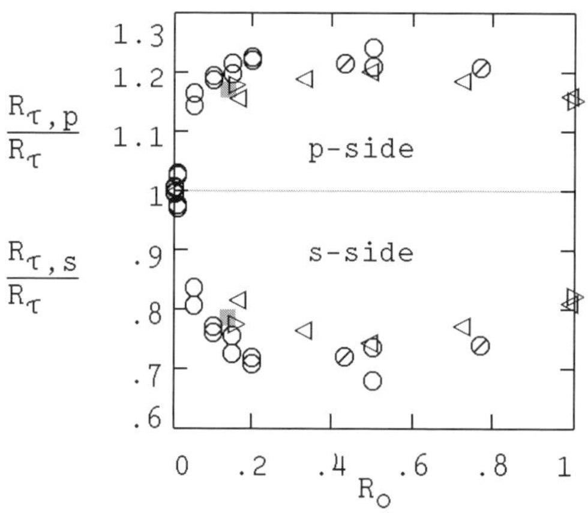

Fig. 13.2.2. Wall friction velocities on the pressure and suction sides of rotating channels, normalized with the r.m.s. average. DNS data, Table 13.2.1.
R_D~4600: ◁, WK2004; ▷, NK1996, EK2003.
R_D~6000: ○, KA1993; ⊘, AJ1999.
R_D = 10,756: ■, WK2004b.

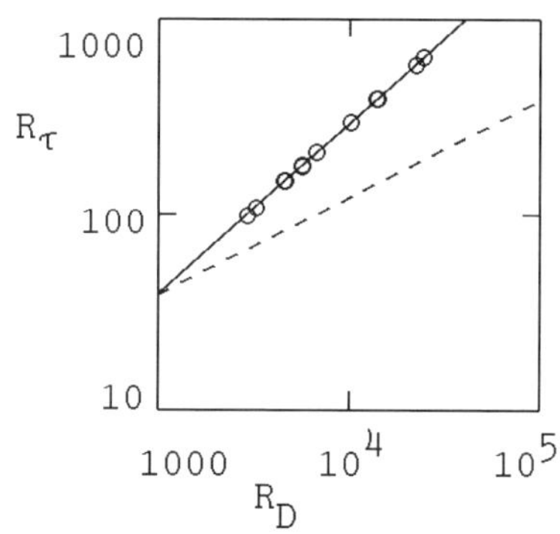

Fig. 13.2.3. Wall friction Reynolds number for non-rotating channel flow.
o, DNS data, Table 6.4.1.
—— fit to DNS data, Eq. 13.2.2.
--- laminar flow.

On the pressure side of the channel, $R_{\tau,p}/R_\tau$ increases with rotation, then levels off, and then begins to fall when the suction side has become laminar. Note that showing both the pressure and suction sides in Fig. 13.2.2 is redundant since $(R_{\tau,p}/R_\tau)^2 + (R_{\tau,s}/R_\tau)^2 = 2$. If the flow becomes completely laminar, either due to low Reynolds number or high rotation, the velocity profile becomes parabolic and $R_{\tau,p}/R_\tau = R_{\tau,s}/R_\tau = 1$.

The variation of R_τ for the WK2004 data, which was at constant R_D, is shown in Fig. 13.2.4a. R_τ rises a little from 150 at zero rotation, and then falls to 128 at $R_o = 1$.

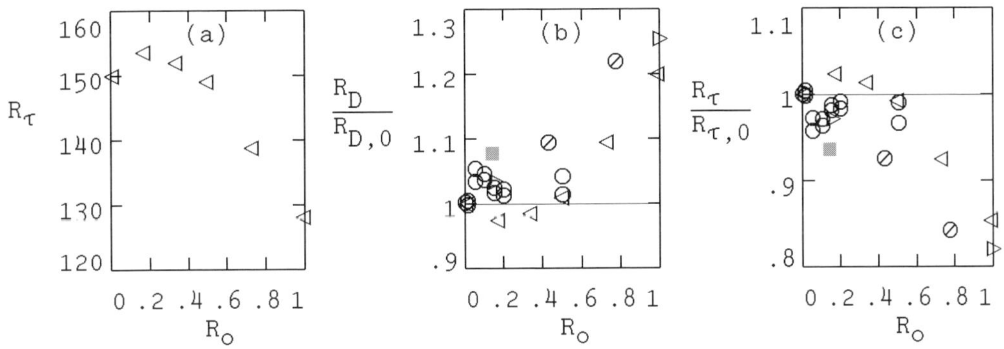

Fig. 13.2.4. Changes in mean friction-velocity and flow Reynolds numbers with rotation number. (a) R_τ for the WK2004 DNS data, R_D=4560. (b) R_D of the rotating channel flow divided by R_D for a non-rotating channel with the same R_τ. (c) R_τ of the rotating channel flow divided by R_τ for a non-rotating channel with the same R_D. DNS data as for Fig. 13.2.2.

The same type of information is shown another way in Fig. 13.2.4b. Here, the flow rate Reynolds number, R_D, for the rotating channel flow, has been divided by the Reynolds number, $R_{D,0}$, for the non-rotating channel flow at the same R_τ. To determine $R_{D,0}$, the DNS data in Fig. 13.2.3 has been fit to the relation,

$$R_{D,0} = 14.89 R_\tau^{1.144} \quad . \tag{13.2.2}$$

This is used to obtain $R_{D,0}$ for all the data in Fig. 13.2.4b, except the KA1993 data, where $R_D / R_{D,0}$ is directly available. For the KA1993 data, a second set of points is also shown to correct for the small variations in $U_{\tau,a}$, ±2 percent. Here the relation

$$R_{D,0} = c R_\tau^{1.144} \tag{13.2.3}$$

was used, where the constant c was determined from the KA1993 zero-rotation data, which gave $c = 14.04$. Overall, the scatter in the data is considerable, about 10 percent, with the trend being to higher $R_D / R_{D,0}$ with increased rotation rate.

In Figure 13.2.4c all the data is shown together as $R_\tau / R_{\tau,0}$, where $R_{\tau,0}$ is the friction-velocity Reynolds number for the channel with zero rotation at the R_D of the data. To convert the data to this common basis, the relation

$$R_D = cR_{\tau,0}^{1.144} \tag{13.2.4}$$

can be used to obtain $R_{\tau,0}$. The results in Figs. 13.2.4b and 13.2.4c are then related by

$$\frac{R_D}{R_{D,0}} = \left(\frac{R_{\tau,0}}{R_\tau}\right)^{1.144} . \tag{13.2.5}$$

Experimental data is available from Moore (M1967), and Halleen and Johnston (HJ1967). Both made measurements on the centerline of channels with aspect ratios of about 7 (in the x_3 direction) to 1 (in the x_2 direction). Moore studied channels of various aspect ratios, and determined that the centerline results at this aspect ratio were relatively unaffected by three-dimensional flows.

The results shown in Fig. 13.2.5 concentrate on two Reynolds numbers, R_D approximately 12,000 and 34,000. The rotation numbers are not as high as for many of the DNS cases. The data was analyzed in the same manner as the KA1993 data above, using Eq. 13.2.3 with c determined from the experimental data. Qualitatively the results are similar to the DNS data. The experimental data at $R_D \approx 12{,}000$ shows a steady fall of $R_{\tau,s} / R_\tau$ with increased rotation number until about $R_o = 0.1$, at which $R_{\tau,s} / R_\tau$ falls rapidly until the suction-side wall layer is laminar.

Halleen and Johnston, using flow visualization, determined that the suction side layer, at $R_D \approx 12{,}000$, begins to lose its fully turbulent character by $R_o \approx 0.11$ (they called the flow "laminar-streaky") and is "completely laminar" for $R_o > 0.15$. The DNS data point at $R_o = 0.137$, Fig. 13.2.5a, shows that the layer is clearly still fully turbulent. Similarly, at lower Reynolds number the flow visualization suggests laminarization of the suction side layer at lower rotation numbers than the DNS data. Halleen and Johnston's laminar flow envelope is shown in Fig. 13.2.6(a). Figure 13.2.6(b) shows, for the DNS data, the maximum turbulent shear stress on the suction side divided by the wall shear stress.

It may be that the experimental data shows early laminarization due to three-dimensional effects. It may also be that the DNS data is biased towards turbulent flow. The DNS calculations impose a repeating boundary (except for reduced pressure) in the x_1 direction. Alvelius (1999) noted that two-point correlations of u_1 still had significant levels with a separation of half the computational box length under some rotating channel flow conditions. In particular, with the two-point correlation defined as 1 at zero separation, he reported two-point correlations at half the box length as high as 0.25 even with a relatively long box of $4\pi D$ (Table 13.2.1) for $R_o = 0.055$ and $R_\tau = 180$. For the WK2004b case, the correlation on the suction side (at $x_2 = 3D/4$) does not fall below 0.4. These may be compared with the corresponding values for non-rotating channel flow which are all below 0.1 for the cases in the Kasagi et al. data base (KHMMN1990). Moin and Mahesh (1998) discuss in some detail numerical issues and spatial resolution requirements to accurately resolve small and large scales in DNS calculations.

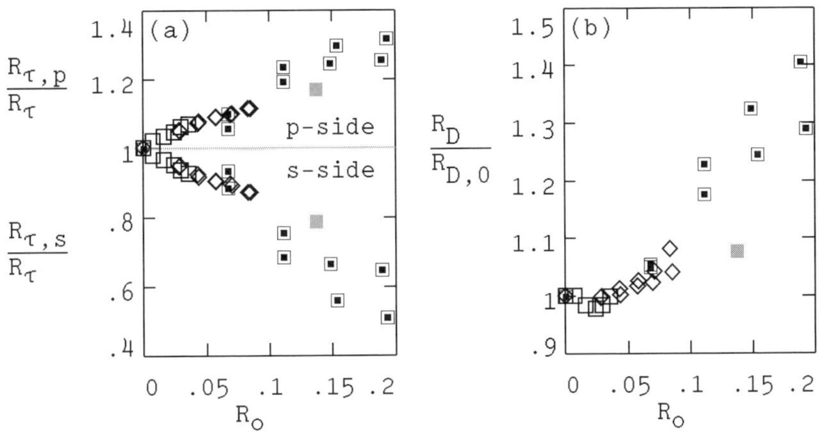

Fig. 13.2.5. Rotating channel data at higher R_D. (a) Wall friction velocities and (b) Reynolds number ratio. Experimental data: □, Moore, M1967, $R_D = 12{,}700$; Halleen and Johnston, HJ1967: ▣, $R_D = 11{,}500$ to 12,800; ◇, $R_D = 32{,}200$ to 36,100. ■ DNS data, WK2004b, $R_D = 10{,}756$ (Table 13.2.1).

Fig. 13.2.6. Evidence of laminarization on the suction side. (a) Visualization data of Halleen and Johnston (HJ1967), S, streaky-laminar; L, completely laminar. (b) Ratio of maximum turbulent shear stress over the wall shear stress for DNS data, symbols as in Fig. 13.2.2.

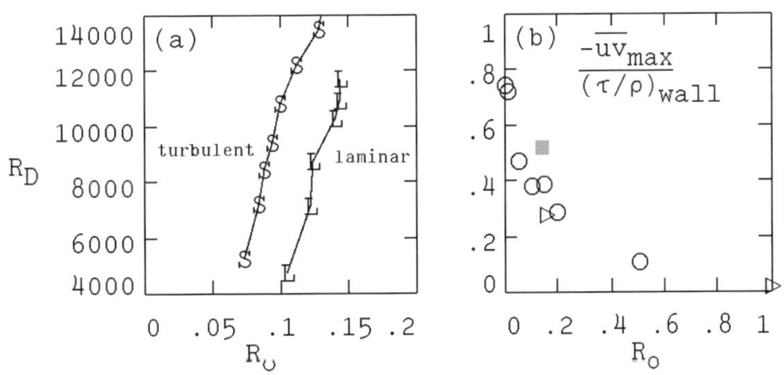

13.3 Constants for the Rotating Wall Modification

In this section we set the constants for the rotating-wall addition to the effective pressure-strain for the MARV model, proposed in Section 12.5.

The addition to the pressure-strain model,

$$\frac{\hat{\phi}_{ij}^{\text{rot. wall}}}{2k} = -\frac{F_{ij}}{2k} f_R = 2 f_R \Omega_k \left(\varepsilon_{ikm} b_{jm} + \varepsilon_{jkm} b_{im} \right) , \tag{13.3.1}$$

adds to the source term for the b_{ij} equations given in Eq. 8.1.5. Otherwise, the turbulence equations for the rotating channel are the same as those for the stationary channel given in Section 8.1. The resulting source term for the b_{ij} equations may be written as

$$\begin{aligned} S_{bij} = & -\frac{1}{2} b_{ij} \left((c_1 - 2)\omega + (2 + \frac{1}{2} c_6) \frac{P}{k} \right) + \frac{1}{2} \left(c_3 - \frac{4}{3} \right) S_{ij} \\ & + \frac{1}{2} (c_4 - 2)(b_{ik} S_{jk} + b_{jk} S_{ik} - \frac{2}{3} \delta_{ij} b_{mn} S_{mn}) \\ & + \frac{1}{2} c_5 (b_{ik} W_{jk} + b_{jk} W_{ik}) - (b_{ik} \omega_{jk} + b_{jk} \omega_{ik}) \\ & -2(1 - f_R) \Omega_k \left(\varepsilon_{ikm} b_{jm} + \varepsilon_{jkm} b_{im} \right) . \end{aligned} \tag{13.3.2}$$

Note that here we have rewritten the vorticity terms using Eq. 12.4.4. If there is a c_5 term in the homogeneous model, it is multiplied by the absolute vorticity (Eq. 3.2.17). The vorticity term arising from production is the relative vorticity term (Eq. 2.2.5). The last line is the rotation term and the addition, $\hat{\phi}_{ij}^{\text{rot. wall}} / 2k$.

The best data, for the purpose of setting the model coefficients, is probably the KA1993 DNS data at $R_o = 0.05$. This is a high enough rotation rate that there are significant rotation effects, and low enough rotation that the DNS and experiments are in agreement that the suction side is turbulent. To set the coefficients, a, b and c, in the expression for f_R (Eq. 12.5.6),

$$f_R = \exp\left(-a \left(\frac{y\omega}{q} \tilde{R}_t^b \right)^c \right) , \tag{13.3.3}$$

series of calculations were made with $R_\tau = 194$ (corresponding to the KA1993 conditions), varying b and c, and setting a to obtain $R_{\tau,p} / R_\tau = 1.141$ when $R_o = 0.05$ (Appendix KA1993).

With $b=1$, values for c of 1, 2, 3 and 4 were tried, and with $c=1$, values for b of 1,2,3 and 4 were tried. Figure 13.3.1a shows $R_{\tau,p}/R_\tau$ and $R_{\tau,s}/R_\tau$ as functions of rotation for all the choices considered; b and c have little effect on these ratios. All the model variations give laminar flow on the suction side for $R_o \geq 0.07$. The choice of b and c, however, does have a strong effect on $R_D/R_{D,0}$, as shown in Figs. 13.3.1b and 13.3.1c. The best model appears to be $c=1$, $b=4$. Therefore for the MARV model we take,

$$f_R = \exp\left(-0.0087\frac{yq}{\nu}\left(\frac{\tilde{R}_t}{100}\right)^3\right). \tag{13.3.4}$$

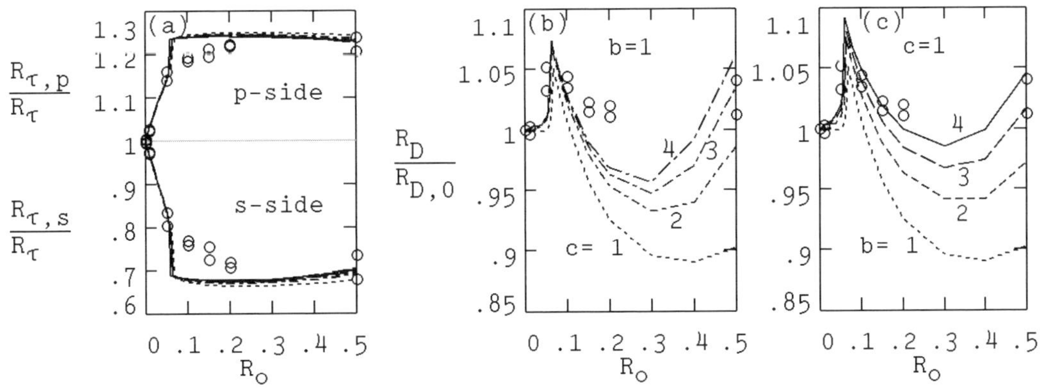

Fig. 13.3.1. Comparison of possible choices for the rotating wall modification to the MARV model, with DNS data at $R_\tau=194$. o, DNS data, KA1993. Lines: model variations with constants b and c, Eq. 13.3.3; —— final model, Eq. 13.3.4.

The rotating-wall parameter, f_R, for $R_o=0.05$ is shown in Fig. 13.3.2. The wall-reflection term affects about 10 percent of the passage width on each side of the channel at this rotation number and Reynolds number. Figure 13.3.2b plots f_R versus the local y^+ on either side of the channel. Through the laminar sublayer, up to about $y^+=10$, the wall reflection essentially cancels out the rotation term. The effect then decays, with little or no effect beyond $y^+=50$.

The physics of the $\hat{\phi}_{ij}^{\text{rot.wall}}$ addition to the model can be seen by considering the momentum equations for U_2 and u_2. For fully developed three-dimensional flow with Ω_3 rotation, Eq. 2.1.10 reduces to

$$U_2\frac{\partial U_2}{\partial x_2}+U_3\frac{\partial U_2}{\partial x_3}+\frac{\partial\overline{u_2u_2}}{\partial x_2}+\frac{\partial\overline{u_2u_3}}{\partial x_3}=-2\Omega_3U_1-\frac{1}{\rho}\frac{\partial\overline{P}_r}{\partial x_2}+\frac{\partial}{\partial x_2}\nu\frac{\partial U_2}{\partial x_2}+\frac{\partial}{\partial x_3}\nu\frac{\partial U_2}{\partial x_3}, \tag{13.3.5}$$

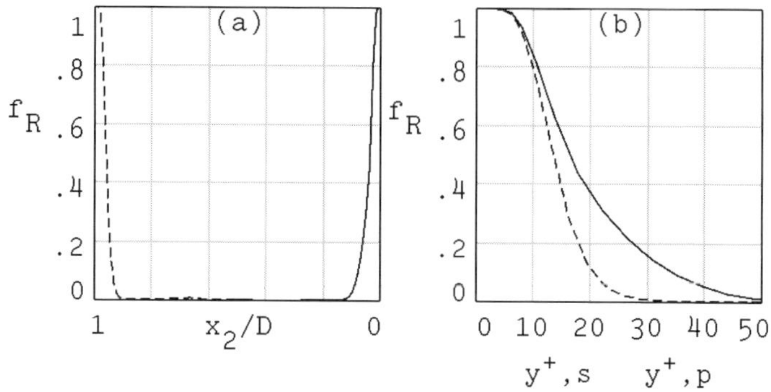

Fig. 13.3.2. MARV model parameter f_R in rotating channel flow, $R_\tau = 194$, $R_o = 0.05$. --- suction side, —— pressure side.

which for two-dimensional flow ($U_3 = 0$, so that by continuity $U_2 = 0$) further reduces to

$$\frac{1}{\rho}\frac{\partial \overline{P}_r}{\partial x_2} = -2\Omega_3 U_1 - \frac{\partial \overline{u_2 u_2}}{\partial x_2}. \tag{13.3.6}$$

The gradient of reduced pressure then balances the Coriolis force term (as well as the small gradient in the Reynolds normal stress). In three-dimensional flow, the secondary flow (U_2 and U_3 velocity) terms provide three-dimensional relief altering the balance between the pressure gradient and Coriolis forces.

The equation for u_2 may be considered is a corresponding way. From Eq. 2.1.11, for constant ν,

$$\underbrace{\frac{\partial u_2}{\partial t}}_{y^2} + \underbrace{u_k \frac{\partial U_2}{\partial x_k}}_{y^3} + \underbrace{U_k \frac{\partial u_2}{\partial x_k}}_{y^3} + \underbrace{\frac{\partial u_2 u_k}{\partial x_k}}_{y^3} - \underbrace{\frac{\partial \overline{u_2 u_k}}{\partial x_k}}_{y^3} = \underbrace{-2\Omega_3 u_1}_{y} - \underbrace{\frac{1}{\rho}\frac{\partial p}{\partial x_2}}_{1} + \underbrace{\frac{\partial}{\partial x_k}\nu\frac{\partial u_2}{\partial x_k}}_{1} . \tag{13.3.7}$$

Under each term is given the Taylor series limiting behavior. Keeping terms of order y and lower, the equation reduces to

$$\frac{1}{\rho}\frac{\partial p}{\partial x_2} = -2\Omega_3 u_1 + \frac{\partial}{\partial x_2}\nu\frac{\partial u_2}{\partial x_2} . \tag{13.3.8}$$

The physical explanation of the $\hat{\phi}_{ij}^{\mathrm{rot.wall}}$ addition is then that, near the wall, there is a contribution to the fluctuating pressure gradient, which balances the fluctuating Coriolis force term. From the empirical constants determined for the model, this appears to occur through the laminar sublayer, beyond which the higher order terms in Eq. 13.3.7 provide time-dependent and three-dimensional relief, paralleling the effects of three-dimensional relief which occur for the mean momentum balance.

13.4 Results from the MARV Model

Comparisons of MARV model results with the DNS data of Kristoffersen and Andersson (KA1993) are shown in Figs. 13.4.1-4. The mean velocity profiles for all the DNS rotation numbers are shown in Fig. 13.4.1. As rotation increases, the peak velocity shifts towards the suction side and a large region develops where the mean velocity gradient is nearly constant, with $\partial U_1/\partial x_2$ approaching $2\Omega_3$. (Dimensionlessly, this slope corresponds to a difference in U/U_{mean} across the channel of $2R_o$.) For R_o = 0.1 to 0.2, the modeled velocity profiles on the suction side of the channel are more laminar than the data, consistent with the laminarization at lower rotation shown in Fig. 13.3.1.

Fig. 13.4.1. Mean velocity profiles in a rotating channel with $R_\tau = 194$.
—— MARV model;
--- KA1993 DNS data.

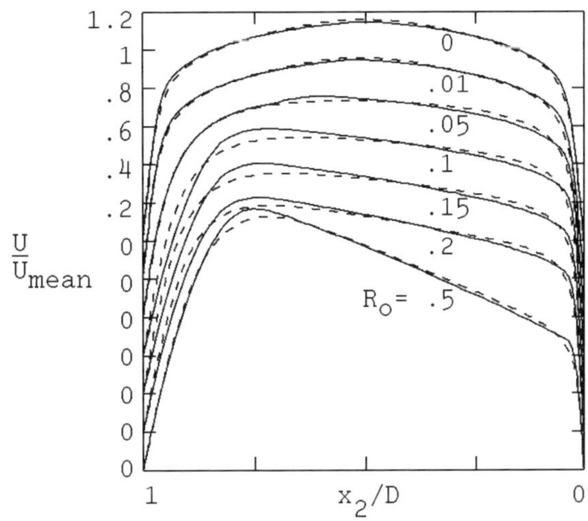

The turbulence kinetic energy is shown in Fig. 13.4.2 as the turbulence velocity scale, $q = \sqrt{k}$. At low rotation numbers, the turbulence increases with rotation on the pressure side and decreases on the suction side, with peaks in turbulence near the walls. At the highest rotation, $R_o = 0.5$, the shape of the q profile changes. There is good agreement between the model and the data, except at $R_o = 0.1$ on the suction side where the model gives lower turbulence.

The anisotropy of the turbulence is shown in Fig. 13.4.3. Overall, the normal anisotropies are well modeled, capturing the decrease in b_{11} and increase in b_{22} over the bulk of the flow as rotation is increased. The shear anisotropies are also well modeled, except near the suction side at $R_o = 0.1$, where the magnitude of b_{12} is smaller, which results in the lower turbulence and a laminar velocity profile.

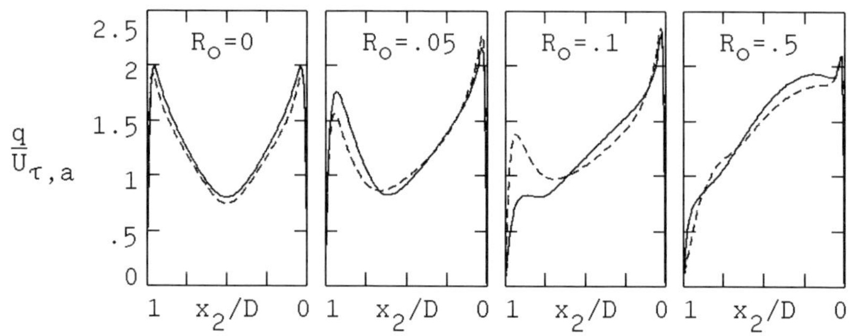

Fig. 13.4.2. Turbulence velocity scale, q, in a rotating channel with $R_\tau = 194$. —— MARV model; - - - KA1993 DNS data.

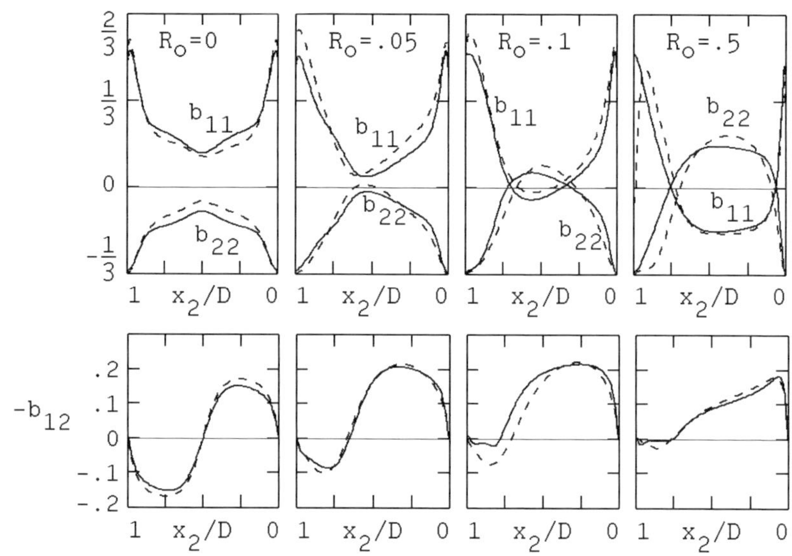

Fig. 13.4.3. Anisotropy of the turbulence in a rotating channel with $R_\tau = 194$. —— MARV model; - - - KA1993 DNS data.

Vorticity and Glyph Rotation

Putting together the turbulence kinetic energy and the turbulence anisotropy gives the Reynolds stresses, some of which are shown as glyphs in Fig. 13.4.4. The largest visual impact of the glyphs is their sizes. At zero rotation, the glyph in the center is a factor of 5 smaller than the near-wall glyphs, reflecting the factor of 5 smaller amount of turbulence kinetic energy. The orientation and shape of the glyphs are also informative. At zero rotation the glyphs are symmetrical relative to each wall.

As the rotation number increases from zero to $R_o = 0.05$, the glyphs are rotated clockwise, opposite to the rotation of the channel. On the suction side this results in a reduction in the magnitude of b_{12}, and on the pressure side an increase (Fig. 13.4.3). Also shown with the glyphs are profiles of $P/\tilde{\varepsilon}$, W/S, and the rotation parameter, R, from Chapter 4, Eq. 4.3.10. In the center of the channel at $R_o = 0.05$, the principal axes of the glyph are at nearly 45 degrees to the x_1, and x_2 directions. Correspondingly W/S is approximately 0.5, and R, –0.5. These values represent the most unstable condition in the bifurcation plots, Figs. 4.3.2 and 4.3.6. However at this location the velocity gradient is small and the turbulence, although larger than that at zero rotation, is still quite small.

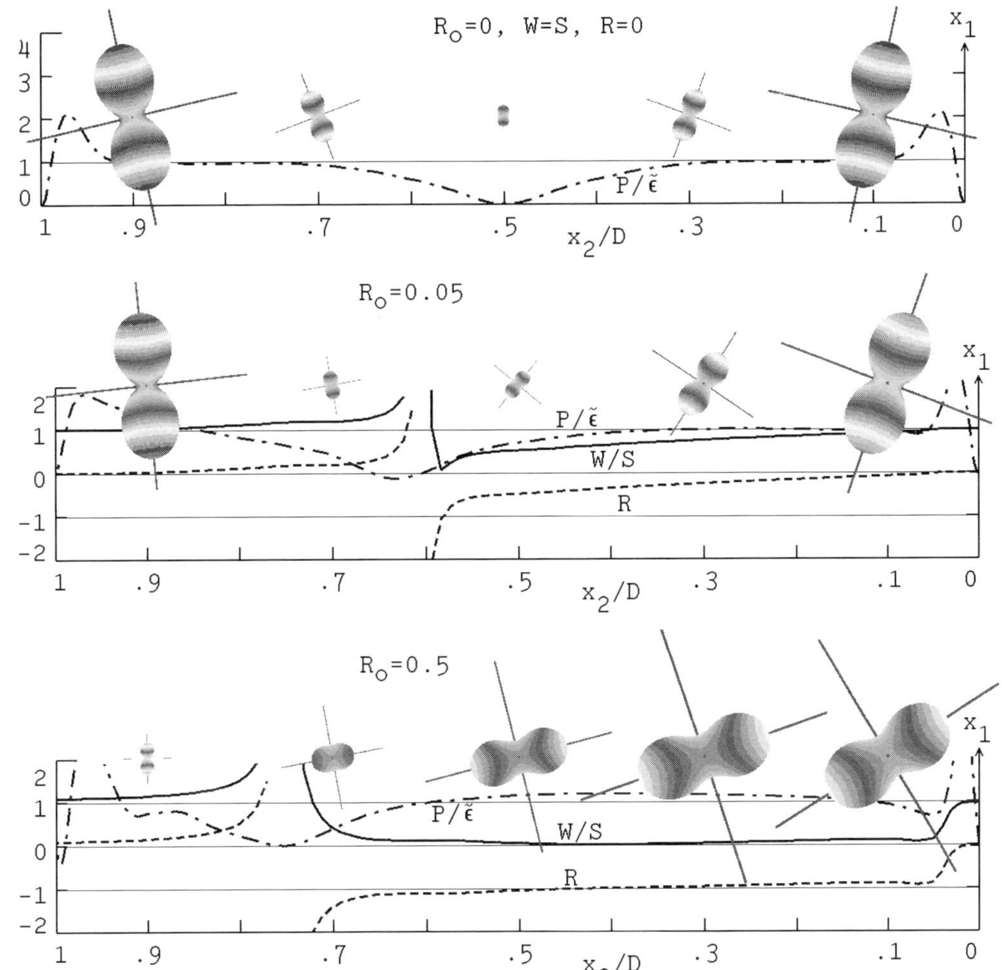

Fig. 13.4.4. Rotating channel with $R_\tau = 194$, MARV model results. – – – $P/\tilde{\varepsilon}$; —— W/S; – – – rotation parameter, R, Eq. 4.3.10. Glyphs of the Reynolds stress tensors, at x_2/D=0.1, 0.3, 0.5, 0.7, and 0.9. Sizes (length of axes) of the glyphs are proportional to k.

Near to this location (still considering the $R_o = 0.05$ case), from $x_2/D = 0.6$ to 0.65, turbulence production is negative ($P/\tilde{\varepsilon} < 0$). $\partial U_1/\partial x_2 = 0$ at $x_2/D = 0.6$, while b_{12}, and correspondingly the shear stress, is zero at $x_2/D = 0.65$. This region of negative production occurs due to the turbulent diffusion of the turbulence anisotropy. The DNS data shows a slightly broader ($x_2/D = 0.57$ to 0.67) but less intense region of negative production at this rotation number.

At the highest rotation number, $R_o = 0.5$, the glyphs are rotated even more, past 45 degrees so that $b_{22} > b_{11}$. The b_{12} potential for creating turbulence is reduced. However, a large region with $W/S \approx 0$, $R \approx -1$, is formed with a significant velocity gradient. The result is fairly high turbulence in the middle of the channel. The small region of negative production on the suction side of the channel is also present in the DNS data.

Reynolds Number Dependence

The Reynolds number dependence of the skin friction velocities is shown in Fig. 13.4.5. Series of calculations were made with the MARV model, keeping the Reynolds number, R_D, constant while increasing R_o. Wall friction-velocity Reynolds numbers, $R_{\tau,p}$ and $R_{\tau,s}$ (Eq. 13.1.2), on either side of the channel, have been normalized as in Fig. 13.2.4c, with $R_{\tau,0}$, the friction-velocity Reynolds number for zero rotation. There are three main features to the Reynolds number dependence:

(1) Near zero rotation at lower Reynolds numbers, increased rotation number causes $R_{\tau,p}/R_{\tau,0}$ to increase more rapidly, and $R_{\tau,s}/R_{\tau,0}$ to fall more rapidly.

(2) On the suction side of the channel, the rotation number at which the velocity profile becomes laminar increases with Reynolds number.

(3) At high rotation numbers, the normalized average friction-velocity Reynolds number (Eq. 13.1.7), $R_\tau/R_{\tau,0}$, is higher at low R_D.

The last feature is due to both higher $R_{\tau,p}/R_{\tau,0}$ at low R_D and higher laminar limiting values for $R_{\tau,s}/R_{\tau,0}$ at low R_D. $R_\tau/R_{\tau,0}$ is shown directly in Fig. 13.4.6.

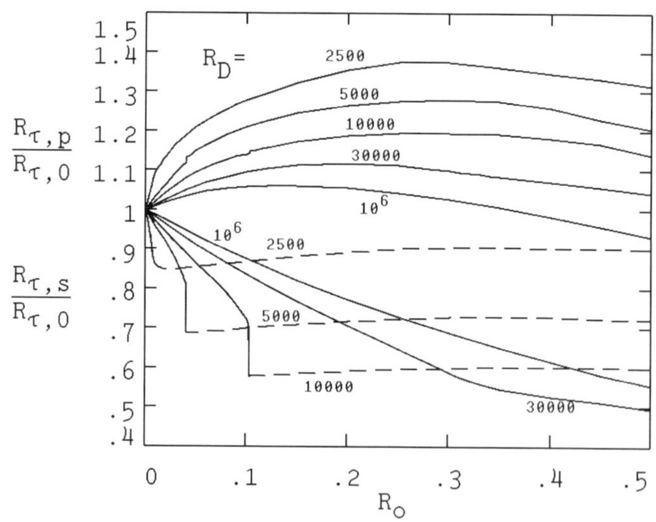

Fig. 13.4.5. Wall friction-velocity Reynolds numbers in rotating channels normalized with zero-rotation values. MARV model, R_D=2500, 5000, 10,000, 30,000, and 10^6.
—— turbulent pressure and suction sides;
– – laminar suction side.

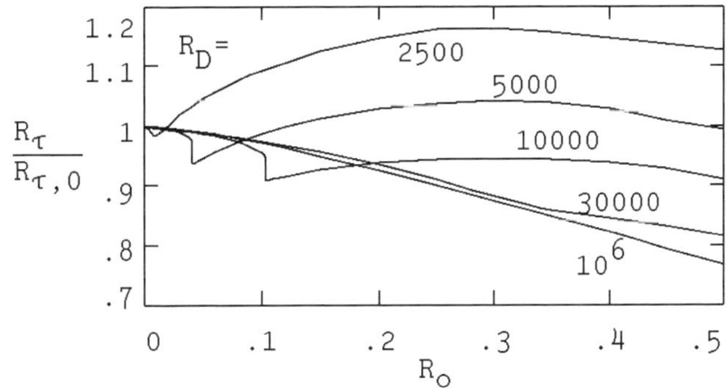

Fig. 13.4.6. Average friction-velocity Reynolds numbers normalized with zero-rotation values. MARV model, R_D=2500, 5000, 10,000, 30,000, and 10^6.

Figures 13.4.7-9 show comparisons with data to quantitatively verify these Reynolds number dependent features. Figure 13.4.7 shows $R_{\tau,s}/R_\tau$ at rotation numbers up to 0.2. $R_{\tau,s}/R_\tau$ instead of $R_{\tau,s}/R_{\tau,0}$ was chosen for the plot to remove errors due to uncertainty in the data values for R_D and therefore $R_{\tau,0}$. For $R_D = 6130$ ($R_{\tau,0} = 194$) the model gives $R_{\tau,s}/R_\tau$ falling with increasing R_o to go through a KA1993 data point at $R_o = 0.05$. The rotating-wall addition constants were chosen to do this. At higher Reynolds numbers, R_D =12,000 and 34,000, the MARV model results are consistent with the less rapid fall of $R_{\tau,s}/R_\tau$ shown in the experimental data.

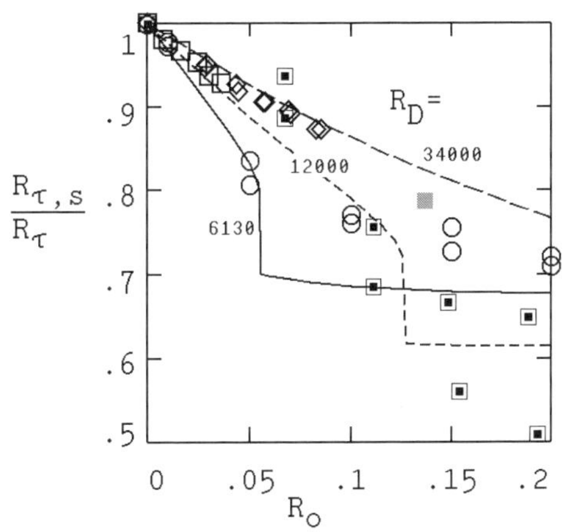

Fig. 13.4.7. Suction-side friction velocity ratio, $R_{\tau,s}/R_\tau$.

R_D	MARV model	data
~ 6130	———	○
~12,000	– – – –	□, ▣, ■
~34,000	— — —	◇

(Data as for Figs. 13.2.2 and 13.2.5.)

The rotating-wall addition constants were also chosen to approximately capture the variation in $R_D/R_{D,0}$ for the KA1993 data (Fig. 13.3.1c). At constant R_D, this is equivalent to capturing the variation in $R_\tau/R_{\tau,0}$. Figure 13.4.8 compares $R_\tau/R_{\tau,0}$ with data over a range of Reynolds numbers. At low rotation numbers the model gives $R_\tau/R_{\tau,0}$ falling gently, followed by a rapid fall as the suction side becomes laminar. This is consistent with the data, except that in Fig. 13.4.8a the experimental data for R_D~12000 falls to lower values of $R_\tau / R_{\tau,0}$. In Fig. 13.4.8b, the model shows $R_\tau / R_{\tau,0}$ increasing after the laminarization of the suction side, then falling, with the levels of $R_\tau / R_{\tau,0}$ being higher for lower R_D in agreement with the DNS data.

Fig. 13.4.8. Variation in friction-velocity Reynolds number ratio, $R_\tau / R_{\tau,0}$.

R_D	MARV	data
~ 4560	----	◁, ▷
~ 6130	———	○, ⊘
~12,000	— — —	□, ▣, (shaded)
~34,000	— — —	◇

(Data as for Figs. 13.2.2 and 13.2.5.)

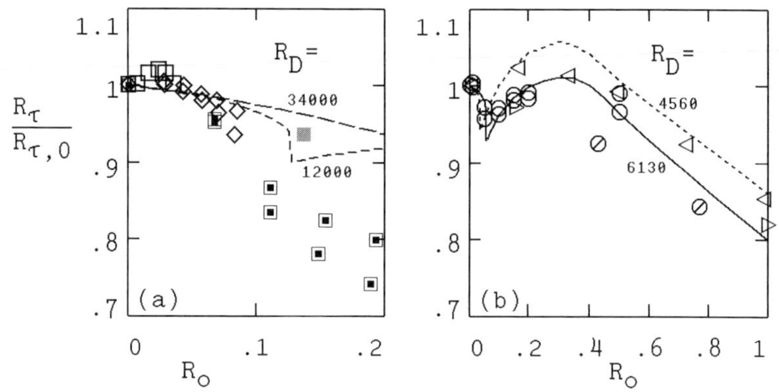

In Fig. 13.4.7, the MARV model shows a rapid fall in $R_{\tau,s} / R_\tau$ at $R_o = 0.056$ for $R_D = 6130$ and at $R_o = 0.128$ for $R_D = 12000$, corresponding to the velocity profiles on the suction side becoming laminar. The rotation at which this occurs is consistent with the experimental data for $R_D \sim 12000$ (▣). Figure 13.4.9 compares modeled laminarization rotation numbers with the visualization data of Halleen and Johnston (HJ1967). The model shows more Reynolds number dependence than the data. However, at zero rotation the model gives transition at $R_D = 1450$, consistent with the data of Patel and Head (1969), which gives transition at $R_D \approx 1464$.

Fig. 13.4.9. Regions of laminar and turbulent flow on the suction sides of rotating channels. Visualization data of Halleen and Johnston (HJ1967): S, streaky-laminar; L, completely laminar.
—— MARV model boundary between laminar and turbulent velocity profiles.
○, laminar limit for zero rotation, data of Patel and Head (1969).

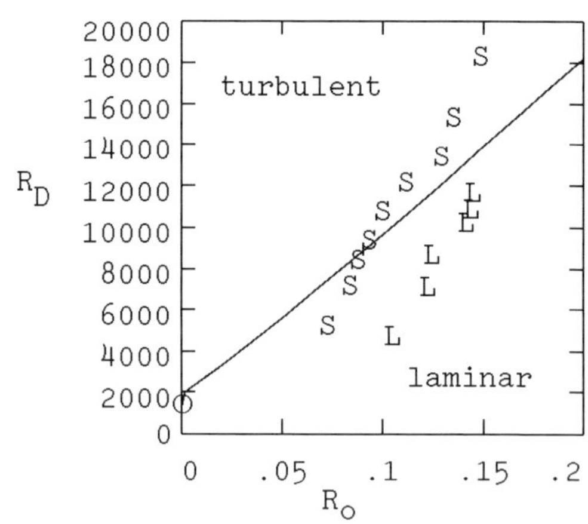

A more complete picture of transition in rotating channels is shown in Fig. 13.4.10, where the skin friction coefficients on each side of the channel are plotted for R_o = 0, 0.01, 0.02 and 0.05. The location of the lowest c_f on the suction side was used to define the laminar limit of the MARV model shown on Fig. 13.4.9. This figure also shows the overall laminar limit, where the pressure side, and therefore the entire flow, is laminar. This occurs at lower R_D with increasing rotation. For example, while the model gives a transition Reynolds number, R_D =1450, at zero rotation, this reduces to 942 at R_o =0.2.

From Fig. 13.4.10, the skin friction coefficients for the laminar suction sides of rotating channels appear to follow an approximately universal curve. This is only valid at low rotation numbers. At higher rotation, the laminar suction-side limit is first above this line then below, as at very high rotation numbers ($R_o > 2$) it approaches the laminar line for zero rotation.

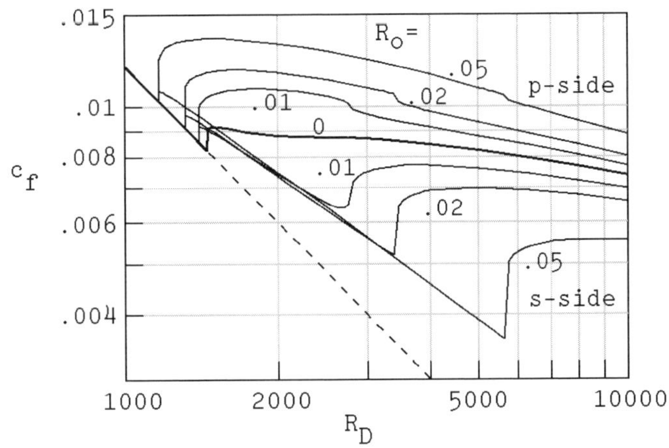

Fig. 13.4.10. Skin friction coefficients on the pressure and suction sides of rotating channels from the MARV model.
——— zero rotation laminar flow.

13.5 Model Variations

In this section we consider which features of the MARV model are important in rotating channel flow, and look at results from the simple MC model.

Parameters in the MARV Model

The importance of the rotating wall modification is shown in Fig. 13.5.1 where results for $R_\tau = 194$ are compared with DNS data. As shown in Fig. 13.5.1a, the suction side becomes laminar at about $R_o = 0.06$, whether or not the modification is included. The big effect is shown in Fig. 13.5.1b. Without the modification, the flow rate falls significantly with rotation, which is inconsistent with the data. Figure 13.5.1c shows the results as a more traditional plot, where the calculations were made at constant R_D, and $R_{\tau,p}$ and $R_{\tau,s}$ have been normalized with $R_{\tau,0}$. Without the rotating wall modification, the friction velocity on the pressure side is very high.

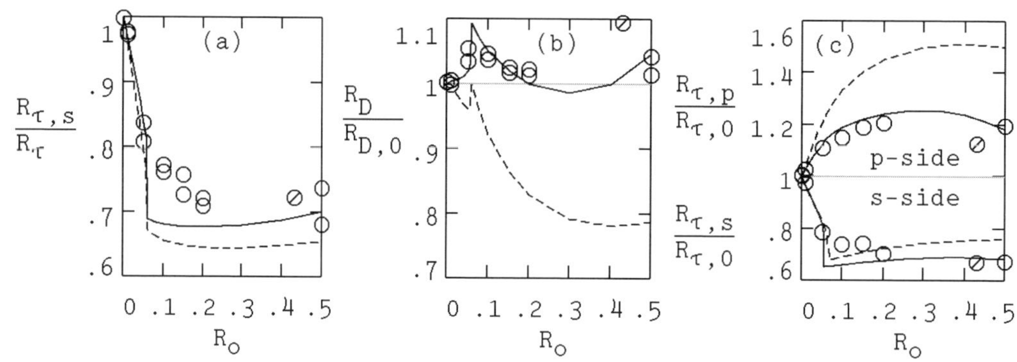

Fig. 13.5.1. Effect of rotating wall parameter, f_R, on rotating channel. (a),(b) with R_τ=194; (c) with R_D=6130. —— MARV model; ---- MARV model with f_R=0 (no rotating wall modification). DNS data, ○, KA1993; ⊘, AJ1999.

At zero rotation, $W = S$ in the channel flow, so that the vorticity parameter f_w is one. As soon as the rotation is present f_w takes the full range of values, zero to one. Some distributions of f_w for $R_o = 0.5$ are shown in Fig. 13.5.2a. At this high rotation number there is a large region with low f_w. The rapid distortion parameter, f_B, is shown in Fig. 13.5.2b. At rotation numbers below $R_o = 0.03$, f_B was zero throughout the channel, but at high rotation numbers, as shown in Fig. 13.5.2b, high values are obtained especially in the region near $\partial U_1/\partial x_2 = 0$. Here, terms in the b_{ij} equations which are proportional to strain rate are small compared to the rotation term, so that in the absolute frame (which is used as the basis for the f_B parameter) b_{ij} is not fully developed.

Fig. 13.5.2. MARV model parameters in a rotating channel with R_o = 0.5: —— R_τ=194; --- R_τ=906.

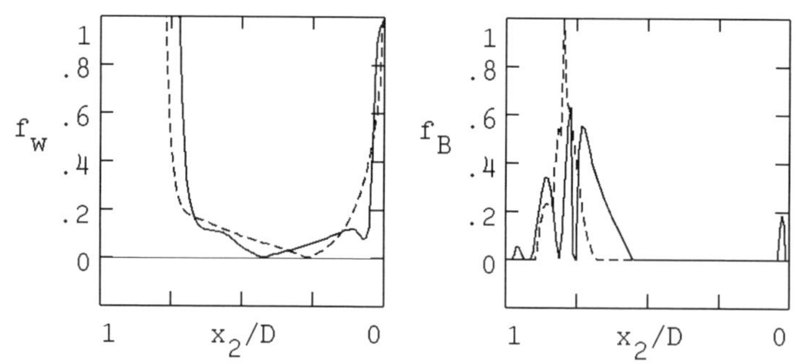

Results of calculations which forced $f_B = 0$, in other words which removed the rapid distortion features of the MARV model, are shown in Fig. 13.5.3 (--- lines). There is very little difference between these results and the full model, so that rapid distortion is not significant in the fully developed rotating channel. Also shown on Fig. 13.5.3 are results from imposing both $f_B = 0$ and $f_w = 1$. This is the MA model but with the rotating-wall addition, the f_R term. The pressure-side results are somewhat affected, which should not be surprising since W/S is less than one in the pressure-side boundary layer.

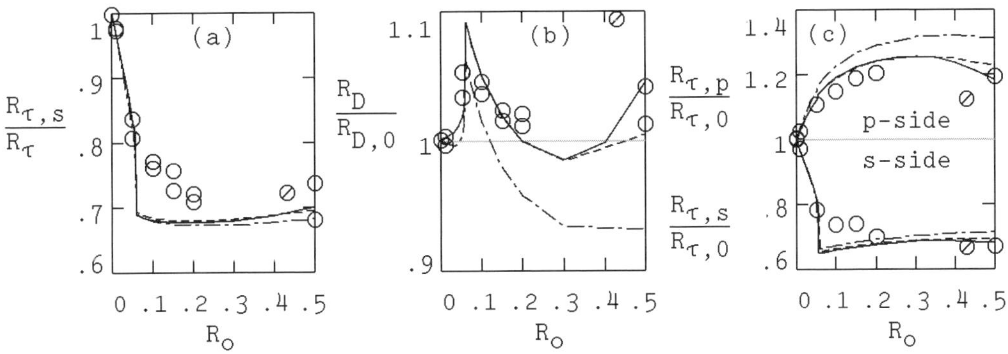

Fig. 13.5.3. Effect of parameters f_B and f_w on rotating channel. (a),(b) with $R_\tau = 194$; (c) with $R_D = 6130$. —— MARV model; ---- MARV model, but with $f_B = 0$; –·–·– MARV model, but with $f_B = 0$ and $f_w = 1$. DNS data, ○, KA1993; ⊘, AJ1999.

Results from the MC Model

Results from the MC model are compared with Kristoffersen and Andersson's (KA1993) data in Figs. 13.5.4-9.

As shown in Fig. 13.5.4a, $R_{\tau,s}/R_\tau$ falls less rapidly than in the MARV model (Fig. 13.5.1) and the MC model gives a more gradual transition to laminar flow, consistent with the DNS data. From Fig. 13.5.4c, $R_{\tau,p}$ is reasonable up to about $R_o = 0.2$, after which it is significantly over-predicted. Correspondingly, Fig. 13.5.4b shows large flow reductions for $R_o > 0.2$, caused by the pressure-side error.

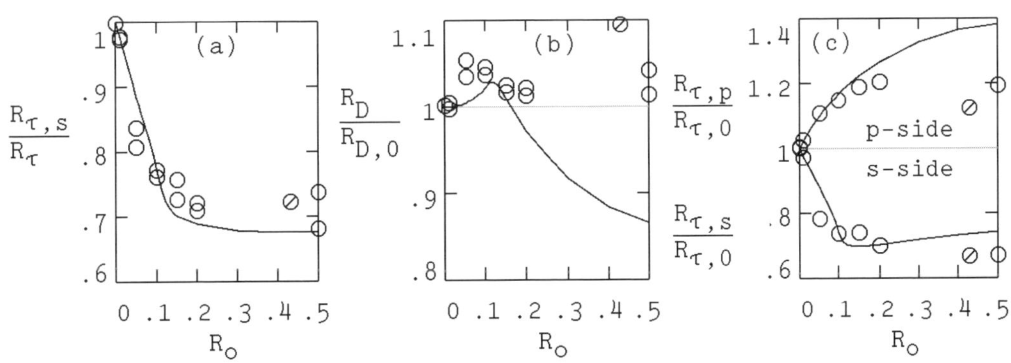

Fig. 13.5.4. Wall friction velocity and Reynolds number ratios from the MC model. (a),(b) with $R_\tau = 194$; (c) with $R_D = 6130$. —— MC model; DNS data, ○, KA1993; ⊘, AJ1999.

Shear anisotropies from the MC model are shown in Fig. 13.5.5. Significantly, the MC model gives wall values for b_{12} which are zero, or nearly zero, for the entire range of rotations, even though it uses the boundary condition $\partial b_{ij}/\partial y = 0$ at the wall. This may be explained using an extension of the two-equation analysis in Chapter 5.

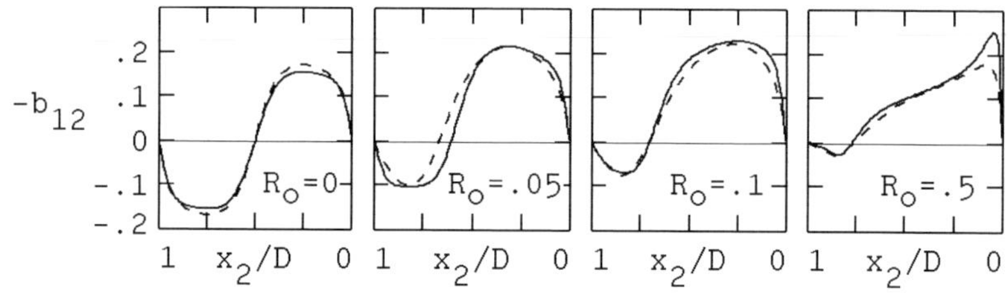

Fig. 13.5.5. Shear anisotropy, b_{12}, in a rotating channel with $R_\tau = 194$.
—— MC model;
--- KA1993 DNS data.

Rewriting Eq. 5.3.8 for rotating flows, a fully developed shear flow with rotation will give b_{12} going to zero as c_1 goes to 2 (which it does at the wall) if

$$\left(\frac{db_{12}}{dD}\right)_{D=0,Eq.5.3.6} \geq \left(\frac{db_{12}}{dD}\right)_{Eq.5.3.2} . \tag{13.5.1}$$

From Eqs. 5.3.6 and 5.3.2, the condition is

$$\frac{-\frac{1}{2}\left(\frac{4}{3} - c_3\right)}{\left\{(4 - c_5)\frac{\Omega_3}{S_{12}} + c_5 - 2\right\}^2 - \frac{1}{3}(c_4 - 2)^2} \geq -\frac{1}{4 + c_6} . \tag{13.5.2}$$

Evaluating this for the MC model (which uses $c_3 = 0.15,\ c_4 = 1,\ c_5 = c_6 = 0$, Eq. 8.2.1) yields the condition (Appendix EQ.13.5.3)

$$\frac{\Omega_3}{S_{12}} \leq \frac{-\sqrt{2.7} + 2}{4} = 0.0892 . \tag{13.5.3}$$

On the suction side of the channel, S_{12} is negative so that this condition is always satisfied. On the pressure side of the channel, this condition may be written as

$$\frac{W}{S} = \frac{S_{12} - \Omega_3}{S_{12}} \geq 0.9108 . \tag{13.5.4}$$

This was satisfied for all the calculations made for the MC model. (For example, for $R_\tau = 194$, $R_o = 0.5$, W/S was 0.977 at the pressure-side wall.) Of course, it takes more than $b_{12} \sim 0$ at the wall to model the shear stress distribution. Looking more closely at the b_{12}'s in Fig. 13.5.5, the over-prediction of b_{12} in the pressure side layer for $R_o = 0.5$, and over-prediction of b_{12} near the suction side for $R_o = 0.05$, both led to U_τ's which were higher than the data (Fig. 13.5.4c).

Interestingly, the velocity profiles and turbulence kinetic energy from the MC model appear to be in reasonable agreement with the data, as shown in Figs. 13.5.6 and 13.5.7. Comparing these two figures with Figs. 13.4.1 and 13.4.2, it would seem that there is little to choose between the MC and MARV models. However, these figures do not reveal some of the near-wall behavior necessary to obtain agreement with wall friction velocities.

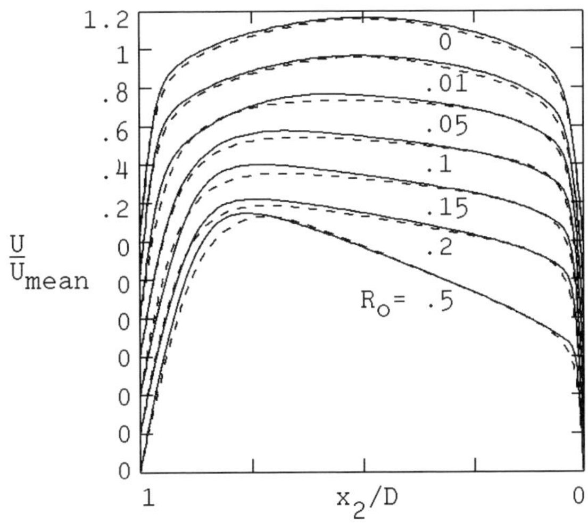

Fig. 13.5.6. Mean velocity profiles in the rotating channel for $R_\tau = 194$.
—— MC model;
--- KA1993 DNS data.

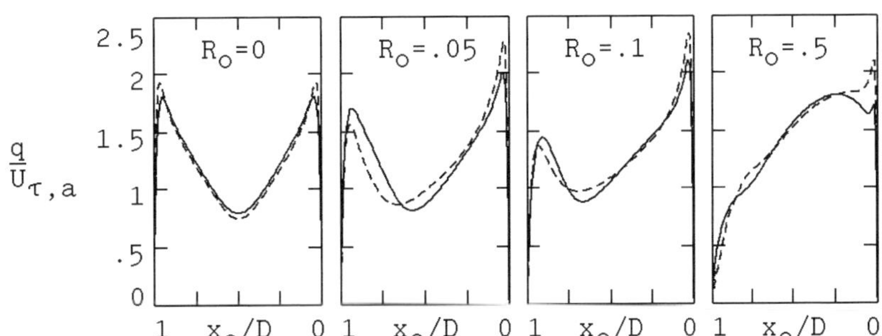

Fig. 13.5.7. Turbulence velocity scale, q, in a rotating channel with $R_\tau = 194$.
—— MC model;
--- KA1993 DNS data.

The normal anisotropies, b_{11} and b_{22}, from the MC model are shown in Fig. 13.5.8. The results are similar to those for the MARV model, Fig. 13.4.3, except near the walls. In particular b_{22} does not go $-1/3$ at the walls. The implication, in terms of near-wall modeling, is that the MC model has a different balance of terms in the b_{12} equation, in particular, production, rotation and pressure-strain.

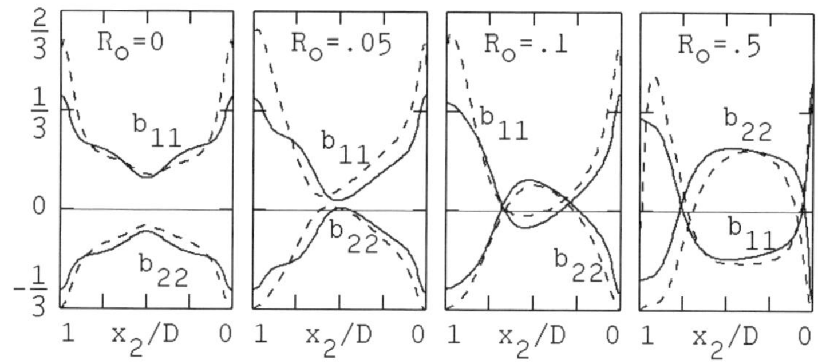

Fig. 13.5.8. Normal anisotropy, rotating channel with $R_\tau = 194$. —— MC model; --- KA1993 DNS data.

The Reynolds number dependence of the MC model is compared with the MARV model in Fig. 13.5.9. While all the same features, as described for the MARV model in connection with Fig. 13.4.5, are present, quantitatively the results are different. For the MC model, the rate at which $R_{\tau,p}/R_{\tau,0}$ and $R_{\tau,s}/R_{\tau,0}$ depart from one with rotation is smaller than for the MARV model, especially at low R_D. The rotation number at which the suction side becomes laminar is about twice that for the MARV model. And, at higher rotation numbers, the MC model gives higher values of $R_{\tau,p}/R_{\tau,0}$, and correspondingly $R_\tau/R_{\tau,0}$, especially at low R_D.

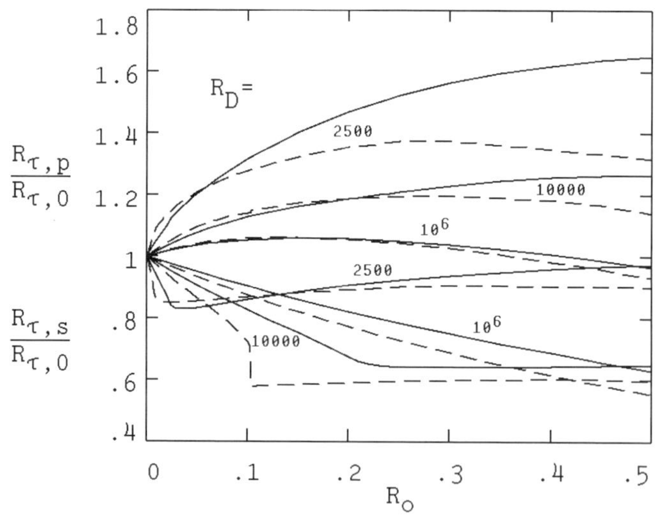

Fig. 13.5.9. Wall friction-velocity Reynolds numbers for R_D=2500, 10000 and 10^6. —— MC model; --- MARV model.

Chapter 14

AXIALLY ROTATING PIPES

In this chapter we consider flow in pipes rotating on their axes. This is an interesting test case, since the Reynolds stress tensor in cylindrical, r,θ,z, coordinates is full with no components which are zero due to symmetry. Consideration is limited to fully developed flow.

The rotation of the pipe wall causes a rotation of the flow, which in turn causes a counter rotation of the Reynolds stress tensor glyph. The result is that skin friction is reduced with increased rotation and the velocity profile takes on a more laminar shape. The glyph rotation also causes a non-zero $\overline{u_r u_\theta}$ Reynolds stress. This in turn, when considering the flow in the rotating frame of the pipe, causes a Reynolds stress driven secondary flow in the θ direction.

14.1 Equations for Fully Developed Flow

The coordinates and geometry for a pipe rotating about its axis are shown in Fig. 14.1.1. We will use the convention that the rotation is positive about the z-axis. Fully developed flow is steady in both the rotating frame of the pipe and in the absolute frame. The mean radial velocity, U_r, is zero and time-averaged properties do not vary in either the θ or z directions, except $\partial \overline{P}/\partial z < 0$. In the absolute frame the wall velocity is $V_{\theta,wall} = \Omega_z D/2$, where Ω_z is the rotation rate of the pipe and D is its diameter. A dimensionless rotation rate, N, is defined using the area-mean axial velocity,

$$N = \frac{V_{\theta,wall}}{U_{mean}} = \frac{1}{2}\frac{\Omega_z D}{U_{mean}} \quad . \tag{14.1.1}$$

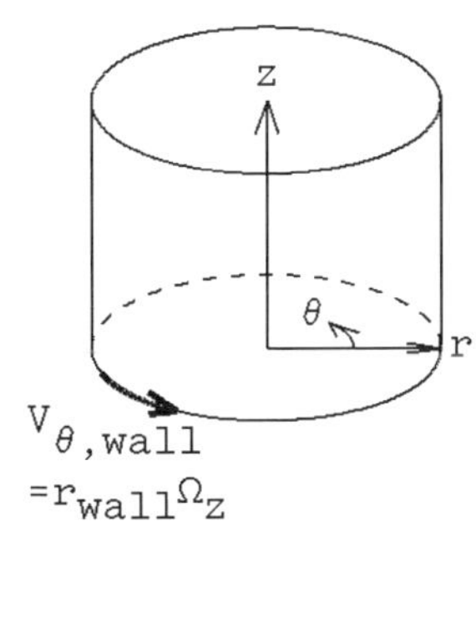

Fig. 14.1.1. Coordinates for axially rotating pipe.

The mean momentum equations for fully developed pipe flow with axial rotation are derived in the relative frame in Appendix D (Eqs. 61-63). The radial momentum equation, here given in the absolute frame,

$$0 = -\frac{1}{r}\frac{\partial r\overline{u_r u_r}}{\partial r} - \frac{1}{\rho}\frac{\partial \overline{P}}{\partial r} + \frac{V_\theta V_\theta}{r} + \frac{\overline{u_\theta u_\theta}}{r} \, , \tag{14.1.2}$$

determines the radial distribution of static pressure, and need not be solved. The tangential momentum equation,

$$0 = -\frac{1}{r}\frac{\partial r\overline{u_\theta u_r}}{\partial r} + \frac{1}{r}\frac{\partial}{\partial r}\nu r\frac{\partial V_\theta}{\partial r} - \frac{\overline{u_r u_\theta}}{r} - \frac{\nu V_\theta}{r^2} \, , \tag{14.1.3}$$

is solved with the boundary conditions of the specified wall velocity, $V_{\theta,wall}$, and $V_\theta = 0$ on the pipe axis. The axial momentum equation,

$$0 = -\frac{1}{r}\frac{\partial r\overline{u_z u_r}}{\partial r} - \frac{1}{\rho}\frac{\partial \overline{P}}{\partial z} + \frac{1}{r}\frac{\partial}{\partial r}\nu r\frac{\partial V_z}{\partial r} \, , \tag{14.1.4}$$

is the same as for the stationary pipe (Eq. 8.1.14).

For comparison with turbulent flow results, it is useful to consider the laminar flow solution, Fig. 14.1.2. The axial velocity profile is parabolic, with a maximum velocity which is twice the area-mean velocity (Eq. 8.1.13). The tangential velocity varies linearly in the absolute frame and is zero in the rotating frame. The absolute and relative tangential velocities are related by

$$V_\theta = U_\theta + r\Omega_z \; . \tag{14.1.5}$$

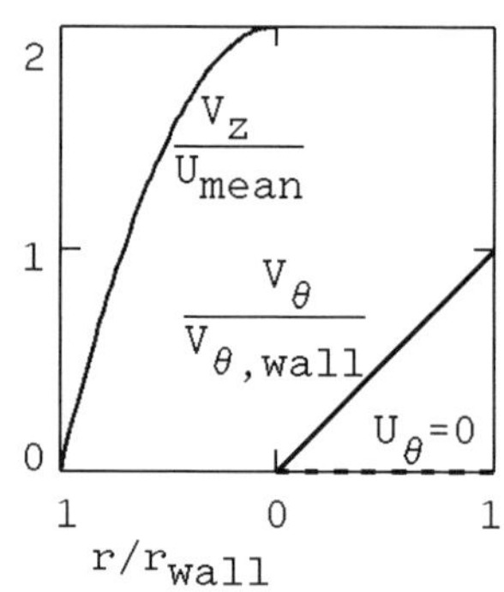

Fig. 14.1.2. Velocity profiles for laminar flow in an axially rotating pipe.

The equations for the scalar turbulence properties, q and ω, are the same as for the stationary pipe, Eq. 8.1.15,

$$0 = \frac{1}{r}\frac{\partial}{\partial r} r\nu \frac{\partial \psi}{\partial r} + \frac{1}{r}\frac{\partial}{\partial r} r\left(C_{rr}\frac{\partial \psi}{\partial r}\right) + S_\psi \ . \qquad (14.1.6)$$

The non-zero components of the strain-rate tensor (Appendix D, Eq. 22) are

$$S_{r\theta} = S_{\theta r} = \frac{1}{2}\left(\frac{\partial V_\theta}{\partial r} - \frac{V_\theta}{r}\right) = \frac{1}{2}\left(\frac{\partial U_\theta}{\partial r} - \frac{U_\theta}{r}\right) \quad \text{and} \quad S_{rz} = S_{zr} = \frac{1}{2}\frac{\partial V_z}{\partial r} \ , \qquad (14.1.7)$$

so that turbulence production (Eq. 2.2.6) and the source terms for q and ω (Eqs. 8.1.3-4) become

$$P = -2kb_{ik}S_{ik} = -4k(b_{r\theta}S_{r\theta} + b_{rz}S_{rz}) \ , \qquad (14.1.8)$$

$$S_q = -2q(b_{r\theta}S_{r\theta} + b_{rz}S_{rz}) - \frac{\omega q}{2} \ , \qquad (14.1.9)$$

and

$$S_\omega = -4\omega(c_{\varepsilon 1} - 1)(b_{r\theta}S_{r\theta} + b_{rz}S_{rz}) - (c_{\varepsilon 2} - 1)\omega^2 \ . \qquad (14.1.10)$$

Note that while the $S_{r\theta}$ term makes a significant contribution to turbulence production in developing flow, the influence of the term on fully developed flow tends to be small, since the wall is no longer imparting tangential momentum to the fluid and $S_{r\theta}$ is zero at the wall (Exercise 14.1 below).

The equations for the turbulence anisotropy are derived for both developing and fully developed flow in Appendix D. Here, the equations are given for fully developed flow in terms of the absolute velocity and vorticity, including the expansion of the source term, S_{bij} (Eq. 13.3.2), for each component. The only non-zero components of the absolute vorticity (Appendix D, Eq. 26) are

$$W_{r\theta} = -W_{\theta r} = \frac{1}{2}\left(-\frac{\partial V_\theta}{\partial r} - \frac{V_\theta}{r}\right) \quad \text{and} \quad W_{rz} = -W_{zr} = \frac{1}{2}\left(-\frac{\partial V_z}{\partial r}\right) \ . \qquad (14.1.11)$$

The differential equations for the turbulence anisotropies are (Appendix EQ.14.1.12-16):

Equation for $b_{rr} - b_{\theta\theta}$,

$$-\frac{1}{r}\frac{\partial}{\partial r} r\left((\nu + C_{rr})\frac{\partial (b_{rr} - b_{\theta\theta})}{\partial r}\right) + \frac{4(\nu + C_{\theta\theta})}{r^2}(b_{rr} - b_{\theta\theta}) + \frac{4}{r}\frac{\partial}{\partial r} C_{r\theta} b_{r\theta} + \frac{4C_{r\theta}}{r}\frac{\partial b_{r\theta}}{\partial r}$$
$$= -\frac{1}{2}(b_{rr} - b_{\theta\theta})\left((c_1 - 2)\omega + (2 + \frac{1}{2}c_6)\frac{P}{k}\right) + (c_4 - 2)b_{rz}S_{rz} + (c_5 - 2)b_{rz}W_{rz} \qquad (14.1.12)$$
$$+2(c_5 - 2)b_{r\theta}W_{r\theta} - 8f_R\Omega_z b_{r\theta} + 4\frac{V_\theta}{r} b_{r\theta} \ .$$

Equation for b_{zz},

$$-\frac{1}{r}\frac{\partial}{\partial r}r\left((\nu + C_{rr})\frac{\partial b_{zz}}{\partial r}\right) = -\frac{1}{2}b_{zz}\left((c_1 - 2)\omega + (2 + \frac{1}{2}c_6)\frac{P}{k}\right)$$
$$+\frac{1}{2}(c_4 - 2)(2b_{rz}S_{rz} + \frac{1}{3}\frac{P}{k}) - (c_5 - 2)b_{rz}W_{rz} \quad . \tag{14.1.13}$$

Equation for b_{rz},

$$-\frac{1}{r}\frac{\partial}{\partial r}r\left((\nu + C_{rr})\frac{\partial b_{rz}}{\partial r}\right) + \frac{(\nu + C_{\theta\theta})}{r^2}b_{rz} + \frac{1}{r}\frac{\partial}{\partial r}C_{r\theta}b_{\theta z} + \frac{C_{r\theta}}{r}\frac{\partial b_{\theta z}}{\partial r}$$
$$= -\frac{1}{2}b_{rz}\left((c_1 - 2)\omega + (2 + \frac{1}{2}c_6)\frac{P}{k}\right) + \frac{1}{2}\left(c_3 - \frac{4}{3}\right)S_{rz}$$
$$-\frac{1}{2}(c_4 - 2)b_{\theta\theta}S_{rz} + \frac{1}{2}(c_5 - 2)(b_{zz} - b_{rr})W_{rz}$$
$$+\frac{1}{2}(c_4 - 2)b_{\theta z}S_{r\theta} + \frac{1}{2}(c_5 - 2)b_{\theta z}W_{r\theta} - 2f_R\Omega_z b_{\theta z} + \frac{V_\theta}{r}b_{\theta z} \quad . \tag{14.1.14}$$

Equation for $b_{r\theta}$,

$$-\frac{1}{r}\frac{\partial}{\partial r}r\left((\nu + C_{rr})\frac{\partial b_{r\theta}}{\partial r}\right) + \frac{4(\nu + C_{\theta\theta})}{r^2}b_{r\theta} - \frac{1}{r}\frac{\partial}{\partial r}C_{r\theta}(b_{rr} - b_{\theta\theta}) - \frac{C_{r\theta}}{r}\frac{\partial(b_{rr} - b_{\theta\theta})}{\partial r}$$
$$= -\frac{1}{2}b_{r\theta}\left((c_1 - 2)\omega + (2 + \frac{1}{2}c_6)\frac{P}{k}\right) + \frac{1}{2}\left(c_3 - \frac{4}{3}\right)S_{r\theta}$$
$$+\frac{1}{2}(c_4 - 2)(b_{\theta z}S_{rz} - b_{zz}S_{r\theta}) + \frac{1}{2}(c_5 - 2)\left(b_{\theta z}W_{rz} - (b_{rr} - b_{\theta\theta})W_{r\theta}\right)$$
$$+2f_R\Omega_z(b_{rr} - b_{\theta\theta}) - \frac{V_\theta}{r}(b_{rr} - b_{\theta\theta}) \quad . \tag{14.1.15}$$

Equation for $b_{\theta z}$,

$$-\frac{1}{r}\frac{\partial}{\partial r}r\left((\nu + C_{rr})\frac{\partial b_{\theta z}}{\partial r}\right) + \frac{(\nu + C_{\theta\theta})}{r^2}b_{\theta z} - \frac{1}{r}\frac{\partial}{\partial r}C_{r\theta}b_{rz} - \frac{C_{r\theta}}{r}\frac{\partial b_{rz}}{\partial r}$$
$$= -\frac{1}{2}b_{\theta z}\left((c_1 - 2)\omega + (2 + \frac{1}{2}c_6)\frac{P}{k}\right) + \frac{1}{2}(c_4 - 2)(b_{r\theta}S_{rz} + b_{rz}S_{r\theta})$$
$$-\frac{1}{2}(c_5 - 2)(b_{r\theta}W_{rz} + b_{rz}W_{r\theta}) + 2f_R\Omega_z b_{rz} - \frac{V_\theta}{r}b_{rz} \quad . \tag{14.1.16}$$

The laminar and turbulent diffusion terms are shown on the left hand side of each of the above equations. Because the MARV model uses an anisotropic diffusion with (Eq. 8.1.12)

$$C_{ij} = c_{t\psi}\frac{q^2}{\omega}\left(\frac{\overline{u_i u_j}}{k}\right)^{n_t}, \qquad (14.1.17)$$

and because the θ direction is not a principal direction of the Reynolds stress tensor (as it is for stationary pipe flow), there are additional diffusion terms involving the $C_{r\theta}$ coefficient.

The last lines of Eqs. 14.1.12 and 14.1.14 are new source terms for the $b_{rr} - b_{\theta\theta}$ and b_{rz} equations, which were not present in stationary pipe flow. In addition, for stationary pipe flow, $b_{r\theta}$ and $b_{\theta z}$ were zero so it was unnecessary to solve Eqs. 14.1.15 and 14.1.16. Since $S_{r\theta}$ is zero at the wall, both the relative tangential velocity and its gradient are zero at the wall. The wall boundary conditions for the b_{ij} equations are then the same as for the stationary pipe.

Exercise 14.1. Rearrange Eq. 14.1.3 into the form $\frac{1}{r^2}\frac{\partial}{\partial r}r^2(2\nu S_{r\theta}) = \frac{1}{r^2}\frac{\partial}{\partial r}r^2\overline{u_r u_\theta}$, and thence show that for fully developed flow in an axially rotating pipe $2\nu S_{r\theta} = \overline{u_r u_\theta}$ in general, and in particular $(S_{r\theta})_{wall} = 0$.

14.2 Data and Flow Physics

The most complete set of experimental data for fully developed pipe flow is that of Imao et al. (IIH1996) at $R_D = 20{,}000$. In addition to measuring wall friction and velocity profiles, Imao et al. used a single-component laser to obtain five of the six Reynolds stress components for $N = 0$, 0.5 and 1. The sixth Reynolds stress, $\overline{u_r u_\theta}$, can be obtained by analysis from the tangential velocity profile and Eq. 14.1.3 (see Exercise 14.1).

Kikuyama et al. (KMNM1983) present wall friction coefficients and mean velocity profiles from their own experiments and those of Yamada and Imao (YI1980). The data is for $R_D = 7{,}500$ to 100,000, with some data to $N = 3$.

The Reich and Beer (1989) experiments on fully developed pipe flow included a heated wall, so that temperature distributions and heat transfer coefficients were measured as well as mean velocities and skin friction. The heated wall appears to have altered the mean velocity profiles, lowering the centerline velocity as shown in Fig. 14.2.1, and therefore making the data unsuitable for comparisons in this chapter.

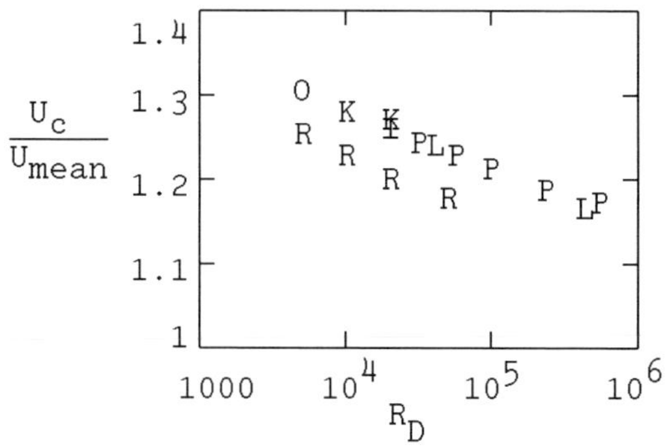

Fig. 14.2.1. Ratio of centerline to average velocity for stationary pipe flow. Data: R, Reich and Beer (1989); K, KMNM1983; I, IIH1996; O, Orlandi and Fatica DNS (1997); L, Laufer (L1954); P, Princeton superpipe, Zagarola (1996).

Orlandi and Fatica (1997) made DNS calculations at $R_D = 4900$, for $N = 0$, 0.5, 1 and 2. Orlandi and Ebstein (2000) extended the calculations to $N = 5$ and 10. Due to ambiguities in the presentation of the data sets, only the tabulated friction coefficients from Orlandi and Fatica will be used here.

Velocity Profiles

Pipe rotation causes the turbulent axial velocity profile to take on a more laminar shape, as shown in Fig. 14.2.2. For $N = 3$ and $R_D = 10{,}000$, the data of Kikuyama et al. (KMNM1983) shows a velocity profile closely resembling laminar flow.

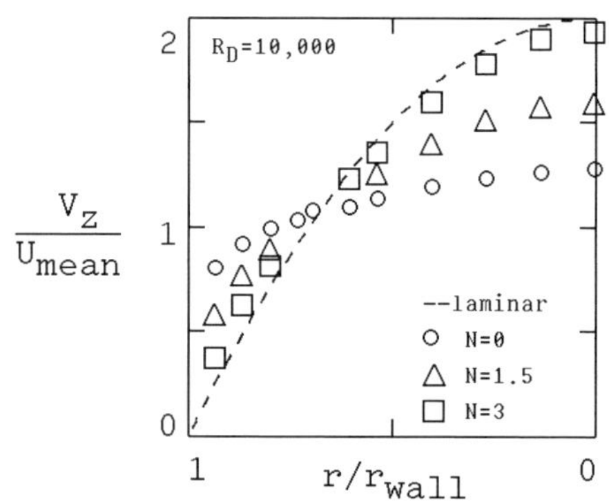

Fig. 14.2.2. Axial velocity profiles data from Kikuyama et al. (KMNM1983) for R_D=10,000: ○, N=0; △, N=1.5; □, N=3. ---laminar flow.

Profiles of tangential velocity are available for $N \le 2$. These appear to have a fairly universal shape when normalized with the wall velocity, as shown in Figs. 14.2.3 and 14.2.4. Figure 14.2.3 shows the absolute tangential velocity and the points have been tagged with rotation number, N. There may be some tendency to more laminar values near the wall for $N > 1$.

The difference between the absolute tangential velocity and the laminar profile, $V_\theta / V_{\theta,wall} = r / r_{wall}$, is the relative tangential velocity. This is shown in Fig. 14.2.4, where the points have been tagged with Reynolds number. There may be some tendency for U_θ to be smaller at low R_D.

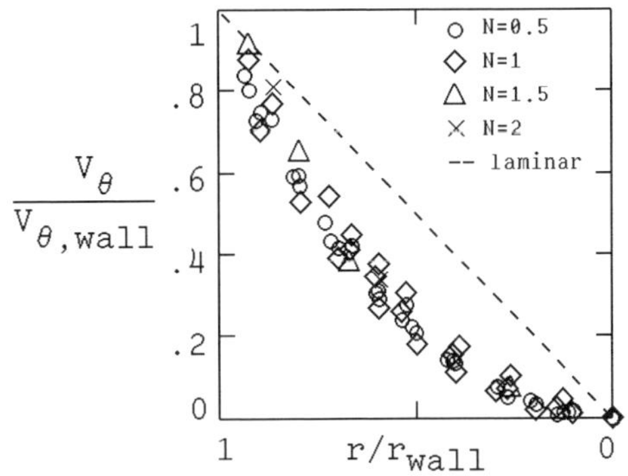

Fig. 14.2.3. Absolute tangential velocity normalized with the wall velocity. Data from YI1980, KMNM1983 and IIH1996, R_D=10,000 to 50,000: o, N=0.5; ◇, N=1; △, N=1.5; ×, N=2. --- laminar flow.

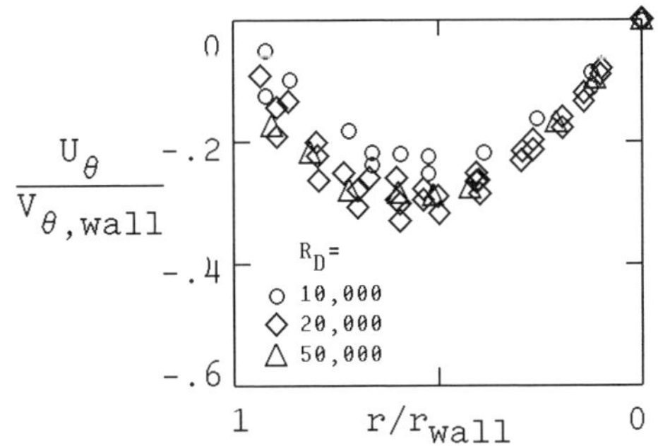

Fig. 14.2.4. Relative tangential velocity normalized with the absolute wall tangential velocity for N=0.5 and N=1. Data: o, R_D=10,000, KMNM1983; ◇, R_D=20,000, KMNM1983 and IIH1996; △, R_D=50,000, YI1980.

The physics of the turbulence which causes apparent laminarization of the axial velocity profile, and relative secondary flows, U_θ, can be seen by considering the V_θ / r source terms in the b_{ij} equations:

Equation	V_θ / r source term
$b_{rr} - b_{\theta\theta}$	$4b_{r\theta}V_\theta / r$
b_{rz}	$b_{\theta z}V_\theta / r$
$b_{r\theta}$	$-(b_{rr} - b_{\theta\theta})V_\theta / r$
$b_{\theta z}$	$-b_{rz}V_\theta / r$.

(14.2.1)

The shear anisotropy, b_{rz}, associated with the axial velocity in the pipe, is positive. This produces negative values of $b_{\theta z}$ in the rotating pipe. These in turn reduce b_{rz}, reducing $\overline{u_r u_z}$, and the axial velocity profile takes on a more laminar appearance. In general $b_{\theta\theta} > b_{rr}$, so that $-(b_{rr} - b_{\theta\theta})$ is positive. This generates positive values of $b_{r\theta}$, and correspondingly positive $\overline{u_r u_\theta}$. These act in the tangential momentum equation, Eq. 14.1.3, to change the shape of the absolute tangential velocity profile and produce secondary flows in the rotating frame.

Glyph Rotation

The modification of turbulence by rotation is visualized with Reynolds stress tensor glyphs in Fig. 14.2.5, using the data of Imao et al. at $r/r_{wall} = 0.8$. The top pictures are *r-z* views. These are similar to glyph views used earlier in that they show the wall normal and primary flow (*z*) directions. For no rotation, Fig. 14.2.5a, the primary principal axis (the direction of the largest normal stress) is tipped towards the wall.

The bottom views look along the axis of rotation. In this view, with $N = 0$, Fig. 14.2.5a, the primary principal axis should appear aligned with the *r* direction. The rotation of the pipe is clockwise in the bottom views, and correspondingly the glyphs in Figs. 14.2.5b and 14.2.5c, for $N = 0.5$ and 1.0, have rotated counterclockwise from the $N = 0$ orientation.

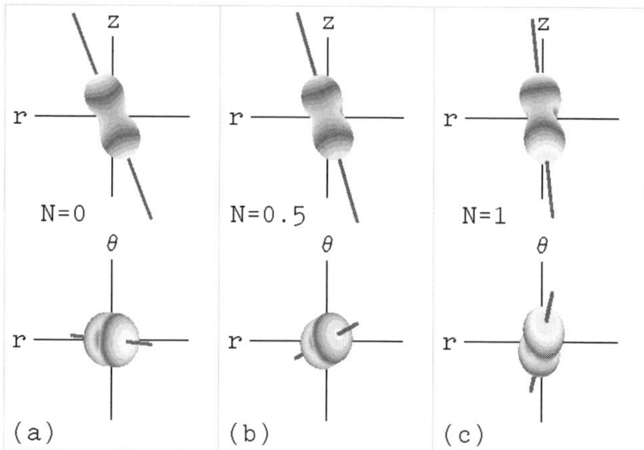

Fig. 14.2.5. Glyphs of the normalized Reynolds stresses, $\overline{u_i u_j}/k$, from the data of Imao et al. (IIH1996) at r/r_{wall}=0.8. The principal axis of the largest normal stress is shown extra long, and the other principal axes have been omitted.

The effect of this counterclockwise rotation on the $\overline{u_r u_z}$ shear stress can be seen in the top views in Fig. 14.2.5. With increased rotation rate the glyphs appear more aligned with the z direction, so that $\overline{u_r u_z}$ is reduced.

14.3 Results from the MARV and MC Models

For fully developed rotating pipe flow there is theoretically less distinction between the MARV and MC models than there was for the rotating channel flows in Chapter 13. $W/S \geq 1$ and the MARV model vorticity parameter, f_w, equals 1 throughout the pipe. Rapid distortion effects are also not important. The f_B parameter remained at zero, or near zero, in these fully developed flow calculations. The main distinctions between the MARV and MC models for this test case are then the anisotropy dependence of the pressure-strain coefficients and the rotating wall modification implemented with the f_R parameter, both of which affect near-wall results.

Velocity Profiles

Mean velocity profiles for $R_D = 20{,}000$ from both the MARV and MC models are compared with data in Fig. 14.3.1. Both models show the increase in U_z/U_{mean} with rotation on the pipe centerline expected from the data, and the corresponding decrease U_z/U_{mean} near the wall, giving the profile a more laminar shape. The tangential velocities are shown in the relative frame and normalized with U_{mean}. Both models give good results at $N = 0.5$, but under-predict the turbulence driven secondary flows at $N = 1$. Note that overall, the turbulence driven secondary flows are small compared with the primary flow.

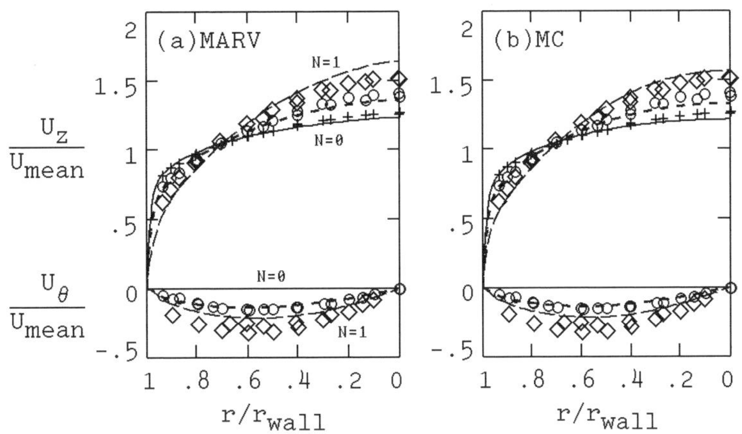

Fig. 14.3.1. Profiles of relative velocity for R_D=20,000. (a) MARV model. (b) MC model. Data from KMNM1983 and IIH1996: +, N=0; o, N=0.5; ◇, N=1. models: —— N=0; - - - N=0.5; — — — N=1.

Absolute tangential profiles for $N = 0.5$, normalized in the traditional way with $V_{\theta,wall}$, are shown in Fig. 14.3.2. The overall profiles in Fig. 14.3.2a show only small changes between the MARV model results for $R_D = 10{,}000$ and 50,000, within the scatter of the data. Very near the wall, Fig. 14.3.2b, the V_θ profiles follow the laminar line consistent with $S_{r\theta} = 0$ at the wall. The modeled results suggest this is a laminar sublayer phenomenon, occurring below $y^+ = 5$.

Components of the strain rate and vorticity are obtained from the mean velocity profiles. Figure 14.3.3 shows the MARV model results for $R_D = 20{,}000$ and $N = 1$. The magnitude of the strain rate associated with the axial velocity, S_{rz}, is much larger than the tangential component, $S_{r\theta}$, as shown in Fig. 14.3.3a. For this case, S_{rz} made dimensionless with r_{wall} and U_{mean} has a value of –9 at the wall (not shown). The pipe rotation has no direct effect on the vorticity associated with the axial velocity, since $W_{zr} = -W_{rz} = S_{rz}$. It does however directly affect $W_{\theta r}$, so that $W_{\theta r} > S_{r\theta}$ and overall $W/S > 1$, as shown in Fig. 14.3.3b. The MARV model's vorticity parameter, f_w, uses a filtered vorticity to strain-rate ratio, $\sqrt{8W_{ij}S_{jk}W_{kn}S_{ni}}/S^2$ (Eq. 10.1.12). As shown in Fig. 14.3.3b, this is less than W/S, but also greater than or equal to one, so that the model uses $f_w = 1$.

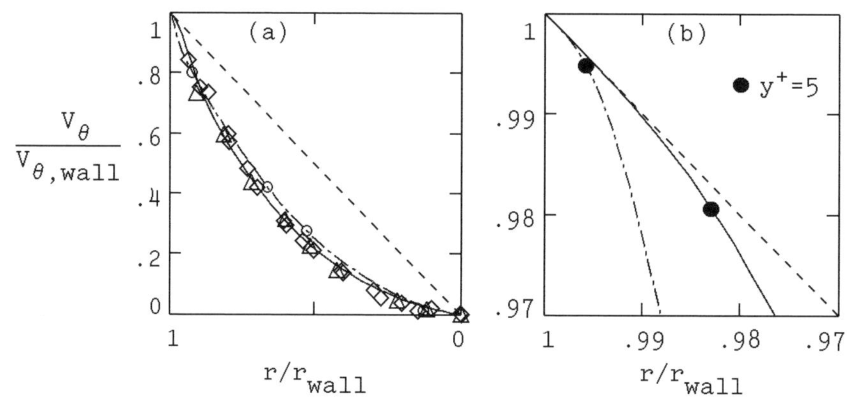

Fig. 14.3.2. Profiles of absolute tangential velocity for N=0.5. Data from YI1980, IIH1996 and KMNM1983: o, R_D=10,000; ◇, R_D=20,000; △, R_D=50,000.
MARV model: —— R_D=10,000; —·— R_D=50,000; ●, y^+ =5. — — — laminar flow.

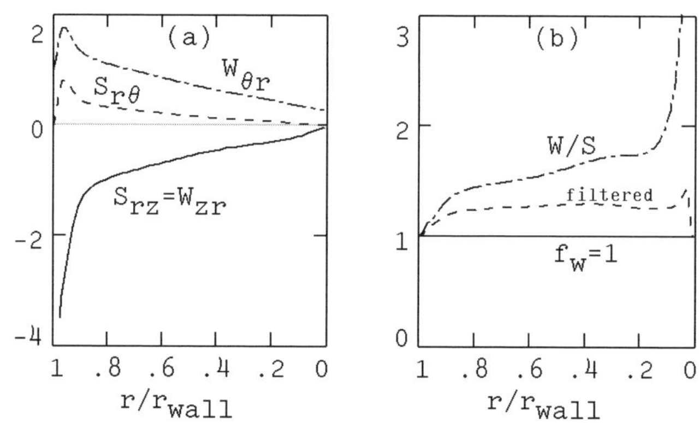

Fig. 14.3.3. MARV model results for R_D=20,000 and N=1. (a) Strain-rate and vorticity components: $S_{ij}\, r_{wall}/U_{mean}$, $W_{ij}\, r_{wall}/U_{mean}$. (b) Vorticity to strain-rate ratios: W/S, f_w and — — filtered = $\sqrt{8W_{ij}S_{jk}W_{kn}S_{ni}}/S^2$.

Skin friction coefficients for $R_D = 20{,}000$ normalized with the laminar flow value, $c_{f,lam} = 16/R_D$, are shown in Fig. 14.3.4. At zero rotation, c_f is between 8 and 9 times the laminar value. Up to $N = 1$, both the MARV and MC models are in reasonable agreement with the data, with $c_f / c_{f,lam}$ falling to about 5 by $N = 1$. By

$N = 1.5$ the MC model gives laminar flow. The data continues to show turbulent flow, as does the MARV model, but the modeled value of c_f is low by a factor of 2.

Fig. 14.3.4. Rotating pipe skin frictions for R_D=20,000 divided by the skin friction for laminar flow. Data: ⊕, YI1980; ⊙, KMNM1983; ○, IIH1996. Models: —— MARV; –·– MC; - - - - MARV but with $f_R = 0$.

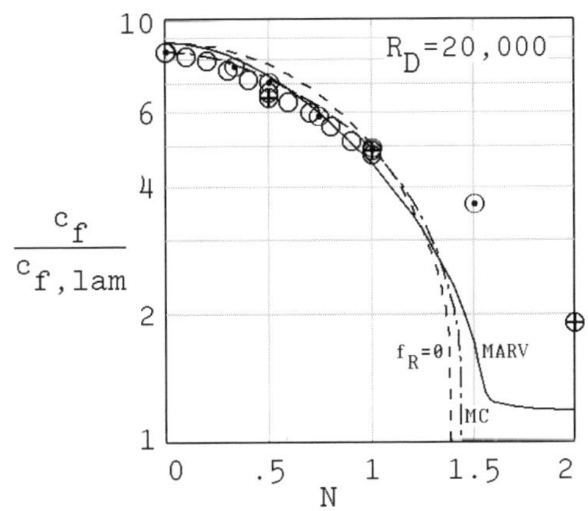

Also shown on Fig. 14.3.4 are results from the MARV model but with the rotating-wall parameter, f_R, set to zero. The c_f values are higher than those from the MARV model at low rotation rates. At higher rotation rates, the modeled flow becomes laminar.

The f_R parameter, which prevents the MARV model results from becoming laminar, is shown in Fig. 14.3.5. For $N = 1$ and 1.5, the rotating wall modification only affects the pressure-strain near the wall. However, for $N = 2$, the modification affects the entire pipe, a consequence of the R_t dependence of f_R (Eq. 13.3.4) and the nearly laminar (low turbulence) state of the modeled flow.

Fig. 14.3.5. Rotating-wall parameter, f_R, from the MARV model for R_D=20,000. —— N=1; –·– N=1.5; – – N=2.

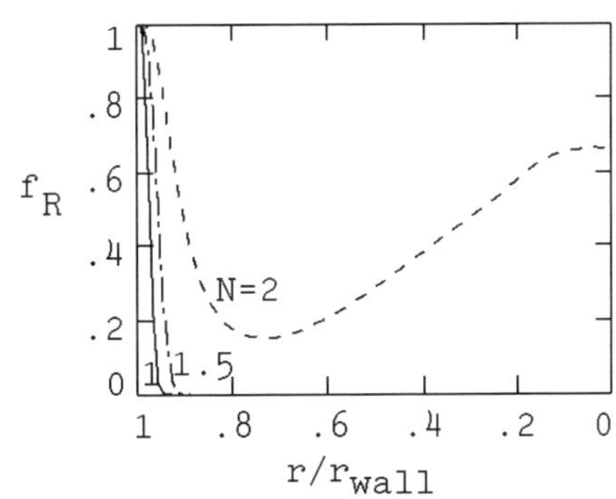

Reynolds Stresses

The modeled profiles of turbulence velocity scale, $q = \sqrt{k}$, are compared with the data of Imao et al. in Fig. 14.3.6. For no rotation, both the MARV and MC models, and the data, show q/U_{mean} increasing from about 0.05 on the centerline to

0.1 at $r/r_{wall} = 0.9$. The data does not go close enough to the wall to capture the near-wall peak the MARV model gets at $r/r_{wall} = 0.97$. As rotation is increased this peak disappears. For the rotating cases, the data shows slightly less turbulence over the bulk of the pipe, while the models give more turbulence.

Fig. 14.3.6. Profiles of turbulence velocity scale, q, for R_D=20,000. (a) MARV model. (b) MC model. Data from IIH1996: +, N=0; o, N=0.5; ◇, N=1. models: —— N=0; - - - N=0.5; — — — N=1.

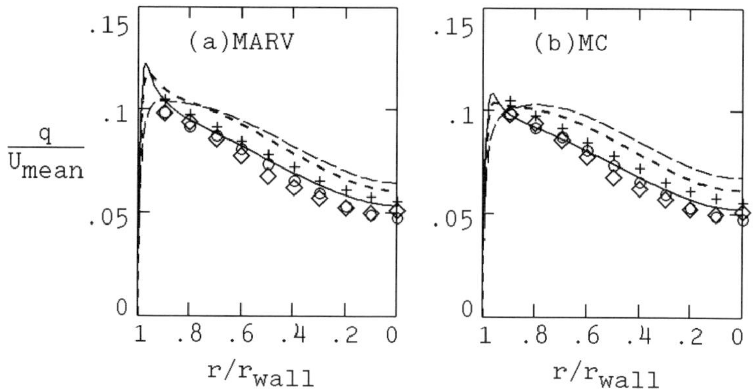

The anisotropies of the Reynolds stresses from the MARV model are shown in Fig. 14.3.7. The model shows little variation with rotation of the r, θ and z components of the normal anisotropies. These are in agreement with the data for $N=0$ and $N=0.5$. For $N=1$, the data shows an increase in $\overline{u_\theta u_\theta}/2k$ and a corresponding decrease in $\overline{u_r u_r}/2k$ in the outer half of the pipe, so that $b_{\theta\theta} - b_{rr}$ has increased. Imao attributed the increase in $\overline{u_\theta u_\theta}$ to low frequency spiral waves observed near the wall. Considering the V_θ/r source terms in the b_{ij} equations (Eq. 14.2.1), this may explain the higher 'data' values for $b_{r\theta}$ in Fig. 14.3.7, which were obtained from the higher measured secondary flows, Fig. 14.3.1.

The model and data agree well on the reduction in b_{rz} with rotation. Over the bulk of the flow, Fig. 14.3.7 shows b_{rz} reduced to about 0.1 for $N=0.5$ and to about 0.05 for $N=1$. However, in the data, the mechanism for the reduction appears to be more complex than the V_θ/r source term, Eq. 14.2.1, since the data shows $b_{\theta z}$ approximately zero in the inner half of the pipe at all rotations.

It is interesting to note that both $S_{r\theta}$ (Figure 14.3.3a) and $b_{r\theta}$ (Fig. 14.3.7), or $\overline{u_r u_\theta}$, are positive. The glyph rotation (Fig. 14.2.5), which caused $b_{r\theta}$ to be positive, is part of the Reynolds stress model. In turbulence models which use the Boussinesq approximation (Eq. 1.3.24), e.g. two-equation models, positive $S_{r\theta}$ would result in negative values for $\overline{u_r u_\theta}$, which through the equation for V_θ would result in negative $S_{r\theta}$! The result is that Boussinesq models give $S_{r\theta}=0$ for fully developed flow, and correspondingly no Reynolds stress driven secondary flow.

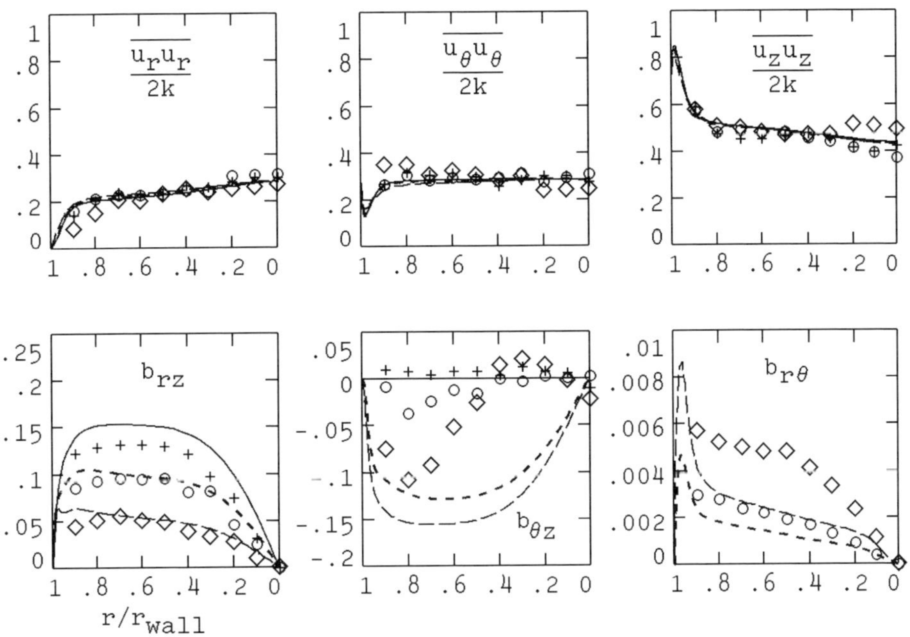

Fig. 14.3.7. Components of the Reynolds stress tensor normalized with $2k$, R_D=20,000. IIH1996 data: +, N=0; o, N=0.5; ◇, N=1. MARV model: —— N=0; - - - N=0.5; — — — N=1.

High Rotation and Reynolds Number Dependence

Skin friction coefficients from the MARV and MC models are compared with data for Reynolds numbers from $R_D = 4900$ to 100,000 in Fig. 14.3.8. As in Fig. 14.3.4, c_f has been normalized with laminar flow values. The reason is to show that the data approaches the laminar limit more slowly at high rotation rates than the modeled results, at all R_D. For $N = 1$, the data is still ordered as for no rotation, with $c_f / c_{f,lam}$ higher at higher R_D. At $N = 2$, the sequencing may be reversed, or to within a reasonable error band for the data, $c_f / c_{f,lam}$ is a constant, between 2 and 2.5, independent of Reynolds number.

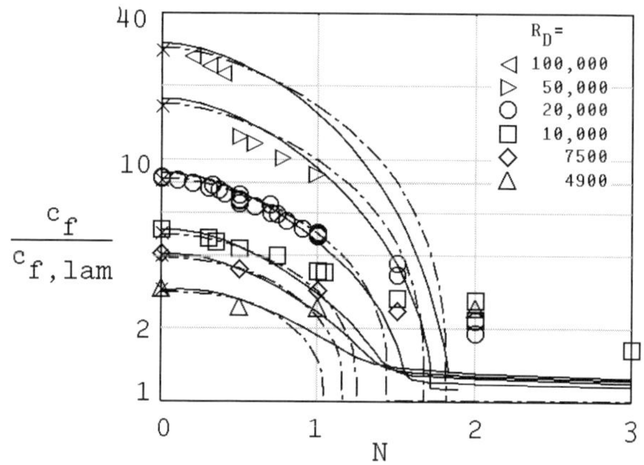

Fig. 14.3.8. Comparison of modeled and measured rotating pipe skin friction ratios, $c_f / c_{f,lam}$, at six Reynolds numbers. Models: —— MARV; – – – MC. Data from YI1980, KMNM1983, IIH1996, and Orlandi and Fatica DNS (1997): ◁, R_D=100,000; ▷, R_D=50,000; ○, R_D=20,000; □, R_D=10,000; ◇, R_D=7500; △, R_D=4900. ×, no rotation, McKeon et al. (MSZDS2004).

Overall, both the MARV and MC models give reasonable results for fully developed rotating pipe flow up to $N = 1$ for $R_D \geq 20{,}000$, and up to $N = 0.5$ at lower Reynolds numbers. At higher rotation rates, the flow experiences high ratios of vorticity to strain rate. The effect of these on the turbulence has not been considered in the model development.

Chapter 15

RETURN TO ISOTROPY

The return-to-isotropy part of the effective pressure-strain models is embodied in the c_1 pressure-strain coefficients in the MC and MARV models. The concepts were introduced in Section 3.2, a homogeneous shear DNS data basis for the models was given in Section 5.1, and the models were finalized in Chapter 8 based on fully developed channel flow.

In this chapter we consider some modeling history, including separate consideration of slow pressure-strain and anisotropy of dissipation. DNS data is used to form a model for anisotropy of dissipation alone. A review of experimental data for the return-to-isotropy of homogeneous turbulence, in the nominal absence of mean velocity gradients, shows varied results. The choice of data used has led to a variety of slow pressure-strain models in the literature.

15.1 Equations for Homogeneous Turbulence

The basic equation for the return to isotropy of homogeneous turbulence, in the absence of mean velocity gradients, was given in Section 3.2 (Eq. 3.2.5),

$$\frac{d\overline{u_i u_j}}{dt} = \Pi_{ij}^{S} - \varepsilon_{ij} \ . \tag{15.1.1}$$

This may be written as an equation for the turbulence kinetic energy,

$$dk/dt = -\varepsilon, \tag{15.1.2}$$

and equations for the anisotropy of the turbulence, b_{ij},

$$2k\frac{db_{ij}}{dt} = \Pi_{ij}^{S} - \varepsilon_{ij} + 2\varepsilon(b_{ij} + \frac{1}{3}\delta_{ij}) \ . \tag{15.1.3}$$

It is modeling and solutions for this latter equation which concern us in this chapter. Using the general modeled algebraic form for slow pressure-strain, Π_{ij}^{S}, and the Reynolds stress dissipation, ε_{ij}, Eqs. 3.2.2 and 3.2.3,

$$\Pi_{ij}^{S} = -c_{1s}\varepsilon b_{ij} + c_{2s}\varepsilon(b_{ik}b_{kj} - \frac{1}{3}b_{mn}b_{mn}\delta_{ij}) , \tag{15.1.4}$$

$$\varepsilon_{ij} = \frac{2}{3}\varepsilon\delta_{ij} + c_{1d}\varepsilon b_{ij} - c_{2d}\varepsilon(b_{ik}b_{kj} - \frac{1}{3}b_{mn}b_{mn}\delta_{ij}) , \tag{15.1.5}$$

the modeled form of Eq. 15.1.3 is

$$2k\frac{db_{ij}}{dt} = -(c_{1s} + c_{1d} - 2)\varepsilon b_{ij} + (c_{2s} + c_{2d})\varepsilon(b_{ik}b_{kj} - \frac{1}{3}b_{mn}b_{mn}\delta_{ij}) . \tag{15.1.6}$$

Whether or not turbulence is returning to isotropy, i.e., becoming more isotropic, is usually described in terms of the invariant, $II = b_{ij}b_{ij}$. If II is becoming smaller, the turbulence is becoming more isotropic. Multiplying Eq. 15.1.6 by b_{ij} yields

$$k\frac{dII}{dt} = -(c_{1s} + c_{1d} - 2)\varepsilon II + (c_{2s} + c_{2d})\varepsilon III . \tag{15.1.7}$$

It is clear that when $c_{1s} + c_{1d} > 2$, the first term will result in a return to isotropy. Whether or not the second term results in a return to isotropy will depend on the sign of both the coefficients and III. An effective rate-of-return coefficient, $c_{1\text{eff}}$, may be defined by

$$k\frac{dII}{dt} = -(c_{1\text{eff}} - 2)\varepsilon II . \tag{15.1.8}$$

Combining this with Eq. 15.1.7 yields

$$c_{1\text{eff}} = c_{1s} + c_{1d} - (c_{2s} + c_{2d})\frac{III}{II} = c_1 - c_2\frac{III}{II} . \tag{15.1.9}$$

The range of anisotropy that turbulence can have in terms of the invariants II and III is shown in Fig. 15.1.1. This is a linearly transformed version of Fig. 1.2.5. II ranges from zero for isotropic turbulence to 1/6 at the two-dimensional axisymmetric contraction limit, to 2/3 at the one-dimensional axisymmetric expansion limit. III ranges from −1/36 at the two-dimensional axisymmetric contraction limit to 2/9 at the one-dimensional limit.

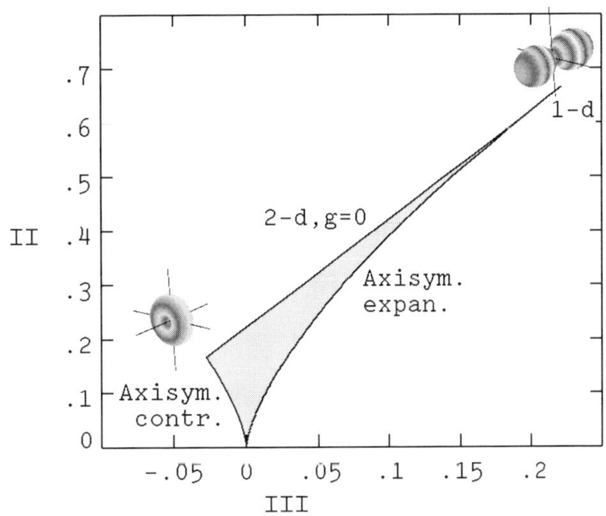

Fig. 15.1.1. Range of turbulence anisotropy in terms of *II* and *III*.

Fourier Components

It is helpful to consider the representations of Π_{ij}^{S} and ε_{ij} in wave-number space. From the equation for development of the Fourier transform of the two-point velocity correlation (Eq. 9.2.8), Π_{ij}^{S} and ε_{ij} can be expressed as integrals in wave-number space,

$$\varepsilon_{ij} = 2\nu \iiint K^2 \Phi_{ij} dK_1 dK_2 dK_3 \quad , \tag{15.1.10}$$

and

$$\Pi_{ij}^{S} = \frac{(2\pi)^3}{V} \iiint \frac{iK_n K_m}{K^2} \iiint \Big(K_j \overline{a_i^*(\mathbf{K}) a_m(\mathbf{K}-\mathbf{K}') a_n(\mathbf{K}')} - K_i \overline{a_j(\mathbf{K}) a_m^*(\mathbf{K}+\mathbf{K}') a_n(\mathbf{K}')} \Big) dK_1' dK_2' dK_3' \, dK_1 dK_2 dK_3 \, , \tag{15.1.11}$$

where a_i is the Fourier transform of the fluctuating velocity.

It can immediately be observed from Eq. 15.1.10 that the dissipation rate is proportional to K^2, so that high wave-number turbulence dissipates more rapidly. From Eq. 15.1.11, the interchange of energy among the components of the Reynolds stress tensor is also higher at high wave numbers since it is proportional to K. This is consistent with spectral measurements in high Reynolds number turbulence, which show that the turbulence in the high wave-number part of the spectrum tends to be “locally isotropic”, with the anisotropy confined to low wave numbers (Batchelor, 1953, p. 110). Since most of the dissipation occurs at high wave numbers,

$$\varepsilon_{ij} = \frac{2}{3}\varepsilon\delta_{ij} \qquad \text{at high } R_t \text{ and low anisotropy.} \tag{15.1.12}$$

This is used in Reynolds stress models which consider only high Reynolds number turbulence, and is used as the starting point for the modeling of Reynolds stress dissipation, Eq. 3.2.3.

As turbulence decays and the turbulence Reynolds number approaches zero, all that remains of the initial turbulence are energies in the lowest wave numbers of the spectrum. The Reynolds stress dissipation is then proportional to the Reynolds stresses (Batchelor, 1953, p. 94).

$$\varepsilon_{ij} \to \varepsilon \frac{\overline{u_i u_j}}{k} \quad \text{as} \quad R_t \to 0 \; . \tag{15.1.13}$$

As the active range of the wave-number spectrum reduces at low R_t, so also does the correlation of the fluctuating velocities in the slow pressure-strain, Eq. 15.1.11, so that

$$\Pi_{ij}^{S} \to 0 \quad \text{as} \quad R_t \to 0 \; . \tag{15.1.14}$$

As turbulence becomes highly anisotropic, approaching the one-dimensional or two-dimensional turbulence limits, the dissipation of the vanishing component also vanishes, e.g.,

$$\varepsilon_{11} \to 0 \quad \text{as} \quad \overline{u_1 u_1} \to 0 \; . \tag{15.1.15}$$

So also must the corresponding component of the slow pressure-strain,

$$\Pi_{11}^{S} \to 0 \quad \text{as} \quad \overline{u_1 u_1} \to 0 \; . \tag{15.1.16}$$

The corresponding limiting values for the model coefficients are therefore

$$R_t = \infty, \; g = 1: \qquad c_{1\mathrm{d}} = 0, \qquad c_{2\mathrm{d}} = 0 \; ; \tag{15.1.17}$$

$$R_t = 0: \qquad c_{1\mathrm{d}} = 2, \quad c_{2\mathrm{d}} = 0, \quad c_{1\mathrm{s}} = 0, \quad c_{2\mathrm{s}} = 0 \; ; \tag{15.1.18}$$

$$g = 0: \qquad c_{1\mathrm{d}} = 2 - c_{2\mathrm{d}}(\frac{1}{3} - II), \qquad c_{1\mathrm{s}} = -c_{2\mathrm{s}}(\frac{1}{3} - II) \; . \tag{15.1.19}$$

For the effective pressure-strain coefficients, $c_1 = c_{1\mathrm{d}} + c_{1\mathrm{s}}$ and $c_2 = c_{2\mathrm{d}} + c_{2\mathrm{s}}$,

$$R_t = 0: \qquad c_1 = 2, \qquad c_2 = 0 \; ; \tag{15.1.20}$$

$$g = 0: \qquad c_1 = 2 - c_2(\frac{1}{3} - II) \; . \tag{15.1.21}$$

The latter is Eq. 3.4.5, and both limits are satisfied by the MARV model.

15.2 Anisotropy of Dissipation

Models in the Literature

Rotta (R1951) combined the concept of isotropic dissipation at high turbulence Reynolds numbers, with dissipation proportional to the Reynolds stress at low R_t, to obtain

$$\varepsilon_{ij} = 2\nu \overline{\frac{\partial u_i}{\partial x_k}\frac{\partial u_j}{\partial x_k}} = \underbrace{\frac{2}{3} c \frac{k^{3/2}}{L}\delta_{ij}}_{\text{high } R_t} + \underbrace{a\nu \frac{\overline{u_i u_j}}{L^2}}_{\text{low } R_t} . \qquad (15.2.1)$$

From Rotta's Eq. 4.7 and Table 1, $a = 5\pi/4 = 3.927$ and $c = 0.228$. Equation 15.2.1 can be contracted to obtain ε, and then solved for Rotta's length scale, L (Appendix R1951). It can then be written in the form of Eq. 15.1.5, yielding

$$\varepsilon_{ij} = \frac{2}{3}\varepsilon\delta_{ij} + c_{1d}\varepsilon b_{ij} , \qquad \text{with} \qquad c_{1d} = \frac{2}{\left(\sqrt{\dfrac{R_t}{302}} + \sqrt{1+\dfrac{R_t}{302}}\right)^2} . \qquad (15.2.2)$$

Rotta's model is then an interpolation between the high turbulence Reynolds number limit, $c_{1d} = 0$, and the low Reynolds number limit, $c_{1d} = 2$, with $c_{2d} = 0$. The mid-value, $c_{1d} = 1$, occurs at $R_t = 38$. Since the c_{1d} values are between 0 and 2, the dissipation, when considered alone, tends to make the turbulence more anisotropic.

Naot, Shavit and Wolfshtein (1973), considering two-point correlation models, described the c_{1d} coefficient (their $2\Lambda_0$) in terms of ratios of the microscales of various spectral functions. Schumann and Patterson (1978), from a numerical study using a spectral method, proposed an interpolation parameter, which in terms of the present nomenclature translates to

$$c_{1d} = \frac{2}{1 + 0.1\sqrt{\dfrac{20}{3}R_t}} . \qquad (15.2.3)$$

The mid-value, $c_{1d} = 1$, occurs at $R_t = 15$.

Hanjalić also considers the modeling of ε_{ij} separate from Π^{S}_{ij}. Hanjalić and Launder (1976) used $c_{1d} = 2/(1 + R_t/10)$, which gives $c_{1d} = 1$ at $R_t = 10$. In a recent review, Hanjalić (1994) discusses ε_{ij} extensively in relation to near-wall flows, and various models in the literature for it. Considering DNS data for fully developed channel flows, Hanjalić and Jakirlić (1992, see Hanjalić, 1994) argue that neither R_t nor turbulence anisotropy is individually suited to model ε_{ij} in these flows.

However, in the recent model of Hadžić and Hanjalić (1999), the R_t dependence of ε_{ij} appears only in wall-normal corrections, so that, neglecting these, ε_{ij} is an implicit function of the turbulence anisotropy alone,

$$c_{1d} = 2 - 2E^2\sqrt{g} \ . \tag{15.2.4}$$

Here E is the measure of the two-dimensionality of ε_{ij}, in the same way as g is the measure of the two-dimensionality of $\overline{u_i u_j}$.

Anisotropy of Dissipation Data

DNS data sets usually tabulate ε_{ij} for each Reynolds stress component, independent of the pressure-strain. This provides the possibility, unavailable or extremely difficult in experimental data, of looking at the components of ε_{ij} separately from pressure-strain. Very near-wall values of ε_{ij}, below $y^+ = 1$, are consistent with the Taylor series analysis results, Eq. 12.2.8, as shown in Fig. 12.2.3, and will not be further considered here.

Figure 15.2.1a shows ε_{ij} from the fully developed channel DNS of Moser et al. (MKM1999) for $R_\tau = 590$. Equation 15.1.5 was used to determine model values for both c_{1d} and c_{2d} from the components of ε_{ij} in the principal stress directions. Note that ε_{ij} can only be modeled exactly when the principal directions of ε_{ij} are the same as the principal directions of b_{ij}. Figure 15.2.1a shows how well the model can do, while Fig. 15.2.1d shows the angle between the principal directions of ε_{ij} and b_{ij}.

Figure 15.2.1b shows c_{1d} approximately equal to 2 up to $y^+ = 10$, then falling to values below 1 by $y^+ = 100$. Figures 15.2.1e and 15.2.1f show that both g and R_t rise through this transition. The values of c_{2d}, Fig. 15.2.1c, are close to zero near the wall and remain low over most of the flow, justifying the standard use of $c_{2d} = 0$. Near the centerline of the channel, both c_{1d} and c_{2d} rise dramatically. Since $III > 0$ in this shear flow, these are competing effects (see Eq. 15.1.7).

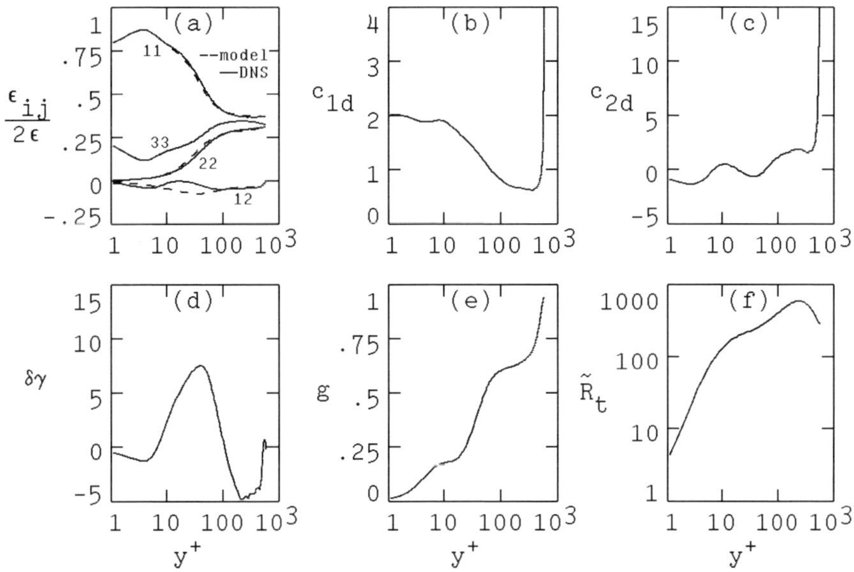

Fig. 15.2.1. Components of Reynolds stress dissipation, a model for it, and related properties from the fully developed channel DNS of Moser et al. (MKM1999) at R_τ=590. $\delta\gamma$, angle (in degrees) between principal directions of ε_{ij} and b_{ij}.

To overcome this difficulty, the data for all three Moser et al. cases ($R_\tau = 180$, 395 and 590) were reanalyzed considering only the c_{1d} coefficient, and using a weighted least-squares fit of ε_{ij} in the principal stress directions (P_1, P_2, P_3),

$$c_{1d} = \frac{\dfrac{\left(\dfrac{\varepsilon_{P_1P_1}}{\varepsilon} - \dfrac{2}{3}\right) b_{P_1P_1}}{(\overline{u_{P_1}u_{P_1}})^2} + \dfrac{\left(\dfrac{\varepsilon_{P_2P_2}}{\varepsilon} - \dfrac{2}{3}\right) b_{P_2P_2}}{(\overline{u_{P_2}u_{P_2}})^2} + \dfrac{\left(\dfrac{\varepsilon_{P_3P_3}}{\varepsilon} - \dfrac{2}{3}\right) b_{P_3P_3}}{(\overline{u_{P_3}u_{P_3}})^2}}{\left(\dfrac{b_{P_1P_1}}{\overline{u_{P_1}u_{P_1}}}\right)^2 + \left(\dfrac{b_{P_2P_2}}{\overline{u_{P_2}u_{P_2}}}\right)^2 + \left(\dfrac{b_{P_3P_3}}{\overline{u_{P_3}u_{P_3}}}\right)^2} . \tag{15.2.5}$$

The results are shown as a function of $\tilde{R}_t$ in Fig. 15.2.2a. The g values along each curve are marked. For $g \geq 0.8$, c_{1d} shows a clear R_t dependence. Between $g = 0.1$ and 0.5, the bulk of the variation in c_{1d} appears to be dependent on g.

Rogers et al. (RMR1986) tabulate the individual components of the dissipation tensor for their DNS calculations of homogeneous shear flows. For these data sets, the principal directions of the dissipation tensor, ε_{ij}, differed from the principal directions of the Reynolds stress tensor by up to 13 degrees. Results of analyzing their data to obtain c_{1d} using Eq. 15.2.5, shown in Fig. 15.2.2b, have similar trends to the channel flow data.

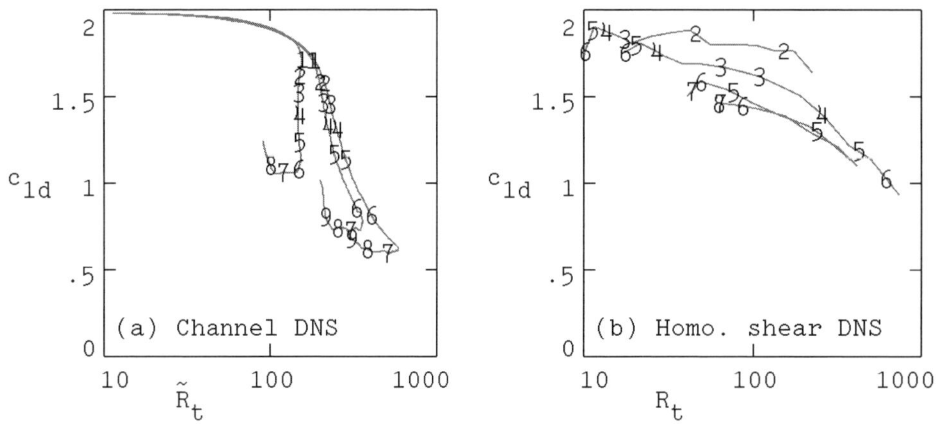

Fig. 15.2.2. c_{1d} as a function of turbulence Reynolds number calculated from DNS data using Eq. 15.2.5. Symbols: 1, $g=0.1$; to 9, $g=0.9$. (a) Fully developed channel DNS of Moser et al. (MKM1999). (b) Homogeneous shear DNS of Rogers et al. (RMR1986).

Model from Data Analysis

An approximate model to fit both sets of data in Fig. 15.2.2 is shown in Fig. 15.2.3. It is

$$c_{1d} = 2 - 2 f_a(g) f_b(\tilde{R}_t) \ , \qquad (15.2.6)$$

with

$$f_a(g) = \frac{\min(g, 0.75)}{0.75} \qquad \text{and} \qquad f_b(\tilde{R}_t) = \frac{\tilde{R}_t^{3/4}}{125^{3/4} + \tilde{R}_t^{3/4}} \ . \qquad (15.2.7)$$

The Reynolds number factor has a mid-value at $R_t = 125$, much higher than for the previously described models in the literature.

This model for the anisotropy of dissipation has no effect on the MC and MARV Reynolds stress models. If for some reason one wishes to consider slow pressure-strain, Π_{ij}^{S}, separate from ε_{ij}, c_{1s} may be obtained from $c_{1s} = c_1 - c_{1d}$. In Chapter 17, this model for c_{1d} is used as part of an algebraic model for the spectrum tensor.

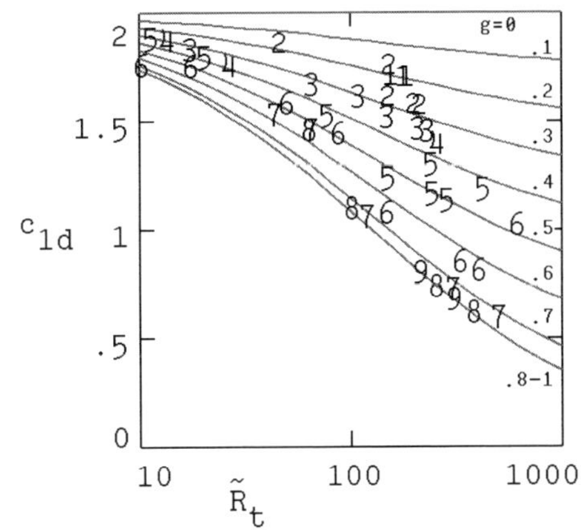

Fig. 15.2.3. Comparison of model and data for c_{1d}.
——— Model, Eq. 15.2.6, for g=0 to 1, by 0.1.
Data as for Fig. 15.2.2:
1, g=0.1; to 9, g=0.9.

15.3 Models and Data for Slow Pressure-Strain

Rotta (R1951) is well known for his formulation of the standard slow pressure-strain model. He proposed

$$\frac{1}{\rho}\overline{p\left(\frac{\partial u_i}{\partial x_j}+\frac{\partial u_j}{\partial x_i}\right)} = -k_p\frac{\sqrt{k}}{L}(\overline{u_i u_j}-\frac{2}{3}k\delta_{ij}) \tag{15.3.1}$$

for the pressure-strain in the absence of mean velocity gradients. Rotta's equation is best known in its high Reynolds number form which, when combined with isotropic dissipation ($\varepsilon_{ij}=\frac{2}{3}\varepsilon\delta_{ij}$), gives the effective slow pressure-strain,

$$\phi_{ij}^{S} = -c_1\varepsilon b_{ij} \quad , \tag{15.3.2}$$

so that c_1 is often referred to as the Rotta coefficient. However, inherent in Eqs. 15.2.1 and 15.3.1 is a turbulence Reynolds number dependent formulation for slow pressure-strain alone. It may be written as (Appendix R1951)

$$\Pi_{ij}^{S} = -c_{1s}\varepsilon b_{ij}\,, \qquad \text{with} \qquad c_{1s} = 2\frac{k_p}{c}\frac{2\sqrt{\dfrac{R_t}{302}}}{\sqrt{\dfrac{R_t}{302}}+\sqrt{1+\dfrac{R_t}{302}}} \quad , \tag{15.3.3}$$

so that at high R_t, $c_{1s}=2k_p/c$, and at low R_t, c_{1s} goes to zero. Rotta determined a value for c/k_p consistent with shear flow (in the log-law region of a turbulent boundary layer) and his rapid pressure-strain model, and obtained $c/k_p = 0.7$ (Rotta, Table 1), so that at high R_t, $c_{1s} = 2.86$. Later, Rotta (1962, p. 48) noted that the measurements of Uberoi (U1957) are consistent with about twice this value.

The remaining models discussed in this section are for the combined effects of the anisotropy of dissipation and slow pressure-strain. In formulating the models, the workers either assumed the dissipation to be isotropic, or consciously modeled the two effects together.

Lumley and Newman (1977) did an extensive analysis of return-to-isotropy data and considered both the effects of turbulence Reynolds number and turbulence anisotropy on the process. Inherent in their model are both c_1 and c_2 coefficients which are functions of R_t, II and III. The model however is difficult to interpret directly in this way. Besides being consistent with the limiting values in Eqs. 15.1.20-21, the model always gives II decreasing in time in the absence of mean velocity gradients. Lumley and Newman thought the magnitude of III should also decrease with time. Their model is consistent with this for $III > 0$.

Lumley (1978) approximated their results in a simplified model for c_1 alone (also given as Eq. 3.5.5, part of the SL1985 model, Table 3.5.1),

$$c_1 = 2 + g\,\mathrm{e}^{-11.65/\sqrt{R_t}}\left(12/\sqrt{R_t} + 8.9\ln(1 + 31.2II + 47.8III)\right) , \tag{15.3.4}$$

which is easier to interpret. The g dependence insures that $c_1 = 2$ in the two-dimensional turbulence limit. The $\sqrt{R_t}$ dependence is based on the dimensional arguments of Tennekes and Lumley (1972). The factor $\mathrm{e}^{-11.65/\sqrt{R_t}}$ is the basic Reynolds number dependence. The constant, 11.65, was determined using data taken by Uberoi (U1956, U1957), and Mills and Corrsin (1959), for return-to-isotropy following axisymmetric contraction. The data was in the range $R_t \sim 40$ to 450 (see Table 15.3.1 below).

The data of Comte-Bellot and Corrsin (CC1966) for the return-to-isotropy of grid generated turbulence was the basis for the term $12/\sqrt{R_t}$. Lumley and Newman saw a Reynolds number dependence in the data which suggested that for nearly isotropic turbulence, c_1 goes to 2 as R_t goes to infinity. The data was also analyzed to determine a II dependence near the isotropic limit. Reynolds (1976) looked at the same Comte-Bellot and Corrsin data and determined $c_1 = 2.5$ to within the scatter of the data, independent of both R_t and II.

The data in question has been reanalyzed (Appendix CC1966) by fitting the data with

$$\frac{d\ln(II)}{d\ln(k)} = c_1 - 2 . \tag{15.3.5}$$

An error band for c_1 is estimated from

$$c_{1,err} = \frac{\left| \ln(II)_{fit} - \ln(II)_{data} \right|_{\max}}{\ln(k)_{start} - \ln(k)_{end}} . \tag{15.3.6}$$

The results are shown in Fig. 15.3.1. We agree with Reynolds that all that is justified is $c_1 = 2.5$. However, all the data is for $g > 0.95$, so that in the MARV model we have taken this as the high Reynolds number isotropic limit,

$$\text{MARV model:} \qquad c_1 = 2.5 \text{ at high } R_t \text{ and } g = 1. \tag{15.3.7}$$

Fig. 15.3.1. c_1 pressure-strain coefficients determined from the data of Comte-Bellot and Corrsin (CC1966).

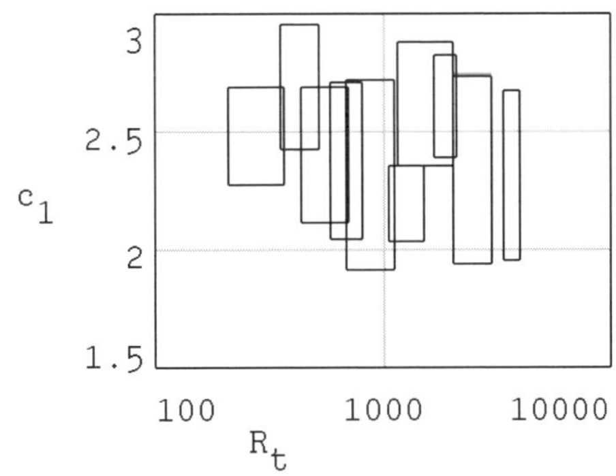

The dependence on III in the Lumley model was not a conscious choice based on data, but simply the result of being consistent with the form of the model that Lumley and Newman chose. The effect of the III dependence in the model is that c_1 is in general higher for axisymmetric expansion and shear flows with $III > 0$, than for axisymmetric contraction where $III < 0$.

Sarkar and Speziale (1990) appear to have based their model on quite different data and considerations. Le Penven et al. (LGC1985) noted that their experimental data (as well as other data) showed that turbulence with $III < 0$ returns to isotropy more rapidly than turbulence with $III > 0$. Sarkar and Speziale observed that this is consistent with c_2 positive (see Eq. 15.1.7). They also noted that if

$$c_2 = 3(c_1 - 2) , \tag{15.3.8}$$

the model will always result in realizable Reynolds stresses (Eq. 3.4.11 with $II = 2/3$, its upper limit). Sarkar and Speziale considered only high Reynolds number turbulence and chose

$$\text{SS model:} \qquad c_1 = 3.4, \qquad c_2 = 3(c_1 - 2) = 4.2 , \tag{15.3.9}$$

which forms part of the SSG1991 model, Table 3.5.1.

Sarkar and Speziale were also concerned with the way in which turbulence returns to isotropy. Following Choi and Lumley (1984), they considered trajectories in ξ,η space, where

$$\xi = \left(\frac{III}{6}\right)^{1/3} \quad \text{and} \quad \eta = \left(\frac{II}{6}\right)^{1/2} . \tag{15.3.10}$$

Note that these variables have been scaled (as did Choi and Lumley) so that axisymmetric expansion acting on isotropic turbulence gives $\xi = \eta$, while axisymmetric contraction gives $\xi = -\eta$.

The Sarkar and Speziale model, Fig. 15.3.2a, shows some trajectories where ξ starts out negative, then becomes positive, before reducing to zero as η goes to zero. The data used by Sarkar and Speziale, Fig. 15.3.2b, is near the axisymmetric expansion and contraction lines. While the data shows the trajectory curvature of their model, none of it crosses the $\xi = 0$ line.

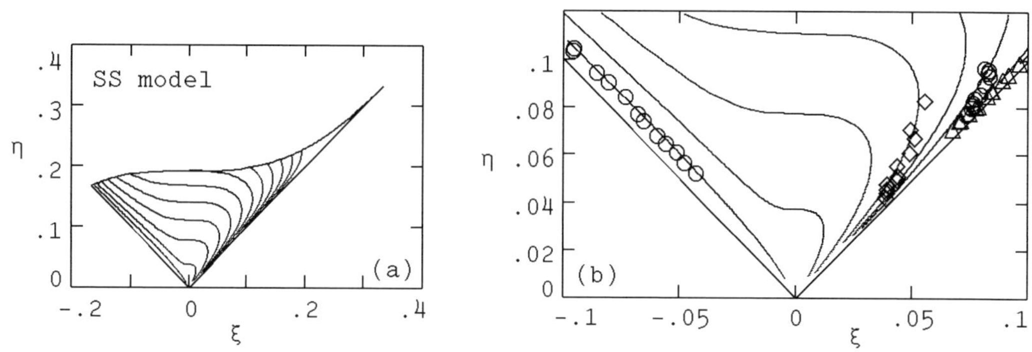

Fig. 15.3.2. (a) Trajectories of ξ,η for turbulence returning to isotropy from the SS model, Eq. 15.3.9. (b) With data: ○, two cases, $\xi < 0$ and $\xi > 0$, Le Penven et al. (LGC1985); from Choi and Lumley (see CL2001): ◇, return after plane strain; △, return after axisymmetric expansion.

The Rotta model, with only the c_1 coefficient, has straight-line trajectories, as shown in Fig. 15.3.3a. Shown with data, in Fig. 15.3.3b, are alternating shaded/clear intervals of $\delta b_m = 0.01$, where b_m is the middle value of the three (principal) anisotropies, b_{11}, b_{22}, and b_{33}. Note the sensitivity to small errors in b_m, in particular the large area covered by $b_m = 0 \pm 0.01$ around $\xi = 0$. Three additional data sets, besides those considered by Sarkar and Speziale have been added to the plot. Case A appears to support the SS model trajectory but case B does not.

Overall, until more is known about ξ,η trajectories, a Rotta-type model would appear to be sufficient. It should be noted, that in experimental situations, it may be very difficult to set up a pure return-to-isotropy that is not contaminated by small strain rates associated with the growth of side wall boundary layers.

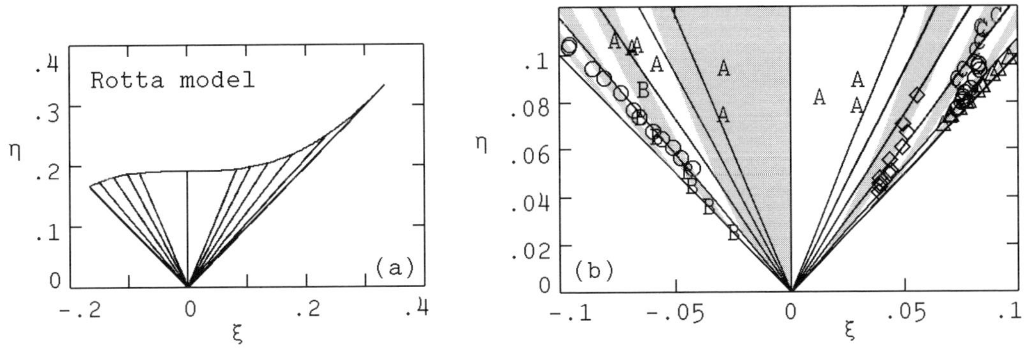

Fig. 15.3.3. (a) Trajectories of ξ,η for turbulence returning to isotropy from Rotta, or c_1 only models. (b) With data as for Fig. 15.3.2, and return after plane strain: A,B, Tucker (T1970, Figs. 16 and 26); C, Gence and Mathieu (GM1980). Alternating shaded/clear, intervals of 0.01 of the middle-valued principal anisotropy.

Figure 15.3.4 compares c_1 values from the MARV model, with results from the analysis of numerous data sets, using Eq. 15.3.5. The data points are tagged by an average g value. The data falls into three groups. Light tone symbols near the $g = 0.5$ model line are from homogeneous shear DNS data on which the model was based (Fig. 5.1.6). $\xi > 1$ for this data. The remaining data is from Fig. 15.3.1 and Table 15.3.1. Data with darker symbols have $\xi < 1$. The light symbols below the $g = 0.9$ model line have $\xi > 1$. Overall, the data from the return-to-isotropy after distortion ducts (contraction, expansion, plane and mixed strain) show slower rates of return than the MARV model predicts. The MC model, which does not include a g dependence of c_1, will predict even more rapid rates of return for these cases.

Fig. 15.3.4. Comparison of the MARV model with effective c_1 coefficients from analysis of experimental and DNS data. Data symbols: 2, $g_{ave} = 0.2 \pm 0.05$ to 9, $g_{ave} = 0.9 \pm 0.05$; a, $g_{ave} > 0.95$. DNS data from Rogers et al. (RMR1986), data from Table 15.3.1, and data from Fig. 15.3.1.

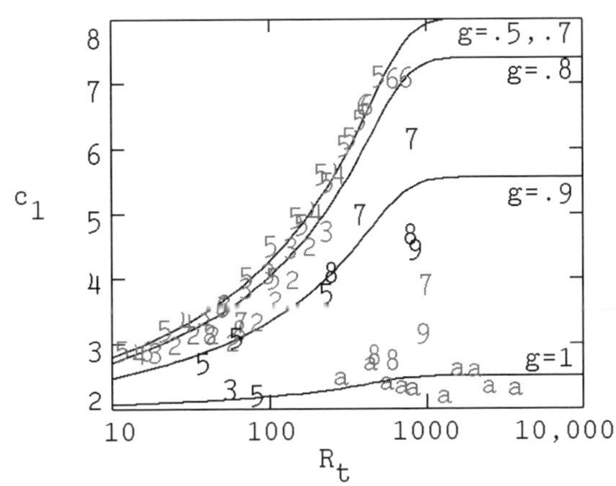

Table 15.3.1. Return-to-Isotropy Data – Analysis Results.

Data Source	No. Pts.	R_t range	g range	$(\xi/\eta)_{initial}$	c_1
Axi-sym. Contraction					
Uberoi (U1956) Fig. 5	7	61, 58	0.36, 0.57	–1	3.08 ±0.19
Fig. 6	3	55, 53	0.31, 0.36	–1	2.31 ±0.03
Fig. 7	2	38, 37	0.43, 0.47	–1	2.69 ± ?
Fig. 10	10	85, 79	0.44, 0.55	–1	2.20 ±0.30
Fig. 11	7	233, 220	0.38, 0.66	–1	3.76 ±0.23
Uberoi (U1957) Fig. 2	14	325, 282	0.59, 0.995	–1	6.54 ±0.75
Mills & Corrsin (1959)	28	462, 446	large scatter, omit		3.67 ±1.52
Choi & Lumley (CL2001)	7	249, 233	0.76, 0.89	–1.00	4.10 ±0.11
Mixed Strain					
Le Penven et al.	12	823, 752	0.66, 0.92	–0.92	4.70 ±0.14
(LGC1985) Fig. 7	12	633, 569	0.78, 0.86	0.84	2.76 ±0.08
Plane Strain					
Tucker (T1970) Fig. 16	10	810, 787	0.62, 0.85	–0.74	6.17 ±1.17
Fig. 26	8	918, 770	0.73, 0.98	–0.69	4.46 ±0.43
Choi & Lumley (CL2001)	11	963, 904	0.83, 0.96	0.68	3.17 ±0.11
Gence & Mathieu (GM1980)	8	970	0.68, 0.81	0.78	3.91 ±0.15
Axi-sym. Expansion					
Choi & Lumley (CL2001)	13	404, 509	0.79 0.88	0.98	2.84 ±0.09

Chapter 16

WALL REFLECTION

As a jet flow approaches a wall, the pressure increases so that the wall normal component stagnates at the wall. To preserve continuity, the flow must also flow outwards, parallel to the wall, two or three dimensionally, depending on the wall and jet geometry. Qualitatively, walls have the same effect on turbulent velocity fluctuations normal to the wall, as they have on mean flow; i.e., the wall normal component of the velocity fluctuations is brought to zero at the wall, due to the impermeability of the wall. If the flow is essentially inviscid, this must occur due to pressure fluctuations, just as the mean flow is brought to zero by the mean pressure.

In this chapter we review the theoretical basis and models in the literature for wall reflection contributions to pressure-strain. The effects of wall reflection on a shear-free boundary layer, and on flow approaching a leading edge are considered.

Note that this chapter on wall reflection is included for completeness. No permanent modifications to the MARV model are made. Demonstration models for the shear-free boundary layer and leading edge flow are given.

16.1 Mathematical Analysis

The pressure fluctuations satisfy the Poisson equation, Eq. 3.1.3,

$$\frac{1}{\rho}\frac{\partial^2 p}{\partial x_i^2} = -2\frac{\partial u_m}{\partial x_i}\frac{\partial V_i}{\partial x_m} - \frac{\partial^2 (u_i u_m - \overline{u_i u_m})}{\partial x_i \partial x_m} \,. \tag{16.1.1}$$

The boundary condition is obtained from the normal component of the fluctuating momentum equation, Eq. 2.1.11, evaluated at the wall. Taking x_2 as the wall normal direction, and with the standard assumptions that the viscosity is constant, and the wall is impermeable ($u_2 = U_2 = 0$ at the wall) and smooth ($\partial U_2/\partial x_1$, $\partial U_2/\partial x_3$, $\partial u_2/\partial x_1$ and $\partial u_2/\partial x_3$ are zero at the wall), the wall boundary condition is (Appendix EQ.16.1.2)

$$\frac{1}{\rho}\frac{\partial p}{\partial x_2} = -2\varepsilon_{2km}\Omega_k u_m + \frac{\partial}{\partial x_k}\nu\frac{\partial u_2}{\partial x_k} \,. \tag{16.1.2}$$

If the wall is considered to be inviscid, with slip wall velocities, then the normal pressure gradient balances the Coriolis term. This is consistent with the conclusion of Section 12.4, that the pressure-strain at the wall cancels out the rotation force term, and the turbulence at the wall is material frame indifferent.

With a viscous wall, all components of the fluctuating velocity at the wall are zero and the normal pressure gradient balances the laminar diffusion of the normal fluctuating velocity. If the wall is also flat, then $\frac{\partial}{\partial x_1}\nu\frac{\partial u_2}{\partial x_1} = \frac{\partial}{\partial x_3}\nu\frac{\partial u_2}{\partial x_3} = 0$, and the boundary condition obtained by Mansour et al. (MKM1988) and Pope (2000) is derived,

$$\frac{1}{\rho}\frac{\partial p}{\partial x_2} = \frac{\partial}{\partial x_2}\nu\frac{\partial u_2}{\partial x_2} \quad . \tag{16.1.3}$$

Since Eq. 16.1.1 is linear in the fluctuating pressure, Mansour et al. considered the pressure in three parts, a 'rapid' part, a 'slow' part and a 'Stokes' part,

$$p = p^{\text{rapid}} + p^{\text{slow}} + p^{\text{Stokes}}. \tag{16.1.4}$$

Decomposing Eq. 3.1.3 (Eq. 16.1.1), and its solution, Eq. 3.1.4, the differential equations, boundary conditions, and solutions for each part are:

Rapid:

$$\frac{1}{\rho}\frac{\partial^2 p^{\text{rapid}}}{\partial x_i^2} = -2\frac{\partial u_m}{\partial x_i}\frac{\partial V_i}{\partial x_m}\ , \qquad \frac{1}{\rho}\frac{\partial p^{\text{rapid}}}{\partial x_2} = 0\ , \tag{16.1.5}$$

$$\frac{1}{\rho}p^{\text{rapid}}(\mathbf{x}) = \frac{1}{4\pi}\iiint\left(2\frac{\partial u_m(\mathbf{x}')}{\partial x_i'}\frac{\partial V_i(\mathbf{x}')}{\partial x_m'}\right)\frac{dVol'}{|\mathbf{x}-\mathbf{x}'|}\ ; \tag{16.1.6}$$

Slow:

$$\frac{1}{\rho}\frac{\partial^2 p^{\text{slow}}}{\partial x_i^2} = -\frac{\partial^2(u_i u_m - \overline{u_i u_m})}{\partial x_i \partial x_m}\ , \qquad \frac{1}{\rho}\frac{\partial p^{\text{slow}}}{\partial x_2} = 0\ , \tag{16.1.7}$$

$$\frac{1}{\rho}p^{\text{slow}}(\mathbf{x}) = \frac{1}{4\pi}\iiint\left(\frac{\partial^2(u_i(\mathbf{x}')u_m(\mathbf{x}') - \overline{u_i(\mathbf{x}')u_m(\mathbf{x}')}\,)}{\partial x_i'\partial x_m'}\right)\frac{dVol'}{|\mathbf{x}-\mathbf{x}'|}\ ; \tag{16.1.8}$$

and Stokes:

$$\frac{1}{\rho}\frac{\partial^2 p^{\text{Stokes}}}{\partial x_i^2} = 0\ , \qquad \frac{1}{\rho}\frac{\partial p^{\text{Stokes}}}{\partial x_2} = \frac{\partial}{\partial x_2}\nu\frac{\partial u_2}{\partial x_2}\ , \tag{16.1.9}$$

$$\frac{1}{\rho}p^{\text{Stokes}}(\mathbf{x}) = \frac{1}{4\pi\rho}\oint\!\!\oint\left(\frac{1}{|\mathbf{x}-\mathbf{x}'|}\frac{\partial p^{\text{Stokes}}(\mathbf{x}')}{\partial n'} - p^{\text{Stokes}}(\mathbf{x}')\frac{\partial}{\partial n'}\frac{1}{|\mathbf{x}-\mathbf{x}'|}\right)dA'\ . \tag{16.1.10}$$

The 'Stokes' part is obviously associated with the wall and the viscous nature of the flow. Mansour et al. (MKM1988) evaluated the pressure-strain associated with each part of the fluctuating pressure from their DNS calculation of fully developed channel flow for $R_\tau = 180$. Their Figs. 13-16 show that the 'Stokes' part is small, in general peaking at about $y^+ = 5$, and fading by about $y^+ = 15$. The turbulence length scale, q^+/ω^+, in this region is about 50 (see e.g. Figs. 8.2.5 and 8.2.6), suggesting that the wall effect extends only a fraction of this length scale. Indeed, the 'Stokes' part appears to be essentially a laminar sublayer phenomenon. It should also be noted, that it is because the wall is viscous, that u_2 varies as y^2 near the wall. Therefore, it could be concluded, any addition to the model to force the y^4 variation of $\overline{u_2 u_2}$ should affect only the laminar sublayer.

Pope (2000, p. 440) mathematically describes wall reflection effects as contributions to the pressure from wall reflected sources, since the normal pressure gradient is zero at the wall. This applies to both the 'slow' and 'rapid' parts of the pressure. Walls can also cause changes to the fluctuating pressure simply due to the overall inhomogeneity the wall creates. In modeling the flow, it is the overall effects the wall has on the pressure-strain, regardless of their mathematical origin, that we seek, and refer to as wall reflection.

16.2 Models in the Literature

Many modelers have included wall reflection pressure-strain terms in their Reynolds stress equations to account for the blockage effect walls have on turbulence. Daly and Harlow (DH1970) modeled the pressure-strain due to the wall as

$$\Pi_{ij}^{\mathrm{W}} = -c_{\mathrm{W}} \frac{\varepsilon}{k}\left(N_{ik}\overline{u_k u_j} + N_{jk}\overline{u_k u_i} - \frac{2}{3} N_{nk}\overline{u_k u_n}\delta_{ij} \right), \tag{16.2.1}$$

where N_{ij} is defined in terms of the unit vector normal to the wall, n_i, and a dimensionless function f to be determined,

$$N_{ij}(\mathbf{x}) = \oiint n_i(\mathbf{x}')n_j(\mathbf{x}')f(\mathbf{x},\mathbf{x}')\frac{dA'}{|\mathbf{x}-\mathbf{x}'|^2}. \tag{16.2.2}$$

Shir (S1973) considered a form appropriate for flat walls, and used a model which may be written as

$$\Pi_{ij}^{\mathrm{W}} = -c_{\mathrm{W}}\omega\left(\overline{u_i u_k} n_k n_j + \overline{u_j u_k} n_k n_i - \frac{2}{3}\overline{u_k u_m} n_k n_m \delta_{ij} \right)\left(\frac{q}{\omega y}\right)^2, \tag{16.2.3}$$

where $q/(\omega y)$ is the ratio of a turbulence length scale, q/ω, to distance from the wall, y. Using x_2 as the coordinate normal to the wall, then

$$\Pi_{22}^{\mathrm{W}} = -\frac{4}{3} c_{\mathrm{w}}\, \omega\, \overline{u_2 u_2} \left(\frac{q}{\omega y}\right)^2 \quad \text{and} \quad \Pi_{11}^{\mathrm{W}} = \Pi_{33}^{\mathrm{W}} = \frac{2}{3} c_{\mathrm{w}}\, \omega\, \overline{u_2 u_2} \left(\frac{q}{\omega y}\right)^2 . \tag{16.2.4}$$

The effect of the wall reflection term is then to take energy out of the normal component, $\overline{u_2 u_2}$, and transfer it equally into the parallel components, $\overline{u_1 u_1}$ and $\overline{u_3 u_3}$. This is consistent with the concept that the wall normal component of the velocity fluctuation is reduced due to pressure effects, as the wall is approached.

Launder, Reece, Rodi (LRR1975) included modifications to the c_1 and c_5 pressure-strain coefficients, as a function of distance to the wall, which may be written as

$$\Pi_{ij}^{\mathrm{W}} = \left(c_{1,\mathrm{w}} \varepsilon b_{ij} - c_{5,\mathrm{w}} k (b_{ik} W_{jk} + b_{jk} W_{ik})\right) \frac{q}{\omega y} . \tag{16.2.5}$$

Gibson and Launder (GL1978) viewed the wall proximity effect for shear flow as "impeding the transfer of energy from the streamwise direction to that normal to the wall." Thus, they viewed Shir's model as a wall modification to slow pressure-strain, and used it in a modified form with the factor $(q/\omega y)^2$ in Eq. 16.2.3 replaced by $q/\omega y$. They also added corresponding terms for other contributions to pressure-strain, for example a rapid pressure-strain term, which may be written as

$$\Pi_{ij}^{\mathrm{W,R}} = -c_{\mathrm{w,R}} \left(\Pi_{ik}^{\mathrm{R}} n_k n_j + \Pi_{jk}^{\mathrm{R}} n_k n_i - \frac{2}{3} \Pi_{km}^{\mathrm{R}} n_k n_m \delta_{ij}\right) \left(\frac{q}{\omega y}\right) . \tag{16.2.6}$$

Gibson and Launder used this form for the log-law region of wall-bounded flows. They placed a heavy reliance on the wall reflection model for the calculation of boundary layer flows in order to counteract the fact that the IP model gives $\overline{u_2 u_2} = \overline{u_3 u_3}$ in shear flow, $U_1(x_2)$ (Launder, 1989).

Craft et al. (CGL1993), when considering jet flow impinging on a flat plate, added the wall reflection term

$$\begin{aligned} \phi_{ij,2}^{\mathrm{W}} = {} & \left(3 c_{2\mathrm{w}} \frac{\partial U_l}{\partial x_m} \overline{u_l u_m} + c_{2\mathrm{w}}'' k \frac{\partial U_l}{\partial x_m} n_l n_m\right) \left(n_i n_j - \frac{1}{3} n_q n_q \delta_{ij}\right) \left(\frac{1}{c_\ell} \frac{q}{\omega y}\right) \\ & - c_{2\mathrm{w}}' k (2 b_{lm}) \left(\frac{\partial U_k}{\partial x_m} n_l n_k \delta_{ij} - \frac{3}{2} \frac{\partial U_i}{\partial x_m} n_l n_j - \frac{3}{2} \frac{\partial U_j}{\partial x_m} n_l n_i\right) \left(\frac{1}{c_\ell} \frac{q}{\omega y}\right) \end{aligned} \tag{16.2.7}$$

to an IP model.

16.3 Experimental Evidence of Wall Reflection

Shear-Free Moving Wall

Thomas and Hancock (TH1977) passed grid produced turbulence over a wall moving at the same speed as the flow. They observed that the wall normal component of the turbulence decreased near the wall, while the wall parallel components remained essentially constant or increased. This is qualitatively consistent with Shir's wall reflection model. The possible extent of the wall reflection effect is shown in Fig. 16.3.1, where the wall normal anisotropy, b_{22}, is plotted against y divided by the free stream turbulence length scale, q_∞/ω_∞. b_{22} begins to fall below $y\omega_\infty/q_\infty = 0.6$, and falls quite linearly towards the wall value of $-1/3$ below $y\omega_\infty/q_\infty = 0.1$.

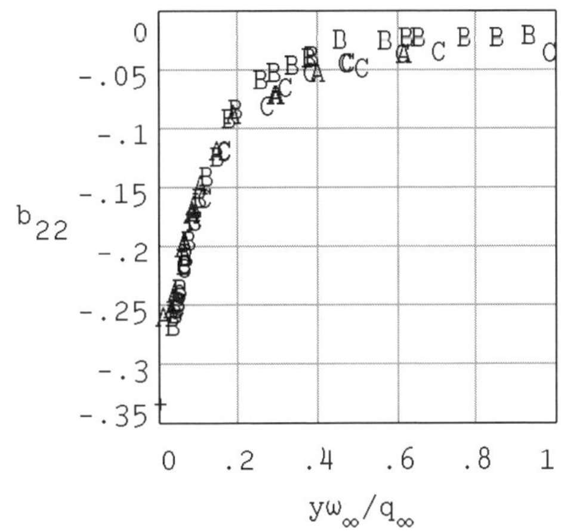

Fig. 16.3.1. Normal turbulence anisotropy from the data of Thomas and Hancock (TH1977) for a moving, shear-free wall. Cases: A-13M; B-22M; C-25M.

To obtain a more quantitative estimate of the extent of the wall reflection effect, a model of the form

$$\Pi_{ij}^{W} = -c_W\omega\left(\overline{u_iu_k}n_kn_j + \overline{u_ju_k}n_kn_i - \frac{2}{3}\overline{u_ku_m}n_kn_m\delta_{ij}\right)f\left(\frac{q}{\omega y}\right), \tag{16.3.1}$$

was added to the MARV model and applied to the Thomas and Hancock test case. It should be noted that since the mean flow is uniform, the rapid pressure-strain is zero and the model elements which influence b_{ij} are the slow pressure-strain, the laminar and turbulent diffusion, the wall boundary condition, and the added wall reflection pressure-strain. The function $f(q/\omega y)$ was tested in the form

$$f\left(\frac{q}{\omega y}\right) = \left(\max\left(\frac{q}{\omega y} - a, 0\right)\right)^b, \tag{16.3.2}$$

to allow the testing of linear and quadratic relations (the models of Gibson and Launder, and Shir), as well as a cut-off length scale, so that when the turbulence length scale becomes less than some value times the distance to the wall, wall reflection no longer applies.

A marching calculation procedure, with uniform conditions at the start of the moving belt, was used to obtain the results shown in Fig. 16.3.2. The distance to the wall has been normalized with the turbulence grid bar diameter, D. With no wall reflection term, b_{22} remains near the free stream value until very near the wall, where it drops to the boundary value of $-1/3$. The linear and quadratic models are the Gibson and Launder (GL1978) and the Shir (S1973) models, respectively. Both give similar results above $y/D = 1.5$, or $y\omega/q = 0.4$, with b_{22} low compared with the data. These models will then be quite active in the log-law region of a turbulent boundary layer where $y\omega/q \approx 0.4$. A best-fit model was obtained using a cutoff on the wall reflection, so that the wall reflection is zero for $y\omega/q > 0.3$. The small reduction in b_{22} above $y/D = 1$ is consistent with the data, and a result of turbulent diffusion in the model.

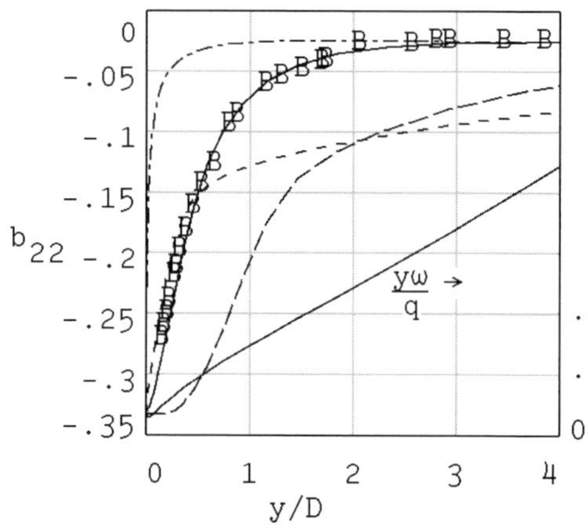

Fig. 16.3.2. Results of wall reflection models for a shear-free moving wall. B- Thomas and Hancock (TH1977) data for case 22M. Model Eqs.: 16.3.1-2.

Models		c_w	a	b
None	– - - -	0		
Linear GL1978	- - -	0.3	0	1
Quad. S1973	– – –	0.143	0	2
Cutoff	———	0.66	3.333	1

Wall Reflection at a Leading Edge

Bearman (B1972) measured the components of the turbulence along the stagnation streamline in flow approaching a bluff body, Fig. 16.3.3. The body was two-dimensional with a flat section of width D on the nose. This is an interesting case because the wall reflection has the opposite effect to the production terms on the turbulence anisotropy. Bearman placed various turbulence grids upstream of the body, so that the integral length scale of the turbulence, L_x, varied between $1.2D$ (grid A) and $2.6D$ (grid C), and noted how the turbulence changed.

As a simple test of wall reflection models, one-dimensional calculations were made of the turbulence on the stagnation streamline using the measured mean velocity distribution. The MARV model includes the effects of rapid distortion which are important in this case where the dimensionless strain rate, s, exceeds 200 close to the body. The MARV model was used with and without wall pressure-reflection terms added.

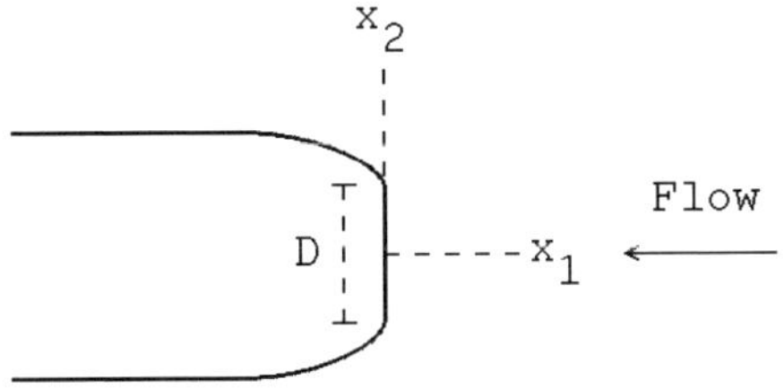

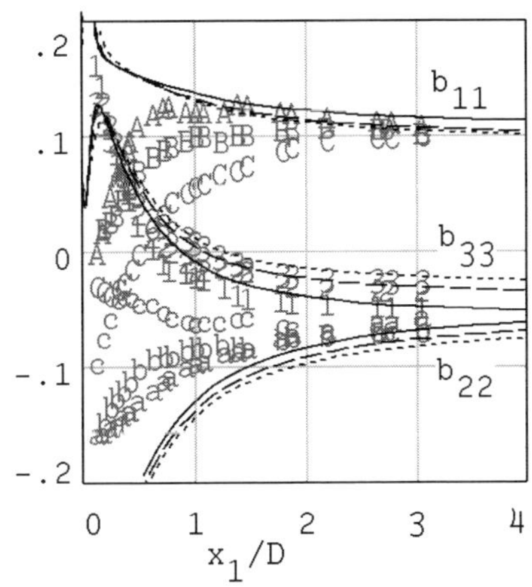

Fig. 16.3.3. Turbulence anisotropy along the stagnation line of a blunt nosed body. b_{11}-A,B,C; b_{22}-a,b,c; b_{33}-1,2,3, data of Bearman (B1972) for L_x/D=1.2, 1.8 and 2.6, respectively. —, – –, - -, corresponding calculations with no wall reflection term.

The data for turbulence grids A, B and C are shown with the calculations with no wall reflection model in Fig. 16.3.3. The flow sees an adverse pressure gradient as the stagnation point is approached, so that $\partial U_1/\partial x_1$ is negative which causes b_{11} to rise. Correspondingly, the calculations with no wall reflection show an increase in b_{11} as the wall is approached. However, since the flow is approaching a wall, b_{11}, which is the normal component, must decline to $-1/3$ at the wall. The data shows that this decline in b_{11} begins approximately one integral length scale, L_x, upstream of the wall.

Three calculations are compared in Fig 16.3.4 with the data for the middle case, grid B. The cutoff model, which successfully modeled the Thomas and Hancock data, gives a reduction in b_{11} below $x_1/D = 0.5$, when compared with no wall reflection, but the amount is insufficient to explain the reduction seen in the data.

Bearman, in qualitatively explaining his data considered two limits. The rapid distortion limit is already included in the MARV model. The other limit was that of quasi-steady flow. For this limit the unsteady velocity in the x_1 direction scales as the mean velocity, so that

$$\frac{u_1 + U_1}{u_{1,\infty} + U_{1,\infty}} = \frac{U_1}{U_{1,\infty}} \quad , \tag{16.3.3}$$

where $u_{1,\infty}$ and $U_{1,\infty}$ are the far upstream values of the fluctuating and mean velocities. If one further makes the assumptions that the unsteadiness of the velocity can be represented by a simple sine wave, $u_1 = a\sin\omega t$, and that at any instant in time total pressure is conserved along the stagnation streamline,

$$\overline{P} + p + \frac{1}{2}\rho(u_1 + U_1)^2 = \text{constant}, \tag{16.3.4}$$

then it can be shown (Appendix EQ.16.3.5) that the component of the velocity–pressure-gradient term, acting on the $\overline{u_1 u_1}$ component of the turbulence, is

$$-2\overline{u_1 \frac{\partial p}{\partial x_1}} = 4\rho\overline{u_1 u_1}\frac{\partial U_1}{\partial x_1} . \tag{16.3.5}$$

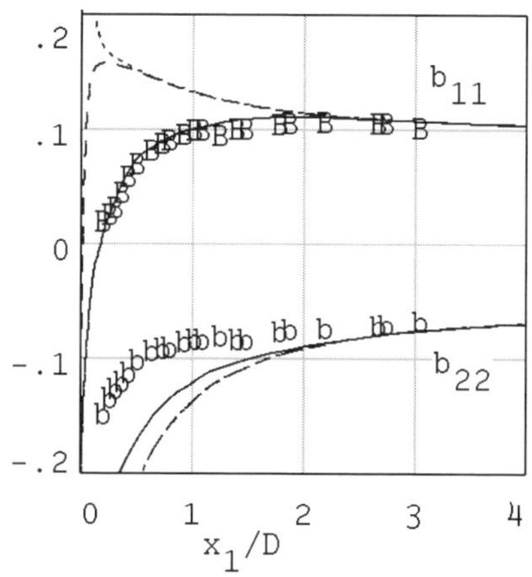

Fig. 16.3.4. Results of wall reflection models. Bearman (B1972) data; B, b_{11}; b, b_{22}. Model Eqs.: 16.3.1, 16.3.2, 16.3.6.

Models:	c_w	a	c_{wn}	a_N
None - - - -	0		0	
Cutoff – – –	0.66	3.333	0	
U-mean ——	0.66	3.333	0.09	1.0

To test the effect of a term of this type in the context of the Bearman test case, an additional wall reflection term proportional to the mean velocity of the flow approaching the wall, U_N, was added in a form similar to Eqs. 16.3.1 and 16.3.2,

$$\Pi_{ij}^{WN} = -c_{wn}\frac{U_N}{y}\left(\overline{u_i u_k}n_k n_j + \overline{u_j u_k}n_k n_i - \frac{2}{3}\overline{u_k u_m}n_k n_m \delta_{ij}\right)\max\left(\frac{q}{\omega y} - a_N, 0\right). \tag{16.3.6}$$

With an appropriate choice of the parameters c_{wn} and a_N, good results can be obtained for the normal anisotropy b_{11}, as shown in Fig. 16.3.4. However, the b_{22} component is still under-predicted by the model.

The individual components of the wall reflection terms along the stagnation line, using Eqs. 16.3.1, 16.3.2 (with $b = 1$) and 16.3.6, are

$$\Pi_{ij}^{W} + \Pi_{ij}^{WN} = \overline{u_1 u_1}\left(c_w \omega \max\left(\frac{q}{\omega y} - a, 0\right) + c_{wn}\frac{U_N}{y}\max\left(\frac{q}{\omega y} - a_N, 0\right)\right)\begin{bmatrix} -4/3 & & \\ & 2/3 & \\ & & 2/3 \end{bmatrix}. \tag{16.3.7}$$

This adds energy equally to the $\overline{u_2u_2}$ and $\overline{u_3u_3}$ components. However, Bearman's quasi-steady analysis for flow around a cylinder suggested that the effect of wall reflection, in this case, is to take energy out of the $\overline{u_1u_1}$ component of the turbulence and put it into the $\overline{u_2u_2}$ component, with the $\overline{u_3u_3}$ component unchanged. This two-dimensional redistribution of the turbulence,

$$\Pi_{ij}^{W} + \Pi_{ij}^{WN} = \overline{u_1u_1}\left(c_w\omega\max\left(\frac{q}{\omega y} - a, 0\right) + c_{wn}\frac{U_N}{y}\max\left(\frac{q}{\omega y} - a_N, 0\right)\right)\begin{bmatrix} -4/3 & & \\ & 4/3 & \\ & & 0 \end{bmatrix}, \tag{16.3.8}$$

is compared with the standard redistribution of Eq. 16.3.7, in Fig. 16.3.5. The two dimensional redistribution works well above $x_1/D = 0.5$, but closer to the wall, b_{22} is so high that the model erroneously predicts $b_{22} > b_{11}$, and therefore negative turbulence production.

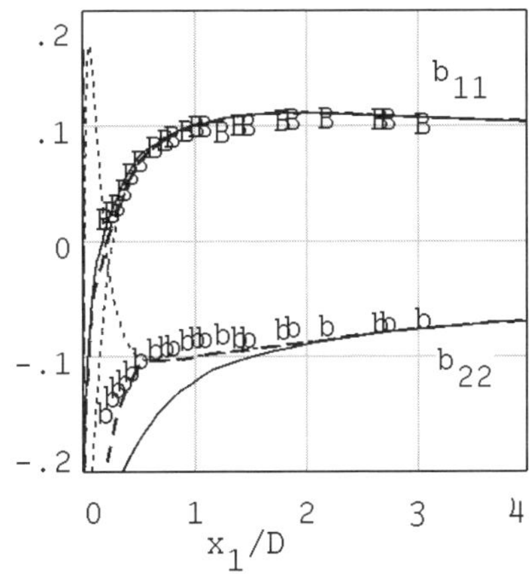

Fig. 16.3.5. Results of various redistribution choices for the U-mean wall reflection model. Bearman (B1972) data; B, b_{11}; b, b_{22}. —— Standard; - - - 2-d; — — — Mixed, 2-d for $x_1/D > 0.5$.

Standard:

$$\Pi_{22}^{W} + \Pi_{22}^{WN} = -\frac{1}{2}\left(\Pi_{11}^{W} + \Pi_{11}^{WN}\right)$$

2-d:

$$\Pi_{22}^{W} + \Pi_{22}^{WN} = -\left(\Pi_{11}^{W} + \Pi_{11}^{WN}\right)$$

Bearman's leading edge is, however, not a cylinder but a flat plate of width D, followed by a streamlined body. Farther from the leading edge, the body will act like a cylinder, but closer to the leading edge it will look like a flat plate. To approximate this, the flow was considered in two regions. For $x_1/D < 0.5$ the wall reflection was taken to act as a flat plate with equal amounts of energy being transferred into $\overline{u_2u_2}$ and $\overline{u_3u_3}$. Farther from the wall all the energy reduction from $\overline{u_1u_1}$ was put into $\overline{u_2u_2}$. The result, the 'mixed' model in Fig.16.3.5, gives a reasonable approximation of the redistribution in Bearman's data.

The 'U-mean' wall reflection mechanism with the 'mixed' redistribution described above, added to the MARV model, explains the turbulence anisotropies observed in Bearman's data, as shown in Fig. 16.3.6.

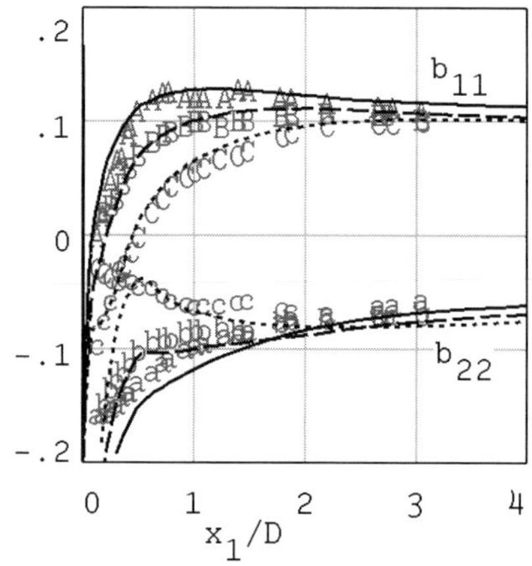

Fig. 16.3.6. Turbulence anisotropy along the stagnation line of a bluff body. b_{11}-A,B,C; b_{22}-a,b,c; data of Bearman (B1972) for L_x/D=1.2, 1.8 and 2.6, respectively. —, – –, - -, corresponding calculations with the U-mean model (Fig. 16.3.4 caption) and the mixed redistribution (Fig. 16.3.5 caption).

16.4 Discussion

Wall reflection is very complex. To model it properly requires both a standard wall reflection model based on the turbulence properties, e.g., Eq. 16.2.3, and an extra wall reflection term to account for flow moving towards the wall. Daly and Harlow's integral formula, Eq. 16.2.2, suggested a method for approximately determining the wall normal direction in a complex geometric configuration. Bearman's data shows that there is also a need to determine how the energy is redistributed into the wall parallel components. Perhaps this could take the form of an effective body diameter as a function of distance to the wall, and geometry. The way energy is redistributed may have implications for concave surfaces, as well as convex leading edges.

However, it is encouraging that wall reflection has very little if any effect on the log-law region of simple boundary layers. (See Appendix A for an extensive review of b_{ij} data.) It is unclear how much of an effect it has on the laminar sublayer and buffer layer regions since the wall normal anisotropy can be explained by the high strain rates in these regions, independent of the wall. Therefore, it is a useful exercise to complete a Reynolds stress model without including wall reflection terms. The simplification, of not including any direct dependency on the wall normal direction (and the effective diameter for redistribution) at every point in the flow, will significantly reduce the difficulty of implementing a Reynolds stress model in complex situations.

Chapter 17

TURBULENCE SPECTRA

Reynolds stress models improve upon two-equation models by using differential equations to say how the energy in the turbulence is directionally oriented. They do not directly consider the frequency, or spectral, distribution of the energy. It may be possible to model this algebraically with the results from a Reynolds stress model. In this chapter we consider the various forms of spectra which are commonly used, and algebraic models for the spectra.

17.1 Common Forms of Turbulence Spectra

The Spectrum Tensor

In Chapter 9, which considered homogeneous turbulence, the wave-number spectral distribution of Reynolds stresses was obtained as the Fourier transform of the two-point correlation (Eq. 9.2.4)

$$\overline{u_i u_j} = R_{ij}(0,0,0) = \iiint \Phi_{ij}(K_1,K_2,K_3)dK_1 dK_2 dK_3 \ . \tag{17.1.1}$$

This is Batchelor's (1953) Eq. 2.4.5. He refers to $\overline{u_i u_j}$ as the energy tensor, and to Φ_{ij} as the energy spectrum tensor. The integral is over all of wave-number space so that more precisely,

$$\overline{u_i u_j} = \int_{-\infty}^{\infty}\int_{-\infty}^{\infty}\int_{-\infty}^{\infty} \Phi_{ij}(K_1,K_2,K_3)dK_1 dK_2 dK_3 \ . \tag{17.1.2}$$

The correlations of derivatives of the fluctuating velocity are related to the energy spectrum tensor by (Appendix EQ.17.1.3)

$$\overline{\frac{\partial u_i}{\partial x_n}\frac{\partial u_j}{\partial x_m}} = \int_{-\infty}^{\infty}\int_{-\infty}^{\infty}\int_{-\infty}^{\infty} K_n K_m \Phi_{ij}(K_1,K_2,K_3)dK_1 dK_2 dK_3 \ . \tag{17.1.3}$$

This leads to the spectral representation of the dissipation tensor,

$$\varepsilon_{ij} = 2\nu\overline{\frac{\partial u_i}{\partial x_n}\frac{\partial u_j}{\partial x_n}} = 2\nu\int_{-\infty}^{\infty}\int_{-\infty}^{\infty}\int_{-\infty}^{\infty} K_n K_n \Phi_{ij}(K_1,K_2,K_3)dK_1 dK_2 dK_3 \ . \tag{17.1.4}$$

The Three-Dimensional or Radial Spectrum

In Section 6.2, a model spectrum for the turbulence kinetic energy was used to shed light on the rate of decay of dissipation. This spectrum takes the form

$$k = \frac{\overline{u_i u_i}}{2} = \int_0^\infty E(K)dK \ , \tag{17.1.5}$$

where K is the magnitude of the wave-number vector. If wave-number space is considered in spherical coordinates, K is the radial coordinate. $E(K)$ is generally referred to as the three-dimensional energy spectrum, because it considers spectral energy in all directions, even though it does so in only an average sense for each wave-number magnitude.

The three-dimensional energy spectrum can be obtained from the energy spectrum tensor. Consider Eq. 17.1.2 in spherical coordinates (Fig. 9.5.1 and Appendix EQ.9.4.7),

$$\overline{u_i u_j} = \int_{-\infty}^{\infty}\int_{-\infty}^{\infty}\int_{-\infty}^{\infty} \Phi_{ij} dK_1 dK_2 dK_3 = \int_0^{2\pi}\int_0^{\pi}\int_0^{\infty} \Phi_{ij} K^2 \sin\psi \, dK d\psi d\theta \ . \tag{17.1.6}$$

Reordering the integration, the turbulence kinetic energy is given by

$$\frac{\overline{u_i u_i}}{2} = \frac{1}{2}\int_0^\infty \left(\int_0^{2\pi}\int_0^{\pi} \Phi_{ii} \sin\psi \, d\psi d\theta \right) K^2 dK , \tag{17.1.7}$$

so that comparing with Eq. 17.1.5,

$$E(K) = \frac{1}{2} K^2 \int_0^{2\pi}\int_0^{\pi} \Phi_{ii} \sin\psi \, d\psi d\theta \ . \tag{17.1.8}$$

Correspondingly, from Eq. 17.1.4, the dissipation is related to the three-dimensional spectrum by

$$\varepsilon = \frac{1}{2}\varepsilon_{ii} = 2\nu \int_0^\infty K^2 E(K) dK \ . \tag{17.1.9}$$

One-Dimensional Spectra

The three-dimensional energy spectrum is so named to distinguish it from one-dimensional spectra, which are measured, for example, by hot-wire anemometers in a flow. Time dependent measurements of velocity at a point in a flow may yield the mean velocity and the fluctuating velocities, $u_1(t)$, $u_2(t)$, and $u_3(t)$. The Reynolds stresses may be obtained by time averaging the correlations of the fluctuating velocity components over a sufficiently long time Δ,

$$\overline{u_i u_j} = \frac{1}{\Delta}\int_0^\Delta u_i(t) u_j(t) dt \ . \tag{17.1.10}$$

The time dependent signals may also be analyzed to give the spectral distribution of the component energies, so that

$$\overline{u_i u_j} = \int_0^\infty \hat{E}_{ij}(f)\, df \ , \tag{17.1.11}$$

where the frequency, f, is the reciprocal of the time to complete one cycle of a sine or cosine wave.

Taking the x_1 direction as the direction of the mean flow at the measurement point, the frequency spectra may be converted to one-dimensional wave-number spectra which are functions of K_1, so that

$$\overline{u_i u_j} = \int_0^\infty E_{ij}(K_1) dK_1 \ . \tag{17.1.12}$$

The corresponding waves in frequency and wave-number representations are

$$\cos(2\pi f t) = \cos(K_1 x_1) \ , \tag{17.1.13}$$

so that $K_1 x_1 = 2\pi f t$, and with an elapsed time, $t = x_1 / U_1$,

$$K_1 = \frac{2\pi f}{U_1} \ , \tag{17.1.14}$$

and

$$E_{ij}(K_1) = \frac{U_1}{2\pi} \hat{E}_{ij}(f) \ . \tag{17.1.15}$$

This correspondence between frequency spectra and K_1 spectra is known as Taylor's hypothesis (Taylor, 1938) and is valid when the fluctuating velocities are much smaller than the mean velocity. (Lumley, 1965, and Wyngaard and Clifford, 1977, have proposed analyses by which the accuracy of the correspondence can be quantitatively assessed as a function of $\overline{u_i u_j} / U_1^2$.)

Comparing Eq. 17.1.12, in which the integration is over K_1 from zero to infinity, with Eq. 17.1.2, in which the integration is from minus infinity to infinity,

$$E_{ij}(K_1) = \int_{-\infty}^{\infty}\int_{-\infty}^{\infty}\left(\Phi_{ij}(\mathbf{K}) + \Phi_{ij}(-\mathbf{K})\right)dK_2 dK_3 = 2\int_{-\infty}^{\infty}\int_{-\infty}^{\infty}\Phi_{ij}(\mathbf{K})\,dK_2 dK_3\,. \qquad (17.1.16)$$

Correspondingly,

$$\overline{\frac{\partial u_i}{\partial x_1}\frac{\partial u_j}{\partial x_1}} = \int_{-\infty}^{\infty} K_1 K_1\left(\int_{-\infty}^{\infty}\int_{-\infty}^{\infty}\Phi_{ij}dK_2 dK_3\right)dK_1 = \int_0^{\infty} K_1^2 E_{ij}(K_1)\,dK_1\,. \qquad (17.1.17)$$

One-dimensional spectra and two-point correlations in the direction of the flow are related. From Eq. 9.2.3,

$$R_{ij}(r_1,0,0) = \int_{-\infty}^{\infty}\left(\int_{-\infty}^{\infty}\int_{-\infty}^{\infty}\Phi_{ij}(K_1,K_2,K_3)\,dK_2 dK_3\right)\exp(iK_1 r_1)\,dK_1\,. \qquad (17.1.18)$$

Since the symmetric part of $\exp(iK_1 r_1)$ is $\cos(K_1 r_1)$, $R_{ij}(r_1,0,0)$ may be obtained from E_{ij} by integrating for each r_1,

$$R_{ij}(r_1,0,0) = \int_0^{\infty} E_{ij}(K_1)\cos(K_1 r_1)\,dK_1\,. \qquad (17.1.19)$$

Isotropic Turbulence

$E(K)$ and $E_{ij}(K_1)$ can always be obtained from Φ_{ij}. They are simply integrations of Φ_{ij} over spherical surfaces and planes, as sketched in Fig. 17.1.1.

In general, however, the reverse conversions may not be made. Isotropic turbulence is special in that Φ_{ij} may be obtained from $E(K)$ using Eq. 9.3.6,

$$\Phi_{ij} = \frac{E(K)}{4\pi K^2}\left(\delta_{ij} - \frac{K_i K_j}{K^2}\right)\,. \qquad (17.1.20)$$

Note that the assumption in this equation is not just that the Reynolds stresses are isotropic, but that the spectral distribution of energy is isotropic at all magnitudes of the wave number, K.

For isotropic turbulence, one-dimensional spectra may be obtained from the three-dimensional spectrum by substituting Eq. 17.1.20 into Eq. 17.1.16. The longitudinal spectrum is given by (Appendix EQ.17.1.21-22)

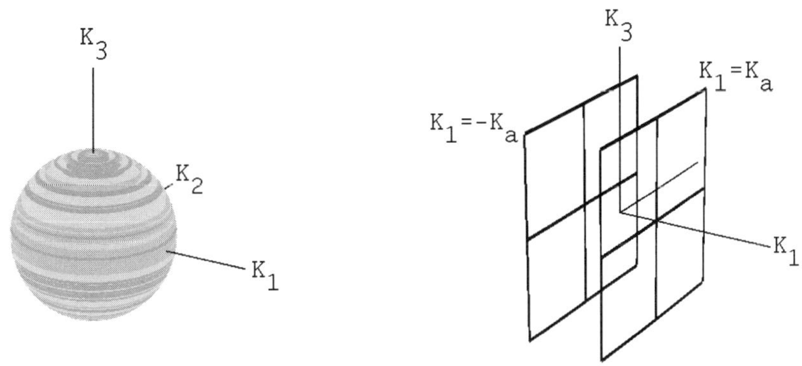

$2E(K)$ = integral of Φ_{ii} over a spherical surface of radius K.

$E_{ij}(K_1 = K_a)$ = integral of Φ_{ij} over infinite planes, $K_1 = \pm K_a$.

Fig. 17.1.1. Sketch of surfaces of integration for $E(K)$ and $E_{ij}(K_1)$.

$$E_{11}(K_1) = \int_{K_1}^{\infty} \frac{E(K)}{K}\left(1 - \frac{K_1^2}{K^2}\right) dK \ , \tag{17.1.21}$$

and the transverse spectra are

$$E_{22}(K_1) = E_{33}(K_1) = \frac{1}{2}\int_{K_1}^{\infty} \frac{E(K)}{K}\left(1 + \frac{K_1^2}{K^2}\right) dK \ . \tag{17.1.22}$$

The reason that $E_{11}(K_1)$ and $E_{22}(K_1)$ are different is sketched in Fig. 17.1.2. If most of the turbulence energy is concentrated at a radius $K = K_b$, say, the K_1-planar average will be strongly influenced by Φ_{ij} on the circle where the plane intersects the sphere. At $K_1 = K_b$, the intersection is a point and the relative magnitudes of $\Phi_{11}:\Phi_{22}:\Phi_{33}$ are 0:1:1, since, by continuity, Φ_{ij} has no component in the direction of the wave vector, **K**. Therefore at high K_1, the planar average gives $E_{22} > E_{11}$. At $K_1 = 0$, the average relative magnitudes of $\Phi_{11}:\Phi_{22}:\Phi_{33}$ on the circle are 2:1:1, so that E_{11} is higher, $E_{11} = 2E_{22}$.

Equation 17.1.21 may be inverted, so that $E(K)$ may be obtained from the measurable $E_{11}(K_1)$ (Appendix EQ.17.1.23-24),

$$E(K) = \frac{1}{2}K^3 \frac{d}{dK}\left(\frac{1}{K}\frac{dE_{11}(K)}{dK}\right) . \tag{17.1.23}$$

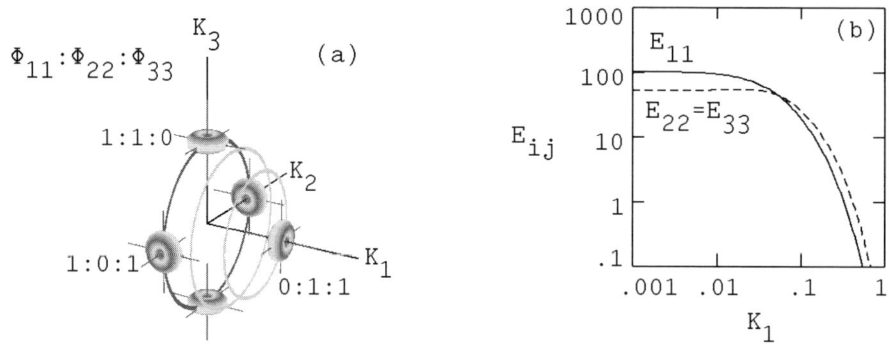

Fig. 17.1.2. Isotropic turbulence. (a) Some glyphs of Φ_{ij} on the intersection of a spherical surface with high Φ_{ii} with K_1 planes of integration. (b) 1-d spectra for $R_t = 100$ ($k = 10$, $\varepsilon = 1$, $\nu = 1$), based on the model spectrum, Eqs. 17.5.7-8, described later.

The transverse spectra may also be obtained from the longitudinal spectrum,

$$E_{22}(K_1) = E_{33}(K_1) = \frac{1}{2}\left(1 - K_1 \frac{d}{dK_1}\right) E_{11}(K_1) \; . \tag{17.1.24}$$

For isotropic turbulence, symmetry gives

$$\overline{\frac{\partial u_1}{\partial x_1}\frac{\partial u_1}{\partial x_1}} = \overline{\frac{\partial u_2}{\partial x_2}\frac{\partial u_2}{\partial x_2}} = \overline{\frac{\partial u_3}{\partial x_3}\frac{\partial u_3}{\partial x_3}} \tag{17.1.25}$$

and

$$\overline{\frac{\partial u_1}{\partial x_2}\frac{\partial u_1}{\partial x_2}} = \overline{\frac{\partial u_1}{\partial x_3}\frac{\partial u_1}{\partial x_3}} = \overline{\frac{\partial u_2}{\partial x_1}\frac{\partial u_2}{\partial x_1}} = \overline{\frac{\partial u_2}{\partial x_3}\frac{\partial u_2}{\partial x_3}} = \overline{\frac{\partial u_3}{\partial x_1}\frac{\partial u_3}{\partial x_1}} = \overline{\frac{\partial u_3}{\partial x_2}\frac{\partial u_3}{\partial x_2}} \; . \tag{17.1.26}$$

Further, by substituting Eq. 17.1.20 into 17.1.3 and integrating in spherical coordinates, it can be shown that for isotropic turbulence (Appendix EQ.17.1.27-28)

$$\overline{\frac{\partial u_1}{\partial x_2}\frac{\partial u_1}{\partial x_2}} = 2\overline{\frac{\partial u_1}{\partial x_1}\frac{\partial u_1}{\partial x_1}} \; . \tag{17.1.27}$$

Using Eq. 17.1.17, the dissipation is then given by

$$\varepsilon = 15\nu\overline{\frac{\partial u_1}{\partial x_1}\frac{\partial u_1}{\partial x_1}} = 15\nu\int_0^\infty K_1^2 E_{11}(K_1)\,dK_1 \, . \tag{17.1.28}$$

17.2 Turbulence Length Scales

Mixing length, two-equation, and Reynolds stress models have length scales associated with the formation of a turbulent viscosity, $\nu_t \propto \text{Velocity} \times \text{Length}$. The mixing length, L, introduced in Eq. 1.3.3, is determined empirically, with the purpose being to return reasonable algebraically modeled shear stresses. In two-equation models, the length scale is formed from turbulence variables, for example $k^{3/2}/\varepsilon$ or $k^{3/2}/\tilde{\varepsilon} = q/\omega$. The new v_{ij} diffusion model, Eq. 7.2.10, uses q/ω as the length scale for the turbulent diffusion of the turbulence properties. These length scales are considered to be characteristic of the size of the energy containing eddies.

In this section we consider additional length scales associated with turbulence spectra. These come in three sizes. The smallest is the Kolmogorov (1941) length scale,

$$\eta = (\nu^3/\varepsilon)^{1/4}. \tag{17.2.1}$$

This is formed from the laminar viscosity and the turbulence dissipation, independent of the turbulence kinetic energy. As such it will be associated with the size of the dissipating eddies.

Integral length scales are a measure of the size of the energy containing eddies, based on two-point velocity correlations. The length scale may be directional, and determined separately for different components of the Reynolds stresses. For homogeneous turbulence, we define integral length scales,

$$L_{ij,k} = \frac{1}{\overline{u_i u_j}} \int_0^\infty \overline{u_i(\mathbf{x}) u_j(\mathbf{x} + r\mathbf{n}_k)}\, dr, \quad \text{with no sum on } i \text{ and } j, \tag{17.2.2}$$

and where $\mathbf{n}_k$ is a unit vector in the x_k direction. In terms of the two-point correlation function, R_{ij},

$$L_{ij,k} = \int_0^\infty \frac{R_{ij}(r\mathbf{n}_k)}{R_{ij}(0)} dr, \quad \text{with no sum on } i \text{ and } j\,, \tag{17.2.3}$$

so that the integral length scale is the area under a normalized two-point correlation function. For example, Fig. 17.2.1a shows (a modeled distribution of) $R_{11}(r_1,0,0)/R_{11}(0,0,0)$ versus r_1. The area under the curve is $L_{11,1}$ $(= L_x)$.

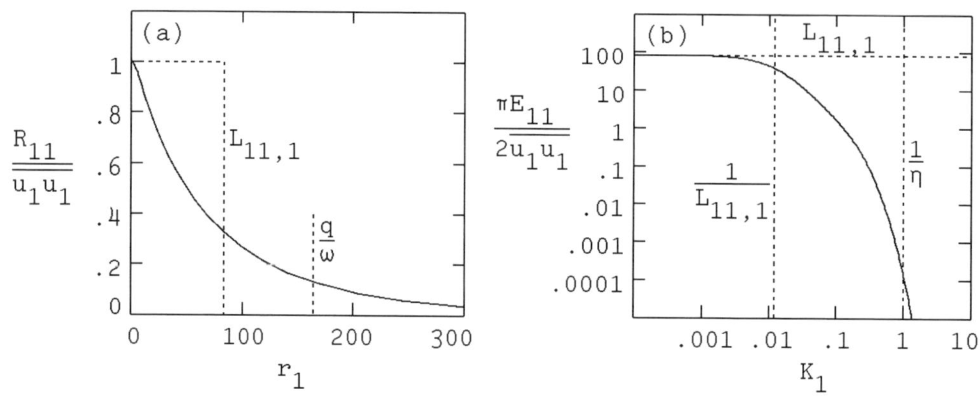

Fig. 17.2.1. The integral length scale, $L_{11,1}$, shown on plots of: (a) $R_{11}(r_1,0,0)/\overline{u_1u_1}$, and (b) $0.5\pi E_{11}(K_1)/\overline{u_1u_1}$. Results for isotropic turbulence with R_t=900 (k=30, ε=1, ν=1) based on the model spectrum, Eqs. 17.5.7-8, described later. (Model gives $L_{11,1}$ =83.03.)

The integral length scales, taken in the direction of the flow, are also related to the one-dimensional spectra. With x_1 as the flow direction (Appendix EQ.17.2.4),

$$L_{ij,1} = \frac{\pi}{2}\frac{E_{ij}(0)}{\overline{u_iu_j}} \qquad \text{with no sum on } i \text{ and } j. \tag{17.2.4}$$

Figure 17.2.1b shows this on a companion distribution of $E_{11}(K_1)$ consistent with the two-point correlation shown in Fig. 17.2.1a. The distributions in Fig. 17.2.1 are based on a modeled spectrum for isotropic turbulence with R_t =900. Also shown for comparison are the length q/ω in Fig. 17.2.1a, and the wave numbers, $1/L_{11,1}$ and $1/\eta$ in Fig. 17.2.1b.

Taylor (1935) defined a microscale, λ, such that

$$\frac{1}{\lambda^2} = \lim_{r_2 \to 0}\left(\frac{1-\dfrac{R_{11}(0,r_2,0)}{R_{11}(0,0,0)}}{r_2^2}\right), \tag{17.2.5}$$

where x_1 is in the direction of the flow. $R_{11}(0,r_2,0)$ is symmetric in r_2, so that it may be expressed as a series in r_2^2. Equation 17.2.5 may alternatively be written as

$$\frac{R_{11}(0,r_2,0)}{R_{11}(0,0,0)} = 1-\left(\frac{r_2}{\lambda}\right)^2 + O(r_2^4)\ . \tag{17.2.6}$$

$R_{11}(0,r_2,0)/\overline{u_1u_1}$ (consistent with the example in Fig. 17.2.1) and the limiting parabola which defines λ are shown in Fig. 17.2.2. The Taylor microscale is, in general, an intermediate length scale, larger than the Kolmogorov scale, η, and smaller than the length scale $L_{11,2}$ – the area under the $R_{11}(0,r_2,0)/\overline{u_1u_1}$ profile.

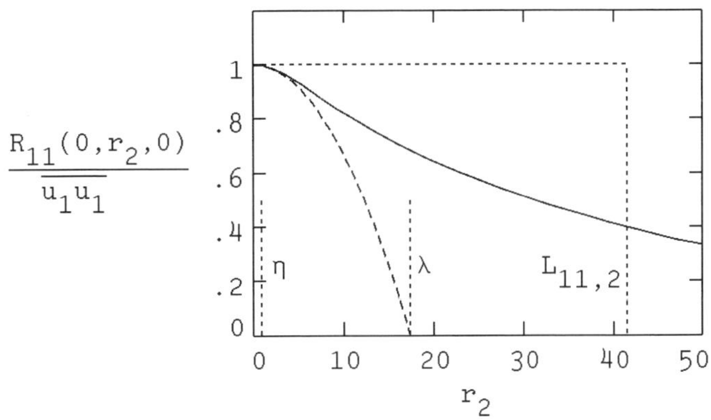

Fig. 17.2.2. Relation between the transverse two-point correlation and the Taylor microscale, λ. Based on the spectrum model for isotropic turbulence with R_t=900 used in Fig. 17.2.1.

Noting that (Appendix EQ.17.1.3)

$$\left(\frac{\partial^2 R_{ij}}{\partial r_k \partial r_n}\right)_{\mathbf{r}=0} = -\overline{\frac{\partial u_i}{\partial x_k}\frac{\partial u_j}{\partial x_n}} , \tag{17.2.7}$$

λ may be evaluated from

$$\lambda = \sqrt{\frac{2\overline{u_1u_1}}{\overline{\frac{\partial u_1}{\partial x_2}\frac{\partial u_1}{\partial x_2}}}} , \tag{17.2.8}$$

and a corresponding Reynolds number is defined by

$$R_\lambda = \frac{\lambda\sqrt{\overline{u_1u_1}}}{\nu} . \tag{17.2.9}$$

We may also define a longitudinal microscale,

$$\lambda_1 = \sqrt{\frac{\overline{u_1u_1}}{\overline{\frac{\partial u_1}{\partial x_1}\frac{\partial u_1}{\partial x_1}}}} . \tag{17.2.10}$$

For isotropic turbulence, Eq. 17.1.27 yields

$$\lambda_1 = \lambda \ , \tag{17.2.11}$$

and λ is related to the dissipation by

$$\varepsilon = \nu \overline{\frac{\partial u_i}{\partial x_k} \frac{\partial u_i}{\partial x_k}} = 15\nu \frac{\overline{u_1 u_1}}{\lambda^2} \ . \tag{17.2.12}$$

17.3 Models for E(K) Spectra

The Universal Range

Kolmogorov (1941) pioneered the concept of a universal range in the spectrum of very large Reynolds number turbulence. His analyses were made considering two-point correlations. Kolmogorov's concepts and hypotheses may be translated to turbulence spectra (see e.g., Batchelor, 1953, Chapter 6).

Consider turbulence which is locally homogeneous, locally isotropic, and locally steady. The word 'locally' has its basis in Kolmogorov's consideration of the difference in fluctuating velocities over a restricted domain with, say, a size much less that the integral length scale of the turbulence. In terms of the spectrum tensor, $\Phi_{ij}(\mathbf{K},t)$, 'locally' means at high wave numbers. This locally homogeneous, locally isotropic, and locally steady turbulence then satisfies (Eq. 9.3.6)

$$\Phi_{ij}(\mathbf{K},t) = \frac{E(K)}{4\pi K^2}\left(\delta_{ij} - \frac{K_i K_j}{K^2}\right) \ , \quad \text{for} \quad K >> 1/L_{11,1} \, . \tag{17.3.1}$$

Reviewing the differential equation for Φ_{ij}, Eq. 9.2.8, local isotropy is likely to occur. The terms labeled turbulent diffusion and slow pressure-strain, will be relatively larger at high wave numbers, than the terms labeled production, Coriolis and rapid pressure-strain, which are independent of the magnitude of the wave number. Since the dissipation term is proportional to K^2, most of the dissipation will occur at high wave numbers, and this region of the spectrum is likely to be locally isotropic.

Kolmogorov's first hypothesis suggests that at high wave numbers there is a locally isotropic region of the spectrum in which all the dissipation occurs and there is a universal high wave-number spectrum (at very high R_t), which is independent of the low wave-number turbulence and the turbulence kinetic energy. For these high wave numbers, $E(K)$ depends only on K, ε and ν. Based on the dimensions of $E(K)$ we then have

$$E(K) = \alpha \varepsilon^{1/4} \nu^{5/4} f(K\eta) \quad \text{for} \quad K >> 1/L_{11,1}, \tag{17.3.2}$$

where η is the Kolmogorov length scale, Eq. 17.2.1, and α is a dimensionless constant. Kolmogorov's second hypothesis suggests that this universal spectrum has an inertial sub-range (at its lower wave-number end) in which there is no dissipation, so that correspondingly $E(K)$ is independent of ν. Writing

$$E(K) = \alpha \varepsilon^{1/4} \nu^{5/4} (K\eta)^a = \alpha \varepsilon^{1/4} \nu^{5/4} \left(\frac{K \nu^{3/4}}{\varepsilon^{1/4}} \right)^a , \tag{17.3.3}$$

this is independent of ν for $a = -5/3$, so that in this inertial sub-range,

$$E(K) = \alpha \varepsilon^{2/3} K^{-5/3} \quad \text{for} \quad 1/\eta >> K >> 1/L_{11,1} . \tag{17.3.4}$$

Several authors have proposed models, which extend the inertial sub-range model, to cover the entire universal range. For the most part these take the form

$$E(K) = \alpha \varepsilon^{2/3} K^{-5/3} \exp\left(-\beta (K\eta)^n\right) \quad \text{for} \quad K >> 1/L_{11,1} , \tag{17.3.5}$$

although some authors (e.g. Aupoix, Eq. 6.2.9) have modified the exponential. In general, values for n ranging from 1 to 2 are used (Martínez et al., MCDKWZ1997). The coefficient β is determined from self consistency, so that the model returns the turbulence dissipation, ε (Eq. 17.1.9). For high Reynolds numbers, β is obtained from

$$\varepsilon = 2\nu \int_0^\infty K^2 E(K) dK = 2\nu \int_0^\infty K^2 \alpha \varepsilon^{2/3} K^{-5/3} \exp\left(-\beta (K\eta)^n\right) dK . \tag{17.3.6}$$

Pao (1965) based his model on the argument that the energy cascade should take the form

$$\sigma E = \int_K^\infty 2\nu K^2 E dK \quad \text{or} \quad \frac{d\sigma E}{dK} = -2\nu K^2 E. \tag{17.3.7}$$

He took the spectral element cascade rate, σ, to be independent of ν so that the dimensions of σ yield $\sigma = C \varepsilon^{1/3} K^{5/3}$. Equations 17.3.5 and 17.3.7 then give $n = 4/3$, and from Eq. 17.3.6, $\beta = 3\alpha/2$ (and $C = \alpha^{-1}$). Pao's model is then (Appendix EQ.17.3.8)

$$E(K) = \alpha \varepsilon^{2/3} K^{-5/3} \exp\left(-1.5\alpha (K\eta)^{4/3}\right), \quad \text{for} \quad K >> 1/L_{11,1} . \tag{17.3.8}$$

This model is shown in Fig. 17.3.1, together with the range in $K\eta$ over which the dissipation occurs. The –5/3rds region extends to about $K\eta = 0.1$, although 11 percent of the dissipation occurs at lower $K\eta$. The dissipation range, where 98 percent of the dissipation occurs, covers about two decades in $K\eta$ from about 0.016 to 1.6.

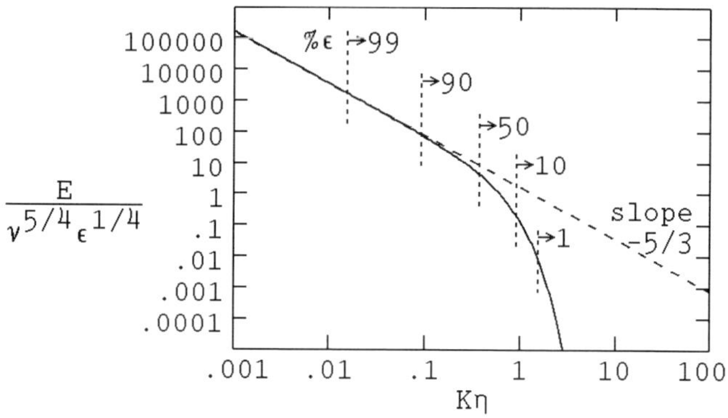

Fig. 17.3.1. The universal range for $E(K)$ at very high R_t. —— Pao's model, Eq. 17.3.8, with α=1.7. — — — inertial sub-range model, Eq. 17.3.4. --- $K\eta$ above which specified amounts of the dissipation occur.

Low Wave Numbers

At $K = 0$, $E(K)$ is also zero, since we require the spatial average of u_i to be zero for homogeneous turbulence. $E(K)$ is not a directly measurable quantity, so that the variation near $K = 0$ is determined from analyses. Synge and Lin (1943, see Saffman, 1967) found $E(K)$ varied as K^2 near $K = 0$. Lin (1947), using an analytic method, found $E(K)$ varied as K^4. Subsequent work has continued to find either a K^2 or a K^4 variation. For example, Chasnov (1995), using numerical simulations, obtained either K^2 or K^4, depending on the impulsive forces used to start the turbulence.

Von Kármán (1948), agreeing with the analysis of Lin (1947), proposed an interpolation formula for high Reynolds number turbulence, combining the K^4 variation at low K and the $K^{-5/3}$ variation at higher K,

$$E(K) = CkL_{vk}\frac{(KL_{vk})^4}{\left(1+(KL_{vk})^2\right)^{17/6}}, \quad \text{for } K << 1/\eta, \tag{17.3.9}$$

where L_{vk} is a length scale of the turbulence. Integrating $E(K)$ to return the turbulence kinetic energy at very high Reynolds number gives (Appendix EQ.17.3.10)

$$C = 2\Gamma(17/6)/\left(\Gamma(5/2)\Gamma(1/3)\right) = 0.9685. \tag{17.3.10}$$

This model is shown in Fig. 17.3.2, together with the range in KL_{vk} of the energy containing eddies.

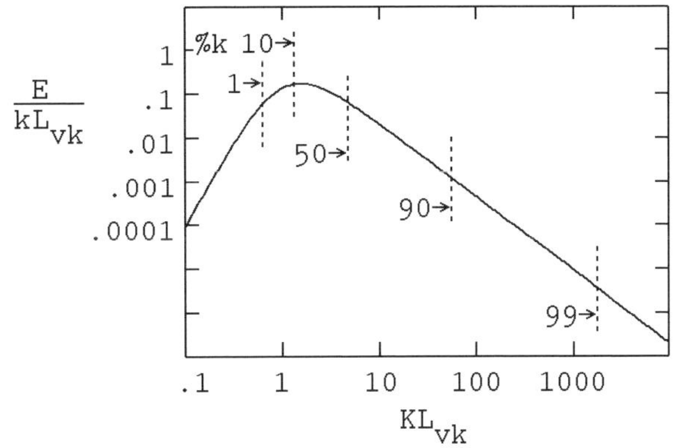

Fig. 17.3.2. The von Kármán spectrum for $E(K)$ at very high R_t, Eq. 17.3.9. --- KL_{vk} below which specified fractions of the turbulence energy occur.

Full Models for E(K)

Pope (2000) has proposed a model for $E(K)$ covering all wave numbers. It is based on the universal $K^{-5/3}$ spectrum, with modification factors for low and high K. It may be written as

$$E(K) = \underbrace{\left(\alpha \varepsilon^{2/3} K^{-5/3}\right)}_{-5/3\text{ region}} \underbrace{\left(\frac{(KL_p)^{5/3}(KL_p)^m}{\left(1+(KL_p)^2\right)^{5/6+m/2}} \right)}_{\text{low } K \text{ factor}} \underbrace{\exp\left(-\beta\left[(K\eta)^4 + c_\eta^4\right]^{1/4} - c_\eta\right)}_{\text{high } K \text{ factor}} . \qquad (17.3.11)$$

α, m, and β are fixed model constants, while the length L_p and the dimensionless parameter c_η are to be adjusted so that the model returns the desired turbulence kinetic energy and dissipation. The low K factor with $m = 4$ is consistent with the von Kármán model. The high K factor is of the form of Eq. 17.3.5 with $n = 1$ when $c_\eta = 0$.

In choosing which model constants to keep fixed, and which should vary with Reynolds number, Pope has defined how he thinks the spectrum should change as Reynolds number is reduced. Pope chose $\alpha = 1.5$, $m = 2$, and $\beta = 5.2$, and his model, over a range of turbulence Reynolds numbers, is shown in Fig. 17.3.3. Also shown are markers for 5 percent and 95 percent of both the turbulence energy and the dissipation. At $R_t = 10{,}000$, there is only a small overlap in the energy and dissipation ranges for $K\eta$, so that this Reynolds number has the separation of scales described by Kolmogorov and a long $K^{-5/3}$ region. At $R_t = 100$, the short $K^{-5/3}$ region corresponds to the overlap region for turbulence energy and dissipation. At $R_t = 1$, there is no $K^{-5/3}$ region. For these three Reynolds numbers, the values of L_p/η are 406, 19, and 2, respectively (Appendix EQ.17.3.11).

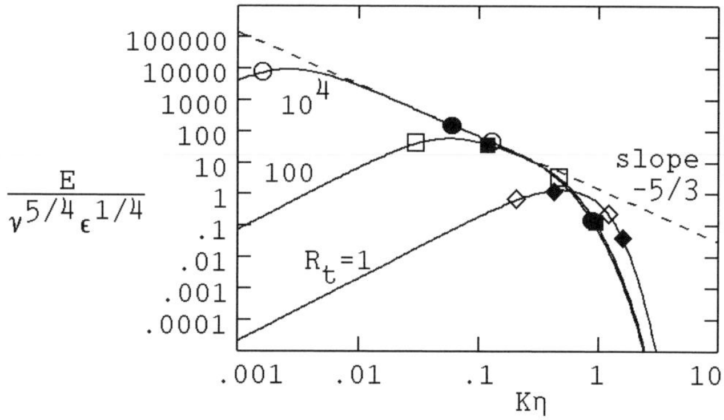

Fig. 17.3.3. Pope's model for $E(K)$, Eq. 17.3.11, for R_t=10,000, 100, 1.
○, □, ◇, mark $K\eta$ for 5% to 95% of the turbulence energy.
●, ■, ◆, mark $K\eta$ for 5% to 95% of the turbulence dissipation.

Exercise 17.3. Show that the consistency requirements, for $E(K)$ models to return both k and ε, are, in dimensionless form,

$$\int_0^\infty \left(\frac{E(K)}{\varepsilon^{1/4}\nu^{5/4}} \right) d(K\eta) = \sqrt{R_t} \quad \text{and} \quad \int_0^\infty (K\eta)^2 \left(\frac{E(K)}{\varepsilon^{1/4}\nu^{5/4}} \right) d(K\eta) = \frac{1}{2} .$$

17.4 Longitudinal Spectra for Nearly Isotropic Turbulence

For isotropic turbulence, the models for $E(K)$ in the previous section may be directly converted into one-dimensional spectra. The converted models for $E_{11}(K_1)$ may also be used for nearly isotropic turbulence, provided care is taken to ensure that the models are consistent and integrate to give $\overline{u_1 u_1}$. Since $E(K)$ is not measured, in this section, we evaluate $E(K)$ models by comparing their corresponding $E_{11}(K_1)$ distributions with experimental data.

von Kármán's Model

Integrating the isotropic turbulence relation, Eq. 17.1.21, using von Kármán's model, Eq. 17.3.9, yields (Appendix EQ.17.4.1-3)

$$E_{11}(K_1) = \frac{18}{55} C k L_{vk} \left(1 + (L_{vk} K_1)^2 \right)^{-5/6} . \tag{17.4.1}$$

Rewriting this equation in terms of $\overline{u_1 u_1}$ instead of k, and the longitudinal length scale, $L_{11,1}$ instead of L_{vk}, gives

$$E_{11}(K_1) = \frac{2}{\pi}\overline{u_1 u_1} L_{11,1}\left(1 + A^2 (L_{11,1}K_1)^2\right)^{-5/6} \quad \text{for} \quad K_1 << 1/\eta, \tag{17.4.2}$$

where

$$A^2 = \left(\frac{L_{vk}}{L_{11,1}}\right)^2 = \left(\frac{\Gamma(1/3)}{\Gamma(5/6)}\frac{1}{\sqrt{\pi}}\right)^2 \approx 1.8 \text{ and } L_{11,1} \approx 0.75 L_{vk}\,. \tag{17.4.3}$$

In this form, the model is internally consistent, so that integrating $E_{11}(K_1)$, the model returns $\overline{u_1 u_1}$ whether or not the turbulence is isotropic.

Bearman (B1972) found that his 1-d spectra measurements of turbulence downstream of a grid agreed well with von Kármán's model. As shown in Fig. 17.4.1a, the data exhibits both the flat region at low $K_1 L_{11,1}$ and an extensive –5/3rds region at high $K_1 L_{11,1}$, consistent with the model. Also shown in Fig. 17.4.1 are measurements Bearman made upstream of his blunt leading edge (Fig. 16.3.3). The data has been normalized with the local values of $\overline{u_1 u_1}$ and $L_{11,1}$. The data farther from the stagnation point, Fig. 17.4.1b, agrees well with von Kármán's model, while the data closer to the body, Fig. 17.4.1c, shows the spectra dropping off after a shorter –5/3rds region.

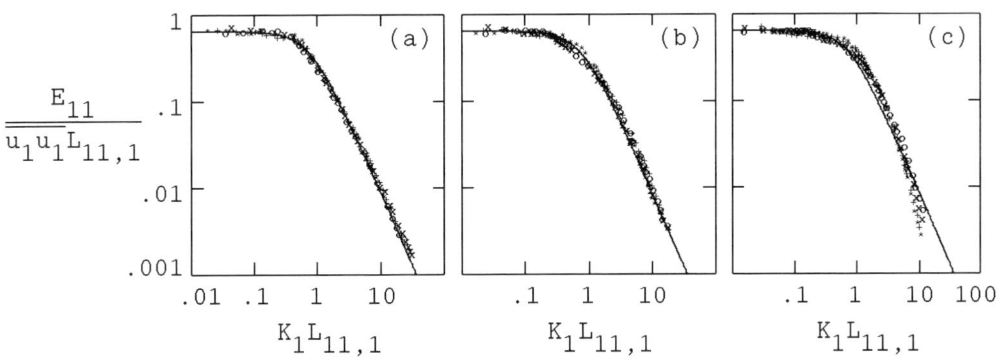

Fig. 17.4.1. One-dimensional longitudinal turbulence spectra. Comparison of the von Kármán model with data of Bearman (B1972). —— model, Eq. 17.4.2; Data: *,+,x,o, turbulence grids A,B,C,D, respectively. (a) At stagnation point, without the body. With body, Fig. 16.3.3: (b) x_1/D=0.624; (c) x_1/D=0.183. $g \geq 0.8$ for all the data sets.

Estimating the Dissipation from Spectrum or Length-Scale Data

At very high Reynolds number, we may obtain the –5/3rds part of $E_{11}(K_1)$ using $E(K) = \alpha\varepsilon^{2/3}K^{-5/3}$ (Eq. 17.3.4), with the isotropic relation, Eq. 17.1.21, to obtain

$$E_{11}(K_1) = \frac{18}{55}\alpha\varepsilon^{2/3}K_1^{-5/3} \quad \text{for} \quad 1/\eta >> K_1 >> 1/L_{11,1}. \tag{17.4.4}$$

If the spectrum for a data set is available, Eq. 17.4.4 may be used with a data point in the –5/3rds region to estimate the dissipation rate, ε. For the Bearman data in Fig. 17.4.1c, this method of estimating ε gives $R_t \sim 380$ to 1100 (using $\alpha = 1.7$).

Alternatively, equating Eq. 17.4.4 with the –5/3rds region of von Kármán's model, Eq. 17.4.2, yields a useful relation which may be used to estimate the dissipation rate from the integral length scale (Appendix Eq. 17.4.5),

$$\varepsilon = \left(\frac{55}{9\alpha\pi}\right)^{3/2}(1.8)^{-5/4}\frac{\overline{u_1u_1}^{3/2}}{L_{11,1}} \approx 0.59\frac{\overline{u_1u_1}^{3/2}}{L_{11,1}} \quad (\text{using } \alpha = 1.7)\ . \tag{17.4.5}$$

Applying this to Bearman's data, in Figs. 17.4.1a,b, gives $R_t \sim 880$ to 10640. The fall of the spectra in Fig. 17.4.1c is then consistent with the lower estimates of turbulence Reynolds number.

Obtaining Consistent 1-d Spectrum Models

Pao's model (Eq. 17.3.8) may be substituted into the isotropic relation, Eq. 17.1.21, and integrated numerically (Pao tabulated his results) to give a 1-d longitudinal spectrum corresponding to the 3-d spectrum. Several authors (e.g. Pao, 1965, Chapman, 1979, and Saddoughi and Veeravalli, SV1994) have compared data for 1-d longitudinal spectra with the result. To do so however, the dissipation, ε, is needed so that η may be evaluated. If the dissipation is isotropic, ε may be found from the longitudinal spectrum using Eq. 17.1.28. Pope's model for $E(K)$, Eq. 17.3.11, contains parameters which also depend on the turbulence Reynolds number, R_t. This too may be found from the longitudinal spectrum if the turbulence is isotropic.

To emphasize that these are assumptions, and to use these models generally for non-isotropic turbulence, as well, we define the parameters,

$$\varepsilon_1 = 15\nu\overline{\frac{\partial u_1}{\partial x_1}\frac{\partial u_1}{\partial x_1}} = 15\nu\int_0^\infty K_1^2 E_{11}(K_1)dK_1, \tag{17.4.6}$$

$$\eta_1 = \nu^{3/4}/\varepsilon_1^{1/4}, \tag{17.4.7}$$

$$R_{t1} = \frac{\left(3\overline{u_1u_1}/2\right)^2}{\nu\varepsilon_1} = \frac{\left(\frac{3}{2}\int_0^\infty E_{11}(K_1)dK_1\right)^2}{\nu\varepsilon_1}, \tag{17.4.8}$$

$$R_{\lambda 1} = \frac{\lambda_1 \sqrt{\overline{u_1 u_1}}}{\nu} = \sqrt{\frac{20}{3} R_{t1}}\,, \qquad (17.4.9)$$

where λ_1 is defined in Eq. 17.2.10. These parameters may be evaluated from the longitudinal spectrum itself, and will equal ε, η, R_t and R_λ, respectively, when the turbulence is isotropic.

Now, if we convert an $E(K)$ model using the isotropic relation, Eq. 17.1.21, and use the values ε_1 and R_{t1}, for ε and R_t, the resulting model for $E_{11}(K_1)$ will be self consistent, and will give

$$\overline{u_1 u_1} = \int_0^\infty E_{11}(K_1)dK_1 \quad \text{and} \quad \overline{\frac{\partial u_1}{\partial x_1}\frac{\partial u_1}{\partial x_1}} = \int_0^\infty K_1^2 E_{11}(K_1)dK_1\,, \qquad (17.4.10)$$

whether or not the turbulence is actually isotropic.

Models of Pao, Pope and Variations

Four models for $E_{11}(K_1)$ are considered in this section. The models are listed in Table 17.4.1 in a common form in terms of dimensionless parameters. Models 1 and 2 use the Pao and Pope $E(K)$ models, respectively, given in Eqs. 17.3.8 and 17.3.11. Model 3 extends the Pao model to low Reynolds numbers by using Pope's low K factor. Consistent with Pao's recommendation, the dissipation in the high Reynolds number model becomes an effective dissipation at low R_t. In essence this introduces R_t dependence for both the α and β parameters in the model. Model 4 is a modified version of Pope's model. Instead of keeping the parameter α fixed and letting c_η vary with R_t, c_η is kept fixed and α varies with R_t.

Grant et al. (GSM1962) measured the longitudinal spectra for turbulence in a tidal channel and reported the results in tabular form. The data is for very high R_t, and is shown in Fig. 17.4.2 with the Pao and Pope models. Both the data and models show a long –5/3rds region in the spectrum before falling more rapidly beyond $K_1\eta_1 = 0.1$. In the –5/3rds region, the ratio of the modeled values is 1.7/1.5, the ratio of values of α used in the models. This difference is nearly indistinguishable on the plot.

To quantitatively determine α, compensated spectra are plotted (see Eq. 17.4.4),

$$\frac{55}{18}\frac{E_{11}(K_1)}{\varepsilon_1^{2/3} K_1^{-5/3}} = \text{"}\alpha\text{"}\,, \qquad (17.4.11)$$

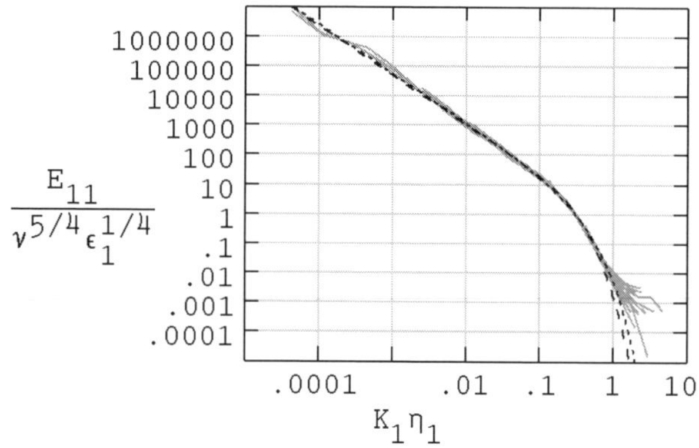

Fig. 17.4.2. E_{11} at high R_t. Comparison of models with data. ——— data of Grant et al. (GSM1962) in a tidal channel, 16 data sets (3Oct59, 1Feb60, 2Feb60, normalized using given values for ε and ν). Models in Table 17.4.1: --- model 1, Pao; — — — model 2, Pope, $R_{t1}=10^8$.

Table 17.4.1. Models for $E_{11}(K_1)$.

Integrate numerically Eq. 17.1.21, $E_{11}(K_1) = \int_{K_1}^{\infty} \frac{E(K)}{K}\left(1 - \frac{K_1^2}{K^2}\right) dK$,

using $E(K) = \alpha \varepsilon_1^{2/3} K^{-5/3} f_{low} f_{high}$,

$$f_{low} = \left(\frac{(K\eta_1)^{5/3+m}}{\left(c_a + (K\eta_1)^2\right)^{5/6+m/2}} \right), \qquad f_{high} = \exp\left(-\beta\left[(K\eta_1)^n + c_\eta^n \right]^{b/n} - c_\eta^b \right).$$

Model 1: Pao, high R_t only, returns dissipation only,
$\alpha = 1.7,\ c_a = 0,\ \beta = 1.5\alpha = 2.55,\ c_\eta = 0,\ b = 4/3$.

Model 2: Pope, with recommended values,
$\alpha = 1.5,\ m = 2,\ \beta = 5.2,\ n = 4,\ b = 1$,
$c_a = c_a(R_{t1})$ and $c_\eta = c_\eta(R_{t1})$ determined to satisfy Eq. 17.4.10.
At high R_t, $c_\eta \approx 0.4$; $c_\eta > 0.4$ at low R_t.

Model 3: Pao, extended to low R_t,
$m = 2,\ c_\eta = 0,\ b = 4/3,\ \alpha = 1.7\left(\varepsilon_{eff}/\varepsilon_1\right)^{2/3},\ \beta = 2.55\left(\varepsilon_{eff}/\varepsilon_1\right)^{-1/3}$,
$c_a = c_a(R_{t1})$ and $\left(\varepsilon_{eff}/\varepsilon_1\right) = \left(\varepsilon_{eff}/\varepsilon_1\right)(R_{t1})$ determined to satisfy Eq. 17.4.10.
At high R_t, $\varepsilon_{eff}/\varepsilon_1 = 1$; $\varepsilon_{eff}/\varepsilon_1 > 1$ at low R_t.

Model 4: Pope, modified,
$m = 2,\ \beta = 5.2,\ n = 4,\ c_\eta = 0.35,\ b = 1$,
$\alpha = \alpha(R_{t1})$ and $c_a = c_a(R_{t1})$ determined to satisfy Eq. 17.4.10.
At high R_t, $\alpha \approx 1.65$; $\alpha > 1.65$ at low R_t.

so that in the –5/3rds region the plot shows α. Figure 17.4.3 shows the compensated spectra for the data and models in Fig. 17.4.2. Within the scatter of the data, both models are acceptable.

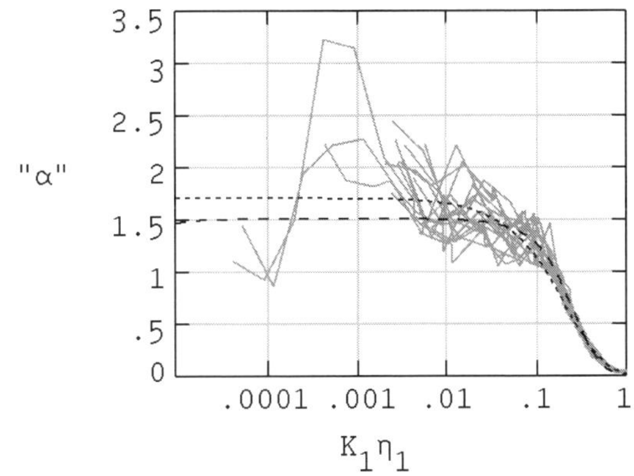

Fig. 17.4.3. Compensated spectra, "α", Eq. 17.4.11. Data and models as for Fig. 17.4.2.

It is interesting to note that the high wave-number factors in the models begin to reduce "α" before $K_1\eta_1 = 0.1$, the apparent end of the –5/3rds region in Fig. 17.4.2. To see the effect of the low and high wave-number factors on "α", the compensated spectra for model 4 in Table 17.4.1 (modified Pope) are shown in Fig. 17.4.4. As R_λ decreases, below say about 300, the peak values of the compensated spectra also decrease.

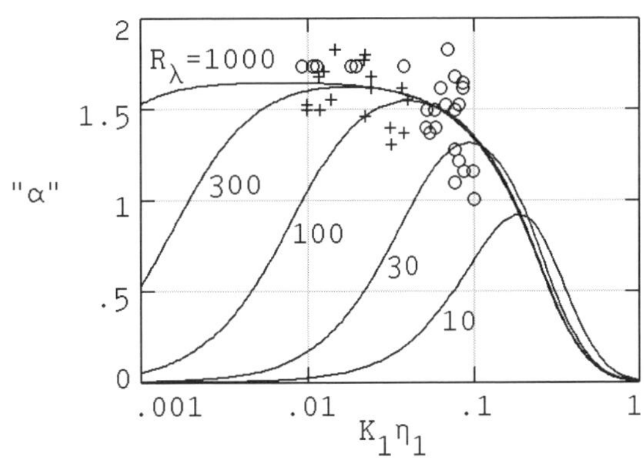

Fig. 17.4.4. Compensated spectra, "α", Eq. 17.4.11. —— model 4, Table 17.4.1 (Pope modified), for $R_{\lambda 1}$ =10, 30, 100, 300, 1000 (R_{t1}=15 to 150,000). Data from Sreenivasan's (1995) compilation (Tables I and II): o, grid turbulence; +, boundary layers. Data points plotted at $K_1\eta_1$ corresponding to location of maximum "α" in the model for the specified R_λ.

Sreenivasan (1995) compiled data from many sources, estimating the Kolmogorov constant, α, from the peak (or at high R_λ, an average) of the compensated spectra. Data from his Tables I and II are shown in Fig. 17.4.4 and cover the range R_λ =28 to 669. Each data point is plotted at the $K_1\eta_1$ location corresponding to the maximum "α" in the model for the specified R_λ. The data forms an envelope around the peak modeled values. Additional data from

Sreenivasan's Tables III and IV, for $R_\lambda > 300$, give an overall average value of $\alpha = 1.64$.

Stewart and Townsend (1951) measured (and tabulated) longitudinal spectra downstream of turbulence grids. Figure 17.4.5 shows the data normalized using ε_1 calculated by integrating the data (Eq. 17.4.6). $R_{\lambda 1}$ varied between 15 and 38 (R_{t1} =36 to 218). Figure 17.4.5a shows the compensated spectra, plotted to emphasize the dissipation region, and in particular the model parameter β, which controls the slope of the decay of log("α") with $K_1\eta_1$. Pope's model which uses $\beta = 5.2$ gives good results while Pao's model has a smaller slope.

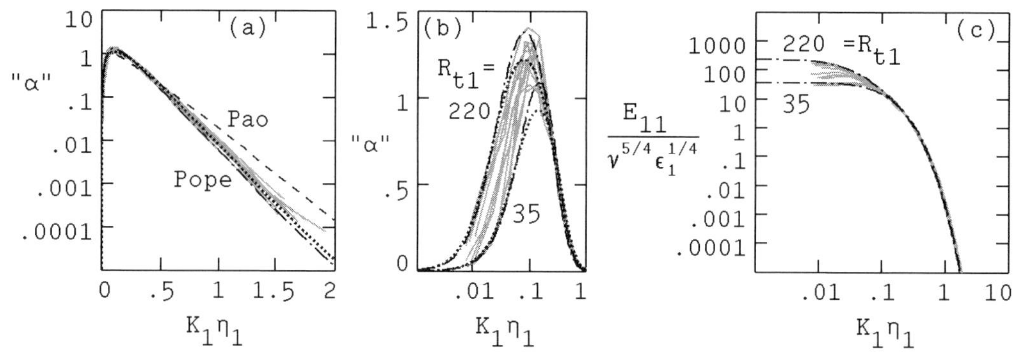

Fig. 17.4.5. Comparison of models for E_{11} with the data of Stewart and Townsend (1951). —— data, 10 cases, R_{t1}=36,,,218 ($R_{\lambda 1}$=15,,,38). Models, Table 17.4.1: ······ model 2 (Pope); - - - - model 3 (Pao extended); — - — model 4 (Pope modified). (a) Compensated spectra, "α", Eq. 17.4.11, models 2, 3 and 4 at R_{t1}=100. (b) "α", models 2 and 4 at R_{t1}=35 and 220. (c) Normalized E_{11}, model 4 at R_{t1}=35 and 220.

Figure 17.4.5b shows the compensated spectra with an emphasis on the region where both energy and dissipation are significant, near the maximum of "α". The modified Pope model (model 4), which allows α to vary with R_t, has higher peak values of "α" consistent with the data. Overall, the spectra from model 4 agree well with the data, as shown in Fig. 17.4.5c.

Figure 17.4.6 shows the same comparisons, only this time with the tabulated grid turbulence data from Comte-Bellot and Corrsin (1971). Again model 4, the modified Pope model, agrees well with the data. In the next section, the $E(K)$ model which is the basis of model 4, will be used as the basis for the energy distribution in a model for the spectrum tensor.

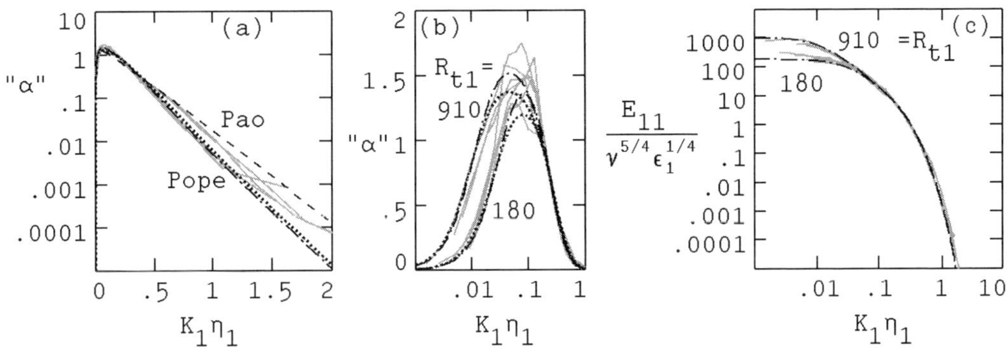

Fig. 17.4.6. Comparison of models for E_{11} with the data of Comte-Bellot and Corrsin (1971). —— data, 7 cases, R_{t1}=178,,,908 ($R_{\lambda 1}$=34,,,78). Models, Table 17.4.1: ······ model 2 (Pope); ---- model 3 (Pao extended); —·— model 4 (Pope modified). (a) Compensated spectra, "α", Eq. 17.4.11, models 2, 3 and 4 at R_{t1}=400. (b) "α", models 2 and 4 at R_{t1}=180 and 910. (c) Normalized E_{11}, model 4 at R_{t1}=180 and 910.

17.5 Modeling the Spectrum Tensor

Models in the Literature

Batchelor's model of the spectrum tensor for isotropic turbulence, Eq. 9.3.6,

$$\Phi_{ij} = \frac{E(K)}{4\pi K^2}\left(\delta_{ij} - \frac{K_i K_j}{K^2}\right), \tag{17.5.1}$$

has made several contributions to this book. It has been used to initialize rapid distortion theory calculations (Section 9.3), and to determine isotropic limits for rapid pressure-strain (Eq. 9.4.7). In the previous section it was used to convert models of 3-d spectra into 1-d spectra for comparison with data. This model for Φ_{ij} is made complete by choosing a model for the 3-d spectrum, $E(K)$.

Reynolds (1989) proposed a model for the spectrum tensor for mildly anisotropic turbulence,

$$\begin{aligned}\Phi_{ij} = C_1(K)\left(\delta_{ij} - \frac{K_i K_j}{K^2}\right) + C_3(K)\left(b_{ij} - \frac{b_{in}K_n K_j + b_{jn}K_n K_i}{K^2} + \frac{b_{nm}K_n K_m K_i K_j}{K^4}\right) \\ + \gamma C_3(K)\left(\frac{b_{nm}K_n K_m}{K^2}\delta_{ij} - \frac{b_{nm}K_n K_m K_i K_j}{K^4}\right)\end{aligned}, \tag{17.5.2}$$

where $C_1(K)$ and $C_3(K)$ need to be chosen to satisfy the constraints

$$4\pi\int_0^\infty K^2 C_1(K)\,dK = k\ , \quad 4\pi\int_0^\infty K^2 C_3(K)\,dK = \frac{30}{7-2\gamma}k\ , \tag{17.5.3}$$

so that integrating Φ_{ij} returns $\overline{u_i u_j}$. γ is a constant to be determined, and b_{ij} is the anisotropy of the turbulence. Shih et al. (SRM1990) discuss the limitations of the model. In particular, the normal components of Φ_{ij} are positive, and therefore realizable, over all wave-number space for only limited anisotropy.

A Complete Spectrum Tensor Model

In this section we describe a way of modeling the spectrum tensor, which is generally applicable at all anisotropies. We will model the spectrum tensor in three parts. The first part is the determination of the energy at each radial value of K, i.e. the three-dimensional spectrum, $E(K)$. The second part is the determination of the anisotropy of the turbulence energy at each radial value of K. The final part specifies $\Phi_{ij}(\mathbf{K})$ at each K; i.e., we obtain an angular distribution of both the energy and the anisotropy on each spherical shell, so that both realizability ($\Phi_{ij} \geq 0$ for $i = j$) and continuity ($K_i\Phi_{ij} = K_j\Phi_{ij} = 0$) are satisfied at each point in wave-number space.

Model Part 1. From the isotropic relation, Eq. 17.5.1, we can extract a dimensionless tensor

Isotropic:
$$\Phi_{ij}^N = \left(\delta_{ij} - \frac{K_i K_j}{K^2}\right) . \tag{17.5.4}$$

This represents isotropic turbulence and the energy is the same at each angular position, $\Phi_{ii}^N = 2$. Integrating Eq. 17.5.4 over a spherical shell,

$$\int_0^{2\pi}\int_0^{\pi} \Phi_{ii}^N \sin\psi\, d\psi\, d\theta = 8\pi\ . \tag{17.5.5}$$

Therefore, writing a more general non-isotropic relation in the form

$$\Phi_{ij} = \frac{E(K)}{4\pi K^2}\Phi_{ij}^N , \tag{17.5.6}$$

we require, for consistency, that Φ_{ii}^{N} satisfy Eq. 17.5.5. The first part of the spectrum tensor model, the $E(K)$ distribution, is now separate from the remaining two parts which are contained in Φ_{ij}^{N}.

For $E(K)$ we use (the generalized form of) Model 4 in Table 17.4.1, which, with the fixed constants included, may be written as

$$\frac{E(K)}{\nu^{5/4}\varepsilon^{1/4}} = \alpha\left(\frac{(K\eta)^2}{\left(c_a + (K\eta)^2\right)^{11/6}}\right)\exp\left(-5.2\left[(K\eta)^4 + (0.35)^4\right]^{1/4} - 0.35\right). \qquad (17.5.7)$$

The dimensionless parameters $\alpha = \alpha(R_t)$ and $c_a = c_a(R_t)$ are determined to satisfy (Exercise 17.3)

$$\int_0^\infty \left(\frac{E(K)}{\varepsilon^{1/4}\nu^{5/4}}\right) d(K\eta) = \sqrt{R_t} \quad \text{and} \quad \int_0^\infty (K\eta)^2 \left(\frac{E(K)}{\varepsilon^{1/4}\nu^{5/4}}\right) d(K\eta) = \frac{1}{2}. \qquad (17.5.8)$$

This model, together with the isotropic relation, Eq. 17.5.1, was used to obtain the profiles in Figs. 17.1.2b, 17.2.1 and 17.2.2.

Model Part 2. Integrating Φ_{ij} to obtain the Reynolds stresses,

$$\overline{u_i u_j} = \int_0^\infty \left(\int_0^{2\pi}\int_0^{\pi} \Phi_{ij} \sin\psi d\psi d\theta\right) K^2 dK = \frac{1}{4\pi}\int_0^\infty \left(\int_0^{2\pi}\int_0^{\pi} \Phi_{ij}^{N} \sin\psi d\psi d\theta\right) E(K) dK, \qquad (17.5.9)$$

we define a dimensionless ratio, $\mathrm{B}_{ij}(K)$, to be modeled, such that $\mathrm{B}_{ii}(K) = 1$,

$$\mathrm{B}_{ij}(K) = \frac{\int_0^{2\pi}\int_0^{\pi} \Phi_{ij}^{N} \sin\psi d\psi d\theta}{\int_0^{2\pi}\int_0^{\pi} \Phi_{nn}^{N} \sin\psi d\psi d\theta} = \frac{\int_0^{2\pi}\int_0^{\pi} \Phi_{ij}^{N} \sin\psi d\psi d\theta}{8\pi}. \qquad (17.5.10)$$

B_{ij} describes the average anisotropy of the spectrum tensor on a spherical shell, K, like $\overline{u_i u_j}/(2k)$ describes the anisotropy of the Reynolds stress tensor.

Substituting back into Eq. 17.5.9, yields the relation that B_{ij} must satisfy so that the spectrum is consistent with the Reynolds stresses,

$$\overline{u_i u_j} = 2\int_0^\infty \mathrm{B}_{ij}(K)E(K)dK \ . \tag{17.5.11}$$

The anisotropy of the dissipation tensor, ε_{ij}, provides an additional constraint on B_{ij}. From Eq. 9.2.9,

$$\varepsilon_{ij} = 4\nu\int_0^\infty K^2 \mathrm{B}_{ij}(K)E(K)dK \, . \tag{17.5.12}$$

Using the dissipation model developed in Chapter 15 (Eqs. 15.1.5, 15.2.6-7), which approximated the anisotropy of the dissipation as being proportional to b_{ij},

$$\frac{\varepsilon_{ij}}{2\varepsilon} - \frac{1}{3}\delta_{ij} = \frac{c_{1\mathrm{d}}}{2} b_{ij} \ , \quad \frac{c_{1\mathrm{d}}}{2} = 1 - \left(\frac{\min(g,0.75)}{0.75}\right)\left(\frac{\tilde{R}_t^{3/4}}{125^{3/4} + \tilde{R}_t^{3/4}}\right) , \tag{17.5.13}$$

both the Reynolds stresses and the dissipation tensor may be returned by taking the anisotropy of the contribution to the turbulence at each K proportional to b_{ij},

$$\mathrm{B}_{ij}(K) - \frac{1}{3}\delta_{ij} = C_{\mathrm{B}}(K) b_{ij} . \tag{17.5.14}$$

Substituting into Eq. 17.5.11 yields

$$\overline{u_i u_j} = 2\int_0^\infty \left(C_{\mathrm{B}}(K) b_{ij} + \frac{1}{3}\delta_{ij}\right) E(K)dK \, , \tag{17.5.15}$$

which may be rewritten as

$$2\int_0^\infty C_{\mathrm{B}}(K)E(K)dK = 2k = 2\int_0^\infty E(K)dK \, , \tag{17.5.16}$$

so that the Reynolds stresses are returned when the $E(K)$ weighted average of C_{B} is one. Substituting Eq. 17.5.14 into the equation for ε_{ij}, Eq. 17.5.12, yields

$$\varepsilon_{ij} = 4\nu\int_0^\infty K^2 \left(C_{\mathrm{B}}(K) b_{ij} + \frac{1}{3}\delta_{ij}\right) E(K)dK \, . \tag{17.5.17}$$

Combining this with the modeled form, $\varepsilon_{ij} = c_{1d}\varepsilon b_{ij} + \frac{2}{3}\varepsilon\delta_{ij}$, Eq. 17.5.17 may be rewritten as

$$c_{1d}\varepsilon = 4\nu\int_0^\infty K^2 C_B(K)E(K)\,dK = c_{1d}2\nu\int_0^\infty K^2 E(K)\,dK, \tag{17.5.18}$$

so that ε_{ij} is returned when the $K^2E(K)$ weighted average of C_B is $c_{1d}/2$.

For $g = 0$ or $\tilde{R}_t = 0$, the model for c_{1d} (Eq. 17.5.13) gives $c_{1d}/2 = 1$. $C_B = 1$ then satisfies both Eqs. 17.5.16 and 17.5.18, and the anisotropy of the contributions to turbulence energy is the same over all spherical shells. Otherwise, $c_{1d}/2 < 1$. Equations 17.5.16 and 17.5.18 are then satisfied by $C_B < 1$ at high K and $C_B > 1$ at low K. This is consistent with the high wave-number part of the spectrum being more isotropic. Correspondingly, the low wave-number part of the spectrum is more anisotropic. Realizability, however, imposes limits. For the modeled Φ_{ij}'s to be realizable, we require that the normal components of B_{ij} are always greater than or equal to zero. This imposes an upper limit on C_B, which may be quantified by considering the b_{ij} corresponding to the smallest principal Reynolds stress, which we will call $b_{P_3P_3}$. In particular, from Eq. 17.5.14, we require that

$$C_B b_{P_3P_3} \geq -\frac{1}{3}, \qquad \text{or since } b_{P_3P_3} \leq 0, \qquad C_B \leq \frac{-1/3}{b_{P_3P_3}} \equiv C_{B,\max}\,. \tag{17.5.19}$$

The model chosen for C_B is

$$C_B = C_{B,0}\exp\left(-\beta_B(K\eta)^2\right), \tag{17.5.20}$$

where the parameters $C_{B,0}$ and β_B are essentially functions of both g and R_t, and are set to satisfy Eqs. 17.5.16 and 17.5.18. While this model form does not guarantee that realizability will be satisfied (we require $C_{B,0} \leq C_{B,\max}$), no problems have been encountered using the c_{1d} model in Eq. 17.5.13. Some examples of $C_B(K\eta)$ are shown in Fig. 17.5.1.

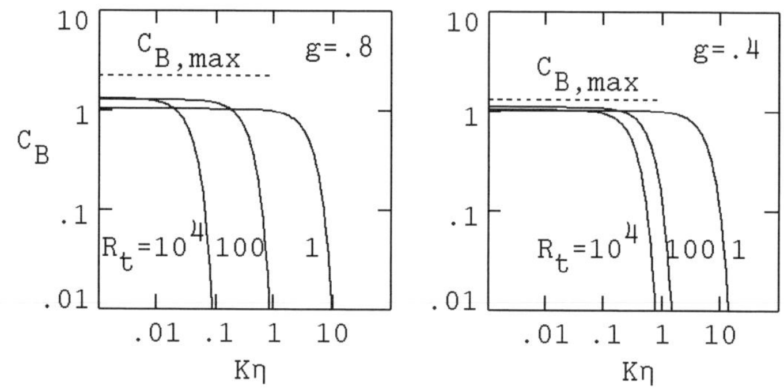

Fig. 17.5.1. Some model results for the anisotropy parameter, C_B: for g=0.8 (using b_{11}=$-b_{22}$=0.149) and for g=0.4 (using b_{11}=$-b_{22}$=0.258).

Model Part 3. Having chosen the anisotropy at each K, the final part of the model is to use B_{ij} to determine Φ_{ij}^N at each K.

Consider forming the spectrum tensor from a vector, $e_i(\mathbf{K},t)$, by writing

$$\Phi_{ij}^N = \overline{\left(e_i - \frac{K_i}{K}\frac{K_n}{K}e_n\right)\left(e_j - \frac{K_j}{K}\frac{K_m}{K}e_m\right)} . \tag{17.5.21}$$

This Φ_{ij}^N satisfies continuity, $K_i\Phi_{ij}^N = 0$, and $K_j\Phi_{ij}^N = 0$, and has the necessary symmetries, $\Phi_{ij}^N = \Phi_{ji}^N$, and $\Phi_{ij}^N(\mathbf{K}) = \Phi_{ij}^N(-\mathbf{K})$. Expanding,

$$\Phi_{ij}^N = \overline{e_ie_j} - \overline{e_je_n}\frac{K_iK_n}{K^2} - \overline{e_ie_m}\frac{K_jK_m}{K^2} + \overline{e_ne_m}\frac{K_iK_jK_nK_m}{K^4} . \tag{17.5.22}$$

Taking $\overline{e_ie_j} = \delta_{ij}$ returns Batchelor's model for isotropic turbulence,

$$\Phi_{ij}^N = \delta_{ij} - \frac{K_iK_j}{K^2} - \frac{K_jK_i}{K^2} + \frac{K_iK_j}{K^2} = \delta_{ij} - \frac{K_iK_j}{K^2} . \tag{17.5.23}$$

Now consider a tensor $\overline{e_ie_j}$ with the only non-zero component being $\overline{e_1e_1}$,

$$\overline{e_ie_j} = \delta_{i1}\delta_{j1}A_1^2 . \tag{17.5.24}$$

Equation 17.5.22 yields,

$$\Phi_{ij}^N = A_1^2\left(\delta_{i1}\delta_{j1} - \delta_{j1}\frac{K_iK_1}{K^2} - \delta_{i1}\frac{K_jK_1}{K^2} + \frac{K_iK_jK_1^2}{K^4}\right) , \tag{17.5.25}$$

or writing out all the terms

$$\Phi_{ij}^{N} = \frac{A_1^2}{K^4}\begin{bmatrix} (K^2-K_1^2)^2 & K_1K_2(K_1^2-K^2) & K_1K_3(K_1^2-K^2) \\ K_1K_2(K_1^2-K^2) & K_2^2K_1^2 & K_2K_3K_1^2 \\ K_1K_3(K_1^2-K^2) & K_2K_3K_1^2 & K_3^2K_1^2 \end{bmatrix}. \qquad (17.5.26)$$

Some glyphs for this Φ_{ij}^{N} distribution are shown in Fig. 17.5.2. The individual glyphs all represent 1-d contributions to the turbulence. At $K_1=0$, the contributions are to $\overline{u_1u_1}$. As K_1/K increases towards 1, the size of the glyphs decreases, representing less energy, and the fraction of the energy that contributes to $\overline{u_1u_1}$ also goes to zero. At $K_1=K$, $\Phi_{ij}^{N}=0$.

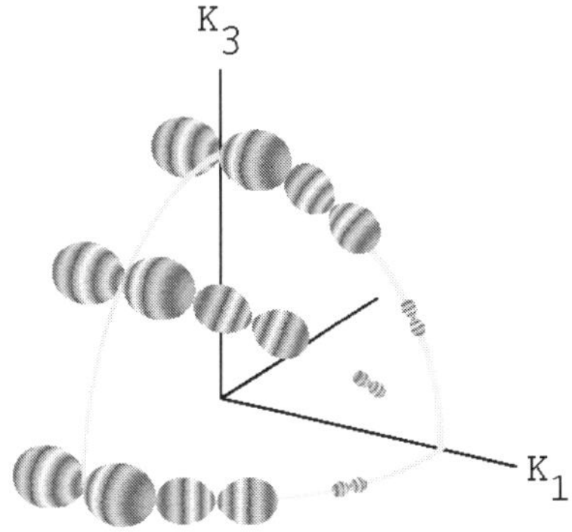

Fig. 17.5.2. Some glyphs for Φ_{ij}^{N} from Eq. 17.5.26.

Integrating Φ_{ij}^{N} over a spherical shell yields (using Appendix EQ.9.4.7, Eqs. 7-10)

$$\int_0^{2\pi}\int_0^{\pi}\Phi_{ij}^{N}\sin\psi d\psi d\theta = A_1^2\begin{bmatrix} 4\pi-2\frac{4\pi}{3}+\frac{4\pi}{5} & 0 & 0 \\ 0 & \frac{4\pi}{15} & 0 \\ 0 & 0 & \frac{4\pi}{15} \end{bmatrix} = \frac{4\pi}{15}A_1^2\begin{bmatrix} 8 & 0 & 0 \\ 0 & 1 & 0 \\ 0 & 0 & 1 \end{bmatrix}. \qquad (17.5.27)$$

There are two things to note about this result. First, the ratio of the largest to smallest normal component is limited to 8:1. A more complex basis function, which removes this limitation is described below. Second, there are no off-diagonal values in the integrated results. Therefore, if Eq. 17.5.25, and similar contributions using $\overline{e_ie_j}=\delta_{i2}\delta_{j2}A_2^2$ and $\overline{e_ie_j}=\delta_{i3}\delta_{j3}A_3^2$, are to be used to form Φ_{ij}^{N}, the analysis needs to be made in the principal coordinates of the Reynolds stress tensor (no off-diagonal stresses).

Working in principal coordinates (subscript P_i), we can attempt to form $\Phi^N_{P_iP_j}$ at each K using as the basis,

$$\overline{e_{P_i}e_{P_j}} = \delta_{i1}\delta_{j1}A_1^2 + \delta_{i2}\delta_{j2}A_2^2 + \delta_{i3}\delta_{j3}A_3^2 . \tag{17.5.28}$$

This yields

$$\Phi^N_{P_1P_1} = A_1^2\left(1 - 2\frac{K_{P_1}^2}{K^2} + \frac{K_{P_1}^2}{K^4}\right) + A_2^2\left(\frac{K_{P_1}^2 K_{P_2}^2}{K^4}\right) + A_3^2\left(\frac{K_{P_1}^2 K_{P_3}^2}{K^4}\right), \tag{17.5.29}$$

and similar expressions for $\Phi^N_{P_2P_2}$ and $\Phi^N_{P_3P_3}$. Integrating each normal component over the spherical surface, yields three equations,

$$\begin{aligned}
\int_0^{2\pi}\int_0^{\pi}\Phi^N_{P_1P_1}\sin\psi d\psi d\theta &= 8\pi \mathrm{B}_{P_1P_1} = \frac{32\pi}{15}A_1^2 + \frac{4\pi}{15}A_2^2 + \frac{4\pi}{15}A_3^2 \\
\int_0^{2\pi}\int_0^{\pi}\Phi^N_{P_2P_2}\sin\psi d\psi d\theta &= 8\pi \mathrm{B}_{P_2P_2} = \frac{4\pi}{15}A_1^2 + \frac{32\pi}{15}A_2^2 + \frac{4\pi}{15}A_3^2 \\
\int_0^{2\pi}\int_0^{\pi}\Phi^N_{P_3P_3}\sin\psi d\psi d\theta &= 8\pi \mathrm{B}_{P_3P_3} = \frac{4\pi}{15}A_1^2 + \frac{4\pi}{15}A_2^2 + \frac{32\pi}{15}A_3^2 ,
\end{aligned} \tag{17.5.30}$$

which may be solved for the three coefficients, A_1^2, A_2^2, A_3^2. If all of the coefficients are positive, the exercise has been successful; $\Phi^N_{P_iP_j}$ will be realizable and give the desired anisotropy at this K. If the turbulence is isotropic at this K, Batchelor's model is recovered.

If any of the three coefficients are negative, a more complex model is needed. Consider the basis function

$$\overline{e_ie_j} = \delta_{i1}\delta_{j1}\frac{K^2A_1^2}{(C_1K)^2 + ((1-C_1)K_1)^2}, \tag{17.5.31}$$

where $0 < C_1 < 1$. This provides a scaling factor (Fig. 17.5.3) for Φ^N_{ij} at each point (K_1, K_2, K_3), so that smaller values of C_1 result in a larger portion of the energy being near $K_1 = 0$. Therefore, from Fig. 17.5.2, a larger portion of the energy will contribute to $\overline{u_1u_1}$.

Fig. 17.5.3. Scaling factor in Eq. 17.5.31,
$f = \dfrac{1}{C_1^2 + (1-C_1)^2 K_1^2/K^2}$
normalized by the value at $K_1/K = 0$.

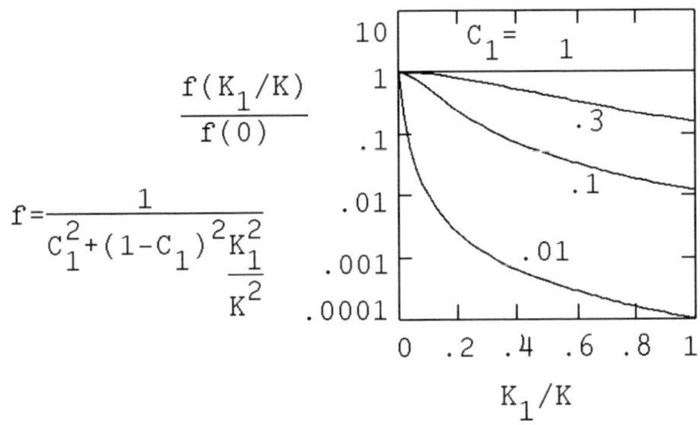

For this scaled basis function, the normal components of Φ_{ij}^N and their integrals are given by (Appendix EQ.17.5.31-36):

$$\Phi_{11}^N = \frac{A_1^2}{K^2\left((C_1K)^2 + ((1-C_1)K_1)^2\right)}\left(K^4 - 2K^2K_1^2 + K_1^4\right),$$

$$\Phi_{22}^N = \frac{A_1^2}{K^2\left((C_1K)^2 + ((1-C_1)K_1)^2\right)}\left(K_1^2K_2^2\right), \tag{17.5.32}$$

$$\int_0^{2\pi}\int_0^{\pi} \Phi_{11}^N \sin\psi d\psi d\theta = A_1^2 C_{cen} \quad \text{and} \quad \int_0^{2\pi}\int_0^{\pi} \Phi_{22}^N \sin\psi d\psi d\theta = A_1^2 C_{out}, \tag{17.5.33}$$

where

$$C_{cen} = \frac{4\pi}{(1-C_1)^2}\left(-\frac{5}{3} - \left(\frac{C_1}{1-C_1}\right)^2 + \left(\frac{1-C_1}{C_1} + 2\frac{C_1}{1-C_1} + \left(\frac{C_1}{1-C_1}\right)^3\right)\tan^{-1}\left(\frac{1-C_1}{C_1}\right)\right) \tag{17.5.34}$$

and

$$C_{out} = \frac{2\pi}{(1-C_1)^2}\left(\frac{2}{3} + \left(\frac{C_1}{1-C_1}\right)^2 - \left(\frac{C_1}{1-C_1} + \left(\frac{C_1}{1-C_1}\right)^3\right)\tan^{-1}\left(\frac{1-C_1}{C_1}\right)\right). \tag{17.5.35}$$

As C_1 approaches zero, C_{cen} approaches $2\pi^2/C_1$ and C_{out} approaches $4\pi/3$, so that the ratio between the largest and smallest integrated components,

$$\lim_{C_1 \to 0} \frac{\int_0^{2\pi}\int_0^{\pi} \Phi_{11}^N \sin\psi d\psi d\theta}{\int_0^{2\pi}\int_0^{\pi} \Phi_{22}^N \sin\psi d\psi d\theta} = \frac{3\pi}{2C_1} , \tag{17.5.36}$$

approaches infinity.

To complete the more complex model, to be used when Eq. 17.5.30 fails to produce a realizable solution, we use the convention that the principal directions are ordered so that

$$\mathrm{B}_{P_1P_1} \geq \mathrm{B}_{P_2P_2} \geq \mathrm{B}_{P_3P_3}, \tag{17.5.37}$$

and we take

$$\overline{e_{P_i}e_{P_j}} = \delta_{i1}\delta_{j1}\frac{K^2A_1^2}{(C_1K)^2 + \left((1-C_1)K_{P_1}\right)^2} + \delta_{i2}\delta_{j2}\frac{K^2A_2^2}{(C_1K)^2 + \left((1-C_1)K_{P_2}\right)^2} . \tag{17.5.38}$$

Integrating over the spherical surface and using Eq. 17.5.10,

$$\begin{aligned}
\int_0^{2\pi}\int_0^{\pi} \Phi_{P_1P_1}^N \sin\psi d\psi d\theta &= 8\pi \mathrm{B}_{P_1P_1} = C_{cen}A_1^2 + C_{out}A_2^2 \\
\int_0^{2\pi}\int_0^{\pi} \Phi_{P_2P_2}^N \sin\psi d\psi d\theta &= 8\pi \mathrm{B}_{P_2P_2} = C_{out}A_1^2 + C_{cen}A_2^2 \\
\int_0^{2\pi}\int_0^{\pi} \Phi_{P_3P_3}^N \sin\psi d\psi d\theta &= 8\pi \mathrm{B}_{P_3P_3} = C_{out}A_1^2 + C_{out}A_2^2,
\end{aligned} \tag{17.5.39}$$

yields three equations for essentially three unknowns, A_1^2, A_2^2, and C_1 which gives both C_{cen} and C_{out} (Eqs. 17.5.34-5). Note that this formulation has the following limiting symmetries: when $\mathrm{B}_{P_2P_2} = \mathrm{B}_{P_3P_3}$, Eq. 17.5.39 yields $A_2^2 = 0$, and when $\mathrm{B}_{P_1P_1} = \mathrm{B}_{P_2P_2}$, Eq. 17.5.39 yields $A_1^2 = A_2^2$.

This completes the definition of a spectrum tensor model, which is generally applicable for all turbulence anisotropies. Once $\Phi_{P_iP_j}^N$ is found at a point in wave-number space, coordinate transformation back to the original Cartesian system will give Φ_{ij}^N. The results may then be integrated over planes of constant K_1 to give the 1-d spectra, $E_{ij}(K_1)$.

17.6 1-d Spectra from the Spectrum Tensor Model

In this section we compare results from the new spectrum tensor model, described in the previous section, with data in the literature. Φ_{ij} was calculated over a 3-d grid in wave-number space and then integrated numerically over planes of constant K_1 (Fig. 17.1.1) to obtain the 1-d spectra, E_{ij}. The input parameters to the spectrum tensor model, b_{ij} and R_t, are listed in Table 17.6.1. In all cases, b_{ij} was obtained directly from the data, either from the Reynolds stresses as given by the authors, or from integrating the spectral data. For the experimental data where only $R_{\lambda 1}$, and correspondingly ε_1, was available, ε, and correspondingly R_t, was determined using the spectrum tensor model results for $\varepsilon_1/\varepsilon$. The cases are considered, and listed in the table, in order of increasing c_{1d}.

Table 17.6.1. Spectrum tensor model parameters. All cases: $b_{13} = b_{23} = 0$, $\tilde{R}_t = R_t$.

Case	b_{11}	b_{22}	b_{12}	R_t	g	c_{1d}	$\varepsilon_1/\varepsilon$	$C_{B,0}$	β_B	$C_{1,min}$
SVd	0.143	–0.108	–0.116	137,000	0.756	0.010	1.012	1.230	11,141	1
SVa	0.191	–0.147	–0.128	15,750	0.635	0.350	1.063	1.066	48.7	1
Tcon	–0.213	0.1065	0	1490	0.629	0.550	0.920	1.160	25.2	0.526
ISKa	0.174	–0.121	–0.141	604	0.664	0.645	1.098	1.224	19.7	1
K.05	0.238	–0.207	–0.131	2160	0.449	0.929	1.169	1.063	8.84	0.419
ISKb	0.565	–0.314	–0.063	144	0.029	1.959	0.482	1.005	0.121	0.062

Case SVd. Saddoughi and Veeravalli (SV1994) measured normal and shear stress spectra in very high Reynolds number boundary layers. Figure 17.6.1 compares data for two of their cases with the spectrum tensor model results. The case here referred to as Case SVd has the highest turbulence Reynolds number and corresponds to their high velocity ($U_e = 50$m/s) mid-layer ($y = 400$mm) case. Shown with the data and the spectrum tensor model results, are results for isotropic turbulence at the same turbulence Reynolds number. For the normal components of E_{ij}, the distinction between the isotropic and anisotropic profiles can only be seen at low K. This is consistent with the dissipation being nearly isotropic. The model gave a very low value for c_{1d}, and ε_1 only one percent higher than ε. The most significant test of the spectrum tensor model is the comparison of the shear component of the 1-d spectra, E_{12}, since for isotropic turbulence $E_{12} = 0$. $-E_{12}$ falls with increasing K, both due to the general fall in energy with increased K, and due to the reduced anisotropy. Overall, the good agreement supports the spectrum tensor model and the sub-model results, c_{1d} and $C_B(K\eta)$.

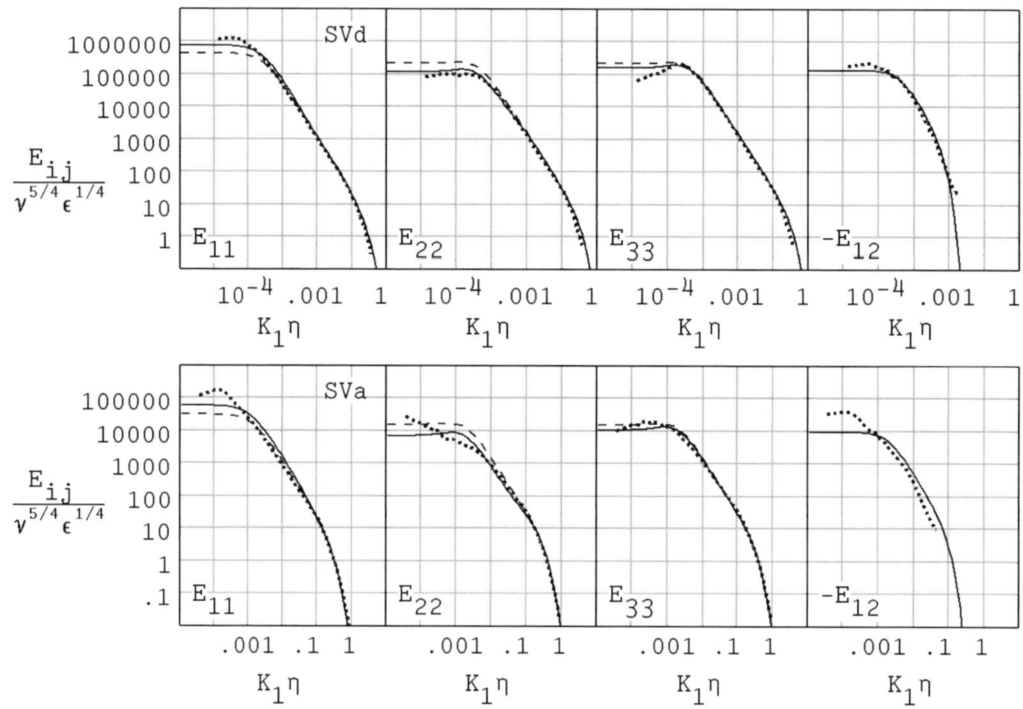

Fig. 17.6.1. Comparison of model results with 1-d spectra data from Saddoughi and Veeravalli (SV1994): Case SVd, high-velocity mid-boundary-layer, U_e=50m/s, y=400mm; Case SVa, low-velocity inner-boundary-layer, U_e=10m/s, y=100mm. •••• data; ——— results from the spectrum tensor model; — — — isotropic turbulence.

Case SVa. Figure 17.6.1 also compares model results with Saddoughi and Veeravalli's low velocity (U_e =10m/s) inner-layer (y=100mm) case, here labeled Case SVa. While the agreement between the data and model is reasonable, it is not as good as for Case SVd. As shown in Table 17.6.1, g is less than 0.75 for Case SVa. Therefore, the model for c_{1d} gives a higher value, c_{1d} =0.350, than would be obtained at the same R_t for $g > 0.75$, c_{1d} =0.052. The value of c_{1d} being too high is consistent with the over-prediction of $-E_{12}$ at high K.

Case Tcon. The data from Tucker (T1970), for E_{11} and E_{22} following an axisymmetric contraction, is shown in Fig. 17.6.2. This is the only non-shear flow case in Table 17.6.1. By symmetry, $E_{33} = E_{22}$, and x_1, x_2 and x_3 are principal directions. The spectrum tensor model agrees well with the data, showing the large reduction in E_{11} (compared to isotropic) at low wave numbers, followed by nearly isotropic dissipation at high wave numbers. The ratios of the principal contributions to the turbulence at low K are obtained from Eq. 17.5.14 with $C_B = C_{B,0}$. This gives

$B_{11}:B_{22}:B_{33}=1:\ 5.3:\ 5.3$, and these ratios were just large enough that the more complex model for Φ_{ij}^{N}, using Eq. 17.5.38, was required at low K. This is indicated in Table 17.6.1 by the value of $C_{1,\min}$ being less than 1, where $\underset{K\to 0}{Lim}\ C_1(K)=C_{1,\min}$. The cases with $C_{1,\min}=1$ used the simpler model, based on Eq. 17.5.28 at all K.

Fig. 17.6.2. Comparison of model results with 1-d spectra data from Tucker (T1970) following axisymmetric contraction. •••• data; ——— results from the spectrum tensor model; – – – isotropic turbulence.

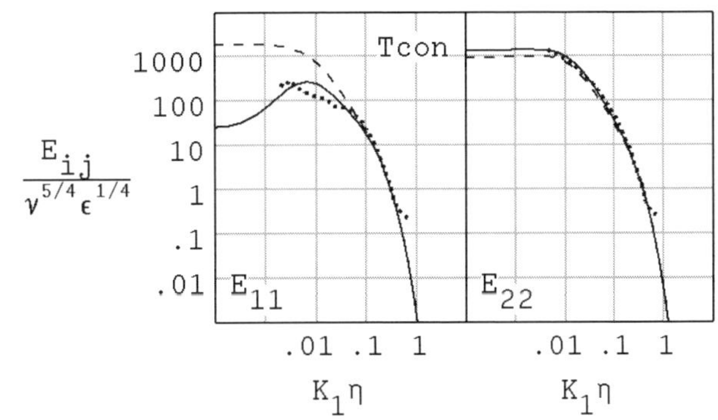

Case ISKa. Test Case ISKa uses the fully developed channel flow DNS from Iwamoto et al. (ISK2002) for $R_\tau=643$, at the mid-layer point, $y/\delta=0.5$. It was chosen because, for a DNS case, it has a high turbulence Reynolds number, $R_t=604$. But since this is a much smaller R_t than for cases SVd and SVa, c_{1d} is higher.

The DNS database (Kasagi et al., KHMMN1990) gives not only the standard 1-d spectra $E_{ij}(K_1)$ for this case, but also spectra in the x_3 direction. Paralleling Eq. 17.1.16,

$$E_{ij}^{Z}(K_3)=2\int_{-\infty}^{\infty}\int_{-\infty}^{\infty}\Phi_{ij}(\mathbf{K})\,dK_1dK_2. \tag{17.6.1}$$

As shown in Fig. 17.6.3, there is good agreement between the spectrum tensor model and all of the $E_{ij}(K_1)$ and $E_{ij}^{Z}(K_3)$ spectra. Note that while the spectrum tensor model gives continuous spectra, the DNS data has only a finite number of spectral components, $K_1\eta=0.00468n$, $n=0,,,143$, and $K_3\eta=0.0117m$, $m=0,,,191$, corresponding to limited box size and limited grid spacing.

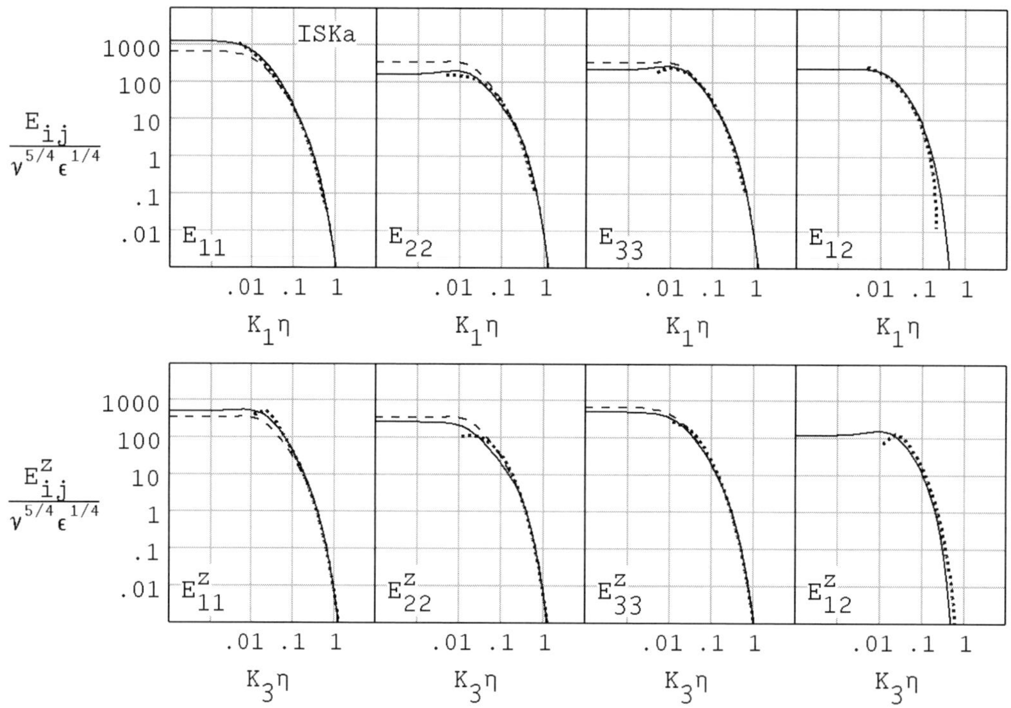

Fig. 17.6.3. Comparison of model results with fully developed channel flow DNS from Iwamoto et al. (ISK2002) for R_τ=643, at mid-layer point, y/δ=0.5. •••• data; ——— results from the spectrum tensor model; – – – isotropic turbulence.

Case K.05. Klebanoff's (K1955) classic boundary layer study included some measurements of E_{11}, E_{22} and E_{12} spectra. Figure 17.6.4 shows his data at y/δ =0.05, and good agreement with the spectrum tensor model results. This case has a higher R_t than case ISKa but a lower g. The result is that c_{1d} is nearly one, so that the dissipation is about half isotropic and half proportional to $\overline{u_i u_j}$. The lower g also resulted in the model using the more complex basis function, Eq. 17.5.38, at low K.

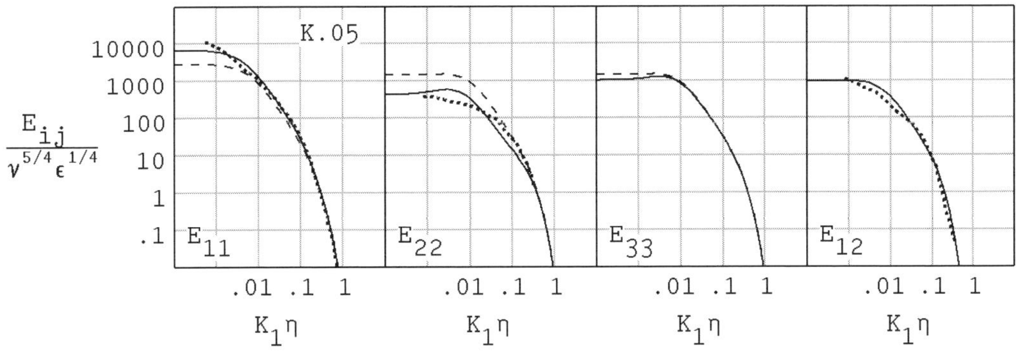

Fig. 17.6.4. Comparison of model results with spectra data from Klebanoff (K1955) at y/δ=0.05 in a boundary layer. •••• data; ——— results from the spectrum tensor model; – – – isotropic turbulence.

Case ISKb. The final case considered is another DNS channel case from Iwamoto et al. (ISK2002), Case ISKb has R_τ =109, and the location is quite near the wall, $y^+ \approx 15$. The turbulence is nearly two-dimensional, $g \approx 0.03$, and the dissipation is primarily proportional to $\overline{u_i u_j}$. The spectrum tensor model, while doing less well than in the earlier cases, as shown in Fig. 17.6.5, provides a reference point for thinking about the turbulence at this location. Because of the high anisotropy of the turbulence (the principal Reynolds stresses are approximately in the ratio 62:6:1), the spectrum tensor model uses the more complex basis function, Eq. 17.5.38, over a large range in K, $K\eta < 1.6$. With the minimum value of C_1 being 0.062, the scaling factor in the model (see Fig. 17.5.3) drops quickly away from the principal planes. The model therefore tends to cluster the contributions to the largest principal stress, $\overline{u_{P_1} u_{P_1}}$, near $K_{P_1} = 0$. Since in this case the K_{P_1} direction differs from the K_1 direction by only 4 degrees, the model essentially clusters the contributions to $\overline{u_1 u_1}$ near $K_1 = 0$. This is consistent with the DNS results, as seen by the peak values in E_{11}, compared to E_{11}^Z. The peak values of E_{22} and E_{12} are also well modeled, but the peak model values of E_{33} and E_{33}^Z appear interchanged relative to the data. A side effect of the model's form, Eq. 17.5.38, clusters the contributions to $\overline{u_3 u_3}$ near $K_3 = 0$. The data is consistent with the $\overline{u_3 u_3}$ energy clustered near $K_1 = 0$ with the $\overline{u_1 u_1}$ energy.

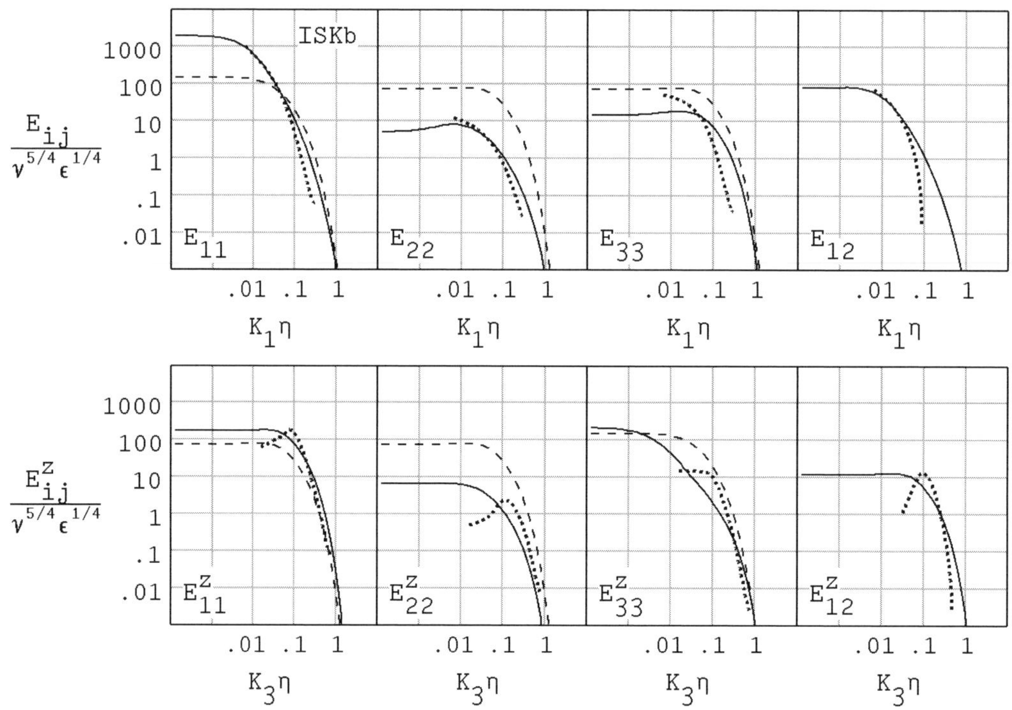

Fig. 17.6.5. Comparison of model results with spectra data from Iwamoto et al. (ISK2002) for $R_\tau = 109$, at $y^+ \approx 15$ in fully developed channel flow. •••• DNS data; ——— results from the spectrum tensor model; – – – isotropic turbulence.

Chapter 18

HIGH ROTATION AND ELLIPTIC FLOWS

In this chapter we review the relationship between high rotation rates, the Taylor-Proudman theorem, material frame indifference, and two-dimensionalization of mean flow and possibly turbulence.

The effect of rotation on turbulence is shown pictorially using rapid distortion theory, and glyphs of the contributions to turbulence at individual wave numbers. This suggests two effects, a scrambling or return-to-isotropy, as well as partial balancing of rotation by pressure-strain. The effect that high rotation has on the decay of dissipation is related to changes in turbulence spectra.

Homogeneous elliptic flow is considered, with and without frame rotation. A parameter, the ratio of rotation rate to relative vorticity, determines whether or not the turbulence may become dominated by elliptical instabilities. However, including the modeling of elliptical instabilities in a Reynolds stress model would require additional complex functionality, which may not be general. Therefore no changes to the MARV model are made.

18.1 Material Frame Indifference

Consider the unsteady momentum equation (Eq. 2.1.5),

$$\frac{\partial \tilde{U}_i}{\partial t} + \tilde{U}_k \frac{\partial \tilde{U}_i}{\partial x_k} = -2\varepsilon_{ikm}\Omega_k \tilde{U}_m + \frac{\partial}{\partial x_i}\left(\frac{\Omega^2 r^2}{2}\right) - \frac{1}{\rho}\frac{\partial \tilde{P}}{\partial x_i} + \frac{\partial}{\partial x_k}\nu\frac{\partial \tilde{U}_i}{\partial x_k}, \tag{18.1.1}$$

and, in particular, the pressure gradient and rotational force terms. For the incompressible flows we are considering here, the centrifugal and pressure gradient terms, may be combined as follows,

$$\frac{\partial \tilde{U}_i}{\partial t} + \tilde{U}_k \frac{\partial \tilde{U}_i}{\partial x_k} = -2\varepsilon_{ikm}\Omega_k \tilde{U}_m - \frac{1}{\rho}\frac{\partial}{\partial x_i}\left(\tilde{P} - \frac{1}{2}\rho\Omega^2 r^2\right) + \frac{\partial}{\partial x_k}\nu\frac{\partial \tilde{U}_i}{\partial x_k}, \tag{18.1.2}$$

so that it is the pressure and not the velocity that is modified by the centrifugal force term.

If the Coriolis term could also be written in gradient form, say

$$2\varepsilon_{ikm}\Omega_k\tilde{U}_m = \frac{1}{\rho}\frac{\partial}{\partial x_i}\left(\tilde{P}_\Omega\right) , \tag{18.1.3}$$

then the unsteady momentum equation could be written as

$$\frac{\partial \tilde{U}_i}{\partial t} + \tilde{U}_k \frac{\partial \tilde{U}_i}{\partial x_k} = -\frac{1}{\rho}\frac{\partial}{\partial x_i}\left(\tilde{P} - \frac{1}{2}\rho\Omega^2 r^2 + \tilde{P}_\Omega\right) + \frac{\partial}{\partial x_k}\nu\frac{\partial \tilde{U}_i}{\partial x_k} . \tag{18.1.4}$$

It then follows that if the boundary and initial conditions for the unsteady relative velocity, $\tilde{U}_i$, are independent of the coordinate rotation rate, Ω_k, then $\tilde{U}_i$ will be independent of Ω_k throughout the flow. The coordinate rotation will affect only the unsteady pressure, $\tilde{P}$. The unsteady relative velocity is then <u>materially frame indifferent</u>, i.e. it is independent of the rotation rate of the coordinate system. Correspondingly then both the mean relative velocity, U_i, and the Reynolds stresses, $\overline{u_i u_j}$, are materially frame indifferent.

The question remains, under what circumstances are these conditions for material frame indifference met.

Two-Dimensional Flow and Turbulence

If the unsteady flow is two-dimensional, such that $\partial \tilde{U}_i / \partial x_3 = 0$, then the continuity equation reduces to

$$\frac{\partial \tilde{U}_1}{\partial x_1} + \frac{\partial \tilde{U}_2}{\partial x_2} = 0, \tag{18.1.5}$$

and the x_1 and x_2 components of the velocity may be written in terms of a scalar stream function, $\tilde{\psi}(x_1,x_2,t)$,

$$\tilde{U}_1 = \frac{\partial \tilde{\psi}}{\partial x_2}, \qquad \tilde{U}_2 = -\frac{\partial \tilde{\psi}}{\partial x_1} . \tag{18.1.6}$$

Noting that $\partial \tilde{\psi} / \partial x_3 = 0$, the Coriolis term may then be written as (Appendix EQ.18.1.7)

$$2\varepsilon_{ikm}\Omega_k\tilde{U}_m = -2\delta_{i3}\Omega_k\frac{\partial \tilde{\psi}}{\partial x_k} + 2\varepsilon_{ik3}\Omega_k\tilde{U}_3 + 2\Omega_3\frac{\partial \tilde{\psi}}{\partial x_i} . \tag{18.1.7}$$

If the rotation is only about the x_3 axis, then this equation reduces to the form of Eq. 18.1.3,

$$2\varepsilon_{ikm}\Omega_k\tilde{U}_m = 2\Omega_3\frac{\partial \tilde{\psi}}{\partial x_i} = \frac{1}{\rho}\frac{\partial}{\partial x_i}\left(\tilde{P}_\Omega\right), \quad \tilde{P}_\Omega = 2\rho\Omega_3\tilde{\psi}, \tag{18.1.8}$$

and the mean flow and the turbulence will be materially frame indifferent if appropriate boundary and initial conditions are met.

Taylor-Proudman Theorem

Consider the situation where the time, convection and viscous terms in the unsteady momentum equation (Eq. 18.1.2) are negligible compared with the Coriolis term,

$$\underbrace{\frac{\partial \tilde{U}_i}{\partial t} + \tilde{U}_k \frac{\partial \tilde{U}_i}{\partial x_k}}_{\text{small}} = -2\varepsilon_{ikm}\Omega_k \tilde{U}_m - \frac{1}{\rho}\frac{\partial}{\partial x_i}\left(\tilde{P} - \frac{1}{2}\rho\Omega^2 r^2\right) + \underbrace{\frac{\partial}{\partial x_k}\nu\frac{\partial \tilde{U}_i}{\partial x_k}}_{\text{small}}, \tag{18.1.9}$$

so that the equation reduces to

$$0 = -2\varepsilon_{ikm}\Omega_k \tilde{U}_m - \frac{1}{\rho}\frac{\partial}{\partial x_i}\left(\tilde{P} - \frac{1}{2}\rho\Omega^2 r^2\right). \tag{18.1.10}$$

Taking the curl of this equation,

$$\varepsilon_{jni}\frac{\partial}{\partial x_n}\left(0 = -2\varepsilon_{ikm}\Omega_k \tilde{U}_m - \frac{1}{\rho}\frac{\partial}{\partial x_i}\left(\tilde{P} - \frac{1}{2}\rho\Omega^2 r^2\right)\right), \tag{18.1.11}$$

and for convenience taking the rotation to be in the x_3 direction, yields three equations (Appendix EQ.18.1.12),

$$\frac{\partial \tilde{U}_1}{\partial x_3} = 0,\ \frac{\partial \tilde{U}_2}{\partial x_3} = 0,\ \frac{\partial \tilde{U}_2}{\partial x_2} + \frac{\partial \tilde{U}_1}{\partial x_1} = 0 = -\frac{\partial \tilde{U}_3}{\partial x_3}. \tag{18.1.12}$$

This turns the problem around, and suggests that when time, convection and viscous terms are small compared with the Coriolis term, the flow will tend to stratify so that there are no variations of velocity in the direction of rotation. This is known as the Taylor-Proudman theorem. It is based on the theoretical work of Proudman (1916) and the theoretical and experimental work of Taylor (1917, 1921).

The concept can be demonstrated by rotating a tank of fluid at constant speed until solid body rotation is achieved, and then slowly towing an obstacle across the bottom of the tank. As the flow goes around the obstacle in the bottom of the tank, it also goes around a column of fluid above the obstacle, as if that fluid were also an obstacle. Greenspan (1968) gives an excellent description of this and another simple experiment (with photographs) to demonstrate the Taylor-Proudman column, in the first ten pages of his book, The Theory of Rotating Fluids.

While the Taylor-Proudman column is a 2-d laminar flow, the Taylor-Proudman theorem is more general. If Eq. 18.1.12 is satisfied, then Eq. 18.1.5 is satisfied and the corresponding expectation for material frame indifference also applies.

Rotation Parameters

Hopfinger (Appendix H1992) discusses the scales and conditions for which the mean flow can be expected to become two-dimensional due to high rotation. He also discusses the scales on which turbulence might reasonably be expected to be affected by high rotation.

One relevant parameter is the Rossby number defined by

$$R_{o,H} = \frac{U}{2\Omega L}, \tag{18.1.13}$$

where Ω is the rotation rate of the rotating reference frame, U is the magnitude of the relative velocity, and L is the scale of the motion normal to the direction of rotation. Hopfinger suggests that the mean flow will become two-dimensional when the height of the fluid, H, in the direction of rotation is limited, in particular,

$$H << \frac{L}{R_{o,H}} = \frac{2\Omega L^2}{U} \quad \text{or} \quad \frac{H}{L} << \frac{2\Omega L}{U}. \tag{18.1.14}$$

Note that limited height implies a wall or other boundary. This provides a mechanism by which three-dimensional flows can rearrange the flow pattern to become two-dimensional.

For the turbulence, Hopfinger defines two rotation length scales, one in terms of the turbulence kinetic energy and the other using the dissipation,

$$L_C = \frac{\sqrt{2k}}{2\Omega} \quad \text{and} \quad L_\Omega = \sqrt{\frac{\varepsilon}{(2\Omega)^3}}. \tag{18.1.15}$$

He then argues that turbulence will be affected by rotation when the integral length scale of the turbulence is greater than these rotation length scales. Approximating the integral length scale by (see Eq. 17.4.5)

$$L_x \approx 0.59(2k/3)^{3/2}/\varepsilon, \tag{18.1.16}$$

these criteria become,

$$\frac{L_x}{L_C} = 0.45\frac{\Omega}{\omega} > 1 \quad \text{and} \quad \frac{L_x}{L_\Omega} = 0.91\left(\frac{\Omega}{\omega}\right)^{3/2} > 1 \tag{18.1.17}$$

Further, he suggests that when these ratios are greater than about 3 to 5, the turbulence is quasi 2-d on the time scales of the turbulence, with inertial waves on the time scales of the rotation.

In terms of the absolute vorticity parameter, $W = \sqrt{2W_{ij}W_{ij}}$, when the relative vorticity is zero, $W = 2\Omega$, so that the conditions are approximately:

$$\text{turbulence affected by rotation,} \quad \frac{W}{\omega} > 3\,; \tag{18.1.18}$$

$$\text{quasi 2-d and inertial waves,} \quad \frac{W}{\omega} > 10\,. \tag{18.1.19}$$

It should be noted that these arguments are made in the absence of mean strain rate. They are not meant to apply in, for example, ordinary boundary layers, where W/ω, which equals S/ω, varies between 3.5 and 30 in the buffer and laminar sublayers (Fig. 5.1.2). There may also be an additional requirement, similar to Eq. 18.1.14, for example,

$$H << \frac{2\Omega L_x^2}{\sqrt{2k}} \approx 0.15\frac{q}{\omega}\frac{\Omega}{\omega} \quad , \quad \text{or} \quad H << 0.07\frac{q}{\omega}\frac{W}{\omega}\,, \tag{18.1.20}$$

to provide a mechanism for the turbulence to become quasi 2-d. This would be consistent with the finding that homogeneous isotropic turbulence, when subject to system rotation alone, stays isotropic ($b_{ij} = 0$, e.g., see Bardina et al., BFR1985).

18.2 Modeling Insights from Rapid Distortion Theory

Significant understanding of the effects of rotation on homogeneous turbulence can be obtained by using rapid distortion theory. Consider the RDT equations (Eqs. 9.3.3 and 9.3.4) for the case of pure rotation, $\partial U_i/\partial x_j = 0$, and for simplicity, without dissipation, $\nu = 0$. Equation 9.3.3 reduces to

$$\frac{dK_k}{dt} = 0, \tag{18.2.1}$$

so that there is no movement of points in wave-number space. Equation 9.3.4 reduces to

$$\frac{d\Phi_{ij}}{dt} = \underbrace{-2\Omega_k\left(\varepsilon_{ikm}\Phi_{mj} + \varepsilon_{jkm}\Phi_{im}\right)}_{\text{Coriolis}} + \underbrace{2\varepsilon_{nkm}\Omega_k\left(\frac{K_iK_n}{K^2}\Phi_{mj} + \frac{K_jK_n}{K^2}\Phi_{im}\right)}_{\text{rapid pressure-strain}}. \tag{18.2.2}$$

Consider a point in wave-number space, which lies on the K_3 axis. Since $K_1 = K_2 = 0$, the pressure-strain terms in the equations for Φ_{11}, Φ_{22}, and Φ_{12} are zero. By continuity, $K_i\Phi_{ij} = 0$ (Eq. 9.2.11), so that

$$\Phi_{33} = \Phi_{13} = \Phi_{23} = 0. \tag{18.2.3}$$

and

$$\frac{d\Phi_{11}}{dt} = -4\Omega_k\Phi_{m1}\varepsilon_{1km} = 4\Omega_3\Phi_{21} \tag{18.2.4}$$

$$\frac{d\Phi_{22}}{dt} = -4\Omega_k\Phi_{m2}\varepsilon_{2km} = -4\Omega_3\Phi_{12} \tag{18.2.5}$$

$$\frac{d\Phi_{12}}{dt} = -2\Omega_k\Phi_{m2}\varepsilon_{1km} - 2\Omega_k\Phi_{1m}\varepsilon_{2km} = 2\Omega_3\left(\Phi_{22} - \Phi_{11}\right). \tag{18.2.6}$$

These equations may be written in the form used in Exercise 4.2,

$$\frac{d}{dt}\begin{bmatrix}\Phi_{11} - \Phi_{22}\\ \Phi_{11} + \Phi_{22}\\ \Phi_{12}\end{bmatrix} = 2\Omega_3\begin{bmatrix}4\Phi_{12}\\ 0\\ \Phi_{22} - \Phi_{11}\end{bmatrix}, \tag{18.2.7}$$

so that the tensor glyph for Φ_{ij} rotates about the K_3 axis at a rate of $-2\Omega_3$, as shown for a sample 1-d glyph in Fig. 18.2.1.

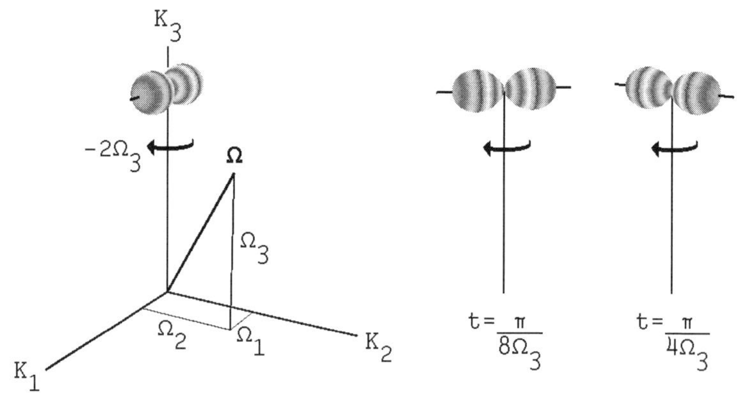

Fig. 18.2.1. Rotation rate of a glyph for Φ_{ij} at K_1=K_2=0 in wave-number space.

Generalizing for any point in wave-number space, the tensor glyph rotates about the vector **K** with a rate which depends on the component of the rotation vector in the **K** direction, $-2K_i\Omega_i/K$. The variation of direction, and rate of rotation, can turn one-dimensional turbulence into three-dimensional turbulence, as shown in Fig. 18.2.2. In terms of the Reynolds stresses, which are the integral of Φ_{ij} over wave-number space, it might be possible to model this effect as a return-to-isotropy type term.

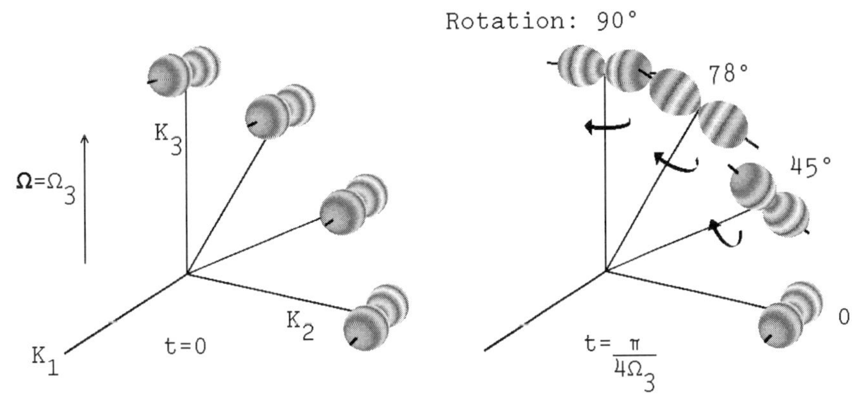

Fig. 18.2.2. Response of some glyphs for Φ_{ij} corresponding to 1-d turbulence ($\overline{u_1u_1}$ only) to rotation about the x_3 axis.

Going back to our original example, with **K** in the K_3 direction, the equations for Φ_{i3} are also of interest. From Eq. 18.2.2,

$$\frac{d\Phi_{i3}}{dt} = -2\Omega_k\Phi_{m3}\left(\varepsilon_{ikm} - \varepsilon_{3km}\frac{K_3K_3}{K^2}\delta_{i3}\right) - 2\Omega_k\Phi_{im}\left(\varepsilon_{3km} - \varepsilon_{3km}\frac{K_3K_3}{K^2}\right). \quad (18.2.8)$$

The equations for Φ_{13} and Φ_{23} contain terms with Φ_{ij}'s which are not initially zero due to continuity,

$$\frac{d\Phi_{13}}{dt} = \ldots\ldots - 2\Omega_1\Phi_{12}\left(\varepsilon_{312} - \varepsilon_{312}\frac{K_3K_3}{K^2}\right) - 2\Omega_2\Phi_{11}\left(\varepsilon_{321} - \varepsilon_{321}\frac{K_3K_3}{K^2}\right),$$

$$(18.2.9)$$

$$\frac{d\Phi_{23}}{dt} = \ldots\ldots - 2\Omega_1\Phi_{22}\left(\varepsilon_{312} - \varepsilon_{312}\frac{K_3K_3}{K^2}\right) - 2\Omega_2\Phi_{21}\left(\varepsilon_{321} - \varepsilon_{321}\frac{K_3K_3}{K^2}\right).$$

For these terms, rotation is balanced by pressure-strain, since with $K_1 = K_2 = 0$, $K^2 = K_3^2$. This maintains continuity by preventing the glyph from tipping towards the **K** vector, as sketched in Fig. 18.2.3. Integrating over wave-number space, it might be possible to model this effect as a c_5 type pressure-strain term.

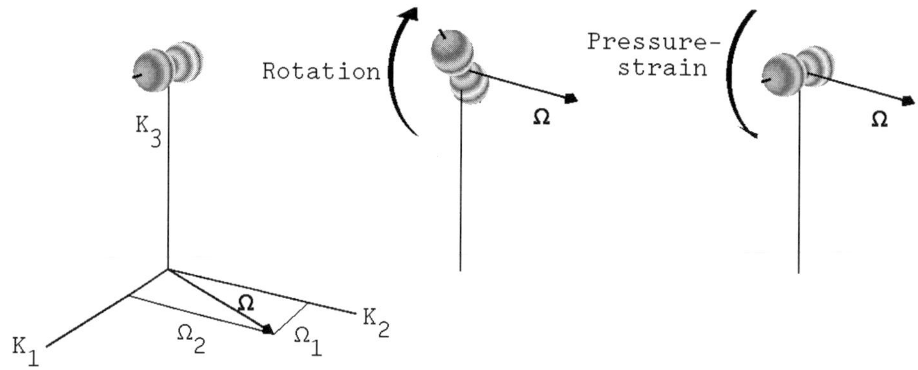

Fig. 18.2.3. Sketch showing pressure-strain balancing rotation when the **K** and Ω vectors are perpendicular.

Taylor-Proudman Limit

In the Taylor-Proudman limit there are no variations of velocity in the direction of rotation (Eq. 18.1.12). Taking the rotation to be in the x_3 direction, the fluctuating velocities then satisfy,

$$\partial u_1 / \partial x_3 = \partial u_2 / \partial x_3 = \partial u_3 / \partial x_3 = 0 \; . \tag{18.2.10}$$

This implies infinitely long length scales in the x_3 direction, so that in wave-number space all the turbulence energy lies in the $K_3 = 0$ plane, as sketched in Fig. 18.2.4. There is then no component of rotation in the direction of the wave vectors, and correspondingly no spinning of the Φ_{ij} glyphs. The rotation only acts normal to the wave-number vector, and this is completely balanced by rapid pressure-strain. In terms of Reynolds stress modeling this corresponds to $c_5 = 4$ (e.g., see Eq. 12.4.5). Note that the Taylor-Proudman limit does not require that the turbulence be two-dimensional. The turbulence may still have $\overline{u_3 u_3} > 0$. It only requires that u_i not be a function of x_3, i.e., $u_i = u_i(x_1, x_2, t)$.

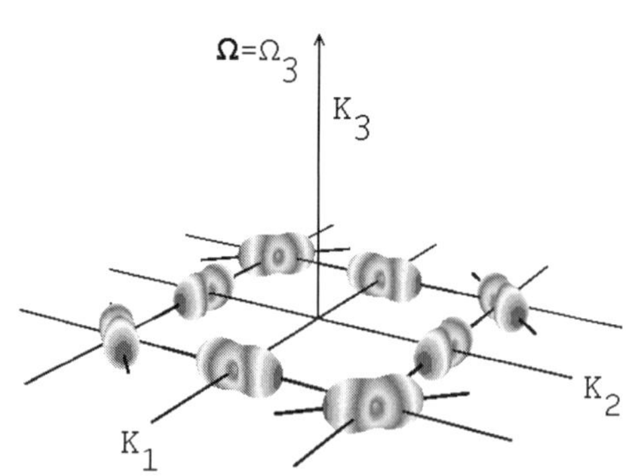

Fig. 18.2.4. An example of the Taylor-Proudman limit with rotation in the x_3 direction. Non-zero Φ_{ij} glyphs lie in the K_3=0 plane.

Exercise 18.2. The glyphs shown in Fig. 18.2.4 sum to give isotropic turbulence, $b_{ij} = 0$. Determine the relative magnitudes of Φ_{11}, Φ_{22} and Φ_{33} for the glyphs lying on the K_1 and K_2 axes.

18.3 Rotating Isotropic Turbulence

The experiments of Wigeland and Nagib (1978) considered nearly isotropic, homogeneous turbulence subjected to rotation. Some of their data was included in the 1980-81 Stanford Conference database (Kline et al., KCL1981), and as shown in Fig. 18.3.1a, the turbulence in the high rotation case decays more slowly than without rotation. The turbulence Reynolds number for the Wigeland and Nagib experiments was low. For the zero rotation case R_t fell from about 22 to 14, and for the high rotation case R_t increased from about 28 to 40. The anisotropy varied little, $b_{11} = 0.08 \pm 0.03$.

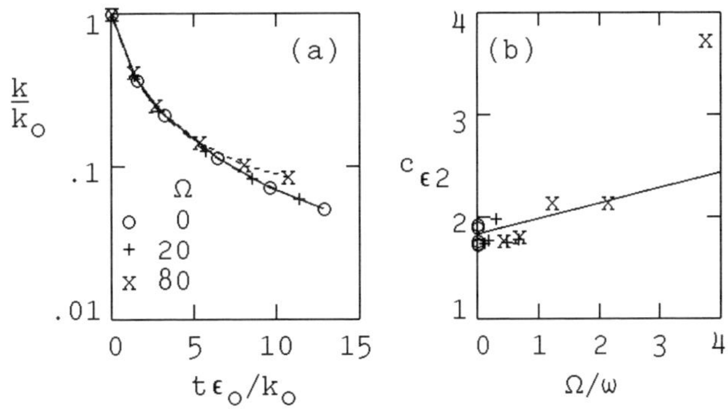

Fig. 18.3.1. Nearly isotropic turbulence subjected to system rotation. Data of Wigeland and Nagib (1978): o, +, x, Ω=0, 20, 80 rad./s, respectively; R_t range, 14-40. (a) Decay of turbulence. (b) $c_{\varepsilon 2}$, —— model of Bardina et al. (BFR1985) Eq. 18.3.2.

Bardina et al. (BFR1985) were able to duplicate Wigeland and Nagib's results starting with isotropic turbulence and using LES (large eddy simulations). They then proposed a model for the increase in the decay rate of turbulence dissipation with increased rotation (see models for $c_{\varepsilon 2}$, Table 6.1.1). Writing the equation for ε, for homogeneous flow, in the absence of mean (relative) velocity gradients as

$$\frac{d\varepsilon}{dt} = -c_{\varepsilon 2}\varepsilon\omega, \tag{18.3.1}$$

the model of Bardina et al. may be written as

$$c_{\varepsilon 2} = \frac{11}{6} + 0.15\frac{\Omega}{\omega} . \tag{18.3.2}$$

As shown in Fig. 18.3.1b, their model is in reasonable agreement with values of $c_{\varepsilon 2}$ deduced from Wigeland and Nagib's data using Eq. 6.2.3. Note, however, that the rotation parameter, Ω/ω, is low. Ω/ω was less than 2, except in the latter half of the high rotation case.

The recent DNS of Morinishi et al. (MNR2001) considers rotating isotropic turbulence at both higher Reynolds numbers, $R_{t,o}=138$, and higher rotation rates, Ω/ω_o up to nearly 100. Like the earlier work they observed that the turbulence anisotropy, b_{ij}, remained essentially zero and that the turbulence decayed less rapidly at high rotation. They explained their results in terms of the turbulence spectra, Fig. 18.3.2. At zero rotation, the turbulence energy both dissipates and cascades to higher wave numbers. At high rotation rates, the transfer of energy to higher wave numbers is reduced so that at very high rotation rates the development of the $E(K)$ spectrum approaches the linear solution, due to dissipation only,

$$\frac{dE(K,t)}{dt} = -2\nu K^2 E(K,t), \qquad E(K,t) = E(K,0)\exp(-2\nu K^2 t) \ . \tag{18.3.3}$$

Fig. 18.3.2. Spectra from the DNS of Morinishi et al. (MNR2001) for rotating isotropic turbulence.
—— initial, (t=0). At $t=3$ ($t\omega_o$=3.05):
--- $\Omega=0$; — — — $\Omega=10$; —— $\Omega=100$;
–·– linear solution, Eq. 18.3.3.
$\varepsilon_o, \omega_o, \eta_o$, values at t=0.

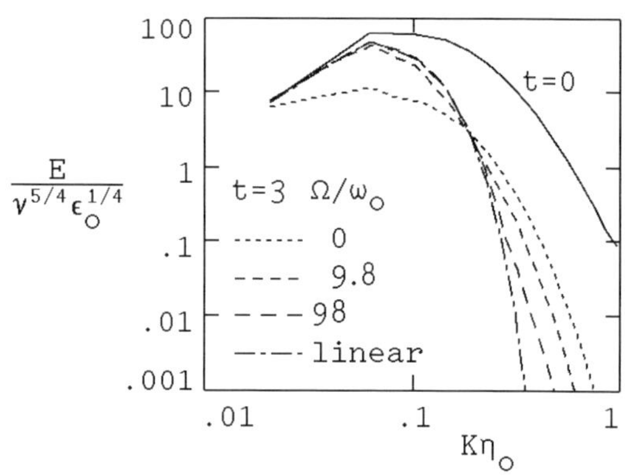

An explanation for the reduction in energy transfer may be obtained by considering the equation for the development of Φ_{ij} (Eq. 9.2.8), or more particularly contracted for Φ_{ii}, in the absence of mean velocity gradients,

$$\underbrace{\frac{\partial \Phi_{ii}}{\partial t} = -2\nu K^2 \Phi_{ii}}_{\text{dissipation}} + \underbrace{\frac{(2\pi)^3}{V} iK_m \int \left(\overline{a_i(\mathbf{K}) a_m^*(\mathbf{K}+\mathbf{K}') a_i(\mathbf{K}')} - \overline{a_i^*(\mathbf{K}) a_m(\mathbf{K}-\mathbf{K}') a_i(\mathbf{K}')} \right) d\mathbf{K}'}_{\text{turbulent diffusion}} . \tag{18.3.4}$$

The term labeled turbulent diffusion is responsible for the movement of energy in wave-number space and, in general, gives rise to the cascade of energy to higher wave numbers. Note that this term relies on there being, on average, a coherence between the Fourier transforms of the fluctuating velocity at three points in wave-number space, $a_i(\mathbf{K})$, $a_m(-\mathbf{K}-\mathbf{K}')$ and $a_i(\mathbf{K}')$ (for the first term). Just as rotation turns glyphs of Φ_{ij} at different rates, depending on their angular location relative to the axis of rotation (Fig. 18.2.2), $a_i(\mathbf{K})$ will also be turned at different rates. Rotation is likely, therefore, to reduce the triple-product correlation and the cascade of energy. An exception will occur if all three points lie in the Taylor-Proudman plane, the plane $K_3 = 0$ for rotation about the x_3 axis (Fig. 18.2.4). Rotation has no effect in this plane, and therefore no effect on this part of the energy cascade. This is consistent with the observations of Morinishi et al., who showed that at high K ($K\eta_o$ =0.61) rotation caused more of a net reduction of energy away from $K_3 = 0$, than near it.

Another observation by Morinishi et al. impacts possible model choices for $c_{\varepsilon 2}$. They observed that while ε initially fell more rapidly at high rotation, it also stopped falling more rapidly, so that at the end of their computations ($t = 3$), ε was higher than for zero rotation. Values for $c_{\varepsilon 2}$ were determined from Morinishi et al.'s data, using $d(1/\omega)/dt = c_{\varepsilon 2} - 1$. For the highest rotation rate case, $c_{\varepsilon 2}$ falls with time, Fig. 18.3.3a, and overall, Fig. 18.3.3b, the levels of $c_{\varepsilon 2}$ are far below those predicted by Eq. 18.3.2.

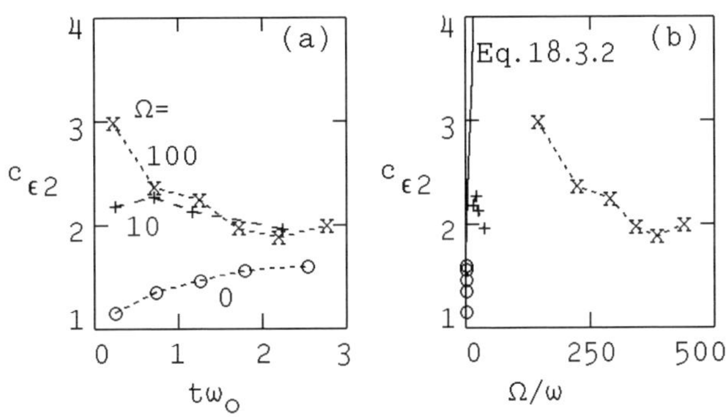

Fig. 18.3.3. Isotropic turbulence subjected to system rotation. DNS of Morinishi et al. (MNR2001): o, +, x, Ω=0, 10, 100, respectively.
(a) $c_{\varepsilon 2}(t\omega_o)$.
(b) $c_{\varepsilon 2}(\Omega/\omega)$, —— model, Eq. 18.3.2.

Overall, the data of Wigeland and Nagib, and Morinishi et al. is consistent with effective $c_{\varepsilon 2}$ values between 2 and 3, for rotating turbulence with Ω/ω greater than about 2, in the absence of any mean strain rate, S_{ij}. Since strain is likely to re-establish the correlation of the triple product, which causes the turbulence cascade, these higher values of $c_{\varepsilon 2}$ are unlikely to be valid when a strain rate is present. Therefore, no changes are made to the Reynolds stress model based on the rotating isotropic turbulence data.

18.4 Models for High Rotation Flows

For simplicity we will limit consideration to homogeneous flows where the strain, vorticity and rotation act in the same plane. As in Chapter 4, we take the rotation to be about the x_3 axis. Then Ω_3, $S_{12}=S_{21}$, and $W_{12}=-W_{21}$ are the only non-zero components of the rotation vector and the strain rate and vorticity tensors. Again considering flows in which b_{13} and b_{23} are initially zero, the turbulence anisotropy develops according to Eq. 4.2.5,

$$b_{13}=0, \qquad b_{23}=0, \qquad \text{and} \tag{18.4.1}$$

$$2k\frac{d}{dt}\begin{bmatrix} b_{11}-b_{22} \\ b_{33} \\ b_{12} \end{bmatrix} = -\left((c_1-2)\varepsilon+(2+\frac{1}{2}c_6)P-\frac{1}{2}c_9 b_{33}P\right)\begin{bmatrix} b_{11}-b_{22} \\ b_{33} \\ b_{12} \end{bmatrix}$$

$$+\left(c_2\varepsilon-\frac{1}{2}c_{10}P+\frac{1}{2}c_{12}b_{33}P\right)\begin{bmatrix} b_{11}^2-b_{22}^2 \\ b_{33}^2-\frac{1}{3}II \\ -b_{12}b_{33} \end{bmatrix}+\left(c_3-\frac{4}{3}\right)kS_{12}\begin{bmatrix} 0 \\ 0 \\ 1 \end{bmatrix}-(c_4-2)kS_{12}\begin{bmatrix} 0 \\ \frac{4}{3}b_{12} \\ b_{33} \end{bmatrix}$$

$$+\left\{c_7 kS_{12}(b_{22}-b_{11})+2k\Omega_3+kW_{12}\left((c_5-2)-c_8 b_{33}+c_{11}(b_{11}b_{22}-b_{12}^2)\right)\right\}\begin{bmatrix} 4b_{12} \\ 0 \\ b_{22}-b_{11} \end{bmatrix}.$$

The differences between the shear flows of Chapter 4 and the more general flows considered here, are the strain-rate and vorticity tensors. In particular, both $\partial U_1/\partial x_2$ and $\partial U_2/\partial x_1$ may be non-zero,

$$S_{12}=\frac{1}{2}\left(\frac{\partial U_1}{\partial x_2}+\frac{\partial U_2}{\partial x_1}\right) \quad \text{and} \quad W_{12}=\frac{1}{2}\left(\frac{\partial U_1}{\partial x_2}-\frac{\partial U_2}{\partial x_1}\right)-\Omega_3 \ . \tag{18.4.2}$$

As for shear flow, turbulence production is given by

$$P=-2kb_{ik}S_{ik}=-4kb_{12}S_{12} \ . \tag{18.4.3}$$

For this limited class of flows, the rapid pressure-strain vorticity terms are (from Eq. 18.4.1 and App. EQ.4.2.5),

$$\begin{bmatrix} \Pi_{11}^{R} \\ \Pi_{22}^{R} \\ \Pi_{33}^{R} \\ \Pi_{12}^{R} \end{bmatrix}_{W\,\text{terms}} = kW_{12}\left(c_5 - c_8 b_{33} + c_{11}(b_{11}b_{22} - b_{12}^2)\right)\begin{bmatrix} 2b_{12} \\ -2b_{12} \\ 0 \\ b_{22} - b_{11} \end{bmatrix} . \quad (18.4.4)$$

Defining an effective c_5,

$$c_{5,eff} = c_5 - c_8 b_{33} + c_{11}(b_{11}b_{22} - b_{12}^2) , \quad (18.4.5)$$

the effect of rotation and vorticity on b_{ij}(from Eq. 18.4.1), may be written as

$$2k\frac{d}{dt}\begin{bmatrix} b_{11} - b_{22} \\ b_{33} \\ b_{12} \end{bmatrix} = \ldots\ldots + \left\{2k\Omega_3 + kW_{12}\left(c_{5,eff} - 2\right)\right\}\begin{bmatrix} 4b_{12} \\ 0 \\ b_{22} - b_{11} \end{bmatrix}$$

$$= \ldots\ldots + \left\{k\Omega_3\left(4 - c_{5,eff}\right) + k\omega_{12}\left(c_{5,eff} - 2\right)\right\}\begin{bmatrix} 4b_{12} \\ 0 \\ b_{22} - b_{11} \end{bmatrix} . \quad (18.4.6)$$

The b_{ij}'s are then independent of the rotation rate, Ω_3, when

MFI limit: $$c_{5,eff} = 4 , \quad (18.4.7)$$

and independent of the relative vorticity, ω_{12}, when $c_{5,eff} = 2$.

Models in the Literature

Ristorcelli et al. (RLA1995) considered the Taylor-Proudman theorem. They developed a rapid pressure-strain model consistent with two-dimensional material frame indifference in the limit of two-dimensional turbulence, in particular, when the Reynolds normal stress in the direction of rotation is zero. Their rapid pressure-strain model is consistent with the general form (Eq. 3.2.17) used to form Eq. 18.4.1, with the coefficients, c_3 through c_{11}, dependent on II and III.

In the two-dimensional turbulence limit, when $g = 0$, the coefficients of the vorticity terms reduce to (App. RLA1995)

$$c_5 = \frac{4}{3(1-1.5II)}, \quad c_8 = \frac{4}{(1-1.5II)} \quad \text{and} \quad c_{11} = \frac{12}{(1-1.5II)} . \quad (18.4.8)$$

In the limited situation being considered in this section, these give

$$c_{5,eff} = \frac{1}{(1-1.5II)}\left(\frac{4}{3} - 4b_{33} + 12(b_{33})^2 - 6II\right), \qquad (18.4.9)$$

so that when $b_{33} = -1/3$, $c_{5,eff} = 4$, and the turbulence is materially frame indifferent. Interestingly, when $b_{33} > -1/3$, and the turbulence is two-dimensional, for example because $b_{22} = -1/3$, then the model gives $c_{5,eff} = 0$ (App. RLA1995, Eq. 26).

Blaisdell and Shariff (BS1996) made DNS calculations for homogeneous elliptic-streamline flows, which have $W/S > 1$. They observed that the turbulence increased much more than expected compared to the LRR1975 and SSG1991 Reynolds stress models. Girimaji (G2000) modified the c_5 coefficient in a partially linearized version of the SSG1991 model, to match the approximate growth of turbulence kinetic energy in the Blaisdell and Shariff DNS. He made c_5 dependent on W/S,

$$c_5 = 0.4, \quad \text{the SSG1991 value,} \quad \text{for } W/S \le 1$$

and

$$c_5 = 2 - (2-0.4)\left(\frac{S}{W}\right)^{1.5}, \qquad \text{for } W/S > 1, \qquad (18.4.10)$$

so that in the limit of high vorticity $c_5 = 2$, and b_{ij} becomes independent of the relative vorticity.

18.5 Homogeneous Elliptic Flows

Consider homogeneous flows where the only non-zero components of the velocity-gradient tensor are $\partial U_1/\partial x_2$ and $\partial U_2/\partial x_1$. Taking the mean velocity as zero at $x_i = 0$, the mean velocity is given by

$$U_1 = x_2 \frac{\partial U_1}{\partial x_2} \quad \text{and} \quad U_2 = x_1 \frac{\partial U_2}{\partial x_1}. \qquad (18.5.1)$$

In this class of flows, we have earlier considered shear flow with $\partial U_2/\partial x_1 = 0$, and plane strain, for which $\partial U_2/\partial x_1 = \partial U_1/\partial x_2$. In this section we consider elliptic flows for which $\partial U_1/\partial x_2$ and $\partial U_2/\partial x_1$ have opposite signs.

To show this, consider the equation for an ellipse,

$$(Ax_1)^2 + (Bx_2)^2 = C^2 . \tag{18.5.2}$$

In transformed cylindrical coordinates we may write

$$Ax_1 = C\cos\theta, \quad Bx_2 = C\sin\theta. \tag{18.5.3}$$

Differentiating with respect to time yields

$$AU_1 = -x_2 B\frac{d\theta}{dt} \quad \text{and} \quad BU_2 = x_1 A\frac{d\theta}{dt}. \tag{18.5.4}$$

Combining with Eq. 18.5.1,

$$\frac{\partial U_1}{\partial x_2} = -\frac{B}{A}\frac{d\theta}{dt} \quad \text{and} \quad \frac{\partial U_2}{\partial x_1} = \frac{A}{B}\frac{d\theta}{dt} , \tag{18.5.5}$$

and we see that the signs of $\partial U_1/\partial x_2$ and $\partial U_2/\partial x_1$ are opposite for an elliptic flow. The fluid particles move along a family of ellipses with aspect ratio, B/A, as sketched in Fig. 18.5.1.

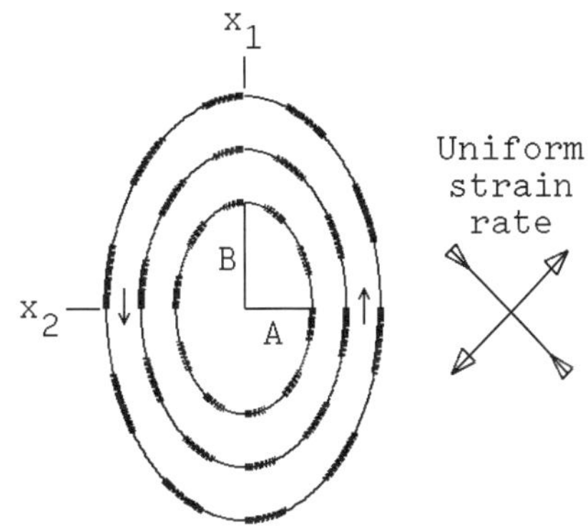

Fig. 18.5.1. Streamlines for elliptic flow. Thick and thin line segments represent equal amounts of time.

The non-zero components of the strain-rate and relative-vorticity tensors are

$$S_{12} = S_{21} = -\frac{1}{2}\left(\frac{B}{A} - \frac{A}{B}\right)\frac{d\theta}{dt} \quad \text{and} \quad \omega_{12} = -\omega_{21} = -\frac{1}{2}\left(\frac{B}{A} + \frac{A}{B}\right)\frac{d\theta}{dt} , \tag{18.5.6}$$

so that for elliptic flow the ratio of the relative vorticity to strain rate is greater than one,

$$\frac{\sqrt{2\omega_{ij}\omega_{ij}}}{S} = \left|\frac{\omega_{12}}{S_{12}}\right| = \left|\frac{B/A + A/B}{B/A - A/B}\right| > 1. \tag{18.5.7}$$

(Without frame rotation, $\sqrt{2\omega_{ij}\omega_{ij}}/S = W/S$.)

The velocity field of the homogeneous elliptic flow may be viewed as solid body rotation on which there has been imposed a uniform strain rate.

Elliptical Instabilities

Flows with elliptical streamlines and a homogeneous constant strain rate in either a stationary or rotating frame are subject to elliptical instabilities. Kerswell (2002) reviewed elliptical instabilities and notes that "the mechanism is always one of resonance in which a normal mode, or pair of normal modes, of oscillation on the undistorted circular flow become tuned to the underlying strain field."

Kerswell also describes a linear theory from which are found the location in wave-number space of the turbulence modes which grow, and their growth rate. The linear theory uses the same approximations as used in rapid distortion theory (without dissipation) with the additional assumption that the streamlines are nearly circular. For elliptic flow where $\partial U_1/\partial x_2$ and $\partial U_2/\partial x_1$ are constants in a frame rotating at Ω_3 about the x_3 axis, the locations in wave-number space of the Fourier components of the fluctuating velocity, a_i, which will grow are dependent on the ratio of the rotation rate to the relative vorticity (Appendix EQ. 18.5.8&10),

$$\frac{K_3}{K} = \pm\frac{1}{2(1+\Omega_E)} \quad , \quad \text{where} \quad \Omega_E = \frac{2\Omega_3}{\dfrac{\partial U_2}{\partial x_1} - \dfrac{\partial U_1}{\partial x_2}} \,. \tag{18.5.8}$$

This is easily confirmed using rapid distortion theory. Results from RDT calculations (without dissipation) for elliptic flows with $\omega_{12}/S_{12} = 10$ are shown in Fig. 18.5.2. The energy distribution, E_{K3}, is defined so that

$$\frac{k}{k_o} = \int_0^1 E_{K3}\, d\left(\frac{|K_3|}{K}\right) . \tag{18.5.9}$$

Three rotation rates are shown, $\Omega_E = 1.5,\ 0,\ -3/8$. Consistent with the linear theory, E_{K3} peaks at $|K_3|/K$ =0.2, 0.5 and 0.8, respectively.

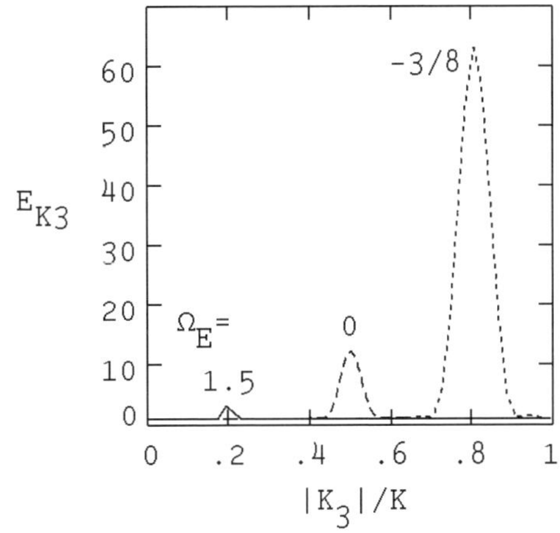

Fig. 18.5.2. Distribution of turbulence energy in wave-number space. RDT (ν=0) results at St=6 for ω_{12}/S_{12}=10. —— Ω_E=1.5; – – – Ω_E= 0; - - - Ω_E= −3/8.

The parameter Ω_E for elliptic flow parallels the parameter R for shear flow, in that $\Omega_E = -1$ and $R = -1$ correspond to zero absolute vorticity, and $\Omega_E = 0$ and $R = 0$ correspond to zero rotation. The Taylor-Proudman limit occurs when $|\Omega_E|$ is very large; from Eq. 18.5.8, energy growth will occur near $K_3/K = 0$, as pictured in Fig. 18.2.4. In the range $-3/2 < \Omega_E < -1/2$, there are no elliptical instabilities, since $|K_3|/K$ cannot be greater than one. At $\Omega_E = -3/2$ and $-1/2$, $|K_3|/K = 1$. These are two-dimensional turbulence limits since all the energy grows on the K_3 axis, and by continuity, Φ_{33} is zero there.

From the linear theory, the Fourier modes of the fluctuating velocity, a_i, which grow most rapidly grow at a rate

$$\sigma = \frac{1}{a^2/2}\left(\frac{d(a^2/2)}{dSt}\right)_{\max} = \frac{1}{4}\left(\frac{1}{2(1+\Omega_E)}+1\right)^2, \quad \text{where} \quad a^2 = a_i a_i \,. \qquad (18.5.10)$$

As shown in Fig. 18.5.3, $\sigma = 1$ at $\Omega_E = -1/2$. σ falls to 9/16 at zero rotation, and at the Taylor-Proudman limit, $\Omega_E = \pm\infty$, $\sigma = 1/4$.

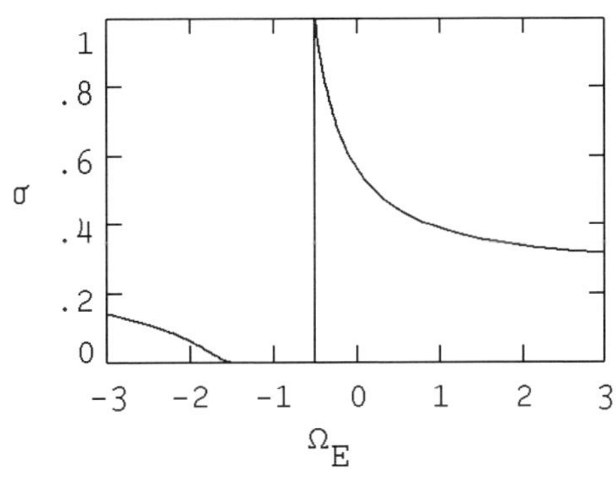

Fig. 18.5.3. Growth rate, σ, of elliptical instabilities from the linear theory, Eq. 18.5.10.

As the elliptical instability grows, so that the most rapidly growing components account for most of the turbulence, σ will become the growth rate of the turbulence kinetic energy.

$$\sigma = \frac{1}{k}\frac{dk}{dSt} \quad \text{at large } St. \tag{18.5.11}$$

Results from RDT calculations, again for $\omega_{12}/S_{12} = 10$, are compared with the theory in Fig. 18.5.4. Figure 18.5.4a shows k/k_o at $St = 2$, 4 and 6 as a function of the rotation parameter Ω_E. k/k_o peaks near $\Omega_E = -0.5$. The growth rate of k, shown in Fig. 18.5.4b for $St = 5$, also peaks at $\Omega_E = -0.5$ as expected, but overall is significantly below the theoretical values. However, the marginal growth rate, evaluated from

$$\sigma = \frac{d\ln(k - k_o)}{dSt}, \tag{18.5.12}$$

is similar in shape to the theory.

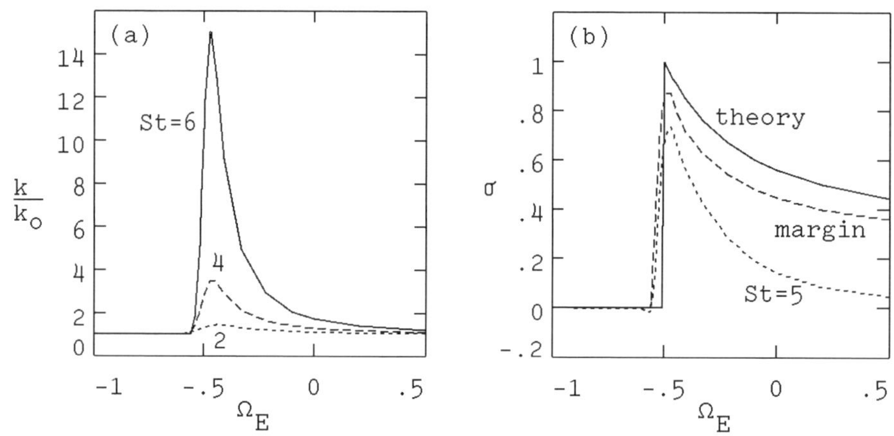

Fig. 18.5.4. Turbulence kinetic energy from RDT calculations without dissipation for ω_{12}/S_{12}=10, as a function of the rotation parameter, Ω_E. (a) k/k_o. (b) Growth rate of k: —— linear theory, Eq. 18.5.10; – – – marginal growth rate at St=5, Eq. 18.5.12; - - - actual growth rate at St=5, Eq. 18.5.11.

Modeling Elliptical Instabilities

Without dissipation, $dk/dt = P = -4b_{12}kS_{12}$, so that Eq. 18.5.11 yields

$$b_{12} = -\frac{\sigma}{2}\frac{S_{12}}{|S_{12}|} \quad \text{at large } St. \tag{18.5.13}$$

Further analysis with the linear theory gives (Appendix EQ.18.5.14)

$$b_{33} = \frac{1}{6} - \frac{1}{8(1+\Omega_E)^2} \quad \text{and} \quad b_{11} = b_{22}, \quad \text{at large } St, \tag{18.5.14}$$

showing that the resulting Reynolds stresses have a Boussinesq orientation (Fig. 2.2.1).

These anisotropies may be used to determine rapid pressure-strain coefficients, which are consistent with the long-term behavior of the elliptical instabilities. Considering only the c_3, c_4, c_5 and c_6 rapid pressure-strain terms, and using Eq. 18.4.1 in the fully developed limit yields $c_3(c_6, \Omega_E)$, $c_4(c_6, \Omega_E)$ and $c_5(\Omega_E)$ (Appendix EQ.18.5.15-17),

$$c_3 = \frac{4}{3} + (2 + \frac{1}{2}c_6)\left(3\left(\frac{1}{6} - \frac{1}{8(1+\Omega_E)^2}\right)^2 - \frac{1}{16}\left(\frac{1}{2(1+\Omega_E)} + 1\right)^4\right), \tag{18.5.15}$$

$$c_4 = 2 + 3(2 + \frac{1}{2}c_6)\left(\frac{1}{6} - \frac{1}{8(1+\Omega_E)^2}\right), \tag{18.5.16}$$

$$c_5 = \frac{2\Omega_E}{1+\Omega_E} + 2\ . \tag{18.5.17}$$

The equation for c_5 gives $c_5 = 4$ in the Taylor-Proudman limit, $\Omega_E = \pm\infty$. It is consistent with Girimaji's model (Eq. 18.4.10) at high W/S and no rotation, $c_5 = 2$ for $\Omega_E = 0$. Equations 18.5.15-17 are consistent with the MARV model for the maximum turbulence production case, $\Omega_E = -1/2$; $c_3 = c_4 = c_5 = c_6 = 0$ for fully developed flow with $g = 0$ (Fig. 10.5.2).

In developing the MARV model, the functionality of the rapid pressure-strain coefficients was limited based on the expectation that rapid pressure-strain in homogeneous flow is independent of the rotation, except through its modification of the absolute vorticity. Equations 18.5.15-17 violate this limitation of the functionality, making the coefficients a function of Ω_E, the ratio of the rotation rate to the relative vorticity in the rotating frame in which the strain rate is homogeneous.

Comparisons with the MARV Model

Results from the MARV model, corresponding to the RDT results in Fig. 18.5.4, are shown in Fig. 18.5.5. With $c_5 = 0$, the growth in turbulence peaks at $\Omega_E = -0.5$, the same as the theory. The MARV model however cuts off quite sharply so that there is essentially no growth in turbulence above $\Omega_E = -0.4$.

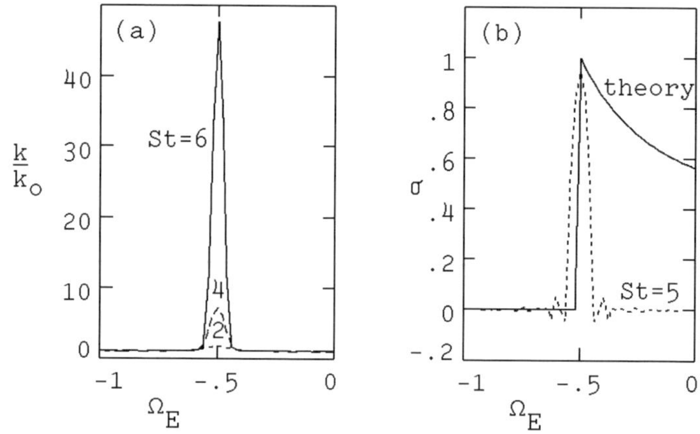

Fig. 18.5.5. Turbulence kinetic energy from the MARV model for ω_{12}/S_{12}=10 and without dissipation. (a) k/k_o. (b) Growth rate of k: —— linear theory, Eq. 18.5.10; --- MARV model at St=5, Eq. 18.5.11.

There are two effects which change this sharp cutoff in turbulence production. At lower ratios of ω_{12}/S_{12}, the range in Ω_E over which turbulence is produced is increased. Figure 18.5.6 compares RDT and MARV model results for elliptic flow with $\omega_{12}/S_{12} = 5/3$. For this flow, which traverses ellipses of aspect ratio 2, the agreement between the RDT and MARV model results is better, since elliptical instabilities are overall less important. Slow pressure-strain, which has been omitted from the calculations in Fig. 18.5.6, also causes a spread in the range of rotations which give increased turbulence. This was shown for shear flow and the D1971 model in Fig. 4.3.3.

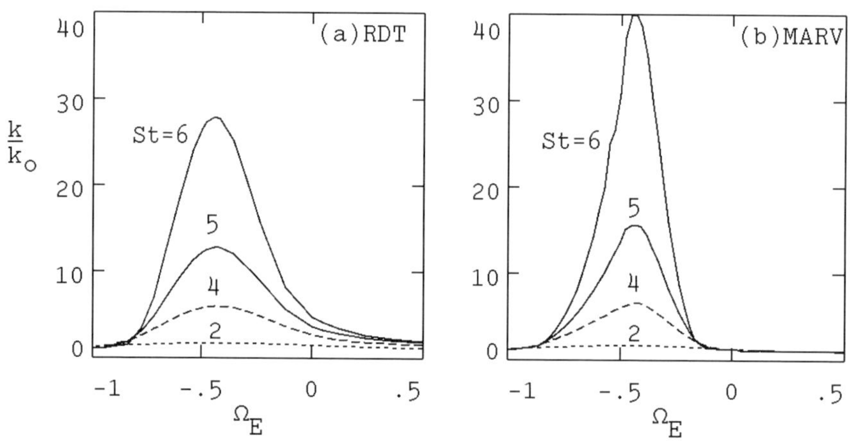

Fig. 18.5.6. Development of turbulence kinetic energy for elliptic flow of aspect ratio 2, $\omega_{12}/S_{12} = 5/3$, as a function of the rotation parameter, Ω_E. (a) From RDT calculations without dissipation. (b) MARV model without dissipation.

18.6 ***Discussion***

In this chapter we have considered flows with high rotation to come to an understanding of the relation between high rotation, the Taylor-Proudman theorem, elliptical instabilities, and turbulence. Rapid distortion theory and glyphs of the individual wave-number contributions to turbulence give a pictorial way of viewing the effects of rotation on turbulence.

The Taylor-Proudman theorem describes a high-rotation limit where the turbulence becomes two-dimensional. This statement however is misleading and has in the past been misinterpreted (since the phrase "two-dimensional" when applied to turbulence can have more than one meaning). If rotation is about the x_3 axis, it is not $\overline{u_3 u_3}$ which tends to zero, but the derivatives, $\partial u_i / \partial x_3$, which tend to zero. If in this rotating frame there is a homogeneous strain-rate, S_{12}, whose magnitude is small compared to the magnitude of the relative vorticity, $|\omega_{12}|$, turbulence may be enhanced and grow from elliptical instabilities. The components of the turbulence which increase are on the $K_3 = 0$ plane in wave-number space (consistent with $\partial u_i / \partial x_3 = 0$), and rapid distortion theory shows us that the turbulence on this plane is unaffected by the Ω_3 rotation, and is therefore materially frame indifferent.

Two Elliptic Flow Examples

Elliptic streamline flows may, or may not, generate turbulence from elliptical instabilities. It depends upon the conditions.

For example, the DNS calculations of Blaisdell and Shariff (BS1996) for elliptic flow were affected by elliptical instabilities. The calculations imposed constant homogeneous mean velocity gradients on an initially isotropic turbulence, and the development of the Reynolds stresses was then calculated over a long time period, from $St = 0$ to about 20. The DNS was made without frame rotation so that the relative and absolute vorticities are equal. It should also be noted that while the initial strain rate was modest, $s_o = Sk_o / \varepsilon_o = 3.37$, the initial turbulence Reynolds number was low, $R_t = 12.75$, so that slow pressure-strain, at least initially, did not play a large role in the turbulence development.

Figure 18.6.1 shows the Reynolds stress tensor glyphs at $St = 1$, 5 and 15, for three cases. At the lowest W/S, 1.25, the glyph begins as one would expect for shear flow. At $St = 1$ the glyph has a near 45 degree orientation due to the initial straining. By $St = 5$ the glyph has rotated clockwise due to flow vorticity. By $St = 15$, however, the glyph is reverting back toward a 45 degree orientation due to the presence of the turbulence generated by elliptical instabilities which have the 45 degree Boussinesq orientation (Fig. 2.2.1). As W/S increases, the vorticity and elliptical instabilities act more quickly, relative to the St time scale. Thus for

$W/S = 2.6$, the glyph at $St = 1$ is rotated clockwise compared with the lower W/S cases, and by $St = 15$, the glyph has nearly returned to the 45 degree orientation.

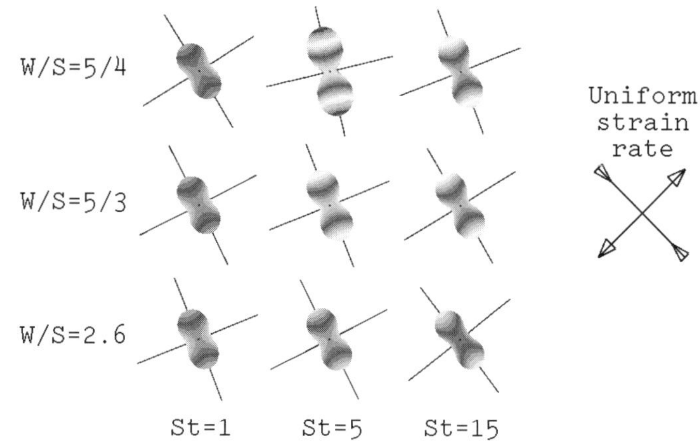

Fig. 18.6.1. Glyphs of $\overline{u_i u_j}/k$ at $St = 1$, 5 and 15 from the elliptic flow DNS of Blaisdell and Shariff (BS1996). Coordinate orientation same as in Fig. 18.5.1.

Figure 18.6.2 shows another example of elliptic streamline flow in a stationary frame. Figure 18.6.2a shows streamlines in a cross-section of one of a pair of counter-rotating vortices measured by Devenport et al. (DZV1997). The outer streamlines have an elliptical shape while the inner streamlines appear circular. However, this vortex more closely resembles the classical vortex in Fig. 4.3.1 than the uniformly-strained solid-body rotation of Fig. 18.5.1. The velocity vectors, Fig. 18.6.2b, increase then decrease with radius, as in Fig. 4.3.1a. The shear component, S_{12}, of the strain rate in the cross-section, Fig. 18.6.2c, alternates between being positive and negative, as for the classical vortex.

The criteria, for the possible generation of the elliptical instabilities analysed in Section 18.5, are twofold. First, in the rotating coordinate system in which the components of the strain-rate tensor are constant along a streamline (or at least uniformly oriented), the ratio of the relative vorticity to strain rate must be greater than one, so that the flow is elliptic (Eq. 18.5.7). Second, the parameter Ω_E (Eq. 18.5.8) must be outside the range $-3/2$ to $-1/2$. Rewriting this criterion as $|\Omega_E + 1| \geq 1/2$, and noting that $W_{21} = \omega_{21} + \Omega_3$, then for elliptical instabilities we require (Appendix Eq. 18.6.1-2)

$$|\Omega_E + 1| = \left|\frac{\Omega_3 + \omega_{21}}{\omega_{21}}\right| = \left|\frac{W_{21}}{\omega_{21}}\right| \geq 1/2\ , \tag{18.6.1}$$

or, using Eq. 18.5.7,

$$\frac{W}{S} \geq \frac{1}{2}\frac{\sqrt{2\omega_{ij}\omega_{ij}}}{S} > \frac{1}{2}\ . \tag{18.6.2}$$

Therefore, for $W/S \leq 0.5$, there will be no elliptical instabilities.

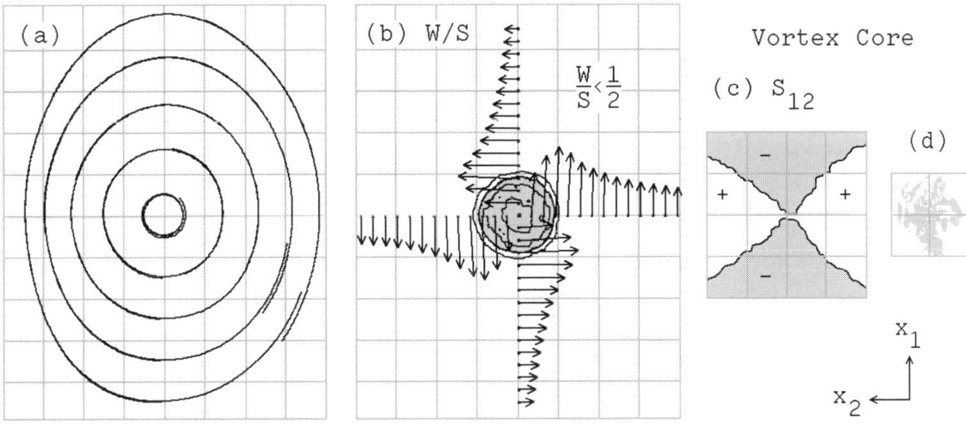

Fig. 18.6.2. Mean flow in a cross-section of one of a pair of counter-rotating vortices, data of Devenport et al. (DZV1997). Results shown in an x_1, x_2 plane, overlaid on a uniform grid. (a) Mean flow streamlines. (b) Selected velocity vectors and contours of W/S, contour interval, 0.5. Shaded core, $W/S > 1$. (c) Contour of $S_{12} = 0$; shaded areas, $S_{12} < 0$. (d) Shaded area, region of possible elliptical instabilities: $|\Omega_E + 1| \geq 1/2$, and $\sqrt{2\omega'_{ij}\omega'_{ij}}/S > 1$.

For Devenport's vortex, contours of W/S are shown in Fig. 18.6.2b. Outside of a core region (approximately coinciding with the maximum velocities), $W/S < 0.5$, and there can be no elliptical instabilities.

Inside the core region, to see if elliptical instabilities are likely, a more complex analysis was carried out for the cross-sectional flow in the x_1, x_2 plane. At each point, a rotation rate, Ω'_3, was determined to minimize the change along the streamline of the dimensionless strain-rate components in the rotating (x'_i) frame (Appendix EQ.18.6.3),

$$\text{minimize} \left(\frac{D(S'_{ij}/S)}{Dt} \right)^2 : \qquad \frac{\partial}{\partial \Omega'_3} \left(\frac{D(S'_{ij}/S)}{Dt} \right)^2 = 0 \,. \tag{18.6.3}$$

Corresponding values of Ω_E $(= \Omega'_3/\omega'_{21})$ and $\sqrt{2\omega'_{ij}\omega'_{ij}}/S$ were obtained. The shaded area in Fig. 18.6.2d corresponds to regions where both criteria for elliptical instabilities, $|\Omega_E + 1| \geq 1/2$, and $\sqrt{2\omega'_{ij}\omega'_{ij}}/S > 1$, are met. The evidence in favor of enhanced turbulence in the vortex core due to elliptical instabilities comes from the Reynolds stress data, which showed peak values of k at the center of the vortex. However, an evaluation of turbulence production rates from the data showed both positive and negative production in the vortex core. This argues against elliptical instabilities, since the Reynolds stresses have not uniformly aligned themselves with the strain rate.

Modeling Comments

The choice of $c_5 = 0$ in the MARV model was based on rapid distortion theory and the initial transient departure from isotropic turbulence. While this does not, in general, capture the increase in turbulence from elliptical instabilities, it does capture the increase in turbulence for the most unstable rotation rate, $\Omega_E = -0.5$ (Fig. 18.5.5).

The analysis in the prior section suggests that more could be done to include turbulence generated from elliptical instabilities in a Reynolds stress model. However, the price might be very high in terms of increased functionality, and the possibility of unwanted side effects would need to be considered.

Chapter 19

SUMMARY OF NEW MODELS

The equations for the new Reynolds stress models developed in this book have been given as they have been developed, and are spread over several chapters. In this final chapter we summarize four models, collecting together the relevant equations. The models have been named to reflect the functionality of the rapid pressure-strain coefficients: C – constant, A – anisotropy, R – rapid distortion, and V – vorticity.

19.1 Shared Model Variables

The Reynolds stress models have differential equations for seven independent turbulence properties, $q = \sqrt{k}$, $\omega = \tilde{\varepsilon}/k$, and 5 independent components of $b_{ij} = \overline{u_i u_j}/(2k) - \delta_{ij}/3$. The transport equations share the common form (Eq. 8.1.1),

$$\frac{\partial \psi}{\partial t} + U_k \frac{\partial \psi}{\partial x_k} = \frac{\partial}{\partial x_k} \nu \frac{\partial \psi}{\partial x_k} + T_\psi + S_\psi , \qquad \text{for} \qquad \psi = q, \omega, b_{ij} , \qquad (19.1.1)$$

where T_ψ, is the turbulent diffusion and S_ψ is the source term.

The source term for the q equation contains no model coefficients (Eq. 8.1.3),

$$S_q = \frac{P}{2q} - \frac{\tilde{\varepsilon}}{2q} = (-b_{ik} S_{ik}) q - \frac{\omega q}{2} , \qquad (19.1.2)$$

while the source term for the ω equation contain the standard production and decay coefficients, $c_{\varepsilon 1}$ and $c_{\varepsilon 2}$ (Eq. 8.1.4),

$$S_\omega = (c_{\varepsilon 1} - 1)(-2 b_{ij} S_{ij}) \omega - (c_{\varepsilon 2} - 1) \omega^2 . \qquad (19.1.3)$$

The source terms for the b_{ij} equations, S_{bij}, are based on production, dissipation, pressure-strain, and (constant) frame rotation. The complexity varies for the different models, and so will be given separately.

Note that the models have been developed and optimized about this choice of variables. While the equations (19.1.1) can be combined and transformed to give equations for other variables, e.g., $\overline{u_i u_j}$ and $\tilde{\varepsilon}$, ignoring the extra terms that arise from the diffusion terms will change the model results. The use of different boundary conditions may also significantly change the results.

19.2 The MC Model

The MC model is so named because the rapid pressure-strain coefficients are constants. The slow pressure-strain coefficient, c_1, and the decay of dissipation coefficient, $c_{\varepsilon 2}$, depend on the turbulence Reynolds number, $\tilde{R}_t = q^2/(\omega\nu)$. The turbulent diffusion of q, ω and b_{ij} is isotropic.

The effective pressure-strain model has one slow and two rapid terms (from Eqs. 3.2.7 and 3.2.17),

$$\frac{\phi_{ij}}{k} = -c_1\omega b_{ij} + c_3 S_{ij} + c_4(b_{ik}S_{jk} + b_{jk}S_{ik} - \frac{2}{3}\delta_{ij}b_{mn}S_{mn}), \tag{19.2.1}$$

so that the source terms for the b_{ij} equations are (from Eq. 8.1.5)

$$\begin{aligned} S_{bij} = &-\frac{1}{2}b_{ij}\left((c_1 - 2)\omega + 2\frac{P}{k}\right) + \frac{1}{2}\left(c_3 - \frac{4}{3}\right)S_{ij} \\ &+\frac{1}{2}(c_4 - 2)(b_{ik}S_{jk} + b_{jk}S_{ik} - \frac{2}{3}\delta_{ij}b_{mn}S_{mn}) \\ &-(b_{ik}W_{jk} + b_{jk}W_{ik}) - \Omega_k\left(\varepsilon_{ikm}b_{jm} + \varepsilon_{jkm}b_{im}\right). \end{aligned} \tag{19.2.2}$$

The pressure-strain model coefficients are (Eq. 8.2.2 and Table 8.2.1)

$$c_1 = 2 + (9-2)\left(1 - \exp(-0.0138\tilde{R}_t^{\,0.75})\right) \tag{19.2.3}$$

and (Eq. 5.2.4)

$$c_3 = 0.15, \qquad c_4 = 1\ . \tag{19.2.4}$$

The coefficients for the production and decay of dissipation are (Eq. 8.2.3 and Table 8.2.1)

$$c_{\varepsilon 1} = 1.55, \qquad c_{\varepsilon 2} = 1.4 + (1.83 - 1.4)\left(1 - \exp(-0.6\tilde{R}_t^{\,0.25})\right)\ . \tag{19.2.5}$$

The turbulent diffusion is isotropic and is given by (Eqs. 8.1.2, 8.2.4, Table 8.2.1)

$$T_\psi = \frac{\partial}{\partial x_k}0.05\frac{q^2}{\omega}\frac{\partial\psi}{\partial x_k} \quad \text{for } \psi = q, \omega, b_{ij}\ . \tag{19.2.6}$$

The wall boundary conditions are (Eqs. 8.2.5, 8.2.6, Table 8.2.1)

$$q = 0, \qquad \omega = 0.07\frac{dq}{dy}, \qquad \frac{db_{ij}}{dy} = 0, \tag{19.2.7}$$

where d/dy is the wall-normal derivative.

19.3 The MR Model

The MR model is the same as the MC model except for the rapid pressure-strain coefficients, which include rapid distortion effects.

The rapid pressure-strain coefficients are (Eqs. 11.5.1-2)

$$c_3 = 0.15(1 - f_B) + 0.8 f_B \tag{19.3.1}$$

and

$$c_4 = (1 - f_B) + \frac{24}{7} f_B , \tag{19.3.2}$$

where

$$f_B(B) = \frac{\max(0, B - 0.7)}{\max(0, B - 0.7) + 1.7} \tag{19.3.3}$$

and

$$B = \frac{k}{(0.9\tilde{\varepsilon} + 0.1\varepsilon)} \frac{b_{ij}}{II} \frac{db_{ij}}{dt} . \tag{19.3.4}$$

Note that db_{ij}/dt for this rapid distortion parameter is not an actual change in b_{ij} with time, but a theoretical change obtained by omitting diffusion and rotation from the b_{ij} equations, so that for the MR model (see Eq. 10.4.2)

$$\begin{aligned} \frac{db_{ij}}{dt} &= -\frac{1}{2} b_{ij}\left((c_1 - 2)\omega + 2\frac{P}{k}\right) + \frac{1}{2}\left(c_3 - \frac{4}{3}\right) S_{ij} \\ &+ \frac{1}{2}(c_4 - 2)(b_{ik}S_{jk} + b_{jk}S_{ik} - \frac{2}{3}\delta_{ij} b_{mn}S_{mn}) - (b_{ik}W_{jk} + b_{jk}W_{ik}) . \end{aligned} \tag{19.3.5}$$

Combining Eqs. 19.3.4 and 19.3.5,

$$B = \frac{k}{(0.9\tilde{\varepsilon} + 0.1\varepsilon) II} \left\{ -\frac{1}{2} II\left((c_1 - 2)\omega + 2\frac{P}{k}\right) + \frac{1}{2}\left(c_3 - \frac{4}{3}\right) b_{ij}S_{ij} + (c_4 - 2) b_{ij} b_{ik} S_{jk} \right\} . \tag{19.3.6}$$

In the limit of isotropic turbulence ($b_{ij}=0$), $f_B=1$. Otherwise, since B depends on c_3 and c_4, it is necessary to iterate for a consistent solution to Eqs. 19.3.1-3 and 19.3.6.

19.4 The MA Model

The MA model is so named because a dependence on anisotropy is included in the model coefficients. This takes several forms: an algebraic dependence on $g=1-4.5b_{ij}b_{ij}+9b_{ik}b_{kn}b_{nj}$ in c_1, c_3, c_4 and $c_{\varepsilon 2}$; a dependence on the gradient of g in c_1, and an anisotropic diffusion model. Like the MC model, c_1 and $c_{\varepsilon 2}$ also depend on the turbulence Reynolds number, $\tilde{R}_t = q^2/(\omega\nu)$.

The MA model was developed to correct three shortcomings of the MC model. With the inclusion of an anisotropy functionality, the MA model can set isotropic limits, and realizable limits for two-dimensional turbulence. The model was also developed to give the strong anisotropy of turbulence observed in laminar sublayers, and extending into the buffer region, of turbulent boundary layers.

The effective pressure-strain model (like the MC model) has one slow and two rapid terms,

$$\frac{\phi_{ij}}{k} = -c_1\omega b_{ij} + c_3 S_{ij} + c_4(b_{ik}S_{jk} + b_{jk}S_{ik} - \frac{2}{3}\delta_{ij}b_{mn}S_{mn}), \tag{19.4.1}$$

so that the source terms for the b_{ij} equations are

$$\begin{aligned} S_{bij} = &-\frac{1}{2}b_{ij}\left((c_1-2)\omega + 2\frac{P}{k}\right) + \frac{1}{2}\left(c_3 - \frac{4}{3}\right)S_{ij} \\ &+\frac{1}{2}(c_4-2)(b_{ik}S_{jk} + b_{jk}S_{ik} - \frac{2}{3}\delta_{ij}b_{mn}S_{mn}) \\ &-(b_{ik}W_{jk} + b_{jk}W_{ik}) - \Omega_k\left(\varepsilon_{ikm}b_{jm} + \varepsilon_{jkm}b_{im}\right). \end{aligned} \tag{19.4.2}$$

The rapid pressure-strain coefficients depend only on g, and are written in terms of the functions f_{g0}, f_{g1} and f_{g3}(Eqs. 5.2.7, 5.2.5 and 5.3.9-10),

$$f_{g0}(g,g_L) = \left(\max\left(0, \frac{g_L - g}{g_L}\right)\right)^2, \tag{19.4.3}$$

$$f_{g1}(g,g_L) = \left(\max\left(0, \frac{g - g_L}{1-g_L}\right)\right)^2, \tag{19.4.4}$$

and

$$f_{g3}(g,g_L,g_H) = 1 - 3\left(f_{lin}(g,g_L,g_H)\right)^2 + 2\left(f_{lin}(g,g_L,g_H)\right)^3, \tag{19.4.5}$$

where

$$f_{lin}(g,g_L,g_H) = \frac{\max\left(0, g_H - g_L - \max(0, g - g_L)\right)}{g_H - g_L} \ . \tag{19.4.6}$$

Using these, c_3 and c_4 are given by (Eq. 8.3.1)

$$c_3 = 0.15 f_{g3}(g,0.05,0.5) + 0.65 f_{g3}(g,0.7,1) \tag{19.4.7}$$

and

$$\begin{aligned} c_4 &= 0.414 - 0.414\left(\max(0, 1 - \sqrt{2g})\right)^2 + 0.086 f_{g3}(g,0,0.5) \\ &+ 0.5 - 0.5 f_{g0}(g,0.5) + \left(\frac{24}{7} - 1\right) f_{g1}(g,0.85) \ . \end{aligned} \tag{19.4.8}$$

The slow pressure-strain coefficient depends on a dimensionless gradient of g (Eq. 8.3.2),

$$G = \frac{q^2}{\omega^2} \frac{\partial g}{\partial x_n} \frac{\partial g}{\partial x_n}, \tag{19.4.9}$$

used to define an effective g (Eq. 8.3.3),

$$g_{eff} = \frac{g^2}{g + 0.3 G^{1.5}} \ . \tag{19.4.10}$$

This is used to evaluate $c_1 = c_1(g_{eff}, \tilde{R}_t)$ (Eq. 8.3.4),

$$c_1 = 2 + \left(c_{1\infty}(g_{eff}) - 2\right) f_{c1}(R_t) \ , \tag{19.4.11}$$

where (Eq. 8.3.6)

$$c_{1\infty}(g_{eff}) = 8 + (2 - 8) f_{g0}(g_{eff},0.5) + (2.5 - 8) f_{g1}(g_{eff},0.7) \tag{19.4.12}$$

and (Eq. 8.3.5 and Table 8.3.1)

$$f_{c1} = 1 - \exp(-3 \times 10^{-6} \tilde{R}_t^{\,2} - 0.045 \tilde{R}_t^{\,0.5}) \ . \tag{19.4.13}$$

The coefficients for the production and decay of dissipation are (Eqs. 8.3.14 and 8.3.9-11)

$$c_{\varepsilon 1} = 1.55, \qquad c_{\varepsilon 2} = 1.4 + (1.83 - 1.4)\left(1 - 0.65 f_{g0}(g, 0.7)\right)\left(1 - \exp(-0.1\tilde{R}_t)\right) . \tag{19.4.14}$$

The turbulent diffusion is a new anisotropic model with the coefficient of diffusion proportional to a turbulent velocity tensor, v_{ij} (Eqs. 8.1.2 and 7.2.7),

$$T_\psi = \frac{\partial}{\partial x_k} c_{t\psi} \frac{q}{\omega} v_{kn} \frac{\partial \psi}{\partial x_n} , \qquad \text{where} \qquad v_{ik} v_{kj} = \overline{u_i u_j} . \tag{19.4.15}$$

The diffusion model constants are (Eq. 8.3.8)

$$\text{for } \psi = q \text{ and for } \psi = b_{ij}, \ c_{t\psi} = 0.0735; \qquad \text{for } \psi = \omega, \ c_{t\psi} = 0.10 . \tag{19.4.16}$$

The wall boundary conditions for q and ω are (Eqs. 8.2.5-6 and 8.3.20)

$$q = 0, \qquad \omega = 0.089 \frac{dq}{dy} , \tag{19.4.17}$$

where d/dy is the wall-normal derivative. The MA model uses fixed values for b_{ij} at the wall. The wall boundary conditions for b_{ij} are defined in terms of a Cartesian coordinate system relative to the point at the wall, such that x_2 is normal to the wall, and x_1 is in the flow direction (the direction of the limiting streamline) parallel to the wall (Fig. 12.1.1). Then, at the wall (Eqs. 8.3.18-19)

$$b_{33} = \frac{1}{6} - \left(0.2 + \frac{1}{6}\right)\left(1 - \exp\left(-0.0384 \frac{1}{\omega} \left|\frac{\partial U_1}{\partial x_2}\right|_{\text{wall}}\right)\right) , \tag{19.4.18}$$

and

$$b_{22} = -1/3, \ b_{11} = -b_{22} - b_{33}, \ b_{12} = 0, \ b_{13} = 0, \text{ and } b_{23} = 0. \tag{19.4.19}$$

19.5 The MARV Model

The MARV model is built around the MA model but with added functionality to include:

(a) the effects of rapid distortion;

(b) the different effects pressure-strain has on simple shear flow and flows without vorticity (e.g. plane strain); and

(c) a wall effect, which makes the anisotropy of the turbulence very near a wall the same, whether or not the wall is rotating.

Same as MA. The effective slow pressure-strain model is the same as for the MA model (from Eq. 19.4.1),

$$\frac{\phi_{ij}^{S}}{k} = -c_1 \omega b_{ij} \,, \tag{19.5.1}$$

so that $c_1 = c_1(g, G, \tilde{R}_t)$ is specified by Eqs. 19.4.9-13 (and 19.4.3-4). The dissipation equation coefficients, $c_{\varepsilon 1}$ and $c_{\varepsilon 2}$, as well as the turbulent diffusion model and the boundary conditions are also the same as for the MA model, Eqs. 19.4.14-19.

Different from MA. The MA and MARV models differ in that the MARV model uses a more complex rapid pressure-strain model (from Eq. 3.2.17),

$$\frac{\Pi_{ij}^{R}}{k} = c_3 S_{ij} + c_4 (b_{ik} S_{jk} + b_{jk} S_{ik} - \frac{2}{3}\delta_{ij} b_{mn} S_{mn}) + c_6 b_{ij} b_{mn} S_{mn} \,, \tag{19.5.2}$$

as well as an additional pressure-strain term to make rotating walls material frame indifferent (Eq. 13.3.1),

$$\frac{\hat{\phi}_{ij}^{\text{rot. wall}}}{k} = 4 f_R \Omega_k \left(\varepsilon_{ikm} b_{jm} + \varepsilon_{jkm} b_{im} \right) . \tag{19.5.3}$$

The parameter f_R is given by (Eq. 13.3.4)

$$f_R = \exp\left(-0.0087 \frac{yq}{\nu} \left(\frac{\tilde{R}_t}{100} \right)^3 \right), \tag{19.5.4}$$

where y is the distance to the nearest wall. Ω_k in Eq. 19.5.3 is the rotation rate of the nearest wall.

If all the walls rotate at the same rate, the source terms for the b_{ij} equations, in the frame rotating with the walls, are (from Eq. 13.3.2)

$$\begin{aligned} S_{bij} = &-\frac{1}{2} b_{ij} \left((c_1 - 2)\omega + (2 + \frac{1}{2} c_6) \frac{P}{k} \right) + \frac{1}{2}\left(c_3 - \frac{4}{3} \right) S_{ij} \\ &+ \frac{1}{2}(c_4 - 2)(b_{ik} S_{jk} + b_{jk} S_{ik} - \frac{2}{3}\delta_{ij} b_{mn} S_{mn}) \\ &-(b_{ik}\omega_{jk} + b_{jk}\omega_{ik}) - 2(1 - f_R)\Omega_k \left(\varepsilon_{ikm} b_{jm} + \varepsilon_{jkm} b_{im} \right) . \end{aligned} \tag{19.5.5}$$

This has been written in terms of the relative vorticity, ω_{ij}, instead of the absolute vorticity, so that it can be seen that at the wall, where $f_R = 1$, the rotation (Ω_k) term is zero.

The rapid pressure-strain coefficients are functions of anisotropy, rapid distortion and vorticity parameters (Eq. 10.1.1),

$$c_n = c_n(g, f_B, f_w), \qquad \text{for} \qquad n = 3,\ 4,\ 6. \tag{19.5.6}$$

f_B and f_w vary between zero and one, and are interpolation parameters between four different limits for each coefficient, c_3, c_4 and c_6 (Eq. 10.1.2),

$$\begin{aligned} c_n(g, f_B, f_w) = {} & (1 - f_w)(1 - f_B)c_{n,fdfW=0} + f_w(1 - f_B)c_{n,fdfW=S} \\ & +(1 - f_w)f_B c_{n,rdtW=0} + f_w f_B c_{n,rdtW=S} \quad . \end{aligned} \tag{19.5.7}$$

The rapid distortion parameter, f_B, is given by (Eq. 10.4.6)

$$f_B(B) = \frac{\max(0, B - 0.7)}{\max(0, B - 0.7) + 1.7}\,, \tag{19.5.8}$$

where (Eq. 10.4.1)

$$B = \frac{b_{ij}}{\omega II}\frac{db_{ij}}{dt} \quad . \tag{19.5.9}$$

db_{ij}/dt for this equation is obtained from (Eq. 10.4.2)

$$\begin{aligned} 2\frac{db_{ij}}{dt} = {} & -b_{ij}\left((c_1 - 2)\omega + (2 + \frac{1}{2}c_6)\frac{P}{k}\right) + \left(c_3 - \frac{4}{3}\right)S_{ij} \\ & +(c_4 - 2)(b_{ik}S_{jk} + b_{jk}S_{ik} - \frac{2}{3}\delta_{ij}b_{mn}S_{mn}) - 2(b_{ik}W_{jk} + b_{jk}W_{ik})\,, \end{aligned} \tag{19.5.10}$$

so that

$$B = \frac{1}{\omega II}\left\{-\frac{1}{2}II\left((c_1 - 2)\omega + (2 + \frac{1}{2}c_6)\frac{P}{k}\right) + \frac{1}{2}\left(c_3 - \frac{4}{3}\right)b_{ij}S_{ij} + (c_4 - 2)b_{ij}b_{ik}S_{jk}\right\} . \tag{19.5.11}$$

The vorticity parameter, f_w, is given by (Eq. 10.1.12)

$$f_w = \left(\min\left(1,\ \max\left(0,\ \frac{8W_{ij}S_{jk}W_{kn}S_{ni}}{S^4}\right)\right)\right)^{1/2} \quad . \tag{19.5.12}$$

The limits for the coefficients are written in terms of the functions f_{g0}, f_{g1} and f_{g3} given in Eqs. 19.4.3-6. For the c_3 coefficient (Eqs. 10.2.8, 10.1.3, 10.1.6 and 10.1.9),

$$c_{3,fdfW=0} = 0.44 - 0.44\left(\max(0,\ 1-\sqrt{4g})\right)^2 + 0.36 f_{g3}(g,0.7,1)\ , \tag{19.5.13}$$

$$c_{3,fdfW=S} = 0.15 f_{g3}(g,0.05,0.5) + 0.65 f_{g3}(g,0.7,1)\ , \tag{19.5.14}$$

$$c_{3,rdtW=0} = 0.6 + 0.2g\ , \tag{19.5.15}$$

$$c_{3,rdtW=S} = 0.1 - 0.1\left(\max\left(0, 1-\sqrt{\frac{g}{0.15}}\right)\right)^2 + 0.7g\ . \tag{19.5.16}$$

For the c_4 coefficient (Eqs. 10.2.7, 10.1.4, 10.1.7 and 10.1.10),

$$c_{4,fdfW=0} = 2.2 - 2.2\left(\max(0,\ 1-\sqrt{4g})\right)^2 + \left(\frac{24}{7} - 2.2\right) f_{g1}(g,0.7)\ , \tag{19.5.17}$$

$$\begin{aligned} c_{4,fdfW=S} = {} & 0.414 - 0.414\left(\max(0,1-\sqrt{2g})\right)^2 + 0.086 f_{g3}(g,0,0.5) \\ & +0.5 - 0.5 f_{g0}(g,0.5) + \left(\frac{24}{7} - 1\right) f_{g1}(g,0.85)\ , \end{aligned} \tag{19.5.18}$$

$$c_{4,rdtW=0} = \frac{24}{7} + 0.8(1-g) + \left(0.1 - \frac{24}{7}\right)\left(\max(0,1-\sqrt{4g})\right)^2\ , \tag{19.5.19}$$

$$c_{4,rdtW=S} = 1.9 - 1.4\left(\max\left(0, 1-\sqrt{\frac{g}{0.15}}\right)\right)^2 - 0.5 f_{g0}(g,0.5) + \left(\frac{24}{7} - 1.9\right) g^2\ . \tag{19.5.20}$$

For the c_6 coefficient (Fig. 10.5.2, Eqs. 10.1.5, 10.1.8 and 10.1.11)

$$c_{6,fdfW=0} = 0\ , \tag{19.5.21}$$

$$c_{6,fdfW=S} = 0\ , \tag{19.5.22}$$

$$c_{6,rdtW=0} = 4.2 - 6\left(\max(0,1-\sqrt{4g})\right)^2\ , \tag{19.5.23}$$

$$c_{6,rdtW=S} = 0\ . \tag{19.5.24}$$

In the limit of isotropic turbulence ($b_{ij} = 0$), $f_B = 1$ and the rdt limits are used. Otherwise, since B depends on c_3, c_4 and c_6, it is necessary to iterate (over Eqs. 19.5.7 and 19.5.11) to obtain consistent values for c_3, c_4 and c_6.

Appendix A

Turbulent Boundary Layers and the Log-Law Region

Simple turbulent boundary layers, by which we refer to flat plate boundary layers with zero pressure gradient and fully developed channel and pipe flows, have many features in common. In this appendix we gather together definitions based on wall friction velocity and commonly accepted values for properties in the log-law region of the boundary layers. In addition, based on data, we consider regions where the turbulence anisotropy is approximately constant. All flows considered are next to smooth walls and without system rotation.

Wall Normalizations using U_τ and ν:

Friction velocity: $$U_\tau = \sqrt{\frac{\tau_w}{\rho}} = \sqrt{\nu\left|\frac{\partial U}{\partial y}\right|_{wall}} \qquad (1)$$

$$y^+ = \frac{yU_\tau}{\nu} \qquad U^+ = \frac{U}{U_\tau} \qquad k^+ = \frac{k}{U_\tau^2} \qquad \varepsilon^+ = \frac{\varepsilon\nu}{U_\tau^4} \qquad q^+ = \frac{\sqrt{k}}{U_\tau} \qquad \omega^+ = \frac{\tilde{\varepsilon}}{k}\frac{\nu}{U_\tau^2} \qquad (2)$$

Very Near Wall

Laminar sublayer: $U^+ = y^+$ valid, $y^+ <$ about 7. (3)

At wall: $$\varepsilon = 2\nu\left(\frac{\partial q}{\partial y}\right)^2 \qquad \varepsilon^+ = 2\left(\frac{\partial q^+}{\partial y^+}\right)^2 \qquad (4)$$

Log-Law Velocity Profiles

The log-law region of a boundary layer is a region where the mean velocity satisfies the relation

$$U^+ = \frac{1}{\kappa}\ln(y^+) + B\ , \qquad (5)$$

where κ and B are constants. Correspondingly,

$$\frac{\partial U^+}{\partial y^+} = \frac{1}{\kappa y^+}\ . \qquad (6)$$

The log-law region begins at about $y^+ = 40$, and based on simple mixing length arguments ends at about $y/\delta = 0.2$. For smooth walls Patel (1965 and 1998) gives

$$B = 5.45, \quad \kappa = 0.418\ . \qquad (7)$$

Comparisons with data are shown in Figs. 1.3.1 and 8.2.8.

Turbulence Properties from Two-Equation Models

Assuming that turbulence kinetic energy is uniform in the log-law region, then turbulence production approximately equals turbulence dissipation. With $P = \varepsilon$, and the turbulent shear stress about equal to the wall shear stress, then (see e.g., Rodi, 1984, pp. 44-45, or Patel, Rodi and Scheuerer, 1985),

$$\frac{k}{U_\tau^2} = \frac{1}{\sqrt{c_\mu}} \qquad k^+ = \frac{1}{\sqrt{c_\mu}} \qquad q^+ = \frac{1}{c_\mu^{1/4}} \tag{8}$$

and

$$\varepsilon = \frac{U_\tau^3}{\kappa y} \qquad \varepsilon^+ = \frac{1}{\kappa y^+} \qquad \omega^+ = \frac{\sqrt{c_\mu}}{\kappa y^+} \,. \tag{9}$$

The two-equation model parameter, $c_\mu = \left(-\overline{uv}/k\right)^2 = (2b_{12})^2 = 0.09,$ (10)
is consistent with $b_{12} = -0.15$.

While (8) and (9) were initially derived to set wall function values for k and ε, they may be used more generally to estimate other log-law region properties.

Dimensionless strain rate: $$s = \frac{k}{\varepsilon}\frac{\partial U}{\partial y} = \frac{1}{\sqrt{c_\mu}} = 3.33\,. \tag{11}$$

Ratio of wall distance to turbulence length scale: $$\frac{y\omega}{q} = \frac{c_\mu^{3/4}}{\kappa} = 0.393 \sim 0.4\,. \tag{12}$$

Turbulence Anisotropy

b_{ij}'s from DNS data for fully developed channel flow and for turbulent boundary layers on a flat plate are shown in Fig. 1. The data is plotted versus both y^+ and y/δ. For the channel, $\delta = D/2$. For the flat plate, $\delta = \delta_{99}$, the location with $U = 0.99U_\infty$. The anisotropies clearly scale on y^+ below $y^+ = 100$. Beyond that, the boundary layers at the higher Reynolds numbers, $R_\tau \geq 298$, have a region where the b_{ij}'s are nearly constant out to about $y/\delta = 0.6$.

Medium and high Reynolds number flat plate data are shown in Fig. 2. The experimental data covers the momentum thickness Reynolds number range, $R_\theta = 2573$ to 57,720. Based on an outer scaling law by Fernholz and Finley (see Appendix FF1996), y/δ is approximated by $4.2y/\Delta$, where Δ is a length scale defined by $\Delta = \int_0^\infty \frac{U_\infty - U}{U_\tau}dy$. Below $y^+ = 100$ there is insufficient experimental data to determine any trends. The plots showing b_{ij} versus y/δ include only data with $y^+ > 100$. There is a fairly flat region in the b_{ij} profiles between $y/\delta = 0.2$ and 0.6.

Below $y/\delta = 0.2$, b_{11} increases to about 0.3 and b_{22} falls to about –0.2, and $-b_{12}$ falls to about 0.1; the results for b_{33} are mixed.

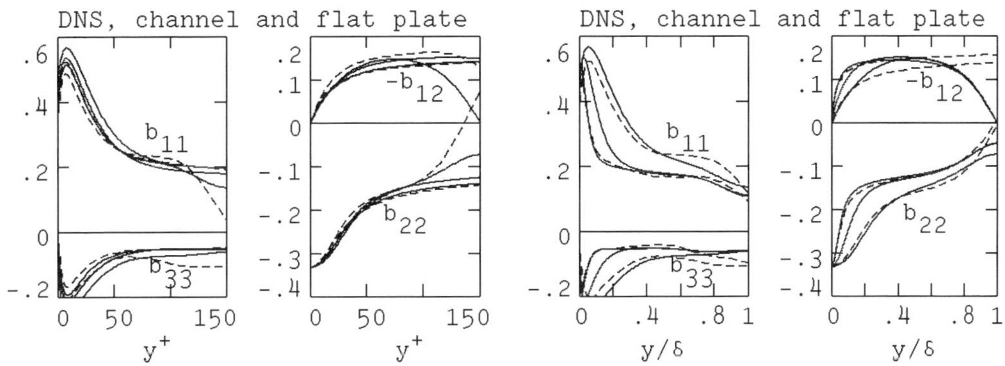

Fig. 1. DNS data: —— fully developed channel flow, Iwamoto et al. (ISK 2002), R_τ = 150, 298 and 643 (R_θ = 236, 514, 1229); ---- flat plate boundary layer, Spalart (1988), R_τ = 150, 650 (R_θ = 300, 1410).

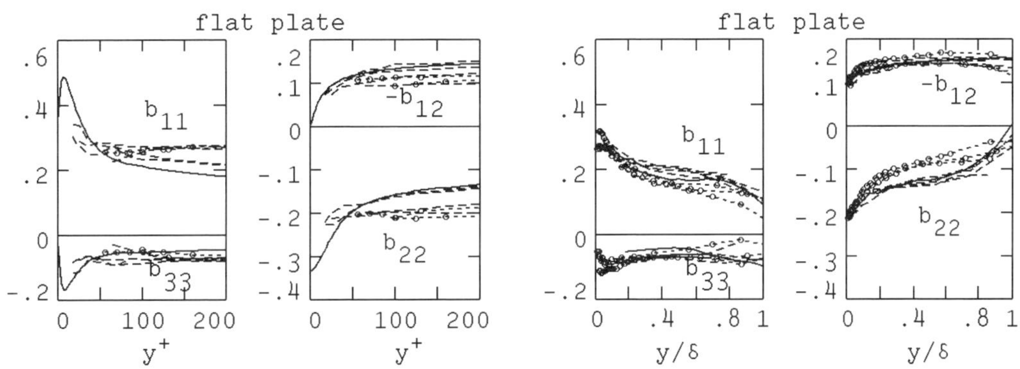

Fig. 2. Flat plate boundary layers with zero pressure-gradient. —— DNS of Spalart (1988), R_τ = 650 (R_θ = 1410). Experimental data: – – Bruns et al. (BDF1992), R_θ = 2573 to 16,080. Fernholz et al. (FKNS1995): o---o R_θ = 20,920 to 57,720. For the y/δ plot, only points with $y^+ > 100$ are shown.

Experimental data for fully developed pipe and channel flows are shown in Fig. 3, together with one DNS case. Most of the experimental channel and pipe data is old and shows significant scatter. However, the same trends, as discussed above, can be observed. When plotted versus y^+, the experimental data shows the same changes in b_{ij} as the DNS below $y^+ = 100$. When the data points for $y^+ > 100$ are plotted versus y/δ, there is a tendency for b_{22} to fall to about –0.2 as y/δ

approaches zero, and a tendency for b_{11} to increase, as for the flat plate data. However, the changes mostly occur more gradually over a larger y/δ range.

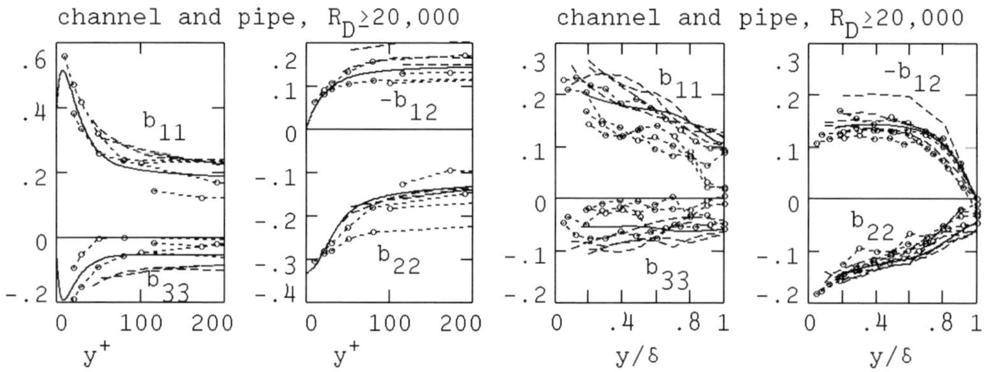

Fig. 3. Fully developed pipe and channel flows. —— DNS channel, Iwamoto et al. (ISK 2002), R_τ = 643 (R_D=24,272, R_θ=1229). Experimental data: – – channel flow, Laufer (L1951), R_D=21,484 to 112,927 (R_θ=800 to 3884); o- - -o pipe flow: Imao et al. (IIH1996), R_D=20,000; Laufer (L1954), R_D=40,285, 428,750; and Lawn (L1971), R_D=88,000. For the y/δ plot, only points with y^+>100 are shown. For the pipe data, $b_{11}=b_{zz}$, $b_{22}=b_{rr}$, $b_{33}=b_{\theta\theta}$ and $-b_{12}=b_{rz}$.

The variation of the data between $y/\delta=0$ and 0.2, but with $y^+>100$ (Figs. 2 and 3), may be evidence of a small amount of wall reflection in the log-law region. Note, however, that the interchange of energy appears to be between b_{11} and b_{22}, with b_{33} staying approximately constant.

Overall for the region $y^+>100$, and $y/\delta<0.6$, the data show the following variations:

	b_{11}		b_{22}		b_{33}		b_{12}		
Channel DNS	0.170	0.206	–0.152	–0.114	–0.057	–0.052	–0.151	–0.136	
Flat plate Exp.	0.134	0.318	–0.212	–0.071	–0.112	–0.047	–0.168	–0.093	
Channel Exp.	0.159	0.267	–0.161	–0.096	–0.106	–0.043	–0.204	–0.125	
Pipe Exp.	0.097	0.229	–0.183	–0.065	–0.076	–0.003	–0.171	–0.106	
Take as standard	**0.20**		**–0.14**		**–0.06**		**–0.15**		(13)

Invariants for Standard Values

The standard values for b_{ij} yield,

$$II=b_{ij}b_{ij}=0.108 \qquad III=b_{ij}b_{jk}b_{ki}=0.00909 \qquad g=0.595\,. \tag{14}$$

Note that the point $(III,g)_{standard}$ lies close to the homogeneous shear points C and T of Champagne et al. (CHC1970) and Tavoularis and Corrsin (TC1981) on Fig. 1.2.5. This boundary layer point is also shown on the functionality cube, Fig. 10.5.3, as the log-law point.

Mohr's Circle and Principal Stresses

Using the standard values, the principal stresses of $\overline{u_i u_j}/(2k)$ can be found from the Mohr's circle for normal and shear stresses in the x_1, x_2 plane. Figure 1.2.1 shows the Mohr's circle for the standard values of $\overline{u_i u_j}/k$.

Mohr's circle for $\overline{u_i u_j}/(2k)$ in x_1, x_2 plane:

$$\text{center at: } 1/3 + 0.5(-0.14+0.2) = 0.3633$$
$$\text{radius} = \sqrt{(0.15)^2 + (.5(0.14+0.2))^2} = 0.2267$$
$$\text{principal stresses}/(2k) = \text{center} + \text{radius},\ \text{center} - \text{radius},\ b_{33} + 1/3$$
$$= 0.5900,\ 0.1367,\ 0.2733\ .$$

Diffusion Model Parameters

For the v_{ij} diffusion used in the MARV model (see Eqs. 7.2.9, 8.1.2 and 8.3.8), it is also of interest to determine the components of $\sqrt{\overline{u_i u_j}/(2k)}$ for the standard log-law anisotropy.

Principal stresses of $\sqrt{\overline{u_i u_j}/(2k)}$ = square root of principal stresses of $\overline{u_i u_j}/(2k)$
$= 0.7681,\ 0.3696,\ 0.5228$.

Mohr's circle for $\sqrt{\overline{u_i u_j}/(2k)}$ in x_1, x_2 plane:

$$\text{center at: } 0.5(0.7681 + 0.3696) = 0.5689$$
$$\text{radius} = 0.5(0.7681 - 0.3696) = 0.1993\ .$$

Principal angles the same for $\sqrt{\overline{u_i u_j}/(2k)}$ and $\overline{u_i u_j}/(2k)$.

Therefore components of $\sqrt{\overline{u_i u_j}/(2k)}$ are:

1,1: $\quad \text{center} + \text{radius} \times (\overline{u_1 u_1} - \overline{u_2 u_2})/\sqrt{(\overline{u_1 u_1} - \overline{u_2 u_2})^2 + (2\overline{u_1 u_2})^2} = 0.7183$

2,2: $\quad \text{center} - \text{radius} \times (\overline{u_1 u_1} - \overline{u_2 u_2})/\sqrt{(\overline{u_1 u_1} - \overline{u_2 u_2})^2 + (2\overline{u_1 u_2})^2} = 0.4195$

3,3: $\quad 0.5228$

1,2: $\quad 2 \times \text{radius} \times \overline{u_1 u_2}/\sqrt{(\overline{u_1 u_1} - \overline{u_2 u_2})^2 + (2\overline{u_1 u_2})^2} = -0.1318\ .$

Some diffusion model parameters:

Daly-Harlow diffusion $\delta_{i2}\delta_{j2}\frac{\overline{u_iu_j}}{2k} = 0.1933$

v_{ij} diffusion $\delta_{i2}\delta_{j2}\sqrt{\frac{\overline{u_iu_j}}{2k}} = 0.4195$.

Therefore in the log-law region, the same turbulent diffusion in the x_2 direction is obtained by:

Isotropic	v_{ij}	Daly-Harlow
$\frac{q^2}{\omega}$	$2.38\,\delta_{i2}\delta_{j2}\sqrt{\frac{\overline{u_iu_j}}{2k}}\,\frac{q^2}{\omega}$	$5.17\,\delta_{i2}\delta_{j2}\frac{\overline{u_iu_j}}{2k}\,\frac{q^2}{\omega}$.

Appendix B

Absolute and Relative Velocity and Vorticity

The equations pertaining to relative and absolute velocity and relative and absolute vorticity are introduced into the text of the book as needed. Here we gather together the definitions and expand upon them for convenience.

The absolute and relative velocities are related by Eq. 2.1.7,

$$V_i = U_i + \varepsilon_{ikm}\Omega_k x_m \ , \qquad (2.1.7)$$

when the origin of the Cartesian, x_m, coordinate system is on the axis of rotation. Differentiating,

$$\frac{\partial V_i}{\partial x_j} = \frac{\partial U_i}{\partial x_j} + \varepsilon_{ikj}\Omega_k \ . \qquad (1)$$

Strain Rate

From (1),

$$\begin{aligned}\frac{\partial V_i}{\partial x_j} + \frac{\partial V_j}{\partial x_i} &= \frac{\partial U_i}{\partial x_j} + \varepsilon_{ikj}\Omega_k + \frac{\partial U_j}{\partial x_i} + \varepsilon_{jki}\Omega_k \\ &= \frac{\partial U_i}{\partial x_j} + \varepsilon_{ikj}\Omega_k + \frac{\partial U_j}{\partial x_i} - \varepsilon_{ikj}\Omega_k = \frac{\partial U_i}{\partial x_j} + \frac{\partial U_j}{\partial x_i} \ ,\end{aligned} \qquad (2)$$

so that the strain-rate tensor, given by Eq. 1.3.20, is

$$S_{ij} = \frac{1}{2}\left(\frac{\partial U_i}{\partial x_j} + \frac{\partial U_j}{\partial x_i}\right) = \frac{1}{2}\left(\frac{\partial V_i}{\partial x_j} + \frac{\partial V_j}{\partial x_i}\right) , \qquad (3)$$

and $S_{ji} = S_{ij}$.

Vorticity

The relative and absolute vorticity tensors are defined by Eqs. 1.3.25 and 3.2.16,

$$\omega_{ij} = \frac{1}{2}\left(\frac{\partial U_i}{\partial x_j} - \frac{\partial U_j}{\partial x_i}\right) \quad \text{and} \quad W_{ij} = \frac{1}{2}\left(\frac{\partial V_i}{\partial x_j} - \frac{\partial V_j}{\partial x_i}\right) , \qquad (4)$$

Note that $\omega_{ji} = -\omega_{ij}$ and $W_{ji} = -W_{ij}$. Summing (3) and (4) gives

$$\frac{\partial U_i}{\partial x_j} = S_{ij} + \omega_{ij} \quad \text{and} \quad \frac{\partial V_i}{\partial x_j} = S_{ij} + W_{ij} \ . \qquad (5)$$

From (1),

$$\frac{\partial V_i}{\partial x_j} - \frac{\partial V_j}{\partial x_i} = \frac{\partial U_i}{\partial x_j} + \varepsilon_{ikj}\Omega_k - \frac{\partial U_j}{\partial x_i} - \varepsilon_{jki}\Omega_k$$
$$= \frac{\partial U_i}{\partial x_j} - \frac{\partial U_j}{\partial x_i} + 2\varepsilon_{ikj}\Omega_k \quad . \tag{6}$$

Comparing with (4),

$$W_{ij} = \omega_{ij} + \varepsilon_{ikj}\Omega_k \ . \tag{7}$$

The relative vorticity, in vector form, is given by Eq. 2.2.10,

$$\omega_i = \varepsilon_{ijk}\frac{\partial U_k}{\partial x_j} \ . \tag{2.2.10}$$

The relations to transform between the relative vorticity vector and the relative vorticity tensor are given by Eq. 2.2.11,

$$w_{jk} = -\frac{1}{2}\omega_i\varepsilon_{ijk} \qquad \text{and} \qquad \omega_i = -\varepsilon_{ijk}\omega_{jk} \ . \tag{2.2.11}$$

Combining this with (7),

$$W_{ij} = \omega_{ij} + \varepsilon_{ikj}\Omega_k = -\frac{1}{2}\omega_k\varepsilon_{kij} + \varepsilon_{ikj}\Omega_k = \frac{1}{2}\varepsilon_{ikj}\omega_k + \varepsilon_{ikj}\Omega_k \ . \tag{8}$$

The component of vorticity representing rotation about the x_3 axis is

$$W_{21} = -W_{12} = \omega_{21} + \Omega_3 = \frac{1}{2}\omega_3 + \Omega_3 = \frac{1}{2}\left(\frac{\partial V_2}{\partial x_1} - \frac{\partial V_1}{\partial x_2}\right) \ . \tag{9}$$

The components of the strain-rate and vorticity tensors in cylindrical coordinates are given in Appendix D, Eqs. (22) and (26).

Fluid Rotation

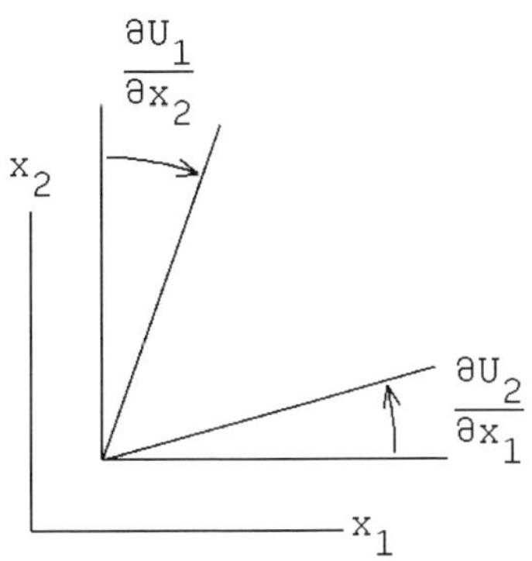

Vorticity = 2 × Average rotation rate of two mutually perpendicular fluid legs;

$$\text{e.g.} \quad \omega_3 = \frac{\partial U_2}{\partial x_1} - \frac{\partial U_1}{\partial x_2} = 2 \times \frac{1}{2}\left(\frac{\partial U_2}{\partial x_1} - \frac{\partial U_1}{\partial x_2}\right) . \tag{10}$$

Also, the average rotation rate of two mutually perpendicular fluid legs is the rate of rotation that the fluid particle would have if it were instantaneously frozen, maintaining its angular momentum. The average rotation rate, r_i, is

$$r_i = \frac{\omega_i}{2} = -\frac{1}{2}\varepsilon_{ijk}\omega_{jk} \, . \tag{11}$$

Equation 2.2.12 then becomes

$$d\overline{u_i u_j}/dt = -2kr_k\left(\varepsilon_{ikm}b_{jm} + \varepsilon_{jkm}b_{im}\right) , \tag{12}$$

compared with Eq. 2.2.15 for the Coriolis term,

$$d\overline{u_i u_j}/dt = F_{ij} = -4k\Omega_k\left(\varepsilon_{ikm}b_{jm} + \varepsilon_{jkm}b_{im}\right) . \tag{2.2.15}$$

Now, in the text examples for

	Eq. 2.2.12	and	Eq. 2.2.15,
	$\omega_3 = -2$	and	$\Omega_3 = 1$,
so	$r_3 = -1$,		
and	$-2r_3 = +2$,	while	$-4\Omega_3 = -4$.

Therefore, in the Coriolis example, Fig. 2.2.5, the glyph rotates at –2 times the rotation rate in the vorticity example, Fig. 2.2.4. The glyph rotation in both cases is opposite to the sense of the driving rotation.

Appendix C

Cayley-Hamilton Theorem and Analysis

Following Lumley (1970, see his Appendix A2.6, pp.156-158), apply Lumley's equations to the 3x3 symmetric tensor, a_{ij}.
Define invariants:

$$I_o = a_{ii}, \quad II_o = \frac{1}{2}(I_o I_o - a_{ij}a_{ji}), \quad III_o = \frac{1}{6}(I_o I_o I_o - 3I_o a_{ij}a_{ji} + 2a_{ij}a_{jk}a_{ki}) . \tag{1}$$

Define the nth power of the tensor as the n factor product:

$$a_{ij}^n = a_{ik}a_{kp}.......a_{pj} . \tag{2}$$

Define the zero power of the tensor:

$$a_{ij}^0 = \delta_{ij} . \tag{3}$$

The eigenvectors, B_i (principal directions), and eigenvalues, λ (principal stresses), are the solutions to

$$B_i a_{ij} = \lambda B_j . \tag{4}$$

The nth power of a_{ij} has the same eigenvectors as a_{ij}, and the eigenvalues are λ^n; i.e. the equation,

$$B_i a_{ij}^n = \lambda^n B_j , \tag{5}$$

is satisfied.

From this one can obtain the Cayley-Hamilton theorem,

$$a_{ij}^n - I_o a_{ij}^{n-1} + II_o a_{ij}^{n-2} - III_o a_{ij}^{n-3} = 0 . \tag{6}$$

Generalized Cayley-Hamilton

Following Rivlin (1955, see his Eqs. 4.21 and 4.22), for 3x3 matrices, A_{ij}, B_{ij}, and C_{ij},

$$\begin{aligned}
&(A_{ik}B_{kn}C_{nj} + A_{ik}C_{kn}B_{nj} + B_{ik}C_{kn}A_{nj} + B_{ik}A_{kn}C_{nj} + C_{ik}A_{kn}B_{nj} + C_{ik}B_{kn}A_{nj}) \\
&- A_{ij}(B_{kn}C_{nk} - B_{kk}C_{nn}) - B_{ij}(C_{kn}A_{nk} - C_{kk}A_{nn}) - C_{ij}(A_{kn}B_{nk} - A_{kk}B_{nn}) \\
&-(B_{ik}C_{kj} + C_{ik}B_{kj})A_{nn} - (C_{ik}A_{kj} + A_{ik}C_{kj})B_{nn} - (A_{ik}B_{kj} + B_{ik}A_{kj})C_{nn} \\
&- \delta_{ij}(A_{kk}B_{nn}C_{mm} - A_{kk}B_{nm}C_{mn} - B_{kk}C_{nm}A_{mn} - C_{kk}A_{nm}B_{mn} \\
&\qquad + A_{kn}B_{nm}C_{mk} + C_{kn}B_{nm}A_{mk}) = 0 .
\end{aligned} \tag{7}$$

Setting $C_{ij} = A_{ij}$,

$$A_{ik}B_{kn}A_{nj} + A_{ik}A_{kn}B_{nj} + B_{ik}A_{kn}A_{nj} - A_{ij}(A_{kn}B_{nk} - A_{kk}B_{nn}) \\ -\frac{1}{2}B_{ij}(A_{kn}A_{nk} - A_{kk}A_{nn}) - (A_{ik}B_{kj} + B_{ik}A_{kj})A_{nn} - A_{ik}A_{kj}B_{nn} \\ -\delta_{ij}\left(A_{kn}A_{nm}B_{mk} - A_{kk}A_{nm}B_{mn} + \frac{1}{2}B_{kk}(A_{nn}A_{mm} - A_{nm}A_{mn})\right) = 0 \quad . \tag{8}$$

Analysis

For the anisotropy tensor, b_{ij}, defining $II = b_{ij}b_{ji}$ and $III = b_{ij}b_{jk}b_{ki}$, the Lumley invariants, defined by (1), are

$$I_o = 0, \qquad II_o = -\frac{1}{2}II \quad \text{and} \quad III_o = \frac{1}{3}III \ . \tag{9}$$

From (6), the equation for b_{ij}^3 is

$$b_{ik}b_{kn}b_{nj} - I_o b_{ik}b_{kj} + II_o b_{ij} - III_o \delta_{ij} = 0 \ , \tag{10}$$

or

$$b_{ik}b_{kn}b_{nj} = \frac{1}{2}IIb_{ij} + \frac{1}{3}III\delta_{ij} \ . \tag{11}$$

Following Ristorcelli et al. (RLA1995), apply (8) with A_{ij} as the anisotropy tensor, b_{ij}, and B_{ij} as the strain-rate tensor, S_{ij}. Since $b_{ii} = 0$ and $S_{ii} = 0$, this gives

$$b_{ik}S_{kn}b_{nj} = -(b_{ik}b_{kn}S_{nj} + S_{ik}b_{kn}b_{nj}) + (b_{kn}S_{nk})b_{ij} + \frac{1}{2}(b_{kn}b_{nk})S_{ij} + (b_{kn}b_{nm}S_{mk})\delta_{ij} \ . \tag{12}$$

Multiplying (12) by b_{jp},

$$b_{ik}S_{kn}b_{np}b_{pj} + b_{ik}b_{kn}S_{np}b_{pj} = -\frac{1}{3}IIIS_{ij} + (b_{kn}S_{nk})b_{ip}b_{pj} + (b_{kn}b_{nm}S_{mk})b_{ij} \ . \tag{13}$$

Multiplying (13) by b_{qi},

$$b_{iq}b_{qk}S_{kn}b_{np}b_{pj} = -\frac{1}{2}IIb_{in}S_{np}b_{pj} - \frac{1}{3}III(b_{ip}S_{pj} + b_{jp}S_{pi}) \\ +(b_{kn}S_{nk})(\frac{1}{2}IIb_{ij} + \frac{1}{3}III\delta_{ij}) + (b_{kn}b_{nm}S_{mk})b_{iq}b_{qj} \quad . \tag{14}$$

Appendix D

Conversion to Cylindrical Coordinates

In the bulk of the text, the velocity, Reynolds stresses and differential equations are presented in Cartesian coordinates. This appendix converts the various expressions to cylindrical, r,θ,z, coordinates. Extra terms arise when there are spatial derivatives of vector or tensor properties, because unit vectors in the radial and tangential directions are not constants. For consistency with Eq. 2.1.3, where r is the distance to the axis of rotation, only rotation Ω_z, about the z axis, is considered.

Unit Vectors

Consider the Cartesian and cylindrical coordinates with a common origin and with $z = x_3$, as shown at right. Unit vectors in the radial, tangential and axial directions, $\underline{m}_r$, $\underline{m}_\theta$, and $\underline{m}_z$, respectively, have the Cartesian components,

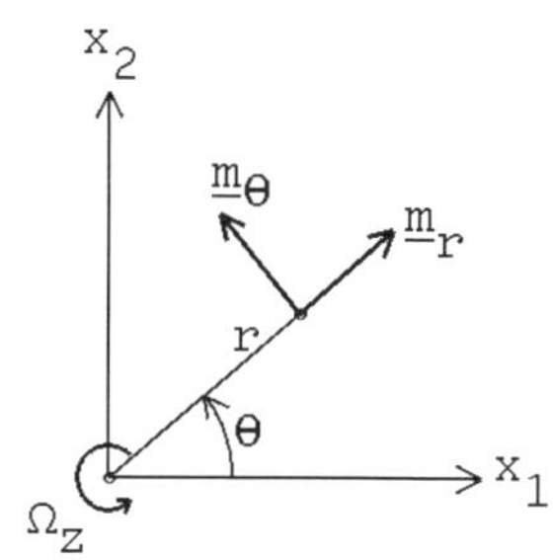

$$m_{r,i} = \cos\theta, \sin\theta, 0$$
$$m_{\theta,i} = -\sin\theta, \cos\theta, 0 \qquad (1)$$
$$m_{z,i} = 0, 0, 1 \quad .$$

Taking the derivatives of $\underline{m}_r$ and $\underline{m}_\theta$ with respect to θ yields

$$\frac{\partial m_{r,i}}{\partial\theta} = \frac{\partial\cos\theta}{\partial\theta}, \frac{\partial\sin\theta}{\partial\theta}, 0 = -\sin\theta, \cos\theta, 0 \quad = m_{\theta,i} \qquad (2)$$

and

$$\frac{\partial m_{\theta,i}}{\partial\theta} = -\cos\theta, -\sin\theta, 0 \quad = -m_{r,i} \quad . \qquad (3)$$

Velocity Components

Using the convention that capital letter subscripts denote r, θ and z components, so that

$$U_I = U_r, U_\theta, U_z, \qquad (4)$$

while lower case subscripts continue to represent Cartesian components, then with

$$\underline{m}_I = \underline{m}_r, \underline{m}_\theta, \underline{m}_z \ , \qquad (5)$$

equations may be written to convert between Cartesian and cylindrical components. For the mean velocity,

$$U_I = m_{I,j} U_j \;, \tag{6}$$

and

$$U_i = m_{J,i} U_J \;, \tag{7}$$

where summation of the three values for J is understood by the repeating subscript.

Note that these relations apply to any vectors, not just velocity. Correspondingly, tensors are transformed using

$$a_{IJ} = m_{I,i} m_{J,j} a_{ij} \;, \tag{8}$$

and

$$a_{ij} = m_{I,i} m_{J,j} a_{IJ} \;. \tag{9}$$

Gradient Transformation

To convert spatial derivatives between Cartesian and cylindrical components, we write

$$\nabla_I = \frac{\partial}{\partial r}, \frac{1}{r}\frac{\partial}{\partial \theta}, \frac{\partial}{\partial z} \;, \tag{10}$$

so that

$$\nabla_I = m_{I,j} \frac{\partial}{\partial x_j} \tag{11}$$

and

$$\frac{\partial}{\partial x_i} = m_{J,i} \nabla_J \;. \tag{12}$$

Velocity Gradient, Strain Rate and Vorticity Tensors

The components of the velocity gradient tensor in cylindrical coordinates, which we will call G_{IJ}, may be found from the Cartesian components using (8),

$$G_{IJ} = m_{I,i} m_{J,j} \frac{\partial U_i}{\partial x_j} \;. \tag{13}$$

Using (7) and (12) to eliminate the Cartesian components,

$$G_{IJ} = m_{I,i} m_{J,j} m_{K,j} \nabla_K m_{N,i} U_N \;, \tag{14}$$

and noting that

$$m_{J,j} m_{K,j} = \delta_{JK} \;, \tag{15}$$

then

$$G_{IJ} = m_{I,i} \nabla_J m_{N,i} U_N = \nabla_J U_I + m_{I,i} U_N \nabla_J m_{N,i} \;. \tag{16}$$

From (2) and (3), the derivatives of the unit vectors may be written as

$$\nabla_J m_{N,i} = \delta_{J\theta} \frac{1}{r} \frac{\partial}{\partial \theta} m_{N,i} = \delta_{J\theta} \frac{1}{r} \left(\delta_{Nr} m_{\theta,i} - \delta_{N\theta} m_{r,i} \right) . \tag{17}$$

Therefore,

$$G_{IJ} = \nabla_J U_I + m_{I,i} U_N \delta_{J\theta} \frac{1}{r} \left(\delta_{Nr} m_{\theta,i} - \delta_{N\theta} m_{r,i} \right) , \tag{18}$$

and using (15),

$$G_{IJ} = \nabla_J U_I + \delta_{J\theta} \frac{1}{r} \left(U_r \delta_{I\theta} - U_\theta \delta_{Ir} \right) . \tag{19}$$

Writing out the individual terms

$$\begin{aligned} G_{rr} &= \frac{\partial U_r}{\partial r} , \quad G_{r\theta} = \frac{1}{r} \frac{\partial U_r}{\partial \theta} - \frac{U_\theta}{r}, \quad G_{rz} = \frac{\partial U_r}{\partial z} \\ G_{\theta r} &= \frac{\partial U_\theta}{\partial r}, \quad G_{\theta\theta} = \frac{1}{r} \frac{\partial U_\theta}{\partial \theta} + \frac{U_r}{r} , \quad G_{\theta z} = \frac{\partial U_\theta}{\partial z} \\ G_{zr} &= \frac{\partial U_z}{\partial r}, \quad G_{z\theta} = \frac{1}{r} \frac{\partial U_z}{\partial \theta}, \quad G_{zz} = \frac{\partial U_z}{\partial z} . \end{aligned} \tag{20}$$

The strain rate in cylindrical coordinates may then be found from

$$S_{IJ} = \frac{1}{2} \left(G_{IJ} + G_{JI} \right) , \tag{21}$$

so that

$$\begin{aligned} &S_{rr} = \frac{\partial U_r}{\partial r} , \quad S_{\theta\theta} = \frac{1}{r} \frac{\partial U_\theta}{\partial \theta} + \frac{U_r}{r} , \quad S_{zz} = \frac{\partial U_z}{\partial z} , \quad S_{r\theta} = S_{\theta r} = \frac{1}{2} \left(\frac{1}{r} \frac{\partial U_r}{\partial \theta} + \frac{\partial U_\theta}{\partial r} - \frac{U_\theta}{r} \right), \\ &S_{rz} = S_{zr} = \frac{1}{2} \left(\frac{\partial U_r}{\partial z} + \frac{\partial U_z}{\partial r} \right), \quad \text{and} \quad S_{\theta z} = S_{z\theta} = \frac{1}{2} \left(\frac{1}{r} \frac{\partial U_z}{\partial \theta} + \frac{\partial U_\theta}{\partial z} \right) . \end{aligned} \tag{22}$$

Correspondingly, the relative vorticity tensor is given by

$$\omega_{IJ} = \frac{1}{2} \left(G_{IJ} - G_{JI} \right) \tag{23}$$

or

$$\begin{aligned} &\omega_{rr} = \omega_{\theta\theta} = \omega_{zz} = 0 , \quad \omega_{r\theta} = -\omega_{\theta r} = \frac{1}{2} \left(\frac{1}{r} \frac{\partial U_r}{\partial \theta} - \frac{\partial U_\theta}{\partial r} - \frac{U_\theta}{r} \right), \\ &\omega_{rz} = -\omega_{zr} = \frac{1}{2} \left(\frac{\partial U_r}{\partial z} - \frac{\partial U_z}{\partial r} \right), \quad \text{and} \quad \omega_{\theta z} = -\omega_{z\theta} = \frac{1}{2} \left(\frac{\partial U_\theta}{\partial z} - \frac{1}{r} \frac{\partial U_z}{\partial \theta} \right) , \end{aligned} \tag{24}$$

The absolute vorticity tensor, W_{IJ}, is obtained by replacing the relative velocity, U, with the absolute velocity, V, in Eq. (24). For rotation Ω_z about the z axis,

$$V_r = U_r, \qquad V_\theta = U_\theta + \Omega_z r, \qquad V_z = U_z, \tag{25}$$

so that

$$W_{rr} = W_{\theta\theta} = W_{zz} = 0 \ , \quad W_{r\theta} = -W_{\theta r} = \frac{1}{2}\left(\frac{1}{r}\frac{\partial U_r}{\partial \theta} - \frac{\partial U_\theta}{\partial r} - \frac{U_\theta}{r} - 2\Omega_z\right),$$

$$W_{rz} = -W_{zr} = \frac{1}{2}\left(\frac{\partial U_r}{\partial z} - \frac{\partial U_z}{\partial r}\right), \quad \text{and} \quad W_{\theta z} = -W_{z\theta} = \frac{1}{2}\left(\frac{\partial U_\theta}{\partial z} - \frac{1}{r}\frac{\partial U_z}{\partial \theta}\right). \tag{26}$$

Equations to be Transformed

The equations to be transformed are the continuity equation (for incompressible flow),

$$\frac{\partial U_i}{\partial x_i} = 0, \tag{27}$$

the mean momentum equation and turbulence property equations. The cylindrical components of the mean momentum equation (Eq. 2.1.10),

$$\frac{\partial U_i}{\partial t} + U_k \frac{\partial U_i}{\partial x_k} + \frac{\partial \overline{u_i u_k}}{\partial x_k} = -2\varepsilon_{ikm}\Omega_k U_m + \frac{\partial}{\partial x_i}\left(\frac{\Omega_z^2 r^2}{2}\right) - \frac{1}{\rho}\frac{\partial \overline{P}}{\partial x_i} + \frac{\partial}{\partial x_k}\nu\frac{\partial U_i}{\partial x_k} + \frac{\partial}{\partial x_k}\nu\frac{\partial U_k}{\partial x_i}\ , \tag{28}$$

are obtained by multiplying the equation by the unit vectors, $m_{I,i}$, for cylindrical coordinates. The equations for the turbulence properties, $\psi = q$, ω and b_{ij}, are given by Eqs. 8.1.1 and 8.1.2 with source terms, S_ψ, described in later equations. Combining Eqs. 8.1.1 and 8.1.2,

$$\frac{\partial \psi}{\partial t} + U_k \frac{\partial \psi}{\partial x_k} = \frac{\partial}{\partial x_k}\nu\frac{\partial \psi}{\partial x_k} + \underbrace{\frac{\partial}{\partial x_k} c_{t\psi}\frac{q^2}{\omega}\left(\frac{\overline{u_k u_n}}{k}\right)^{n_t}\frac{\partial \psi}{\partial x_n}}_{T_\psi} + S_\psi\ . \tag{29}$$

To obtain the cylindrical components of the b_{ij} equations, Eq. (29) is multiplied by the unit vectors, $m_{I,i}m_{J,j}$, for cylindrical coordinates.

For all the equations, (7) and (12) are then used to eliminate references to Cartesian components.

Continuity

For the continuity equation,

$$0 = \frac{\partial U_i}{\partial x_i} = m_{I,i}\nabla_I m_{K,i}U_K \quad = m_{I,i}m_{K,i}\nabla_I U_K + m_{I,i}U_K\nabla_I m_{K,i}$$
$$= \nabla_I U_I + m_{I,i}U_K \frac{\delta_{I\theta}}{r}\left(\delta_{Kr}m_{\theta,i} - \delta_{K\theta}m_{r,i}\right) \;= \frac{\partial U_r}{\partial r} + \frac{1}{r}\frac{\partial U_\theta}{\partial \theta} + \frac{\partial U_z}{\partial z} + \frac{U_r}{r} \ ,$$

or

$$\frac{1}{r}\frac{\partial r U_r}{\partial r} + \frac{1}{r}\frac{\partial U_\theta}{\partial \theta} + \frac{\partial U_z}{\partial z} = 0 \ . \tag{30}$$

For the remaining equations to be transformed, the algebra to derive the terms is omitted here but given in Appendix D.CylCoor (on CD). The equation numbering used here is compatible with the full appendix.

Convection Terms

The convection term for scalars is

$$U_k\frac{\partial \psi}{\partial x_k} = U_r\frac{\partial \psi}{\partial r} + \frac{U_\theta}{r}\frac{\partial \psi}{\partial \theta} + U_z\frac{\partial \psi}{\partial z} \ . \tag{31}$$

For the momentum equation,

$$m_{I,i}U_k\frac{\partial U_i}{\partial x_k} = U_r\frac{\partial U_I}{\partial r} + \frac{U_\theta}{r}\frac{\partial U_I}{\partial \theta} + U_z\frac{\partial U_I}{\partial z} - \delta_{Ir}\frac{U_\theta U_\theta}{r} + \delta_{I\theta}\frac{U_\theta U_r}{r} \ . \tag{32}$$

For the b_{ij} equations, the terms are straightforward, except for the term containing $\partial/\partial\theta$,

$$\begin{aligned} m_{I,i}m_{J,j}U_k\frac{\partial b_{ij}}{\partial x_k} &= U_r\frac{\partial b_{IJ}}{\partial r} + \frac{U_\theta}{r}\frac{\partial b_{IJ}}{\partial \theta} + U_z\frac{\partial b_{IJ}}{\partial z} \\ &\quad + \frac{U_\theta}{r}\left(-\delta_{Ir}b_{\theta J} - \delta_{Jr}b_{I\theta} + \delta_{I\theta}b_{rJ} + \delta_{J\theta}b_{Ir}\right) \ . \end{aligned} \tag{33}$$

Diffusion Terms

The terms, for both the laminar and turbulent diffusion of turbulence properties, can be written in the form $\frac{\partial}{\partial x_k}C_{kn}\frac{\partial \psi}{\partial x_n}$, with

$$C_{kn} = \nu\delta_{kn} \quad \text{for laminar diffusion,} \tag{34}$$

and

$$C_{kn} = c_{t\psi}\frac{q^2}{\omega}\left(\frac{\overline{u_k u_n}}{k}\right)^{n_t} \quad \text{for turbulent diffusion.} \tag{35}$$

Converting to cylindrical coordinates,

$$\frac{\partial}{\partial x_k} C_{kn} \frac{\partial \psi}{\partial x_n} = \frac{1}{r}\frac{\partial}{\partial r} r\left(C_{rr} \frac{\partial \psi}{\partial r} + C_{r\theta} \frac{1}{r}\frac{\partial \psi}{\partial \theta} + C_{rz} \frac{\partial \psi}{\partial z} \right) + \frac{1}{r}\frac{\partial}{\partial \theta}\left(C_{\theta r} \frac{\partial \psi}{\partial r} + C_{\theta\theta} \frac{1}{r}\frac{\partial \psi}{\partial \theta} + C_{\theta z} \frac{\partial \psi}{\partial z} \right) + \frac{\partial}{\partial z}\left(C_{zr} \frac{\partial \psi}{\partial r} + C_{z\theta} \frac{1}{r}\frac{\partial \psi}{\partial \theta} + C_{zz} \frac{\partial \psi}{\partial z} \right) \quad . \tag{36}$$

For laminar diffusion, this reduces to

$$\frac{\partial}{\partial x_k} \nu \frac{\partial \psi}{\partial x_k} = \frac{1}{r}\frac{\partial}{\partial r} r\nu \frac{\partial \psi}{\partial r} + \frac{1}{r}\frac{\partial}{\partial \theta}\frac{\nu}{r}\frac{\partial \psi}{\partial \theta} + \frac{\partial}{\partial z} \nu \frac{\partial \psi}{\partial z} \quad . \tag{37}$$

The laminar diffusion term for the momentum equation becomes

$$m_{I,i} \frac{\partial}{\partial x_k} \nu \frac{\partial U_i}{\partial x_k} = \frac{1}{r}\frac{\partial}{\partial r} \nu r \frac{\partial U_I}{\partial r} + \frac{1}{r}\frac{\partial}{\partial \theta}\frac{\nu}{r}\frac{\partial U_I}{\partial \theta} + \frac{\partial}{\partial z} \nu \frac{\partial U_I}{\partial z} - \delta_{Ir} \frac{1}{r^2}\left(\nu \frac{\partial U_\theta}{\partial \theta} + \frac{\partial \nu U_\theta}{\partial \theta} + \nu U_r \right) + \delta_{I\theta} \frac{1}{r^2}\left(\nu \frac{\partial U_r}{\partial \theta} + \frac{\partial \nu U_r}{\partial \theta} - \nu U_\theta \right) \quad . \tag{38}$$

Considering the diffusion of b_{ij} in the general form

$$m_{I,i} m_{J,j} \frac{\partial}{\partial x_k} C_{kn} \frac{\partial b_{ij}}{\partial x_n} = m_{I,i} m_{J,j} \frac{1}{r}\frac{\partial}{\partial r} r\left(C_{rr} \frac{\partial b_{ij}}{\partial r} + C_{r\theta} \frac{1}{r}\frac{\partial b_{ij}}{\partial \theta} + C_{rz} \frac{\partial b_{ij}}{\partial z} \right) + m_{I,i} m_{J,j} \frac{1}{r}\frac{\partial}{\partial \theta}\left(C_{\theta r} \frac{\partial b_{ij}}{\partial r} + C_{\theta\theta} \frac{1}{r}\frac{\partial b_{ij}}{\partial \theta} + C_{\theta z} \frac{\partial b_{ij}}{\partial z} \right) + m_{I,i} m_{J,j} \frac{\partial}{\partial z}\left(C_{zr} \frac{\partial b_{ij}}{\partial r} + C_{z\theta} \frac{1}{r}\frac{\partial b_{ij}}{\partial \theta} + C_{zz} \frac{\partial b_{ij}}{\partial z} \right),$$

there are 5 terms with θ derivatives. Defining $E_{D,IJ}$ as the additional terms due to these derivatives,

$$m_{I,i} m_{J,j} \frac{\partial}{\partial x_k} C_{kn} \frac{\partial b_{ij}}{\partial x_n} = \frac{1}{r}\frac{\partial}{\partial r} r\left(C_{rr} \frac{\partial b_{IJ}}{\partial r} + C_{r\theta} \frac{1}{r}\frac{\partial b_{IJ}}{\partial \theta} + C_{rz} \frac{\partial b_{IJ}}{\partial z} \right) + \frac{1}{r}\frac{\partial}{\partial \theta}\left(C_{\theta r} \frac{\partial b_{IJ}}{\partial r} + C_{\theta\theta} \frac{1}{r}\frac{\partial b_{IJ}}{\partial \theta} + C_{\theta z} \frac{\partial b_{IJ}}{\partial z} \right) + \frac{\partial}{\partial z}\left(C_{zr} \frac{\partial b_{IJ}}{\partial r} + C_{z\theta} \frac{1}{r}\frac{\partial b_{IJ}}{\partial \theta} + C_{zz} \frac{\partial b_{IJ}}{\partial z} \right) + E_{D,IJ} \quad , \tag{44}$$

then

$$E_{D,rr} = -\frac{2}{r}\left(\frac{\partial}{\partial r}C_{r\theta}b_{r\theta} + \frac{1}{r}\frac{\partial}{\partial\theta}C_{\theta\theta}b_{r\theta} + \frac{\partial}{\partial z}C_{z\theta}b_{r\theta} + C_{\theta r}\frac{\partial b_{r\theta}}{\partial r} + C_{\theta z}\frac{\partial b_{r\theta}}{\partial z}\right) + \frac{2C_{\theta\theta}}{r^2}\left(-\frac{\partial b_{r\theta}}{\partial\theta} - b_{rr} + b_{\theta\theta}\right) \tag{45}$$

$$E_{D,\theta\theta} = -E_{D,rr}, \qquad E_{D,zz} = 0 \tag{46,47}$$

$$E_{D,r\theta} = \frac{1}{r}\left(\frac{\partial}{\partial r}C_{r\theta}(b_{rr} - b_{\theta\theta}) + \frac{1}{r}\frac{\partial}{\partial\theta}C_{\theta\theta}(b_{rr} - b_{\theta\theta}) + \frac{\partial}{\partial z}C_{z\theta}(b_{rr} - b_{\theta\theta})\right) + \frac{C_{\theta r}}{r}\frac{\partial(b_{rr} - b_{\theta\theta})}{\partial r} + \frac{C_{\theta z}}{r}\frac{\partial(b_{rr} - b_{\theta\theta})}{\partial z} + \frac{C_{\theta\theta}}{r^2}\left(\frac{\partial(b_{rr} - b_{\theta\theta})}{\partial\theta} - 4b_{r\theta}\right) \tag{48}$$

$$E_{D,rz} = -\frac{1}{r}\left(\frac{\partial}{\partial r}C_{r\theta}b_{\theta z} + \frac{1}{r}\frac{\partial}{\partial\theta}C_{\theta\theta}b_{\theta z} + \frac{\partial}{\partial z}C_{z\theta}b_{\theta z} + C_{\theta r}\frac{\partial b_{\theta z}}{\partial r} + C_{\theta z}\frac{\partial b_{\theta z}}{\partial z}\right) - \frac{C_{\theta\theta}}{r^2}\left(\frac{\partial b_{\theta z}}{\partial\theta} + b_{rz}\right) \tag{49}$$

$$E_{D,\theta z} = \frac{1}{r}\left(\frac{\partial}{\partial r}C_{r\theta}b_{rz} + \frac{1}{r}\frac{\partial}{\partial\theta}C_{\theta\theta}b_{rz} + \frac{\partial}{\partial z}C_{z\theta}b_{rz} + C_{\theta r}\frac{\partial b_{rz}}{\partial r} + C_{\theta z}\frac{\partial b_{rz}}{\partial z}\right) + \frac{C_{\theta\theta}}{r^2}\left(\frac{\partial b_{rz}}{\partial\theta} - b_{\theta z}\right) . \tag{50}$$

Momentum Equation

The terms other than convection and laminar diffusion (considered above) in the momentum equation (26) are:

$$m_{I,i}\frac{\partial U_i}{\partial t} = \frac{\partial U_I}{\partial t}, \tag{51}$$

$$m_{I,i}\frac{\partial\overline{u_i u_k}}{\partial x_k} = \frac{1}{r}\frac{\partial r\overline{u_I u_r}}{\partial r} + \frac{1}{r}\frac{\partial\overline{u_I u_\theta}}{\partial\theta} + \frac{\partial\overline{u_I u_z}}{\partial z} - \delta_{Ir}\frac{\overline{u_\theta u_\theta}}{r} + \delta_{I\theta}\frac{\overline{u_r u_\theta}}{r} . \tag{52}$$

Considering only rotation about the z axis,

$$m_{I,i}(-2\varepsilon_{ikm}\Omega_k U_m) = -2\delta_{I\theta}\Omega_z U_r + 2\delta_{Ir}\Omega_z U_\theta . \tag{53}$$

$$m_{I,i}\frac{\partial}{\partial x_i}\left(\frac{\Omega_z^2 r^2}{2}\right) = \delta_{Ir}\Omega_z^2 r . \tag{54}$$

$$m_{I,i}\left(-\frac{1}{\rho}\frac{\partial \overline{P}}{\partial x_i}\right) = -\frac{1}{\rho}\nabla_I \overline{P} \; . \tag{55}$$

$$\begin{aligned} m_{I,i}\frac{\partial}{\partial x_k}\nu\frac{\partial U_k}{\partial x_i} &= \nabla_K \nu \nabla_I U_K + \nu\frac{1}{r}\nabla_I U_r + \nu\frac{1}{r}\left(\delta_{I\theta}\nabla_r U_\theta - \delta_{Ir}\nabla_\theta U_\theta\right) \\ &+\delta_{I\theta}\frac{1}{r}\left(\nabla_\theta \nu U_r - \nabla_r \nu U_\theta\right) - \delta_{Ir}\nu U_r \frac{1}{r^2} \; . \end{aligned} \tag{56}$$

Collecting terms for the momentum equation, using (28), (32), (38), and (51)-(56),

$$\begin{aligned} \frac{\partial U_I}{\partial t} + U_r\frac{\partial U_I}{\partial r} + \frac{U_\theta}{r}\frac{\partial U_I}{\partial \theta} + U_z\frac{\partial U_I}{\partial z} &= -\frac{1}{r}\frac{\partial r\overline{u_I u_r}}{\partial r} - \frac{1}{r}\frac{\partial \overline{u_I u_\theta}}{\partial \theta} - \frac{\partial \overline{u_I u_z}}{\partial z} \\ &+\frac{1}{r}\frac{\partial}{\partial r}\nu r\frac{\partial U_I}{\partial r} + \frac{1}{r}\frac{\partial}{\partial \theta}\frac{\nu}{r}\frac{\partial U_I}{\partial \theta} + \frac{\partial}{\partial z}\nu\frac{\partial U_I}{\partial z} \\ &+\delta_{Ir}\left(-\frac{1}{\rho}\frac{\partial \overline{P}}{\partial r} + \frac{U_\theta U_\theta}{r} + \frac{\overline{u_\theta u_\theta}}{r} + 2\Omega_z U_\theta + \Omega_z^2 r - \frac{\nu U_r}{r^2} - \frac{\nu}{r^2}\frac{\partial U_\theta}{\partial \theta} - \frac{1}{r^2}\frac{\partial \nu U_\theta}{\partial \theta}\right) \\ &+\delta_{I\theta}\left(-\frac{1}{\rho r}\frac{\partial \overline{P}}{\partial \theta} - \frac{U_\theta U_r}{r} - \frac{\overline{u_r u_\theta}}{r} - 2\Omega_z U_r - \frac{\nu U_\theta}{r^2} + \frac{\nu}{r^2}\frac{\partial U_r}{\partial \theta} + \frac{1}{r^2}\frac{\partial \nu U_r}{\partial \theta}\right) - \delta_{Iz}\frac{1}{\rho}\frac{\partial \overline{P}}{\partial z} \\ &+\nabla_K \nu \nabla_I U_K + \frac{\nu}{r}\nabla_I U_r - \frac{\nu\delta_{Ir}}{r}\left(\nabla_\theta U_\theta + \frac{U_r}{r}\right) + \frac{\delta_{I\theta}}{r}\left(\nabla_\theta \nu U_r - \nabla_r \nu U_\theta + \nu \nabla_r U_\theta\right). \end{aligned} \tag{60}$$

Note that the last line, which came from the term, $\frac{\partial}{\partial x_k}\nu\frac{\partial U_k}{\partial x_i}$, is zero for incompressible flow with constant ν.

For steady fully-developed incompressible pipe flow with constant ν, where $U_r = 0$, and the velocity does not vary in either the θ or z directions, the equations for r, θ, and z momentum yield:

$$0 = -\frac{1}{r}\frac{\partial r\overline{u_r u_r}}{\partial r} - \frac{1}{\rho}\frac{\partial \overline{P}}{\partial r} + \frac{U_\theta U_\theta}{r} + \frac{\overline{u_\theta u_\theta}}{r} + 2\Omega_z U_\theta + \Omega_z^2 r \tag{61}$$

$$0 = -\frac{1}{r}\frac{\partial r\overline{u_\theta u_r}}{\partial r} + \frac{1}{r}\frac{\partial}{\partial r}\nu r\frac{\partial U_\theta}{\partial r} - \frac{\overline{u_r u_\theta}}{r} - \frac{\nu U_\theta}{r^2} \tag{62}$$

$$0 = -\frac{1}{r}\frac{\partial r\overline{u_z u_r}}{\partial r} - \frac{1}{\rho}\frac{\partial \overline{P}}{\partial z} + \frac{1}{r}\frac{\partial}{\partial r}\nu r\frac{\partial U_z}{\partial r} \; . \tag{63}$$

Equation for Scalar Turbulence Properties

The transformed equation for the scalar turbulence properties, (29) with $\psi = q$ and ω, using (31), (36) and (37), becomes,

$$\frac{\partial\psi}{\partial t}+U_r\frac{\partial\psi}{\partial r}+\frac{U_\theta}{r}\frac{\partial\psi}{\partial\theta}+U_z\frac{\partial\psi}{\partial z}=\frac{1}{r}\frac{\partial}{\partial r}r\left((\nu+C_{rr})\frac{\partial\psi}{\partial r}+C_{r\theta}\frac{1}{r}\frac{\partial\psi}{\partial\theta}+C_{rz}\frac{\partial\psi}{\partial z}\right)$$
$$+\frac{1}{r}\frac{\partial}{\partial\theta}\left(C_{\theta r}\frac{\partial\psi}{\partial r}+(\nu+C_{\theta\theta})\frac{1}{r}\frac{\partial\psi}{\partial\theta}+C_{\theta z}\frac{\partial\psi}{\partial z}\right) \tag{64}$$
$$+\frac{\partial}{\partial z}\left(C_{zr}\frac{\partial\psi}{\partial r}+C_{z\theta}\frac{1}{r}\frac{\partial\psi}{\partial\theta}+(\nu+C_{zz})\frac{\partial\psi}{\partial z}\right)+S_\psi \quad ,$$

where

$$C_{IJ}=c_{t\psi}\frac{q^2}{\omega}\left(\frac{\overline{u_I u_J}}{k}\right)^{n_t} \quad . \tag{65}$$

For steady fully-developed pipe flow, where $U_r = 0$, and the properties do not vary in either the θ or z directions, (60) reduces to

$$0=\frac{1}{r}\frac{\partial}{\partial r}r\nu\frac{\partial\psi}{\partial r}+\frac{1}{r}\frac{\partial}{\partial r}r\left(C_{rr}\frac{\partial\psi}{\partial r}\right)+S_\psi \, . \tag{66}$$

Equations for Turbulence Anisotropy

Collecting the terms for the transformed b_{IJ} equations, using (29), (33) and (44)-(50),

$$\frac{\partial b_{IJ}}{\partial t}+U_r\frac{\partial b_{IJ}}{\partial r}+\frac{U_\theta}{r}\frac{\partial b_{IJ}}{\partial\theta}+U_z\frac{\partial b_{IJ}}{\partial z}=\frac{1}{r}\frac{\partial}{\partial r}r\left((\nu+C_{rr})\frac{\partial b_{IJ}}{\partial r}+C_{r\theta}\frac{1}{r}\frac{\partial b_{IJ}}{\partial\theta}+C_{rz}\frac{\partial b_{IJ}}{\partial z}\right)$$
$$+\frac{1}{r}\frac{\partial}{\partial\theta}\left(C_{\theta r}\frac{\partial b_{IJ}}{\partial r}+(\nu+C_{\theta\theta})\frac{1}{r}\frac{\partial b_{IJ}}{\partial\theta}+C_{\theta z}\frac{\partial b_{IJ}}{\partial z}\right) \tag{67}$$
$$+\frac{\partial}{\partial z}\left(C_{zr}\frac{\partial b_{IJ}}{\partial r}+C_{z\theta}\frac{1}{r}\frac{\partial b_{IJ}}{\partial\theta}+(\nu+C_{zz})\frac{\partial b_{IJ}}{\partial z}\right)+S_{bIJ}+E_{B,IJ} \quad ,$$

where $C_{IJ}=c_{t\psi}\frac{q^2}{\omega}\left(\frac{\overline{u_I u_J}}{k}\right)^{n_t}$ and E_B is all the extra terms from convection, and laminar and turbulent diffusion. Note also that the normal coefficients have become $\nu + C_{rr}$, $\nu + C_{\theta\theta}$, and $\nu + C_{zz}$, accounting for both laminar and turbulent diffusion.

$$E_{B,rr} = \frac{2U_\theta}{r} b_{r\theta} - \frac{2}{r}\frac{\partial}{\partial r} C_{r\theta} b_{r\theta} - \frac{2}{r^2}\frac{\partial}{\partial \theta}(\nu + C_{\theta\theta}) b_{r\theta} - \frac{2}{r}\frac{\partial}{\partial z} C_{z\theta} b_{r\theta}$$
$$- \frac{2C_{\theta r}}{r}\frac{\partial b_{r\theta}}{\partial r} - \frac{2C_{\theta z}}{r}\frac{\partial b_{r\theta}}{\partial z} + \frac{2(\nu + C_{\theta\theta})}{r^2}\left(-\frac{\partial b_{r\theta}}{\partial \theta} - b_{rr} + b_{\theta\theta}\right) \quad (68)$$

$$E_{B,\theta\theta} = -E_{B,rr}\,, \qquad E_{B,zz} = 0 \quad (69,70)$$

$$E_{B,r\theta} = \frac{U_\theta}{r}(b_{\theta\theta} - b_{rr}) + \frac{1}{r}\frac{\partial}{\partial r} C_{r\theta}(b_{rr} - b_{\theta\theta}) + \frac{1}{r^2}\frac{\partial}{\partial \theta}(\nu + C_{\theta\theta})(b_{rr} - b_{\theta\theta})$$
$$+ \frac{1}{r}\frac{\partial}{\partial z} C_{z\theta}(b_{rr} - b_{\theta\theta}) + \frac{C_{\theta r}}{r}\frac{\partial(b_{rr} - b_{\theta\theta})}{\partial r} + \frac{C_{\theta z}}{r}\frac{\partial(b_{rr} - b_{\theta\theta})}{\partial z}$$
$$+ \frac{(\nu + C_{\theta\theta})}{r^2}\left(\frac{\partial(b_{rr} - b_{\theta\theta})}{\partial \theta} - 4b_{r\theta}\right) \quad (71)$$

$$E_{B,rz} = \frac{U_\theta}{r} b_{\theta z} - \frac{1}{r}\frac{\partial}{\partial r} C_{r\theta} b_{\theta z} - \frac{1}{r^2}\frac{\partial}{\partial \theta}(\nu + C_{\theta\theta}) b_{\theta z} - \frac{1}{r}\frac{\partial}{\partial z} C_{z\theta} b_{\theta z}$$
$$- \frac{C_{\theta r}}{r}\frac{\partial b_{\theta z}}{\partial r} - \frac{C_{\theta z}}{r}\frac{\partial b_{\theta z}}{\partial z} - \frac{(\nu + C_{\theta\theta})}{r^2}\left(\frac{\partial b_{\theta z}}{\partial \theta} + b_{rz}\right) \quad (72)$$

and

$$E_{B,\theta z} = -\frac{U_\theta}{r} b_{rz} + \frac{1}{r}\frac{\partial}{\partial r} C_{r\theta} b_{rz} + \frac{1}{r^2}\frac{\partial}{\partial \theta}(\nu + C_{\theta\theta}) b_{rz} + \frac{1}{r}\frac{\partial}{\partial z} C_{z\theta} b_{rz}$$
$$+ \frac{C_{\theta r}}{r}\frac{\partial b_{rz}}{\partial r} + \frac{C_{\theta z}}{r}\frac{\partial b_{rz}}{\partial z} + \frac{(\nu + C_{\theta\theta})}{r^2}\left(\frac{\partial b_{rz}}{\partial \theta} - b_{\theta z}\right) \,. \quad (73)$$

For steady fully-developed pipe flow, the equations for $b_{rr} - b_{\theta\theta}$, b_{zz}, $b_{r\theta}$, b_{rz} and $b_{\theta z}$ are

$$0 = \frac{1}{r}\frac{\partial}{\partial r} r\left((\nu + C_{rr})\frac{\partial(b_{rr} - b_{\theta\theta})}{\partial r}\right) + S_{brr} - S_{b\theta\theta} + \frac{4U_\theta}{r} b_{r\theta} - \frac{4}{r}\frac{\partial}{\partial r} C_{r\theta} b_{r\theta} - \frac{4C_{\theta r}}{r}\frac{\partial b_{r\theta}}{\partial r}$$
$$+ \frac{4(C_{\theta\theta} + \nu)}{r^2}(b_{\theta\theta} - b_{rr}) \quad (74)$$

$$0 = \frac{1}{r}\frac{\partial}{\partial r} r\left((\nu + C_{rr})\frac{\partial b_{zz}}{\partial r}\right) + S_{bzz} \quad (75)$$

$$0 = \frac{1}{r}\frac{\partial}{\partial r} r\left((\nu + C_{rr})\frac{\partial b_{r\theta}}{\partial r}\right) + S_{br\theta} + \frac{U_\theta}{r}(b_{\theta\theta} - b_{rr})$$
$$+\frac{1}{r}\frac{\partial}{\partial r} C_{r\theta}(b_{rr} - b_{\theta\theta}) + \frac{C_{\theta r}}{r}\frac{\partial(b_{rr} - b_{\theta\theta})}{\partial r} - \frac{4(\nu + C_{\theta\theta})}{r^2} b_{r\theta} \tag{76}$$

$$0 = \frac{1}{r}\frac{\partial}{\partial r} r\left((\nu + C_{rr})\frac{\partial b_{rz}}{\partial r}\right) + S_{brz} + \frac{U_\theta}{r} b_{\theta z} - \frac{1}{r}\frac{\partial}{\partial r} C_{r\theta} b_{\theta z} - \frac{C_{\theta r}}{r}\frac{\partial b_{\theta z}}{\partial r} - \frac{(\nu + C_{\theta\theta})}{r^2} b_{rz} \tag{77}$$

$$0 = \frac{1}{r}\frac{\partial}{\partial r} r\left((\nu + C_{rr})\frac{\partial b_{\theta z}}{\partial r}\right) + S_{b\theta z} - \frac{U_\theta}{r} b_{rz} + \frac{1}{r}\frac{\partial}{\partial r} C_{r\theta} b_{rz} + \frac{C_{\theta r}}{r}\frac{\partial b_{rz}}{\partial r} - \frac{(\nu + C_{\theta\theta})}{r^2} b_{\theta z} \,. \tag{78}$$

Note that the source terms for the turbulence properties may be simply evaluated in terms of the cylindrical components of the strain and vorticity, since they contain no other derivatives of vectors or tensors. From Eqs. 8.1.3 to 8.1.5,

$$S_q = \frac{P}{2q} - \frac{\tilde{\varepsilon}}{2q} = (-b_{IJ}S_{IJ})q - \frac{\omega q}{2} \,, \tag{79}$$

$$S_\omega = (c_{\varepsilon 1} - 1)(-2b_{IJ}S_{IJ})\omega - (c_{\varepsilon 2} - 1)\omega^2 \,, \tag{80}$$

$$S_{bIJ} = -\frac{1}{2} b_{IJ}\left((c_1 - 2)\omega + (2 + \frac{1}{2}c_6)\frac{P}{k}\right) + \frac{1}{2}\left(c_3 - \frac{4}{3}\right)S_{IJ}$$
$$+\frac{1}{2}(c_4 - 2)(b_{IK}S_{JK} + b_{JK}S_{IK} - \frac{2}{3}\delta_{IJ} b_{MN}S_{MN}) \tag{81}$$
$$+\frac{1}{2}(c_5 - 2)(b_{IK}W_{JK} + b_{JK}W_{IK}) - \Omega_z(\varepsilon_{IzM} b_{JM} + \varepsilon_{JzM} b_{IM}) \,.$$

Appendix E

List of Appendices on CD

The CD in the pocket on the inside of the back cover contains appendices, exercise solutions and figures. The appendices and exercise solutions are in pdf file format, and are listed below. The figures are the original versions, mostly in color, before conversion to grayscale, which are then printed in halftone in the book. The files are named Fig.1.2.1.gif, etc., reflecting their format. To begin, open the file read.me.html on the CD.

Long Versions of Book Appendices

C.CH.thm Cayley-Hamilton Theorem and Analysis
D.CylCoor Conversion to Cylindrical Coordinates

Analysis of Data and Models in the Literature

A1987 Aupoix's spectra method for decay of dissipation
AS1980 Abu-Ghannam and Shaw transition data
B1972 Bearman leading edge stagnation data and 'Hunt' limits
BC1978 Bennett and Corrsin, grid turbulence, $c_{\varepsilon 2}$
BS1996 Blaisdell and Shariff, elliptic flow DNS
BT1948 Batchelor and Townsend, grid turbulence, $c_{\varepsilon 2}$
CC1966 Comte-Bellot and Corrsin, grid turbulence, $c_{\varepsilon 2}$ and c_1
CGL1993 Craft et al., wall reflection models for impinging flow
CHC1970 Champagne et al., homogeneous shear data
CL2001 Choi and Lumley, return-to-isotropy data
D1971 see DR1969
D1978 Dean, channel skin friction coefficients
DH1970 Daly and Harlow, model diffusion, pressure-strain and walls
DR1969 Donaldson and Rosenbaum, pressure-strain model (also D1971, D1973)
DW1928 Davies and White, channel skin friction coefficients
FF1996 Fernholz and Finley, flat plate boundary layer data
G2000 Girimaji, pressure-strain model for homogeneous elliptic flows
GL1978 Gibson and Launder, pressure-strain and wall reflection models
GM1980 Gence and Mathieu, return-to-isotropy data
H1992 Hopfinger, notes on rotation scales
HG1972 Hall and Gibbings, transition data
HGC1977 Harris et al., homogeneous shear data
HJ1967 Halleen and Johnston, rotating channel data
IIH1996 Imao et al., rotating pipe data

K1955 Klebanoff, spectra in a boundary layer
KA1993 Kristoffersen and Andersson, rotating channel DNS
KMNM1983 Kikuyama et al., rotating pipe data (also YI1980)
L1951 Laufer, channel data
L1954 Laufer, pipe data, velocity and Reynolds stresses
L1971 Lawn, pipe data, Reynolds stresses
LGC1985 Le Penven et al., return-to-isotropy, $III > 0$ and $III < 0$
LKM1990 Lee et al., homogeneous shear DNS at high s
LR1985 Lee and Reynolds, plane strain DNS
LRR1975 Launder, Reece, Rodi, modeling notes
M1967 Moore, rotating channel data
R1951 Rotta, modeling notes
RB1990 Roach and Brierley, ERCOFTAC transition data
RLA1995 Ristorcelli et al., pressure-strain model with MFI limit
RMR1986 Rogers et al., homogeneous shear DNS
S1973 Shir, pressure-strain and wall reflection models, L equation
SS1948 Schubauer and Skramstad, transition data
SV1994 Saddoughi and Veeravalli, spectra data
T1970 Tucker, plane strain, return-to-isotropy, and spectra data
TC1981 Tavoularis and Corrsin, homogeneous shear data
TH1977 Thomas and Hancock, moving-wall data
TK1989 Tavoularis and Karnik, homogeneous shear data
U1956 Uberoi, return-to-isotropy data
U1957 Uberoi, return-to-isotropy data
W1967 Wells, transition data
WK2004 Wu and Kasagi, rotating channel DNS
YI1980 see KMNM1983
ZK1969 Zysina-Molozhen and Kuznetsova, transition data

Derivation of Equations

EQ.2.1.12 Derive the Reynolds stress equations
EQ.3.1.4 References for deriving integral equation for p
EQ.3.2.17 Derive algebraic form for rapid pressure-strain
EQ.3.3.4 Isotropic turbulence limits for pressure-strain coefficients
EQ.4.2.5 Equations for homogeneous shear flow with in-plane rotation
EQ.4.3.4 Velocity gradients in a vortex
EQ.4.3.11 Equations for bifurcation analysis
EQ.4.3.18 Relate So's equation to Donaldson's model
EQ.5.3.1-6 Derive equations for two-equation analysis for homogeneous shear
EQ.6.1.3 Derive transport equation for dissipation
EQ.7.1.18 Relation between turbulent diffusion of ε and ω
EQ.9.1.8 Fourier transform of fluctuating continuity and momentum equations

Exercise Solutions

Exercise 5.3	High strain rate properties of modified SSG models
Exercise 6.2	Range of $c_{\varepsilon 2}$ for decaying turbulence Reynolds number
Exercise 7.1	2-d turbulence limits of diffusion models
Exercise 9.3	Φ_{ij} for isotropic turbulence, and limits imposed by continuity
Exercise 14.1	$S_{r\theta}$ at the wall in fully developed rotating pipe flow
Exercise 17.3	Dimensionless consistency requirements for *E(K)*
Exercise 18.2	Glyphs of Φ_{ij} for isotropic turbulence in Taylor-Proudman limit

REFERENCES

Notation: Author(s), date, reference. (Sections or Appendices, where mentioned or data used)
Note: some references have corresponding appendices on CD, which include discussion or analysis of data or models. These appendices are named with author initials and year.

Abu-Ghannam, B.J., and Shaw, R., 1980, "Natural Transition of Boundary Layers – The Effects of Turbulence, Pressure Gradient and Flow History," Journal of Mechanical Engineering Science, Vol. 22, pp. 213-228. (11.1, 11.3, App. AS1980)

Alvelius, K, 1999, "Studies of Turbulence and Its Modelling through Large Eddy- and Direct Numerical Simulation," Doctoral thesis, Dept. of Mechanics, Royal Institute of Technology, Stockholm, Sweden. (13.2)

Alvelius, K, and Johansson, A.V., 1999, "Direct Numerical Simulation of Rotating Channel Flow at Various Reynolds Numbers and Rotation Numbers," in Doctoral thesis, Technical Report 1999:09, ISSN 0348-467X, Dept. of Mechanics, Royal Institute of Technology, Stockholm, Sweden, see Wallin and Johansson, 2002. (13.2, 13.4-5)

Aupoix, B.,1987, "Homogeneous Turbulence, Two-Point Closures and Applications to One-Point Closures," AGARD Report No. 755, "Special Course on Modern Theoretical and Experimental Approaches to Turbulent Flow Structure and Its Modelling," Paper 3. (6.2-3, 8.2, 17.3, App. A1987)

Bardina, J., 1983, "Improved Turbulence Models Based on Large Eddy Simulation of Homogeneous, Incompressible, Turbulent Flows," Ph.D. thesis, Dept. of Mechanical Engineering, Stanford Univ. (4.3, 9.6)

Bardina, J., Ferziger, J.H., and Rogallo, R.S., 1985, "Effect of Rotation on Isotropic Turbulence: Computation and Modelling," Journal of Fluid Mechanics, Vol. 154. pp. 321-336. (6.1, 18.1, 18.3)

Batchelor, G.K., 1953, The Theory of Homogeneous Turbulence, Cambridge Univ. Press. (9.1, 9.3, 15.1, 17.1, 17.3, 17.5)

Batchelor, G.K., and Townsend, A.A., 1947, "Decay of Vorticity in Isotropic Turbulence," Proc. Roy. Soc. A, Vol. 190, pp. 534-550. (App. CC1966)

Batchelor, G.K., and Townsend, A.A., 1948, "Decay of Turbulence in the Final Period," Proc. Roy. Soc. A, Vol. 194, pp. 527-543. (6.2, 8.2-3, App. BT1948)

Bearman. P.W., 1972, "Some Measurements of the Distortion of Turbulence Approaching a Two-Dimensional Bluff Body," Journal of Fluid Mechanics, Vol. 53, pp. 451-467. (16.3-4, 17.4, App. B1972)

Beer, F.P., and Johnston, E.R., 1981, Mechanics of Materials, McGraw-Hill. (1.2)

Bennett, J.C., and Corrsin, S., 1978, "Small Reynolds Number Nearly Isotropic Turbulence in a Straight Duct and a Contraction," Physics of Fluids, Vol. 21, pp. 2129-2140. (6.2, 8.2-3, App. BC1978)

Bertoglio, J.P., 1982, "Homogeneous Turbulent Field within a Rotating Frame," AIAA Journal, Vol. 20, pp. 1175-1181. (4.3, 5.2, 9.3)

Bilanin. A.J., Teske, M.E., Donaldson, C.duP, and Snedeker, R.S., 1977, "Viscous Effects in Aircraft Trailing Vortices," NASA SP-409. (4.3)

Blaisdell, G.A., and Shariff, K., 1996, "Simulation and Modeling of the Elliptic Streamline Flow," Center for Turbulence Research, Proc. of the Summer Program 1996, Stanford Univ., pp. 433-446. (18.4, 18.6, App. BS1996, App. G2000)

Boltz, F.W., Kenyon, G.C., and Allen, C.Q., 1960, "The Boundary-Layer Transition Characteristics of Two Bodies of Revolution, a Flat Plate, and an Unswept Wing in a Low-Turbulence Wind Tunnel," NASA TN D-309. (11.1, 11.3)

Boussinesq, J., 1877, "Essai sur la Théorie des Eaux Courantes," L'Académie des Sciences de L'Institut de France Mémoires présenté par divers savants, Vol. 23, pp. 24-61. (1.3)

Bradshaw, P., 1969, "The Analogy Between Streamline Curvature and Buoyancy in Turbulent Shear Flow," Journal of Fluid Mechanics, Vol. 36, pp. 177-191. (4.3)

Bradshaw, P., 1973, "Effects of Streamline Curvature on Turbulent Flow," AGARDograph No. 169. (4.3)

Bruns, J., Dengel, P., and Fernholz, H. H., 1992, "Mean Flow and Turbulence Measurements in an Incompressible Two-Dimensional Turbulent Boundary Layer. Part I: Data." Institutsbericht Nr. 02/92, Herman-Föttinger-Institut für Thermo- und Fluiddynamik, Technische Universität Berlin. (App. A, App. FF1996)

Champagne, F.H., Harris, V.G., and Corrsin, S., 1970, "Experiments on Nearly Homogeneous Turbulent Shear Flow," Journal of Fluid Mechanics, Vol. 41, pp. 81-139. (1.2, 3.5, 5.1-3, 6.3, 8.4, App. A, App. CHC1970)

Chapman, D.R., 1979, "Computational Aerodynamics Development and Outlook," AIAA Journal, Vol. 17, pp. 1293-1313. (6.2, 17.4)

Chasnov, J.R., 1995, "The Decay of Axisymmetric Homogeneous Turbulence," Physics of Fluids, Vol. 7, pp. 600-605. (17.3)

Chen, C.-J., and Jaw, S.-Y., 1998, Fundamentals of Turbulence Modeling, Taylor and Francis. (2.2, 6.3)

Choi, K.-S., and Lumley, J.L., 1984, "Return to Isotropy of Homogeneous Turbulence Revisited," Turbulence and Chaotic Phenomena in Fluids, T. Tatsumi (ed), Elsevier, pp. 267-272. (15.3)

Choi, K.-S., and Lumley, J.L., 2001, "The Return to Isotropy of Homogeneous Turbulence," Journal of Fluid Mechanics, Vol. 436, pp. 59-84. (15.3, App. CL2001)

Chou, P.Y., 1945, "On Velocity Correlations and the Solutions of the Equations of Turbulent Fluctuation," Quarterly of Applied Mathematics, Vol. 3, pp. 38-54. (1.0, 2.2, 3.1, App. EQ.3.1.4)

Cleak, J.G.E., Gregory-Smith, D.G., and Birch, N.T., 1991, "Experimental Verification of a 3D Turbulent Flow Calculation in an Axial Turbine Cascade," AGARD Conf. Proc. No. 510, Paper No. 8, San Antonio, Texas. (1.3)

Coakley, T.J., and Hsieh, T.,1985, "A Comparison between Implicit and Hybrid Methods for the Calculation of Steady and Unsteady Inlet Flows," AIAA Paper No. 85-1125. (1.3, 6.1)

Comte-Bellot, G., and Corrsin, S., 1966, "The Use of a Contraction to Improve the Isotropy of Grid-Generated Turbulence," Journal of Fluid Mechanics, Vol.25, pp. 657-682. (6.2,

8.2-3, 15.3, 17.4, App. CC1966)

Craft, T.J., Graham, L.J.W., and Launder, B.E., 1993, "Impinging Jet Studies for Turbulence Model Assessment – II. An Examination of the Performance of Four Turbulence Models," Int. J. Heat Mass Transfer, Vol. 36, pp. 2685-2697. (16.2, App. CGL1993)

Craft , T.J., Launder, B.E., and Suga, K., 1993, "Extending the Applicability of Eddy Viscosity Models Through the Use of Deformation Invariants and Non-Linear Elements," Proc. of Int. Symposium on Refined Flow Modelling and Turbulence Measurements, pp 125-132, Paris. (5.1)

Craft , T.J., Launder, B.E., and Suga, K., 1996, "Development and Application of a Cubic Eddy-Viscosity Model of Turbulence," Int. J. Heat and Fluid Flow, Vol. 17, pp. 108-115. (1.3)

Crow, S.C., 1968, "Viscoelastic Properties of Fine-Grained Incompressible Turbulence," Journal of Fluid Mechanics, Vol. 33, pp. 1-20. (3.3, 3.5, 9.4)

Daly, B.J., and Harlow, F.H., 1970, "Transport Equations in Turbulence", Physics of Fluids, Vol. 13, pp. 2634-2649. (7.1-2, 8.1-2, 16.2, 16.4, App. A, App. DH1970)

Davies, S.J., and White, C.M., 1928, "An Experimental Study of the Flow of Water in Pipes of Rectangular Section," Proc. Roy. Soc. A, Vol. 119, pp. 92-107. (8.2-3, App. DW1928)

da Vinci, Leonardo, ca. 1510, "Studies of an Old Man Seated and of Swirling Water," Windsor, Royal Library, No. 12,579 recto; see A. E. Popham, The Drawings of Leonardo da Vinci, Jonathan Cape Paperback, JCP 14, Ill. No. 282 and p. 155, London, 1964. (1.0)

Dean, R.B., 1978, "Reynolds Number Dependence of Skin Friction and Other Bulk Flow Variables in Two-Dimensional Rectangular Duct Flow," Journal of Fluids Engineering, Vol. 100, pp. 215-223. (8.2-3, App. D1978)

Demuren, A.O., and Sarkar, S, 1993, "Perspective: Systematic Study of Reynolds Stress Closure Models in the Computations of Plane Channel Flows," Journal of Fluids Engineering, Vol. 115, pp. 5-12. (3.5)

Devenport, W.J., Zsoldos, J.S., and Vogel, C.M., 1997, "The Structure and Development of a Counter-Rotating Wing-Tip Vortex Pair," Journal of Fluid Mechanics, Vol. 332, pp. 71-104. (18.6)

Donaldson, C.duP., 1969, "A Computer Study of an Analytical Model of Boundary-Layer Transition," AIAA Journal, Vol. 7, pp. 271-278. (7.1)

Donaldson, C.duP., 1971, "A Progress Report on an Attempt to Construct an Invariant Model of Turbulent Shear Flows," AGARD CP-93, Turbulent Shear Flows, London. (3.5, 4.3, 5.1-3, 18.5, App. DR1969)

Donaldson, C.duP., 1973, "The Relationship Between Eddy-Transport and Second-Order Closure Models for Stratified Media and for Vortices," NASA SP-321, Free Turbulent Shear Flows, Volume I – Conference Proceedings. (3.5, 4.3, App. DR1969)

Donaldson, C.duP., and Bilanin, A.J., 1975, "Vortex Wakes of Conventional Aircraft," AGARDograph No. 204. (3.5, 4.3, 7.1)

Donaldson, C.duP., and Rosenbaum, H., 1969, "Calculation of Turbulent Shear Flows Through Closure of the Reynolds Equations by Invariant Modeling," NASA SP 216. (3.5, App. DR1969)

Dryden, H.L., 1936, "Air Flow in the Boundary Layer Near a Plate," NACA Report No.

562. (11.1, 11.3)

Durbin, P.A., 1993, “A Reynolds Stress Model for Near-Wall Turbulence,” Journal of Fluid Mechanics, Vol. 249, pp. 465-498. (12.2, 12.5)

Durbin, P.A., 1994, “Turbulence Modeling for Separated Flow,” NACA-CR-197352, pp. 97-105. (1.3)

Durbin, P.A., 1996, “On the k-ε Stagnation Point Anomaly,” Int. J. Heat and Fluid Flow, Vol. 17, pp. 89-90. (1.3)

Dutoya, D., and Michard, P., 1981, “A Program for Calculating Boundary Layers Along Compressor and Turbine Blades,” Numerical Methods in Heat Transfer, R.W. Lewis, K. Morgan, and O.C. Zienkiewicz (eds.), Wiley, New York. (6.1)

El-Samni, O., 2001, "Heat and Momentum Transfer in Turbulent Rotating Channel Flow," Ph. D. Thesis. The University of Tokyo, Tokyo, Japan. (13.2, 13.4-5, App. WK2004)

El-Samni, O., and Kasagi, N., 2000, "Heat and Momentum Transfer in Rotating Turbulent Channel Flow," Proc. 4th JSME-KSME Thermal Engineering Conference, Kobe, Vol. 3, pp. 517-522. (13.2, 13.4-5, App. WK2004)

El-Samni, O., and Kasagi, N., 2000, "Passive Scalar Transport in Rotating Turbulent Channel Flows," Proc. 37th National Heat Transfer Symposium, Kobe, pp. 559-560. (13.2, 13.4-5, App. WK2004)

El-Samni, O., and Kasagi, N., 2001, "The Effects of System Rotation with Three Orthogonal Rotating Axes on Turbulent Channel Flow," 7th International Conference of Fluid Dynamics and Propulsion, Cairo, Egypt, CD ROM (ASME). (13.2, 13.4-5, App. WK2004)

Eskinazi, S., and Erian, F.F., 1969, “Energy Reversal in Turbulent Flows,” Physics of Fluids, Vol. 12, pp. 1988-1998. (4.3)

Eskinazi, S., and Yeh, H., 1956, “An Investigation on Fully Developed Turbulent Flows in a Curved Channel,” Journal of the Aeronautical Sciences, Vol. 23, pp. 23-34, 75; see Eskinazi and Erian, 1969, Fig, 2. (4.3)

Fernholz, H.H., and Finley, P. J., 1996, “ The Incompressible Zero-Pressure-Gradient Turbulent Boundary Layer: an Assessment of the Data”, Prog. Aerospace Sci., Vol. 32, pp. 245-311. (App. A, App. FF1996)

Fernholz, H. H., Krause, E., Nockemann, M., and Schober, M., 1995. “Comparative Measurements in the Canonical Boundary Layer at $Re_{\delta 2} \leq 6 \times 10^4$ on the Wall of the German-Dutch Windtunnel,” Physics of Fluids Vol. 7, pp. 1275-1281. (App. A, App. FF1996)

Gatski, T.B., and Speziale, C.G., 1993, “On Explicit Algebraic Stress Models for Complex Turbulent Flows,” Journal of Fluid Mechanics, Vol. 254, pp. 59-78. (5.3)

Gence, J.N., and Mathieu, J., 1980, “The Return to Isotropy of an Homogeneous Turbulence Having Been Submitted to Two Successive Plane Strains,” Journal of Fluid Mechanics, Vol. 101, pp. 555-566. (15.3, App. GM1980)

Gibson, M.M., and Launder, B.E., 1978, “Ground Effects on Pressure Fluctuations in the Atmospheric Boundary Layer,” Journal of Fluid Mechanics, Vol. 86, pp. 491-511; see Rodi, 1984, Table 4. (3.5, 4.3, 5.1-3, 16.2-3, App. GL1978)

Gilbert, N., 1993, "Turbulence Model Data Derived from Direct Numerical Simulations", DLR Internal Report IB 221-93 A 14. (6.4)

Gilbert, N., and Kleiser, L., 1991, "Turbulence Model Testing with the Aid of Direct Numerical Simulation Results", Proc. of 8th Symposium on Turbulent Shear Flows, Paper 26-1, Munich, Sept. 9-11. (6.4, 8.2-3, 13.2)

Girimaji, S.S., 2000, "Pressure-Strain Correlation Modelling of Complex Turbulent Flows," Journal of Fluid Mechanics, Vol. 422, pp. 91-123. (18.4, App. BS1996, App. G2000)

Grant, H.L., Stewart, R.W., and Moilliet, A., 1962, "Turbulence Spectra from a Tidal Channel," Journal of Fluid Mechanics, Vol. 12, pp. 241-268. (17.4)

Greenspan, H.P., 1968, The Theory of Rotating Fluids, Cambridge Univ. Press. (18.1)

Hadžić, I., and Hanjalić, K., 1999, "Modelling of Separation-Induced Transition to Turbulence with a Second-Moment Closure," Engineering Turbulence Modelling and Experiments - 4, W. Rodi and D. Laurence (eds), Elsevier Science Ltd., pp. 587-596. (15.2)

Hall, A.A., and Hislop, G.S., 1938, "Experiments on the Transition of the Laminar Layer on a Flat Plate, " British Aeronautical Research Committee, Reports and Memoranda No. 1843. (11.1, 11.3, App. HG1972)

Hall, D.J., and Gibbings, J.C., 1972, "Influence of Stream Turbulence and Pressure Gradient upon Boundary Layer Transition," Journal of Mechanical Engineering Science, Vol. 14, pp. 134-146. (11.1, 11.3, App. HG1972)

Halleen, R.M., and Johnston, J.P., 1967, "The Influence of Rotation on Flow in a Long Rectangular Channel – An Experimental Study," Report MD-18, Thermosciences Division, Dept. of Mechanical Engineering, Stanford Univ. (13.2, 13.4, App. HJ1967)

Hanjalić, K., 1994, "Advanced Turbulence Closure Models: A View of Current Status and Future Prospects," Int. J. Heat and Fluid Flow, Vol. 15, pp. 178-203. (6.1, 15.2)

Hanjalić, K., and Jakirlić, S., 1992, "A Model of Stress Dissipation in Second-Moment Closures," Proc. 4th Eur. Turbulence Conf., Delft, F. Niewstadt et al. (eds.), Kluwer Academic Press, Dordrecht, Holland. (15.2)

Hanjalić, K., and Launder, B.E., 1972, "A Reynolds Stress Model of Turbulence and Its Application to Thin Shear Flows," Journal of Fluid Mechanics, Vol. 52, pp. 609-638. (7.1-2)

Hanjalić, K., and Launder, B.E., 1976, "Contribution Towards a Reynolds Stress Closure for Low-Reynolds-Number Turbulence," Journal of Fluid Mechanics, Vol. 74, pp. 593-610. (6.1-2, 15.2)

Harris, V.G., Graham, J.A.H., and Corrsin, S., 1977, "Further Experiments in Nearly Homogeneous Turbulent Shear Flow," Journal of Fluid Mechanics, Vol. 81, pp. 657-687. (5.1, App. HGC1977)

Hildebrand, F.B., 1962, Advanced Calculus for Applications, Prentice-Hall. (App. EQ.9.4.7, App. EQ.17.1.23-24)

Hislop, G.S., 1940, "The Transition of a Laminar Boundary Layer in a Wind Tunnel," Ph.D. Thesis, Cambridge University. (11.1, 11.3, App. HG1972)

Hopfinger, E.J., 1992, Sections I.1 and II.1 of Rotating Fluids in Geophysical and Industrial Applications, E. J. Hopfinger (ed.), Springer-Verlag, pp. 3-25. (18.1)

Imao, S., Itoh, M., and Harada, T., 1996, "Turbulent Characteristics of the Flow in an Axially Rotating Pipe," Int. J. Heat and Fluid Flow, Vol. 17, pp. 444-451. (14.2-3, App. A, App. IIH1996)

Iwamoto, K., 2002, "Database of Fully Developed Channel Flow," THTLAB Internal Report, No. ILR-0201, THTLAB, Dept. of Mech. Eng., the Univ. of Tokyo. (6.4)

Iwamoto, K., Suzuki, Y., and Kasagi, N., 2002. "Reynolds Number Effect on Wall Turbulence: Toward Effective Feedback Control," Int. J. Heat and Fluid Flow, Vol. 23, pp. 678-689. (6.4, 7.2, 8.2-3, 12.2-3, 13.2, 17.6, App. A)

Jeffreys, H., and Jeffreys, B.S., 1962, Methods of Mathematical Physics, Cambridge Univ. Press. (App. EQ.3.1.4)

Johnston, J.P., Halleen, R.M., and Lezius, D.K., 1972, "Effects of Spanwise Rotation on the Structure of Two-Dimensional Fully Developed Turbulent Channel Flow," Journal of Fluid Mechanics, Vol. 56, pp. 533-557. (4.3)

Jones, W.P., and Launder, B.E., 1972, "The Prediction of Laminarization with a Two-Equation Model of Turbulence," Int. J. Heat Mass Transfer, Vol. 15, pp. 301-314. (6.1)

Kasagi, N., Horiuti, K., Miyake, Y., Miyauchi, T., and Nagano, Y., 1990, "Establishment of the Direct Numerical Simulation Data Bases of Turbulent Transport Phenomena", Co-operative Research No. 02302043, Ministry of Education, Science and Culture, Japan. (6.4, 8.2-3, 13.2, 13.4-5, 17.6, App. WK2004)

Kerswell, R.R., 2002, "Elliptical Instability," Annual Review of Fluid Mechanics, Vol. 34, pp. 83-113. (18.5)

Kikuyama, K., Murakami, M., Nishibori, K., and Maeda, K., 1983, "Flow in an Axially Rotating Pipe (A Calculation of Flow in the Saturated Region)," Bulletin of the JSME, Vol. 26, pp. 506-513. (14.2-3, App. KMNM1983)

Kim, J., Moin, P., and Moser, R., 1987, "Turbulence Statistics in Fully Developed Channel Flow at Low Reynolds Number," Journal of Fluid Mechanics, Vol. 177, pp. 133-166. (1.2, 5.1)

Klebanoff, P.S.,1955, "Characteristics of Turbulence in a Boundary Layer with Zero Pressure Gradient", NACA Report 1247. (17.6, App. K1955)

Kline, S.J., Cantwell, B.J., and Lilley, G.M. (eds), 1981, 1980-81 AFOSR-HTTM-Stanford Conference on Complex Turbulent Flows, Volume I, Stanford Univ., California. (18.3)

Kolmogorov, A.N., 1941, "The Local Structure of Turbulence in Incompressible Viscous Fluid for Very Large Reynolds Numbers," Dokl. Akad. Nauk SSSR, Vol. 30, pp. 301-305. Translation by V. Levin, 1991, Proc. Roy. Soc. A, Vol. 434, pp. 9-13. (17.2-3)

Kristoffersen, R., and Andersson, H.I., 1993, "Direct Simulations of Low-Reynolds-Number Turbulent Flow in a Rotating Channel," Journal of Fluid Mechanics, Vol. 256, pp. 163-197. (13.0, 13.2-5, App. KA1993)

Kuroda, A., 1990, "Direct Numerical Simulation of Couette-Poiseuille Flows," Dr. Eng. Thesis, the Univ. of Tokyo, Dec. (6.4)

Kuroda, A., Kasagi, N. and Hirata, M., 1990, (a):"A Direct Numerical Simulation of the Turbulent Flow Between Two Parallel Walls: Turbulence Characteristics Near the Wall Without Mean Shear," 5th Symposium on Numerical Simulation of Turbulence, IIS of the Univ. of Tokyo, pp. 1-5. (b): "Investigation of Dynamical Effects of the Mean Shear Rate on the Wall Turbulence via Direct Numerical Simulation," 27th National Heat Transfer Symposium of Japan, Nagoya, pp.46-48. (c): "A Direct Numerical Simulation of the Fully Developed Turbulent

Channel Flow," International Symposium on Computational Fluid Dynamics, Nagoya, 1989, pp. 1174-1179; also in Numerical Methods in Fluid Dynamics, M. Yasuhara et al. (eds.), Vol. 2, Jap. Soc. Comp. Fluid Dyn., pp. 1012-1017. (6.4)

Laufer, J., 1951, "Investigation of Turbulent Flow in a Two-Dimensional Channel," NACA Report 1053, supersedes NACA TN 2123. (App. A, App. L1951)

Laufer, J., 1954, "The Structure of Turbulence in Fully Developed Pipe Flow," NACA Report No. 1174. (14.2, App. A, App. L1954)

Launder, B.E., 1989. "Second-Moment Closure: Present … and Future?" Int. J. Heat and Fluid Flow, Vol. 10, pp. 282-300. (16.2)

Launder, B.E., 1993, "An Introduction to Single-Point Closure Methodology," Lecture Series 1993-02, Introduction to the Modeling of Turbulence, von Karman Inst. for Fluid Dynamics. (3.5)

Launder, B.E., Reece, G.J., and Rodi, W., 1975, "Progress in the Development of a Reynolds Stress Turbulence Closure," Journal of Fluid Mechanics, Vol. 68, pp. 537-566. (3.2, 3.5, 5.1-3, 6.1, 16.2, 18.4)

Launder, B.E., and Reynolds, W.C., 1983, "Asymptotic Near-Wall Stress Dissipation Rates in a Turbulent Flow," Physics of Fluids, Vol. 26, p. 1157. (12.2)

Launder, B.E., and Spalding, D.B., 1974, "The Numerical Computation of Turbulent Flows," Computer Methods in Applied Mechanics and Engineering, Vol. 3, pp. 269-289. (1.3)

Lawn, C.J., 1971, "The Determination of the Rate of Dissipation in a Turbulent Pipe Flow," Journal of Fluid Mechanics, Vol. 48, pp. 477-505. (App. A, App. L1971)

Lee, M.J., Kim, J., and Moin, P., 1990, "Structure of Turbulence at High Shear Rate," Journal of Fluid Mechanics, Vol. 216, pp. 561-583. (1.2, 5.1-3, 6.3, 8.4, 10.5-6, App. LKM1990)

Lee, M.J., and Reynolds, W.C., 1985, "Numerical Experiments on the Structure of Homogeneous Turbulence," Stanford Univ. Report No. TF-24. (10.2-6, App. LR1985).

Lee, M.J., and Reynolds, W.C., 1987, "On the Structure of Homogeneous Turbulence," Turbulent Shear Flows 5, Springer-Verlag, pp 54-66. (App. LR1985)

Le Penven, L., Gence, J.N., and Comte-Bellot, G., 1985, "On the Approach to Isotropy of Homogeneous Turbulence: Effect of the Partition of Kinetic Energy Among the Velocity Components," Frontiers of Fluid Mechanics, S.H. Davis and J.L. Lumley (eds), Springer-Verlag, Berlin, pp. 1-21. (15.3, App. LGC1985)

Lewellen, W.S., and Teske, M., 1973, "Prediction of the Monin-Obukhov Similarity Function from an Invariant Model of Turbulence," J. of the Atmospheric Sciences, Vol. 30, pp. 1340-1345. (3.5)

Lin, C.C., 1947, "Remarks on the Spectrum of Turbulence," 1st Symposium in Applied Mathematics, Brown Univ.; 1949, Proc. of Symposia in Applied Math, Vol. 1, pp. 81-86. (17. 3)

Lumley, J.L., 1965, "Interpretation of Time Spectra Measured in High-Intensity Shear Flows," Physics of Fluids, Vol. 8, pp. 1056-1062. (17.1)

Lumley, J.L., 1970, Stochastic Tools in Turbulence, Academic Press. (App. C)

Lumley, J.L., 1975, "Prediction Methods for Turbulent Flows," VKI Lecture Series No. 76. (9.6)

Lumley, J.L., 1978, "Computational Modeling of Turbulent Flows," Advances in Applied

Mechanics, Vol. 18, pp. 123-176. (3.5, 5.2, 15.3)

Lumley, J.L., and Newman, G.R., 1977, "The Return to Isotropy of Homogeneous Turbulence," Journal of Fluid Mechanics, Vol. 82, pp. 161-178. (3.2, 6.2, 15.3)

Mansour, N.N., Kim, J., and Moin, P., 1988, "Reynolds-Stress and Dissipation-Rate Budgets in a Turbulent Channel Flow," Journal of Fluid Mechanics, Vol.194, pp. 15-44. (16.1)

Martínez, D.O., Chen, S., Doolen, G.D., Kraichnan, R.H., Wang, L.-P., and Zhou, Y., 1997, "Energy Spectrum in the Dissipation Range of Fluid Turbulence," Journal of Plasma Physics, Vol. 57, pp. 195-201. (17.3)

Mayle, R.E., 1991, "The Role of Laminar-Turbulent Transition in Gas Turbine Engines," Trans. ASME, Journal of Turbomachinery, Vol. 113, pp. 509-537. (11.1)

McKeon, B.J., Swanson, C.J., Zagarola, M.V., Donnelly, R.J., and Smits, A.J., 2004, "Friction Factors for Smooth Pipe Flow, " Journal of Fluid Mechanics, Vol. 511, pp. 41-44. (1.3, 8.2-3, 14.3)

Menter, F.R., 1994, "Two-Equation Eddy-Viscosity Turbulence Models for Engineering Applications," AIAA Journal, Vol. 32, pp. 1598-1605. (7.1)

Mills, R.R., and Corrsin, S., 1959, "Effect of Contraction on Turbulence and Temperature Fluctuations Generated by a Warm Grid," NASA Memo 5-5-59W. (15.3)

Moin, P., and Mahesh, K., 1998, "Direct Numerical Simulation: A Tool in Turbulence Research," Annual Review of Fluid Mechanics, Vol. 30, pp. 539-578. (13.2)

Moore, J., 1967, "Effects of Coriolis on Turbulent Flow in Rotating Rectangular Channels," S.M. Thesis, Dept. of Mechanical Engineering, Massachusetts Institute of Technology. (13.2, 13.4-5, App. M1967)

Moore, J., and Moore, J.G., 1985, "Performance Evaluation of Linear Turbine Cascades Using Three-Dimensional Viscous Flow Calculations," Journal of Engineering for Gas Turbines and Power, Vol. 107, pp. 969-975. (1.3)

Moore, J., and Moore, J.G., 1991, "A Computational Study of Tip Leakage Flow and Losses in a Linear Turbine Cascade," AGARD Conf. Proc. No. 510, Paper No. 4, San Antonio, Texas. (1.3)

Moore, J., and Moore, J.G., 1995, "Osborne Reynolds: Energy Methods in Transition and Loss Production: A Centennial Perspective," Journal of Turbomachinery, Vol. 117, pp. 142-153. (1.3)

Moore, J., and Moore, J.G., 1998, "Turbulence Energy and Spectra Downstream of an Axial Compressor Rotor," AIAA Paper No. 98-2550. (1.3)

Moore, J., Moore, J.G., and Timmis, P.H., 1984, "Performance Evaluation of Centrifugal Compressor Impellers Using Three-Dimensional Viscous Flow Calculations," Journal of Engineering for Gas Turbines and Power, Vol. 106, pp. 475-481. (1.3)

Moore, J.G., and Moore, J., 1984, "Calculation of Horseshoe Vortex Flow Without Numerical Mixing," ASME Paper No. 84-GT-241. (1.3)

Moore, J.G., and Moore, J., 1997, "Controlling Over-Production of Turbulence in Two-Equation Models by Limiting the Anisotropy of the Reynolds Normal Stresses," ASME Publication FED-Vol. 239, Pumping Machinery (Paper No. FEDSM97-3356). (1.3)

Moore, J.G., and Moore, J., 1999, "Realizability in Two-Equation Turbulence Models,"

AIAA Paper No. 99-3779. (1.3)

Moore, J.G., Schorn, S.A., and Moore, J., 1996, "Methods of Classical Mechanics Applied to Turbulence Stresses in a Tip Leakage Vortex," Journal of Turbomachinery, Vol. 118, pp. 622-629. (1.2)

Morinishi, Y., Nakabayashi, K., and Ren, S.Q., 2001, "A New DNS Algorithm for Rotating Homogeneous Decaying Turbulence," Int. J. Heat and Fluid Flow, Vol. 22, pp. 30-38. (18.3)

Morkovin, M.V., 1969, "On the Many Faces of Turbulence," Viscous Drag Reduction, C.S. Wells (ed.) Plenum Press, New York, pp. 1-31. (11.1)

Moser, R.D., Kim, J., and Mansour, N.N., 1999, "Direct Numerical Simulation of Turbulent Channel Flow up to Re_τ = 590," Physics of Fluids, Vol. 11, pp. 943-945. (1.3, 5.3, 6.4, 8.2-3, 13.2, 15.2)

Naot, D., Shavit, A., and Wolfshtein, M., 1970, "Interactions between Components of the Turbulent Velocity Correlation Tensor due to Pressure Fluctuations," Israel J. of Technology, Vol. 8, pp. 259-269. (3.5)

Naot, D., Shavit, A., and Wolfshtein, M., 1973, "Two-Point Correlation Model and the Redistribution of Reynolds Stresses," Physics of Fluids, Vol. 16, pp. 738-743. (15.2)

Nishimura, M., and Kasagi, N., 1996, "Direct Numerical Simulation of Combined Forced and Natural Turbulent Convection in a Rotating Plane Channel," Proc. 3rd KSME-JSME Thermal Engineering Conference, Kyongju, Korea, Vol. 3, pp. 77-82. (13.2, 13.4-5, App. WK2004)

Orlandi, P, and Ebstein, D., 2000, "Turbulent Budgets in Rotating Pipes by DNS," Int. J. Heat and Fluid Flow, Vol. 21, pp. 499-505. (14.2)

Orlandi, P., and Fatica, M., 1997, "Direct Simulations of Turbulent Flow in a Pipe Rotating About Its Axis," Journal of Fluid Mechanics, Vol. 343, pp. 43-72. (14.2-3)

Pao, Y.-H., 1965, "Structure of Turbulent Velocity and Scalar Fields at Large Wavenumbers," Physics of Fluids, Vol. 8, pp. 1063-1075. (6.2, 17.3-4)

Patankar, S.V., and Spalding, D.B., 1968, Heat and Mass Transfer in Boundary Layers, C.R.C.Press, Cleveland. (11.2)

Patel, V.C., 1965, "Calibration of the Preston Tube and Limitation on Its Use in Pressure Gradients," Journal of Fluid Mechanics, Vol. 23, pp. 185-208. (App. A)

Patel, V.C., 1998, "Perspective: Flow at High Reynolds Number and Over Rough Surfaces – Achilles Heel of CFD," Journal of Fluids Engineering, Vol. 120, pp. 434-444. (App. A)

Patel, V.C., and Head, M.R., 1969, "Some Observations on Skin Friction and Velocity Profiles in Fully Developed Pipe and Channel Flows," Journal of Fluid Mechanics, Vol. 38, pp. 181-201. (8.2-3)

Patel, V.C., Rodi, W., and Scheuerer, G., 1985, "Turbulence Models for Near-Wall and Low Reynolds Number Flows: A Review," AIAA Journal, Vol. 23, pp. 1308-1319. (1.3, 6.1, App. A)

Pope, S.B., 1975, "A More General Effective-Viscosity Hypothesis," Journal of Fluid Mechanics, Vol. 72, pp. 331-340. (1.3)

Pope, S.B., 2000, Turbulent Flows, Cambridge Univ. Press. (1.2, 16.1, 17.3-4, App. EQ.3.1.4)

Poppleton, E.D., 1971, “Exploratory Measurements of the Flow in the Wing Tip Vortices of a Lockheed Hercules”, Sydney Univ. Aero. Dept., Report ATN-7104, and STAR N72-28277. (4.3)

Poroseva, S.V., Kassinos, S.C., Langer, C.A., and Reynolds, W.C., 2000, “Computation of Turbulent Flow in a Rotating Pipe using the Structure-Based Model,” Center for Turbulence Research, Annual Research Briefs 2000, Stanford Univ., pp. 279-290. (1.3)

Prandtl, L., 1925, “Über die ausgebildete Turbulenz,” ZAAM, 5, pp. 136-139. (1.0, 1.3)

Proudman, J., 1916, “On the Motion of Solids in a Liquid Possessing Vorticity,” Proc. Roy. Soc. A, Vol. 92, pp. 408-424. (18.1)

Reich, G., and Beer, H., 1989, “Fluid Flow and Heat Transfer in an Axially Rotating Pipe – I. Effect of Rotation on Turbulent Pipe Flow,” Int. J. Heat Mass Transfer, Vol. 32, pp. 551-562. (14.2)

Reynolds, Osborne, 1895, “On the Dynamical Theory of Incompressible Viscous Fluids and the Determination of the Criterion,” Phil. Trans. Roy. Soc. A, Vol. 186, pp. 123-164. (1.0)

Reynolds, W.C., 1974, “Computation of Turbulent Flows,” AIAA Paper No. 74-556. (6.2)

Reynolds, W.C., 1976, “Computation of Turbulent Flows,” Annual Review of Fluid Mechanics, Vol. 8, pp. 183-208. (15.3)

Reynolds, W.C., 1987, “Fundamentals of Turbulence for Turbulence Modeling and Simulation,” AGARD Report No. 755, “Special Course on Modern Theoretical and Experimental Approaches to Turbulent Flow Structure and Its Modelling.” (3.2, App. EQ.3.2.17)

Reynolds, W.C., 1989, “Effects of Rotation on Homogeneous Turbulence,” Tenth Australasian Fluid Mechanics Conf. – Univ. of Melbourne, Paper KS-2. (17.5)

Ristorcelli, J.R., Lumley, J.L., and Abid, R., 1995, “A Rapid-Pressure Covariance Representation Consistent with the Taylor-Proudman Theorem Materially Frame Indifferent in the Two-Dimensional Limit,” Journal of Fluid Mechanics, Vol. 292, pp. 111-152. (3.2, 18.4, App. C, App. EQ.3.2.17, App. RLA1995)

Rivlin, R.S., 1955, “Further Remarks on the Stress-Deformation Relations for Isotropic Materials,” Journal of Rational Mechanics and Analysis, Vol. 4, pp. 681-702. (App. C)

Roach, P.E., and Brierley, D.H., 1990, "The Influence of a Turbulent Free Stream on Zero Pressure Gradient Transitional Boundary Layer Development. Part 1: Test Cases T3A and T3B," Numerical Simulation of Unsteady Flows and Transition to Turbulence, O. Pironneau, W. Rodi, and I.L. Ryhming (eds.), Cambridge University Press. (11.1, 11.3-5)

Rodi, W., 1984, Turbulence Models and Their Application in Hydraulics, A.A.Balkema Publishers. (1.3, App. A)

Rogers, M.M., Moin, P., and Reynolds, W.C., 1986, “The Structure and Modeling of the Hydrodynamic and Passive Scalar Fields in Homogeneous Turbulent Shear Flow,” Report No. TF–25, Thermosciences Div., Dept. of Mech. Eng., Stanford Univ. (5.1-3, 6.3, 8.2-4, 10.6, 15.2-3, App. RMR1986)

Rotta, J., 1951, “Statistische Theorie Nichthomogener Turbulenz,” Zeitschrift für Physik,

Bd. 129, S. 547-572. English translation by W. Rodi, 1968, Report No. TWF/TN/39. Dept. of Mech. Eng., Imperial College, London Univ. (1.0, 3.1-2, 15.2-3, App. R1951)

Rotta, J.C., 1962, "Turbulent Boundary Layers in Incompressible Flow," Progress in Aeronautical Sciences, Vol. 2, A. Ferri et al. (eds), Pergamon Press. (15.3)

Ryhming, I.L., 1990, "Test Case Specifications," Numerical Simulation of Unsteady Flows and Transition to Turbulence, O. Pironneau, W. Rodi, and I.L. Ryhming (eds.), Cambridge University Press. (11.1)

Saddoughi, S.G., and Veeravalli, S.V., 1994, "Local Isotropy in Turbulent Boundary Layers at High Reynolds Numbers," Journal of Fluid Mechanics, Vol. 268, pp. 333-372. (17.4, 17.6, App. SV1994)

Saffman, P.G., 1967, "The Large-Scale Structure of Homogeneous Turbulence," Journal of Fluid Mechanics, Vol. 27, pp. 581-593. (17.3)

Salhi, A., and Cambon, C., 1997, "An Analysis of Rotating Shear Flow using Linear Theory and DNS and LES Results," Journal of Fluid Mechanics, Vol. 347, pp. 171-195. (9.3, 10.3)

Sarkar, S., and Speziale, C.G., 1990, "A Simple Nonlinear Model for the Return to Isotropy in Turbulence," Physics of Fluids, Vol. A2, pp. 84-93. (15.3)

Savill, A.M., 1992, "Transition Modelling for Turbomachinery," Proceedings of the ERCOFTAC Turbomachinery Special Interest Group Seminar and Workshop on 3D Turbomachinery Flow Prediction, Part I. (11.4)

Schlichting, H., 1979, Boundary-Layer Theory, McGraw-Hill. (8.2, 11.1)

Schubauer, G.B., and Klebanoff, P.S., 1956, "Contributions on the Mechanics of Boundary-Layer Transition," NACA Report No. 1289. (11.1, 11.3)

Schubauer, G.B., and Skramstad, H.K., 1948, "Laminar-Boundary-Layer Oscillations and Transition on a Flat Plate," NACA Report No. 909. (11.1, 11.3, App. SS1948)

Schumann, U., and Patterson, G.S., 1978, "Numerical Study of Pressure and Velocity Fluctuations in Nearly Isotropic Turbulence," Journal of Fluid Mechanics, Vol. 88, pp. 685-709. (15.2)

Shabbir, A., and Shih, T.-H., 1993, "Critical Assessment of Reynolds Stress Turbulence Models Using Homogeneous Flows," AIAA Paper No. 93-0082. (10.2, App. LR1985)

Shih, T.-H., and Lumley, J.L., 1985, "Modeling of Pressure Correlation Terms in Reynolds Stress and Scalar Flux Equations," Technical Report No. FDA-853, Cornell Univ.; see Demuren and Sarkar, 1993, and Shih and Lumley, 1993. (3.5, 4.2, 5.1)

Shih, T.-H., and Lumley, J.L., 1993, "Critical Comparison of Second-Order Closures with Direct Numerical Simulations of Homogeneous Turbulence," AIAA Journal, Vol. 31, pp. 663-670.

Shih, T.-H., Reynolds, W.C., and Mansour, N.N., 1990, "A Spectrum Model for Weakly Anisotropic Turbulence," Physics of Fluids, Vol. 2, pp. 1500-1502. (17.5)

Shir, C.C., 1973, "A Preliminary Numerical Study of Atmospheric Turbulent Flows in the Idealized Planetary Boundary Layer," J. of the Atmospheric Sciences, Vol. 30, pp. 1327-1339. (3.5, 5.3, 6.1, 16.2-3, App. S1973)

So, R.M.C., 1975, "A Turbulence Velocity Scale for Curved Shear Flows," Journal of

Fluid Mechanics, Vol. 70, pp. 37-57. (4.3)

So, R.M.C., and Mellor, G.L., 1972, "An Experimental Investigation of Turbulent Boundary Layers Along Curved Surfaces," NASA CR-1940. (4.3)

Spalart, P. R., 1988, "Direct Simulation of a Turbulent Boundary Layer up to R_θ =1410," Journal of Fluid Mechanics, Vol. 187, pp. 61-98. (App. A)

Speziale, C.G., 1991, "Analytical Methods for the Development of Reynolds –Stress Closures in Turbulence," Annual Review of Fluid Mechanics, Vol. 23, pp. 107-157. (App. EQ.3.2.17)

Speziale, C.G., Sarkar, S., and Gatski, T.B., 1991, "Modelling the Pressure-Strain Correlation of Turbulence: An Invariant Dynamical Systems Approach," Journal of Fluid Mechanics, Vol. 227, pp. 245-272. (3.5, 4.2-3, 5.1-3, 6.1, 6.3, 9.6, 15.3, 18.4, App. G2000)

Sreenivasan, K.R., 1995, "On the Universality of the Kolmogorov Constant," Physics of Fluids, Vol. 7, pp. 2778-2784. (17.4)

Stewart, R.W., and Townsend, A.A., 1951, "Similarity and Self-Preservation in Isotropic Turbulence," Phil. Trans. Roy. Soc. A, Vol. 243, pp. 359-386. (17.4)

Synge, J.L., and Lin, C.C., 1943, " On a Statistical Model of Isotropic Turbulence," Trans. Roy. Soc. Canada, Section III, Vol. 37, pp. 45-79. (17.3)

Tavoularis, S., and Corrsin, S., 1981, "Experiments in Nearly Homogeneous Turbulent Shear Flow with a Uniform Mean Temperature Gradient. Part 1," Journal of Fluid Mechanics, Vol. 104, pp. 311-347. (1.2, 3.5, 5.1-3, 6.3, 8.4, App. A, App. TC1981)

Tavoularis, S., and Karnik, U., 1989, "Further Experiments on the Evolution of Turbulent Stresses and Scales in Uniformly Sheared Turbulence," Journal of Fluid Mechanics, Vol. 204, pp. 457-478. (5.1-3, 6.3, 8.4, App. TK1989)

Taylor, G.I., 1917, "Motion of Solids in Fluids when the Flow is not Irrotational," Proc. Roy. Soc. A, Vol. 93, pp. 99-113. (18.1)

Taylor, G.I., 1921, "Experiments with Rotating Fluids," Proc. Roy. Soc. A, Vol. 100, pp. 114-121. (18.1)

Taylor, G.I., 1935, "Statistical Theory of Turbulence, Part I," Proc. Roy. Soc. A, Vol. 151, pp. 421-444. (17.2)

Taylor, G.I., 1938, "The Spectrum of Turbulence," Proc. Roy. Soc. A, Vol. 164, pp. 476-490. (17.1)

Tennekes, H., and Lumley, J.L., 1972, A First Course in Turbulence, MIT Press. (6.1, 15.3)

Thomas, N.H., and Hancock, P.E., 1977, "Grid Turbulence Near a Moving Wall", Journal of Fluid Mechanics, Vol. 82, pp. 481-496. (16.3, App. TH1977)

Townsend, A.A., 1976, The Structure of Turbulent Shear Flow, Cambridge Univ. Press. (3.3, 9.0, 9.1, 9.3, 10.3)

Tucker, H.J., 1970, "The Distortion of Turbulence by Irrotational Strain," Ph.D. thesis, Dept. of Mechanical Engineering, McGill University, Montreal. (10.2, 10.5-6, 15.3, 17.6, App. T1970)

Uberoi, M.S., 1956, "Effect of Wind-Tunnel Contraction on Free-Stream Turbulence," Journal of the Aeronautical Sciences, pp. 754-764. (15.3, App. U1956)

Uberoi, M.S., 1957, "Equipartition of Energy and Local Isotropy in Turbulent Flows," Journal of Applied Physics, Vol. 28, pp. 1165-1170. (15.3, App. U1957)

van Driest, E.R., 1956, "On Turbulent Flow Near a Wall," Journal of the Aeronautical Sciences, Vol. 23, pp. 1007-1011, 1036. (1.3)

von Kármán, T., 1948, "Progress in the Statistical Theory of Turbulence," Proc. Natl. Acad. Sci. USA, Vol. 34, pp. 530-539. (17.3-4)

Wallin, S., and Johansson, A.V., 2002, "Modelling Streamline Curvature Effects in Explicit Algebraic Reynolds Stress Turbulence Models," Int. J. Heat and Fluid Flow, Vol. 23, pp. 721-730. (13.2)

Wells, C.S., 1967, "Effects of Freestream Turbulence on Boundary-Layer Transition," AIAA Journal, Vol. 5, pp. 172-174. (11.1, 11.3, App. W1967)

Wigeland, R.A., and Nagib, H.M., 1978, "Grid-Generated Turbulence With and Without Rotation About the Streamwise Direction," IIT Fluids and Heat Transfer Report No. R78-1, Illinois Inst. of Tech., Chicago, Illinois. (18.3)

Wilcox, D.W., 1993, Turbulence Modeling for CFD, DCW Industries, Inc, CA. (1.3)

Wu, H., and Kasagi, N., 2004, "Effects of Arbitrary Directional System Rotation on Turbulent Channel Flow," Physics of Fluids, Vol. 16, pp. 979-990. (13.2, 13.4-5, App. WK2004)

Wu, H., and Kasagi, N., 2004b, "Turbulent Heat Transfer in a Channel Flow with Arbitrary Directional System Rotation," Int. J. Heat Mass Transfer, Vol. 47, pp. 4579-4591. (13.2, 13.4-5, App. WK2004)

Wyngaard, J.C., and Clifford, S.F., 1977, "Taylor's Hypothesis and High-Frequency Turbulence Spectra," J. of the Atmospheric Sciences, Vol. 34, pp. 922-929. (17.1)

Yamada, Y., and Imao, S., 1980, Trans. JSME, 46-409, B(1980), 1662 (in Japanese). (14.2, 14.3, App. KMNM1983)

Younis, B.A., 1984, Ph.D. Thesis, Faculty of Engineering, Univ. of London; see Launder, 1993. (3.5, 5.3)

Zagarola, M.V., 1996, "Mean Flow Scaling of Turbulent Pipe Flow," Ph.D. Thesis, Dept. of Mechanical and Aerospace Engineering, Princeton Univ. (1.3, 8.2-3, 14.2)

Zysina-Molozhen, L.M., and Kuznetsova, V.M., 1969, "Investigation of Transition Conditions in a Boundary Layer," Teploenergetika, Vol 16, No. 7, pp. 16-20. (11.1, 11.3, App. ZK1969)

INTERNET RESOURCES

(availability checked February, 2006)

Devenport et al., DZV1997, counter-rotating tip vortex experimental data.
http://www.aoe.vt.edu/research/groups/flowdata/

Kasagi et al., KIIMMN1990, extensive DNS data, including fully developed stationary and rotating channel flows (see Tables 6.4.1 and 13.2.1).
http://www.thtlab.t.u-tokyo.ac.jp/

Kristoffersen and Andersson, 1993, DNS data, 7th ERCOFTAC/IAHR Workshop on Refined Turbulence Modelling, Case 7.1: Fully developed flow in a plane channel in orthogonal mode rotation with rotation numbers up to 0.5.
http://tmdb.ws.tn.tudelft.nl/workshop7/workshop.html

Moser et al., MKM1999, DNS data for turbulent channel flow.
http://www.tam.uiuc.edu/faculty/moser/channel/

Roach and Brierley, 1990, experimental data for flat plate transitional boundary layers, ERCOFTAC "Classic Collection" Database, Case C.20, by Coupland, 1990.
(Registration and password required.)
http://cfd.me.umist.ac.uk/ercoftac/

Spalart, 1988, DNS data for constant pressure boundary layer, ERCOFTAC "Classic Collection" Database, Case C.33, by Spalart, 1987.
(Registration and password required.)
http://cfd.me.umist.ac.uk/ercoftac/

Zagarola, 1996, experimental velocity profiles in fully developed pipe flow.
http://gasdyn.princeton.edu/data/e247/zagarola_data.html

AUTHOR INDEX

FIGURE INDEX

SUBJECT INDEX

elliptical instability...

rotating plane channel flow…